T0202184

Victorians and Numbers

To the number 3:

Alexander, Jonathan, and Saskia

'this I know - that Numbers teach us whether the world is well or ill-governed'
(William Farr to Florence Nightingale, c.17 July 1857)
B.M. Add Ms. 43398, f.14

OXFORD
UNIVERSITY PRESS

Great Clarendon Street, Oxford, OX2 6DP,
United Kingdom

Oxford University Press is a department of the University of Oxford.
It furthers the University's objective of excellence in research, scholarship,
and education by publishing worldwide. Oxford is a registered trade mark of
Oxford University Press in the UK and in certain other countries

Published in the United States of America by Oxford University Press
198 Madison Avenue, New York, NY 10016, United States of America

British Library Cataloguing in Publication Data
Data available

Library of Congress Control Number: 2021943497

ISBN 978-0-19-284774-4

DOI: 10.1093/oso/9780192847744.001.0001

Printed and bound by
CPI Group (UK) Ltd, Croydon, CR0 4YY

Links to third party websites are provided by Oxford in good faith and
for information only. Oxford disclaims any responsibility for the materials
contained in any third party website referenced in this work.

Victorians and Numbers

Statistics and Society in Nineteenth Century Britain

LAWRENCE GOLDMAN

OXFORD
UNIVERSITY PRESS

Acknowledgments

Victorians and Numbers was largely written in 2020–21 during the Covid-19 pandemic that 'locked-down' Britain and much of the world for months on end. The statistics were unrelenting: the billions of pounds borrowed to pay for the disruption to working lives, the millions of people who contracted the virus, the many thousands who died. Every day ended with a tally of fatalities. We compared death rates in different countries. We followed closely 'the reproduction number R', the average number of secondary infections caused by someone who already had the disease. The rates of infection in different parts of the country were studied closely, and, when allowed out of our homes, we planned our lives accordingly. The numbers vaccinated were counted each day. The reported efficacy of the different vaccines manufactured to combat the virus were compared and contrasted. As the data and statistics came thick and fast, the struggles of many of the figures examined in this book—James Kay-Shuttleworth and William Farr dealing with cholera, John Simon counting the victims of environmental disease, William Guy assessing the threats to occupational health, Florence Nightingale seeking a uniform method for the reporting of sickness—seemed all the more compelling. Indeed, the very subject itself, the history of statistics, took on a new relevance. I had always thought that my fascination with Victorian statistics, and with the attempts of our nineteenth-century ancestors to picture and explain their world numerically, was an essentially private passion. Now I discovered that everyone, with good reason, was interested 'in the numbers'. The task I had set myself of explaining a formative period in the history of statistics in Britain no longer seemed obscure and academic. I found that people understood immediately and intuitively why this book was worth writing. I hope I have answered some of the questions they asked me.

It was also composed at a time of heightened racial awareness in Britain that has prompted the re-evaluation of many leading historical figures known to have held racist views. It is central to the argument developed here that Victorian statisticians before the 1880s—before Francis Galton, in fact—were distinguished by their social and political liberalism. They took an environmental approach to social problems, including poverty, disease, and ignorance, which, they believed, could only be solved by changing the conditions in which people lived. They valued numbers as instruments that could help in the investigation and solution of such ills. As William Farr wrote to Florence Nightingale in 1864, 'And what are figures worth if they do no good to men's bodies or souls [?]'[1] Galton, and some of the leading statisticians who followed him, like Karl Pearson and R. A. Fisher, were eugenicists, however. They believed in the primacy of 'nature over nurture', a term that was coined, like 'eugenics' itself, by Galton. They used the sophisticated mathematical statistics which they themselves developed from the 1880s to differentiate between classes and races on the basis of supposedly inherited mental and physical attributes and characteristics, and to build a case against liberal intervention and reform. Some groups and races were worthier than others; some should be assisted while others discouraged, deterred, or controlled. By the time I came to write about Galton's remarkable intellectual achievements and also his racist and socially-divisive attitudes and views, the rooms and buildings named after him, and also after Karl Pearson, at the university most closely associated with their work, University College, London, had been renamed.[2] Meanwhile, in Cambridge, a memorial window dedicated in 1989 to the academic achievements of Ronald Fisher by Gonville and Caius College, where Fisher had studied as an undergraduate before the First World War and where later he held his chair, had been removed.[3] These developments led to public discussion about freedom of conscience in academic life, about the relationship between intellectual distinction and political conviction, and about judging the past by the values of the present. They also focused attention on the political and racial connotations of the discipline of statistics at a particular stage in its history. They have made this book more necessary than it might otherwise have been, therefore, for it demonstrates that the history of statistics, in Britain and elsewhere, had a liberal and genuinely reformist lineage before the coming of eugenics. From the early nineteenth century, people from many different backgrounds struggled valiantly to collect and interpret numerical social data as a basis for environmental reform and in furtherance of that key Victorian concept, 'improvement'.

Yet if written recently at an exceptional time, this book had a long gestation. In some ways it began with my undergraduate Special Subject in Cambridge on the history of geology from the seventeenth to the nineteenth centuries, taught by the late Roy Porter, historian of science, medicine, and so much more. He kindled an interest in intellectual history and the history of disciplines—academic subjects.

---

[1] William Farr to Florence Nightingale, 2 September 1864, B.M. Add Mss 43,399, f. 154.
[2] 'UCL Renames Three Facilities that Honoured Prominent Eugenicists', *The Guardian*, 19 June 2020, https://www.theguardian.com/education/2020/jun/19/ucl-renames-three-facilities-that-honoured-prominent-eugenicists.
'UCL Denames Buildings Named After Eugenicists', *UCL News*, 19 June 2020, https://www.ucl.ac.uk/news/2020/jun/ucl-denames-buildings-named-after-eugenicists.
[3] 'Cambridge College to Remove Window Commemorating Eugenicist', *The Guardian*, 27 June 2020' https://www.theguardian.com/education/2020/jun/27/cambridge-gonville-caius-college-eugenicist-window-ronald-fisher. A.W.F. Edwards, 'Statisticians in Stained Glass Houses', *Significance* (Royal Statistical Society), 3 (4), 182–3, https://rss.onlinelibrary.wiley.com/doi/pdf/10.1111/j.1740-9713.2006.00203.x. Richard J. Evans, 'R A Fisher and the Science of Hatred', *New Statesman*, 28 July 2020, https://www.newstatesman.com/international/science-tech/2020/07/ra-fisher-and-science-hatred.

Boyd Hilton and Gareth Stedman Jones gave valuable support to my research in the early years. I was fortunate to participate in the seminars of the 'History of Political Economy' project in the King's College, Cambridge Research Centre in the 1980s when a postgraduate. My debt to Vic Gatrell, and my appreciation of his work, are evident in the chapter in this book on the Manchester Statistical Society in the 1830s and 1840s. Joanna Innes, David Eastwood, and Simon Szreter were early companions in pursuit of the history of numbers and we have all gone on to publish on the subject. Iowerth Prothero gave help and encouragement as I searched for the artisans who, in the 1820s, founded the London Statistical Society.

In Oxford, I was part of a group of historians of the Victorian period the like of which, as the History syllabus broadens and is internationalized, will never be equalled again in any university. Some of us came together to teach the undergraduate Further Subject entitled 'Victorian Intellect and Culture' and I pay tribute to the late John Prest, Colin Matthew, and John Burrow, as well as to Jane Garnett, Peter Ghosh, Jose Harris, William Thomas, William Whyte, and Brian Young, among others, for making it so interesting and rewarding. Over the years they tried hard to instruct me in the history of Victorian religion. I hope this book will assist them in teaching the history of Victorian social science, another of the components of the course. Trinity College, Cambridge, and Kellogg and St Peter's Colleges, Oxford provided support and good fellowship for four decades. In St Peter's I was privileged to have Henry Mayr-Harting, Henrietta Leyser, and the late, deeply lamented Mark Whittow, as close colleagues in the teaching of History. I enjoyed many stimulating conversations on the history of his subject with the late John Bithell, Fellow and Tutor in Statistics in St Peter's. Geoff Nicholls, his successor, helped me identify the statistical concept that was discussed at the inaugural meeting of the Statistical Movement in 1833. Inflicting my interest in the history of social science on undergraduates in St Peter's elicited rather varied responses over the years. I'm grateful for their patience, and for what they taught me in return. Rebecca Ryan deserves special thanks for providing me again with her first-class undergraduate dissertation on Manchester in 1840, *Chaos vom Fabriken*. This book would have been much more difficult to research and write without the assistance of dozens of articles in the *Oxford Dictionary of National Biography*. I pay tribute to my former full-time colleagues working at the *ODNB* and to the more than 11,000 contributors whose essays have made the Dictionary such a valuable resource for historians of Britain.

Sections of this book were also written as a Visiting Fellow in the History Department of the Australian National University whose hospitality I've enjoyed greatly over the years, and also as a Senior Fellow of the Kinder Institute for Constitutional Democracy in the University of Missouri. Professors Melanie Nolan, the Editor of the *Australian Dictionary of Biography*, and Kim Sterelny, were, as always, especially welcoming and supportive in Canberra and in the bush. I thank Kim for his help with Natural Theology and Charles Darwin. In Missouri I'm grateful to my former student, Professor Jay Sexton, and to Professors Justin Dyer and Jeff Pasley, for their support and comradeship, and for the very special culture they have created on the top floor of Jesse Hall. For their hospitality on my research trips to the United States, I thank David Shribman and Cindy Skrzycki, Jerry Kleiner and Kristin Warbasse, and Fred and Jean Leventhal.

David Feldman was there at the beginning and at the end to hear me talk about numbers and statistics over numerous curries and pints, and he kindly read and commented on the introduction to this book. This would be a less satisfactory work without the many suggestions from the anonymous reader chosen to comment on my typescript by Oxford University Press whose level of engagement with the text, profusion of ideas for its improvement, and overall assessment of its argument, exemplified all the best features of peer review. Cathryn Steele, Commissioning History Editor at OUP, grasped the point of the book from the outset and has been a most helpful and sympathetic guide through the processes of publication. My wife, Madeleine, has listened patiently to talk of arithmeticians, statisticians, political economists, physicians, sanitarians and eugenicists on walks across the beaches, heaths, and woodlands of Suffolk. Poor her. But she has been the very best of companions, the wisest of advisors, and a fine translator of nineteenth-century German and Florence Nightingale's French.

The chapters to follow incorporate material from conference papers, and some reworked passages from published academic articles. For the reuse of this material I am grateful as follows: to Cambridge University Press for material included in 'The Origins of British "Social Science". Political Economy, Natural Science and Statistics, 1830–35', *The Historical Journal*, 26, no. 3, 1983, 587–616, and 'Experts, Investigators and The State in 1860: British Social Scientists through American Eyes', in M. Lacey and M. O. Furner (eds), *The State and Social Investigation in Britain and the United States* (1993), 95–126; to Oxford University Press for 'Statistics and the Science of Society in Early Victorian Britain: An Intellectual Context for the General Register Office', *Social History of Medicine*, 4, 3, Dec. 1991, 415–34, and 'Victorian Social Science: From Singular to Plural', in Martin Daunton (ed.), *The Organisation of Knowledge in Victorian Britain* (2005), 87–114. A version of Chapter 6 on the London Statistical Society was given at a conference on the history of social investigation at the Open University in December 2004. Chapter 14 on the International Statistical Congress was presented to a colloquium on the history of the social sciences at the Centre for Research in the Arts, Social Sciences and Humanities (CRASSH) in Cambridge in September 2008.

I'm indebted to the librarians and archivists of many different institutions for their help and support over the years. These include the University Library, Cambridge, the Bodleian Library, Oxford, and the British Library, London; the Wren Library, Trinity College, Cambridge; the libraries of the Royal Society and the Royal Statistical Society; the Goldsmith's Library, University of London; the

Royal Archives, Windsor; the Houghton Library and the Francis A. Countway Library of Medicine, Harvard University; the Concord Free Public Library, Concord, Massachusetts; the Académie Royale de Belgique; the Liverpool Record Office; the Manchester Central Library; the D. M. S. Watson Library, University College, London; the British Library of Political and Economic Science, London School of Economics; the Wellcome Library, London; the Oxford Museum of the History of Science; and the library of St Peter's College, Oxford.

My greatest debts are to the historians who have written already on this subject. Among many, I would pick out Michael Cullen who opened up these themes in the 1970s; Richard Stone for his study of the political arithmeticians; John Eyler, author of the magnificent biography of William Farr; Susan Faye Cannon who first defined and understood the Cambridge Network; Jack Morrell and Arnold Thackray on the natural scientists of the 1830s; Ian Hacking and Peter Bernstein on the history of chance and risk; Anthony Hyman and Doron Swade for their work on Charles Babbage; Andrea Wulf on Alexander von Humboldt; Vic Gatrell for his work on the Manchester middle class; Royston Lambert for his study of Dr John Simon and mid-Victorian medicine; Eddie Higgs and Simon Szreter for work on the General Register Office and on Victorian vital statistics and social reform; Marion Diamond and Mervyn Stone for their work on Florence Nightingale; and the historians of statistics, Donald Mackenzie, Theodore Porter, Stephen Stigler, Alain Desrosières, Libby Schweber, and Eileen Magnello. It was a reading of Stephen J. Gould's great study, *The Mismeasurement of Man*, which helped to bring together both the argument and the constituent elements of this book. I'm also grateful to Plamena Panayotova for her work on the history of British sociology and the conference on that subject which she organized in the University of Edinburgh in 2018, now published as essays in *The History of Sociology in Britain. New Research and Revaluation*. A version of the argument of this book is among them.[4]

This can only be a short and therefore defective list of some of the works that have been of the greatest help and influence. There are many more that could be mentioned, and more besides that I've failed to read or even locate. In Victorian Studies, widely defined, we have reached a point where reading it all, 'keeping up', may be beyond any scholar, and I apologize in advance for my lapses and failings in this regard, as in many others.

[4] Lawrence Goldman, 'Victorians and Numbers: Statistics and Social Science in Nineteenth-Century Britain' in Plamena Panayotova (ed.), *The History of Sociology in Britain. New Research and Revaluation* (Cham, Switzerland, 2019), 71–100.

Contents

PART IV: STATISTICS AT MID-CENTURY

PART V: LIBERAL DECLINE AND REINVENTION

List of Abbreviations

BAAS	British Association for the Advancement of Science
GRO	General Register Office
IRC	Industrial Remuneration Conference
ISC	International Statistical Congress
JSSL	Journal of the Statistical Society of London
JRSS	Journal of the Royal Statistical Society
LSS	London Statistical Society
MOH	Medical Officer of Health
MSS	Manchester Statistical Society
ODNB	Oxford Dictionary of National Biography
PCMO	Privy Council Medical Office/Officer
SDF	Social Democratic Federation
SSA	Social Science Association
SSofL	Statistical Society of London
TMSS	Transactions of the Manchester Statistical Society

Introduction

Victorians and Numbers

In the last week of June 1833, the University of Cambridge was host to the third annual meeting of the British Association for the Advancement of Science, a new organization that had been formed to unite all those engaged in the natural sciences in Britain. In rooms in Trinity College being used temporarily by the Rev. Richard Jones, then Professor of Political Economy at King's College, London, a small group of savants who were attending the BAAS met together. Eight in number, they included not only Jones but a much more famous founder of economics, the Rev. Thomas Robert Malthus, author of the *Essay on the Principle of Population*; the mathematician Charles Babbage, then the Lucasian Professor of Mathematics in Cambridge, the chair held by Isaac Newton long before him and by Stephen Hawking in our own time; and the Belgian mathematician and astronomer Adolphe Quetelet who would soon become the most famous and influential statistician in Europe following publication in 1835 of his book *Sur L'Homme et le Développement de ses Facultés: Physique Sociale*. They heard Quetelet present a 'statistical budget' on criminal behaviour in Belgium and France. Babbage followed, offering material on vital statistics—the term used for basic demographic data on life and death—in Scandinavia where records were plentiful and accurate, and also some calculations he had made on the number of inebriates—drunks— arrested in London in each month of the year. This was the kind of teasing humour at which Babbage excelled, but it was also part of a highly individual exploration of the uses and limits of social statistics. It was not the only seemingly eccentric statistical exercise in which Babbage delighted, as we shall see.

Babbage and Quetelet were already friends and would remain in contact for the rest of their lives. Quetelet was the first person Babbage told in 1835 about his work devising an 'Analytical Engine'—often described as the first computer. Quetelet published the letter, making it a foundational text in the history of computing.[1] The two men were never central to the day-to-day operations of what became known as 'the statistical movement' in Britain. That role was left to their close mutual friend, 'the great William Farr' as the historian Eric Hobsbawm once

---

[1] See pp. 112–13 below.

Victorians and Numbers: Statistics and Society in Nineteenth Century Britain. Lawrence Goldman, Oxford University Press.
© Lawrence Goldman 2022. DOI: 10.1093/oso/9780192847744.001.0001

called him.[2] But Quetelet was the greatest intellectual influence over the statistical movement in the early and mid-Victorian eras; and Babbage was, in retrospect, its most innovative and creative figure, the person who, more than any other, links the 1830s with the present day and who will thereby knit together this book. Their presence with Malthus in that college room gives the meeting on 27 June 1833 a special significance: this is where the systematic study of social statistics may be said to have begun in Britain.

Deciding that this irregular grouping should be made permanent, Babbage, a member of the BAAS's Council, served as their spokesman in negotiations with the Association over the next couple of days. Meanwhile, more members and attendees heard about this embryonic 'statistical section' and came to further irregular statistical meetings in Trinity College. The tide could not be held back and wary members of the Council had to accept this *fait accompli*: the BAAS had a new statistical section, Section F. The Association's President for the year, Adam Sedgwick, Professor of Geology in Cambridge and one of Charles Darwin's teachers, was unimpressed, and he warned his colleagues that they had allowed 'the foul Daemon of discord' into their Eden of science.[3] As events were to show, he was not wrong. The creation of Section F then set off a national movement. Within three months, a statistical society had been established in Manchester. In February of the following year the Statistical Society of London was founded in Babbage's home in London. It held its first public meeting in the following month. Since 1887 it has been the Royal Statistical Society. In subsequent years statistical societies were set up in many provincial cities in Britain. From the early 1830s statistics found a home in central government as well. The Statistical Department of the Board of Trade, founded in 1832, and the General Register Office, established in 1837, are the ancestors of today's Office for National Statistics which combines within it all the social statistical functions of the contemporary British state.

Historians like to debate the date when, and the place where 'modernity' began. Where does modern history start? If we look to the history of ideas, 'the creation of the modern world' according to one notable historian who taught the author of this book, begins with the 'British Enlightenment' of the eighteenth century.[4] A more specific answer to the question might locate modernity in the English midlands in the 1770s if we take the Industrial Revolution as the starting point. Or it could be at roughly the same time in the American colonies with the revolution against British rule if the essence of modernity is representative government and liberal freedom. Or it could be in Paris in 1789 with the overthrow that summer

---

[2] Eric Hobsbawm, *The Age of Capital. Europe 1848–1875* (London, 1975), 307.
[3] See p. 42 below.
[4] Roy Porter, *The Creation of the Modern World. The Untold Story of the British Enlightenment* (New York, 2000).

of monarchy and feudalism. Or it could be across the 'imperial meridian' at some
time between the 1780s and 1830s as the British established a durable global
empire under effective administration.[5] Or it could be in the sequence of events
between 1815, the end of the Napoleonic Wars, and 1830, which established the
dominance of the Atlantic powers for the next two centuries.[6] Set against these
grand historical events, what one historian has called 'the avalanche of numbers'
after about 1820 may be small beer. Yet 'between 1820 and 1840 there was an
exponential increase in the number of numbers that were being published. The
enthusiasm for numbers became almost universal' and gave rise 'to a new way of
talking and doing'.[7] This was an age 'becoming numerical and measured in every
corner of its being',[8] an 'era of statistical enthusiasm'.[9] For the first time 'the world
was constituted by measurements or countings'.[10] As one of those who collected
and published those numbers wrote in 1851, 'The materials . . . which relate to the
occurrences of the present century, are vastly superior in amount and value to
those that are to be collected from any existing records of an earlier date'.[11]

The profusion of numbers was not a cause of modernity, but rather, a notable
consequence of some of its features: the enhanced requirements of the modern
state, better and more professional public administration, a growing interest in
science and experimentation, the expansion of independent intellectual life, the
growth of commercial transactions. As nations grew in political sophistication, as
their populations increased, as their economies became more diverse and their
trade more complex, so more numbers were available, more were collected, and
more were needed for better regulation of these new societies: 'large-scale
statistics . . . became available, at the beginning of the nineteenth century, as a by-
product of the rapidly expanded census activities undertaken by various government agencies'.[12] The famous exponent of the concept of the 'scientific revolution',
Thomas Kuhn, has argued that there was a 'second scientific revolution' between
1800 and 1850, one that focused on measurement and the application of mathematics to physical phenomena.[13] In this period 'people came to regard much of

[5] Christopher Bayly, *Imperial Meridian. The British Empire and the World 1780–1830* (London, 1989).
[6] Paul Johnson, *The Birth of the Modern. World Society 1815–1830* (New York, 1991).
[7] Ian Hacking, 'How should we do the History of Statistics?', *I & C*, Spring 1981, no. 8, 18, 19, 20, 24.
[8] Ian Hacking, *The Taming of Chance* (Cambridge, 1990), 61.
[9] Harald Westergaard, *Contributions to the History of Statistics* (London, 1932), title to chapter 13.
[10] Ian Hacking, 'Prussian Numbers 1860–1882', in Lorenz Krüger, Lorraine J. Daston, and Michael Heidelberger (eds), *The Probabilistic Revolution* (2 vols, Cambridge, MA), vol. 1, *Ideas in History* (Cambridge, Mass., 1987), 392.
[11] G. R. Porter, *The Progress of the Nation in its Various Social and Economical Relations* (3rd edn, 1851), 2.
[12] Paul F. Lazarsfeld, 'Notes on the History of Quantification in Sociology – Trends, Sources and Problems', *Isis*, 52, 2, June 1961, 278.
[13] Thomas Kuhn, 'The Function of Measurement in Modern Physical Science', *Isis*, 52, 2, June 1961, 190.

nature as essentially numerical and measurable'.[14] Numbers seemed to offer, as well, a sure means by which to interrogate and understand the range of social developments that we now see as 'modern', and some of the earliest attempts to understand the economics and sociology of modern societies depended on the analysis of social statistics. Numbers were generated by social and political changes from the late eighteenth century and by about 1830 they were being used to explain those changes, as well. Many different social groups, across all classes and even some religious denominations, sought to appropriate numbers to their cause, to use them to construct or interrogate reality in their interests. 'A statistical way of looking at the world gave them confidence they could control it.'[15] Charles Babbage, meanwhile, at the very heart of the new statistical culture in Britain, looked ahead to the mechanical manipulation of numbers, and laid down the idea of, and a basis for, the invention that now most symbolizes modernity, the computer. This book is about the quest to understand social change in the nineteenth century and control it more effectively, whatever that implied and involved, using the data generated as the modern era began.

It concerns Britain for several reasons, pre-eminently because by several of the definitions of 'the modern' set out above, Britain was the place where modern history had its first home. The expansion of population, industrialization, representative government and empire were all aspects of British history from the mid-eighteenth century. Britain also generated a varied and complex statistical culture to account for these developments, though in this respect it was only one of several European countries in the period to foster institutions for the collection and analysis of social statistics. Nevertheless, as Theodore Porter has observed, 'Nowhere else was statistics pursued with quite the level of enthusiasm as in Britain.'[16] At the end of the nineteenth century it was in Britain, in the work of Francis Galton, Karl Pearson, Francis Ysidro Edgeworth, W. F. R. Weldon, and George Udny Yule among others, that the new discipline of mathematical statistics was initiated and refined, an intellectual achievement of the first order, 'a statistical revolution in social science akin to that which Laplace had launched in astronomy'.[17] This book focuses on the collection and use of data in Britain in the preceding era, however, which is less well understood. But if its focus is British, the history to be related has many international connections. At its founding meeting in Cambridge in 1833 it was agreed that Section F and its progeny 'would not be exclusively confined to the British Isles. It would be a part of their duty to

---

[14] Ian Hacking, 'Was there a Probabilistic Revolution 1800–1930?', in Krüger, Daston, and Heidelberger (eds), *The Probabilistic Revolution*, vol. 1, 51–2.
[15] William Lubenow, *Liberal Intellectuals and Public Culture in Modern Britain 1815–1914* (Woodbridge, Suffolk, 2010), 140.
[16] Theodore Porter, *The Rise of Statistical Thinking 1820–1900* (Princeton, NJ, 1986), 37.
[17] Stephen Stigler, 'The Measurement of Uncertainty in Nineteenth-Century Social Science', in ibid, 291. Stigler refers specifically to Galton, Edgeworth, and Pearson. See also Donald A. Mackenzie, *Statistics in Britain, 1865–1930: The Social Construction of Scientific Knowledge* (Edinburgh, 1981).

collect statistical materials from all parts of the world.'[18] The most significant intellectual influences on the founders of the statistical movement in Cambridge in the 1830s were a Belgian, Adolphe Quetelet, and a German, Alexander von Humboldt, the most famous nineteenth-century naturalist before Darwin. Charles Babbage gave the most enduring exposition of the mathematical ideas embodied in his calculating engines to an audience of Italian savants in Turin in 1840. The International Statistical Congress, which attempted to unite the statisticians of the world, was founded in London in 1851. When the artisans who created the London Statistical Society needed help in compiling their statistical critique of British life in the 1820s, they turned to a French diplomat and statistician then in London.

It has been argued that the 'statistical systems' that developed in the major states of the nineteenth and early twentieth centuries 'shared certain common intellectual origins, certain technical preconditions' but were 'distinguished by their relationship to politics'.[19] Comparative studies have certainly demonstrated differences between national statistical cultures in this period.[20] As one historian has put it, 'the French counted criminals, the British paupers, and the Prussians foreigners'.[21] Quetelet's ideas and example were deeply influential in Britain; in Germany, however, his work was received critically and was never endorsed.[22] British and French statisticians developed an 'atomistic' conception of social statistics, focused on individuals and their behaviours; in Germany the approach was more societal and 'holistic', in line with an older tradition of the compilation of descriptive and historical statistics in the German states in the eighteenth century.[23] In comparison with the descriptive work of French demographers, the instrumental and problem-orientated approach to the study of population and disease that developed in Britain generated more, and more usable, data. It more readily incorporated the work of cognate subjects including actuarial science and economics, and it was better placed, therefore, to support social reform and public policy.[24] In contrast to the centralization of statistical collection in both France and Prussia—indeed, of bureaucratic functions in general in both countries—the absence of a central statistical bureau in Britain and the reliance on learned societies and lone individuals to do some of the most intensive work with

[18] Ibid, f. 4.
[19] J. Adam Tooze, Statistics and the German State, 1900–1945. The Making of Modern Economic Knowledge (Cambridge, 2001), 39.
[20] Alain Desrosières, The Politics of Large Numbers. A History of Statistical Reasoning (1993) (Cambridge and London, 1998), 147–209. Libby Schweber, Disciplining Statistics. Demography and Vital Statistics in France and England, 1830–1885 (Durham N.C. and London, 2006), 220–2.
[21] Lorraine J. Daston, 'Introduction', in Krüger, Daston, and Heidelberger (eds), The Probabilistic Revolution, vol. 1, 3.
[22] Theodore Porter, 'Lawless Society: Social Science and the Reinterpretation of Statistics in Germany, 1850–1880', in ibid., 356–61.
[23] Ian Hacking, 'Prussian Numbers 1860–1882', in ibid., 377, 382–3.
[24] Schweber, Disciplining Statistics, esp. pp. 128–9, 214–15.

numbers made Britain something of an outlier.[25] Ian Hacking has contrasted 'the British way with numbers' that 'reflected a resistance to centralized management' with the establishment in Prussia of 'an office of numbers-in-general', a Bureau of Statistics, used by all departments of government and under government control.[26]

In the nineteenth-century world, the collection, publication, and use of statistics defined nations and, in the process, exemplified their cultural and institutional differences. Statistics 'helped to form a sense of what it meant to be British or American, or French, Prussian, Saxon or Italian'.[27] In Prussia (and in Germany after 1871, therefore) data collection and analysis came under the strong influence of state structures that were unsympathetic to the liberal internationalism of mid-nineteenth century statistics, a theme that will arise later when we consider the demise of the International Statistical Congress. In Britain, on the contrary, the very political independence of the statistical culture, which was never fully integrated into official bureaucratic structures, may have encouraged not only political independence but intellectual freedom and remarkable technical innovation. Whatever we may think of Francis Galton, who dominates the final section of this book, he made the most outstanding contribution to the development of statistics in the nineteenth century and was an independent English gentleman of means, without any connection to the British state, who was educated, as it happens, at Trinity College, Cambridge.

Soon after the Statistical Society of London began work, it characterized statistics as a new type of language: 'the spirit of the present age has an evident tendency to confront the figures of speech with the figures of arithmetic'.[28] This was entirely appropriate in a period that had to find a way of measuring, conceptualizing and discussing the new industrial civilization that had grown up in its midst. Only numbers could express the new scale and speed of mechanized production and the growth of the towns and cities in which that production was taking place. In this sense, the growing interest in, and use of, statistics expressed the new realities of the nineteenth century. We can also think of statistics as a new type of technology offering fresh and surprising methods for the analysis and understanding of society. Like many technologies, it was developed subsequently in ways that could not have been envisaged. Everyone, and every social group in the 1830s, wished to use statistics to make their case more effectively, whether Cambridge dons who saw in statistics a new and exciting scientific technique, the Whig political grandees and professional men who founded the Statistical Society of London as a centre for the origination of social policy, the Manchester

[25] Alain Desrosières, *The Politics of Large Numbers*, 147–77.
[26] Ian Hacking, *The Taming of Chance*, 33.
[27] Theodore M. Porter, 'Statistics and the Career of Public Reason. Engagement and Detachment in a Quantified World', in Tom Crook and Glen O'Hara (eds), *Statistics and the Public Sphere. Numbers and the People in Modern Britain, c. 1800–2000* (New York and Abingdon, 2011), 40.
[28] 'Prospectus', *Journal of the Statistical Society of London*, 1, 1839, 8.

industrialists who used statistics to count their economic contribution and to promote their political and religious causes, or the London artisans who used statistics to hold the governing class to account and to advance an alternative political economy in the interests of the working classes.

Contestation over social statistics began long before the 1830s but became acute in that decade as different social interests, now better organized, tried to advance their views by marshalling numbers and using them in their defence. This was the decade that witnessed the first cracks in the aristocratic dominance of British politics, the political emergence of the British middle classes with the Great Reform Act of 1832, and the gathering of working-class protest into Chartism, the movement for manhood suffrage. It was the era in which the 'language of class' developed to express and define the new social divisions in the first industrial society.[29] Everyone reading this book will be used to political and social arguments over statistics in contemporary society: numbers are central to understanding and governing ourselves, and a battleground over which rival ideologies and policies contend. In Shakespeare's *Othello*, Iago is contemptuous of Cassio who would fight a battle by numbers:

> And what was he?
> Forsooth, a great arithmetician...
> That never set a squadron in the field
> Nor the division of a battle knows
> More than a spinster;

But though Iago scoffs that 'mere prattle without practise is all his soldiership', Cassio's military skills win Othello's favour.[30] Even in the sixteenth century there was no place in command for the innumerate. On today's 'digital battlefield' no general can be without the latest technology and the contemporary equivalent of the best arithmeticians.

The argument of this book is that the collection and analysis of statistics was an increasingly important aspect of the liberal social reforms of the early and mid-nineteenth century, although at this stage statistics as a discipline was relatively underdeveloped as a mode of social and scientific procedure. Paradoxically perhaps, its transformation from the late-Victorian period onwards into a much more sophisticated and powerful analytical technique was inspired by Francis Galton, his protégé Karl Pearson, and other statisticians influenced by eugenics. They rejected key principles of Victorian liberalism in favour of the authoritarian, socially-divisive, and racist ideas and applications of eugenics. In the 1830s and

---

[29] Asa Briggs, 'The Language of "Class" in Early Nineteenth-Century England', in A. Briggs and J. Saville, *Essays in Labour History* (London, 1960), 43–73.
[30] *Othello*, Act 1, scene 1, lines 19–28.

1840s, however, during the so-called 'Age of Reform', the emerging discipline of statistics offered support to a broad liberal environmentalism—'environmental ameliorationism' as one historian has termed it—which then characterized all notable movements for social change.[31] According to Babbage's friend and biographer, H. Wilmot Buxton, Babbage 'believed that the greater portion of the misery of which the human race is heir to, is caused by ignorance of the means of modifying and adapting nature to the wants of man'. Accurate and reliable social data would rectify that ignorance. Babbage valued mathematics 'so far as it was subservient to practical purposes, and on account of its usefulness to the ends of life'.[32]

It was widely believed that statistics would demonstrate social regularities and fixed patterns of behaviour.[33] At Cambridge in 1833 the stated aim of the new statistical section was to collect 'facts relating to communities of men which are capable of being expressed by numbers, and which promise when sufficiently multiplied, to indicate general laws'.[34] Those general laws gave hope, indeed evidence, that human behaviour was not random and unpredictable, but repetitive and consistent, and therefore amenable to good influences and to change. According to Quetelet, even the predictability of criminal behaviour

> so discouraging at first sight becomes, on the contrary, consolatory, when examined more nearly, by showing the possibility of ameliorating the human race, by modifying their institutions, their habits, the amount of their information, and generally, all which influences their mode of existence.[35]

It is the cry, the belief, of all liberal reformers: change the environment, the background, the context, and behaviour can be changed as well. By studying statistics and appreciating the patterns in social behaviour, social pathologies may be ameliorated. One notable historical demographer has described 'the opportunity which quantitative data offered to test consistency, to make inferences about missing information, to note the existence of relationships or correlations, in short to find structure' in this period.[36] Whigs at Westminster, liberals in Manchester, and radicals from Clerkenwell in London, all centres of statistical culture, were seeking this 'structure' and all believed that the social problems of

---

[31] Simon Szreter, *Fertility, Class and Gender in Britain 1860–1940* (Cambridge, 1996), 86.
[32] H. W. Buxton, *The Life and Labours of the Late Charles Babbage* (1872–80) (ed. Anthony Hyman) (Cambridge, Mass., 1988), 353.
[33] Llewellyn Woodward, *The Age of Reform 1815–1870* (Oxford, 1963).
[34] *Report of the Third Meeting of the British Association; Held at Cambridge in 1833* (London, 1834), xxxvii.
[35] L. A. J. Quetelet, *Sur L'Homme et le Développement de ses Facultés. Physique Sociale* (Brussels, 1835). *A Treatise on Man and the Development of his Faculties* (Edinburgh, 1842) 'Introductory', p. 6.
[36] E. A. Wrigley, 'Comments', in Richard Stone, *Some British Empiricists in the Social Sciences 1650–1900* (Cambridge, 1997), 423.

early Victorian Britain would be revealed, and could be understood and treated, with the help of accurate statistics.

The symbol of this belief, and the very zenith of the statistical movement in Britain, was the fourth meeting of the International Statistical Congress in London in 1860. It was opened by Albert, Prince Consort, another of the great Victorian reformers, and almost all the key figures in this story, from Babbage to Florence Nightingale, were present. Indeed, the ISC itself was the brainchild of Quetelet and Babbage, united once again at the Great Exhibition of 1851 in London where they hatched the plan for the internationalization of statistics. Because the Congress offers the best possible illustration of the hopes invested in statistics in Victorian Britain, this book opens with it, then moves backwards to 1830 to explain why such distinguished company came together in a festival of statistics of all things, and then goes forward, beyond 1860, to consider the decline of this liberal movement and its resurgence in a different guise.

By the 1870s, there was a sense that the project of understanding and reforming society through numbers was losing impetus. The case studies of three notable physicians, William Farr, William Guy, and John Simon demonstrate how, in different ways—intellectual, practical, bureaucratic, and political—the statistical movement was losing ground or simply being taken for granted, having lost its crusading zeal and the sense that it would, alone, change social conditions and human health for the better. Farr's 'laws of disease' could not explain epidemics; Simon's department for the study of environmental effects on health was sidelined and disbanded; Guy lost faith that statistics really were the science of society. The International Statistical Congress was undermined by conflicts across Europe caused by the unifications of Italy and Germany in the 1860s, and then by the rise of nationalism. Meanwhile, the claim by Henry Thomas Buckle that he could write a systematic history of humanity based on statistical regularities, which had taken the nation by storm in 1857, was, under sustained historical and philosophical scrutiny, coming apart. As the liberal project faltered, so statistics took on a new guise. Under the impact of Darwinian ideas, and in the work of someone who did not follow liberal norms, Francis Galton revolutionized the way in which numbers could be interpreted and applied. He founded modern mathematical statistics in the 1870s and 1880s, but used the new discipline to advocate change of a quite different nature that would divide and differentiate between social and racial groups. In London in 1860 men and women spoke often of the unity of nations and mankind that would be achieved through statistical collection and the reforms thus made possible. Yet within a generation, as a consequence of the development of mathematical statistics, statistics had become an integral aspect of eugenics, the new pseudo-science of human perfectibility and Darwinian social ethics.

The term 'Statistical Movement' has been applied in this book to all of the history it covers up to Galton. It captures the sense of a multi-faceted social development, unfolding over time and including within itself many different elements, some of them only barely compatible. In the word 'movement' it conveys the almost evangelical belief of men and women like Quetelet, Farr, and Nightingale in the reformative power of numbers. By design, it is a capacious and none-too-definite characterization of the way in which numbers touched many different aspects of Victorian society, political life, and letters. To have applied any more formal definition would have required excluding elements that rightly belong. One of the first and best historians of the statistical movement argued that by the mid-1840s 'as a broadly based interconnected national endeavour the statistical movement had collapsed.'[37] That may have been the conclusion reached by limiting a definition of the movement to the institutions founded in its first years, many of them small provincial statistical societies established in the mid-1830s which had indeed disappeared within a decade. It may also have stemmed from thinking about 'statistics' only, rather than conceiving of the broader category of 'numbers' and the multiple roles that they played in the Victorian period: as a medium for the representation of social and economic life; as a method in the natural and social sciences; as a way of making social policy; as a tool by which to understand and eradicate disease; as raw material for the critique of the social and political status-quo; as a way of thinking about social relationships in general.

> Charles Babbage was the
> first to range
> Lone analytic heights, through pathways dim,
> By lettered sign and symbol quaint and strange.[38]

He believed that everything could be reduced to 'number' and he used that word as a generic term for all his mathematical exploits.[39] He looked forward to the enormous range of scientific and practical advantages that would flow from 'the complete control which mechanism now gives us over number'.[40] As they worked together to explain his great invention to the world, he described his friend and collaborator, Ada, Countess of Lovelace, as 'the Enchantress of Number'.[41] For her part, Ada described her dismay 'at having got into so amazing a quagmire and

---

[37] Michael J. Cullen, *The Statistical Movement in Early Victorian Britain. The Foundations of Empirical Social Research* (Hassocks, Sussex, 1975), 132.

[38] 'Burgoyne – Murchison – Babbage', *Punch*, 4 Nov. 1871.

[39] Maboth Moseley, *Irascible Genius. A Life of Charles Babbage, Inventor* (London, 1964), 23. Jenny Uglow, 'Introduction', in F. Spufford and J. Uglow (eds), *Cultural Babbage. Technology, Time and Invention* (London, 1996), 2.

[40] [C. Babbage] 'On the Mathematical Powers of the Calculating Engine', 26 Dec. 1837, Mss draft. Buxton Collection, Oxford Museum of the History of Science.

[41] Charles Babbage to Ada Lovelace, 9 Sept. 1843, quoted in Moseley, *Irascible Genius*, 191.

botheration with these *numbers*'.[42] But she persevered because she recognized that 'without numbers it is not given to us to raise the veil which envelopes the mysteries of nature'.[43] According to Babbage's Italian disciple and expositor, Luigi Frederigo Menabrea, later a prime minister of the new Italian state, Babbage's calculating engine 'may be considered as a real manufactory of numbers'.[44]

Taken in this broader sense as an intellectual and cultural response to the proliferation of social data from the 1830s, the statistical movement was flourishing in the late 1840s and long after that as well. When Carlyle, Dickens, and Ruskin took aim at statistics and the related trends of exactitude and precision, they were opposing a tendency in the thought and culture of the age which had already installed itself in heart as well as head—in both Victorian sensibility and mentality. Numbers had entered into the consciousness of the era and these authors found that deeply unsettling. Yet, to many others, numbers seemed to offer a new means for both understanding society and curing it of its ills. The London meeting of the ISC in 1860 was the symbol of confident practitioners convening at a time of international optimism, a forum for a movement in full advance. The term 'statistical movement' only lost its validity and applicability as that very liberal optimism lost its potency in the following generation. By this time the term 'statistics' was no longer simply interchangeable with 'numbers'. It denoted the mathematical framework required to understand those very numbers. It was the academic subject, in its own right, by which numbers could be studied and applied across all disciplines.[45] The development of mathematical statistics at this point, in the 1880s, and their use by some of the most gifted statisticians to advocate various types of social manipulation and control, brought to an end the sense of a social tendency and a reformative campaign as captured in the name 'statistical movement'.

The different definition and chronology of the statistical movement as used in this book are also consequences of a different approach to its history: one that focuses more on the intellectual aspirations of the movement than the study of institutions and the role of statistics in the public sphere.[46] Important agencies and bureaucracies that collected and used statistics like the decennial national census and the system of registration are not covered in any detail here, not least

---

[42] Ada Lovelace to Charles Babbage, ND (? 21 July 1843), Babbage Papers, British Library, Add. Mss, 37192, f. 382.

[43] Ada Lovelace, 'Sketch of the Analytical Engine Invented by Charles Babbage by L. F. Menebrea. With Notes upon the Memoir by the Translator, Ada Augusta, Countess of Lovelace', *Taylor's Scientific Memoirs*, 3, 1843, 666–731. Quotation from Note A.

[44] Ibid., 690.

[45] Mackenzie, *Statistics in Britain*, 7. Martin Brookes, *Extreme Measures. The Dark Visions and Bright Ideas of Francis Galton* (London, 2004), 228.

[46] For coverage of this theme, see the excellent essays in Crook and O'Hara (eds), *Statistics and the Public Sphere*, referred to at fn. 29 above.

because they are the subject of excellent pre-existing work.[47] Similarly, though the commercial applications of statistics, especially in the insurance business, are discussed in the context of Babbage's polymathic interests and activities, this subject has also been covered recently by historians.[48] While the origin of the different statistical societies in Cambridge, London, Manchester, and Clerkenwell in the 1820s and 1830s is the subject of the book's opening section, their subsequent institutional history is of less importance than the examination of the different intellectual and political projects they embodied. The case studies of Babbage, Whewell, Jones, Quetelet, von Humboldt, Buckle, Nightingale, Galton, Booth et al., not to mention Farr, Guy and Simon, the mid-Victorian physicians who used numbers to try to understand disease, show how many of the finest minds of the era sought to use numbers in the study of society as well as nature. This common approach and methodology, as well as the mutual relations, collaborations and arguments between some of the key figures discussed in the chapters to follow, unite many of the case studies in the book. Taken together, these studies cannot represent all the many applications of statistics in nineteenth century Britain, and the book is not attempting to do that. But I hope I have exposed some of the key themes and issues that make it possible to insert the history of Victorian statistics into the wider social and intellectual history of the age.

According to the great astronomer, John Herschel, who was intimate with many of the intellectual founders of nineteenth-century statistics, 'What astronomical records or meteorological registers are to a rational explanation of the movements of the planets or of the atmosphere, statistical returns are to social and political philosophy.'[49] When taken together, these examples constitute an overlooked thread, or stream, or tradition in Victorian culture: collecting, counting and calculating would, it was believed, reveal the fundamental structures of both the natural and social worlds. A single approach, a unitary method, united research across many different disciplines at this time. This was the methodology

---

[47] D. V. Glass, *Numbering the People: The Eighteenth-Century Population Controversy and the Development of Census and Vital Statistics in Britain* (Farnborough, 1973); R. Lawton, *The Census and Social Structure* (London, 1978); M. Nissel, *People Count. A History of the General Register Office* (London, 1987); S. Szreter (ed.), 'The General Register Office of England and Wales and the Public Health Movement 1837–1914. A Comparative Perspective', Special Issue, *Social History of Medicine*, vol. 4, no. 3, Dec. 1991; Edward Higgs, *Making Sense of the Census. The Manuscript Returns 1801–1901* (London, 1989) and *A Clearer Sense of the Census: The Victorian Censuses and Historical Research* (London, 1996).

[48] Timothy L. Alborn, 'A Calculating Profession: Victorian Actuaries among the Statisticians', *Science in Context*, 7, 3 (1994), 433–68, and *Regulated Lives: Life Insurance and British Society, 1800–1914* (Toronto, 2009); Barry Supple, *The Royal Exchange Assurance: A History of British Insurance 1720–1970* (Cambridge, 1970); Clive Trebilcock, *Phoenix Assurance and the Development of British Assurance. Volume 1: 1780–1870; Volume 2: The Era of Insurance Giants, 1870–1984* (Cambridge, 1985, 1998).

[49] [Sir John Herschel], 'Quetelet on Probabilities', *Edinburgh Review*, xcii (July–Oct. 1850), 41.L. A. J. Quetelet, *Letters addressed to HRH the grand-duke of Saxe-Coburg and Gotha on the theory of probabilities as applied to the moral and physical sciences* (London, 1849).

made famous by the great German naturalist, Alexander von Humboldt, who was known to, and admired by, several of the leading figures in this story. The collection of statistics, whether of the poor in London, the uneducated in Manchester, or of the movement of the tides across the world's oceans, were collaborative enterprises stemming from a shared hope in the 1830s that the world might be understood and set right through numbers.

Many of the key figures in this story were natural scientists widely defined: Charles Babbage was a mathematician and engineer; Quetelet and Herschel were both astronomers; William Farr trained and practised briefly as a physician; William Whewell, a polymathic physicist, actually coined the very word 'scientist' in 1833 at the aforementioned meeting of the British Association in Cambridge.[50] This is hardly surprising: one of the great intellectual enterprises of the nineteenth century was the application of natural scientific methods and concepts to the study of society. The term 'social science', dating from the era of the French Revolution, was in use from the 1830s, coinciding with the development of social statistics in that decade and with the widespread claim that the collection of numerical data would provide a truly scientific approach to the study of society.[51] In a famous statement that defined the nature of the first scientific revolution of the seventeenth century, Galileo had written in 1623 that

> Philosophy is written in this grand book — I mean the universe — which stands continually open to our gaze, but it cannot be understood unless one first learns to comprehend the language in which it is written. It is written in the language of mathematics, and its characters are triangles, circles, and other geometric figures, without which it is humanly impossible to understand a single word of it; without these, one is wandering about in a dark labyrinth.[52]

Two centuries later, though the intellectual task had moved on, it relied on a similar language: many Victorian statisticians believed that society and humanity were also best understood mathematically through the analysis of social data, for which similar skills as implied by Galileo were required. As one of those most taken with this methodology, the historian Henry Thomas Buckle, was to write,

> It is this deep conviction, that changing phenomena have unchanging laws and that there are principles of order to which all apparent disorder may be referred, - it is this, which, in the seventeenth century, guided in a limited field Bacon,

[50] Laura J. Snyder, *The Philosophical Breakfast Club. Four Remarkable Friends Who Transformed Science and Changed the World* (New York, 2011), 1–3.
[51] Brian W. Head, 'The Origins of "La Science Sociale" in France', *Australian Journal of French Studies*, vol. xix, 2, May–Aug 1982, 115–32. See below, pp. 133–4.
[52] Galileo Galilei, *Il Saggiotore* (The Assayer) (1623), quoted in R. H. Popkin, *The Philosophy of the Sixteenth and Seventeenth Centuries* (1966), 65.

Descartes and Newton; which in the eighteenth century was applied to every part of the material universe; and which it is the business of the nineteenth century to extend to the history of the human intellect.[53]

Among practitioners as well as some historians, the view is sometimes encountered 'that the period before about 1880 is generally regarded within the social sciences as pre-history and not as history proper of their fields'.[54] As this book will show, nineteenth-century statistics was never just a 'dry Victorian discipline concerned with the manual manipulation of columns and numbers'[55] but, in the minds of its intellectual leaders, a struggle to uncover the deep structures of the social and natural worlds through the numbers they generated, which could be collected and analysed. To discover and demonstrate the order of nature has been an intellectual quest since ancient times. It has always been one of the goals of mathematics itself, which is characterized by 'the quest for order... to build systems of knowledge... to form patterns of explanation'.[56] As Stephen Stigler has argued, the development of nineteenth century statistics can be understood to have comprised two main phases: first, the search for regularity and order within large masses of data in the age of Quetelet, and then, from the 1880s, the refinement of techniques for understanding the nature and causes of variation within those masses in the era of Galton, Pearson, Edgeworth, and those who came after.[57] The quest for order goes on: a new sub-branch of mathematics is focused on 'nature's numbers', the 'deep mathematical regularities that can be detected in natural forms' from the structure of petals on a flower, and snowflakes falling to earth, to the stripes on a tiger.[58] We live in a universe which 'organises itself along mathematical lines'.[59] To many Victorians, that universe was social as well as physical, and could be studied in the same way.

This book will also argue, however, that during the first phase, the 'age of Quetelet', the quest was unsuccessful for three broad reasons. First, because the statistical techniques and applications available to the social-statistical movement for the half century or so between the 1820s and 1870s were, by our standards, simple and crude. If deeper structures and patterns existed they could not be teased-out merely through the collection of data and their subsequent contemplation in the privacy of the library. It took the development of mathematical statistics

---

[53] Henry Thomas Buckle, *History of Civilization in England* (2 vols) (London, 1857, 1861), I, 807.
[54] Theodore M. Porter, 'Statistics and the Career of Public Reason', in Crook and O'Hara (eds), *Statistics and the Public Sphere,* 38.
[55] David J. Hand, *Statistics. A Very Short Introduction* (Oxford, 2008), 1.
[56] Morris Kline, *Mathematics. The Loss of Certainty* (Oxford, 1980), 353.
[57] Stephen Stigler, 'Francis Ysidro Edgeworth, Statistician', *JRSS,* series A, 141 (1976), 309–10.
[58] Ian Stewart, *Nature's Numbers. Discovering Order and Pattern in the Universe* (London, 1995) (1996 edn), 165. See also Ian Stewart, *Life's Other Secret. The New Mathematics of the Living World* (New York, 1998). Work of this type takes inspiration from D'Arcy Thompson's evidence for the mathematical determination of biological forms in his seminal study *On Growth and Form* (1917).
[59] 'The Nature of Numbers', *The Times,* 29 June 1998, 15 (interview with Ian Stewart).

from the 1880s to provide statisticians with the tools required for sophisticated social analysis. Second, social statistics on their own might illustrate and exemplify trends and behaviours, but they could not explain them adequately. It was Darwin's theory of natural selection that ignited Galton's interest in numbers and prompted his remarkable technical virtuosity in founding mathematical statistics. Darwinism gave him a way of seeing the world, albeit one that led him to embrace profoundly anti-social conclusions. Third, William Farr spent his whole life searching for theories of disease in the statistics generated by the periodic epidemics that swept across Victorian Britain, but the microbiological causes of illness and death were uncovered and explained by quite other means in the laboratory at the end of a microscope. Ironically, it was in the year of Farr's death, 1883, that Robert Koch linked cholera, the focus of Farr's statistical research, to the waterborne organic vector of the disease, the comma bacillus. That undermined all theories of disease in this era premised on the vague idea of 'bad air' and dangerous miasmas, as it also 'reduced the relative importance of statistical epidemiology' as Farr, John Simon, and other physicians discussed in the third section of this book had practised it.[60]

This study thus records a complete cycle of intellectual history, examining the rise of statistics from the 1830s in relationship to liberal reforms in public administration, trade, public health, education, and other aspects of Victorian life. But from the 1870s this ideological and intellectual complex was in decline as many of its assumptions were undermined. The final chapters of the book examine this decline through studies of the International Statistical Congress, founded in London in 1851, which was weakened by conservative nationalism in Europe after 1870; by considering the failure of 'statistical medicine' to explain disease causation and transmission; by the cessation of creative medical research sponsored by central government, which was a consequence of bureaucratic reorganization in the early 1870s; and most important of all, by explaining the advent of mathematical statistics led by Francis Galton, who revolutionized the techniques of statistical analysis and then used them for conservative social ends. This latter transition is examined through the study of Galton himself, including his public criticism of the statistical movement as it existed in the late 1870s, his presentation of new mathematical methods at the Royal Statistical Society in 1885, and his interaction with Florence Nightingale in 1891. It was when she applied to him for help in establishing a chair of statistics in Oxford that the differences between the old social statistics, of which Nightingale was a passionate devotee, and the new mathematical statistics became evident.

I have entitled the Nightingale/Galton section 'The end of the Statistical Movement'. It was the final act in a process that extended over two decades. There

---

[60] Desrosières, *The Politics of Large Numbers*, 170.

is no generally recognized moment in this period when one intellectual and ideological formation gave way to another, but Nightingale's unsuccessful application to Galton has symbolic significance. Whether we assign any specific year or event as the endpoint, or focus instead on a range of related changes over a longer period, there is a natural and acknowledged transformation in the 1880s—acknowledged by contemporaries and subsequent historians, both. If the older 'descriptive statistics' were employed to understand the characteristics of whole populations and were closely associated with the introduction of liberal reforms between the 1830s and the 1870s, mathematical statistics, so useful for understanding the *differences* between phenomena, were generated, taken up, and used by eugenicists and Social Darwinists at the end of the century to try to achieve discriminatory social outcomes. These interrelated technical and ideological changes, which were mutually reinforcing, established statistics on a new academic and professional basis, and they mark a break with the 'social counting' which had preceded the 1880s, little of which had gone beyond the plotting of distributions and the calculation of averages.

It is important to recognize, however, that this transition did not result in the eradication of earlier types of statistical work, which is the point of the book's final case study of the 1885 Industrial Remuneration Conference at which the conventional social problems of the decade were addressed by distinguished participants from all levels of society in conventional ways: by collecting descriptive data on wages, piece-rates, and prices, and arguing about them. The IRC thus links back to the research and the views of the artisan social statisticians of the 1820s, the subject of Chapter 4, and points forward to enduring debates, up to the very present, about distribution and social equity in our own society. These issues have been central to the development of British statistics for two centuries: in every generation statistics have been used, and statisticians have come to the fore, in the discussion of social justice. Galton, Pearson, and their collaborators may have reconfigured statistics at a much higher level of mathematical and technical sophistication, and harnessed this reformulated discipline to eugenics, but many citizens, civil servants, and reformers remained both unimpressed and uninfluenced by these innovations, whether technical or ideological. The Liberal governments between 1906 and 1914 pursued traditional environmental reforms in measures like the Children's Act of 1908, free school meals, old age pensions, and national insurance, though social innovations of this type were now contested by those who would use statistics in support of illiberal policies and ends.

Great social theorists of the nineteenth century like Tocqueville, Mill, and Marx certainly used data published by the statistical movement. Tocqueville attended a meeting of the Manchester Statistical Society and Mill went to the Statistical Society of London in 1835 and 1860 respectively, while Marx heaped praise on the work of British statisticians and the veracity and profusion of the social information they provided. But social statistics were for these thinkers, and for other students of society, the material from which theory might emerge, not the theory already formulated. In short, in studying the statistical movement in this period we are examining a stage in intellectual history when the profusion of numerical information seemed to offer the prospect of a science of society *in itself*. The movement ended, roughly in the 1880s, by which time far more powerful intellectual paradigms had emerged, like Darwinism and Marxism, that offered apparently coherent, and to their devotees, highly attractive and persuasive explanations of social structure and function. It is only in our own age, in the current generation, that numbers themselves, in the form of so-called Big Data, have once again opened up the possibility of the structural approaches to human behaviour, health, and welfare that were glimpsed, but never realized, by some Victorians.

Peter Laslett, the historian of both political thought and population, once remarked that 'tracing the ultimate origins of the statistical preoccupations of the Fabians and the Webbs back to the earliest work of Graunt, Petty, and King stands as a challenge to intellectual historians. It would have to be pursued largely beneath the surface for much of its length, with vagaries and discontinuities conspicuous at every stage.'[61] This book, though it is focused on the nineteenth century, is a response to that challenge. To explain *Victorians and Numbers* it has certainly been necessary to go back to the origins of 'political arithmetic' as it was known from the late seventeenth century, and the narrative does indeed end with the very first public outing of the Fabian Society, members of which came down from Hampstead, where the society first met in 1884, to Piccadilly in January 1885, to participate in the Industrial Remuneration Conference in London.[62] It covers the period from Graunt's work on London's vital statistics in the 1660s until Charles Booth's great investigative project on *The Life and Labour of the People of London* at the end of the Victorian period, to which Beatrice Webb was an early and key contributor.[63] But little of this is, or needs to be, subterranean. If some of the figures examined here are not as well-known as they might be, such as Drs. Guy, Simon, and Jarvis, many of those who feature in this story are extremely significant: Malthus, Whewell, Herschel, Prince Albert, Francis Galton, et al. Among them are those who were transfixed by the potential and poetry of numbers: Babbage, his collaborator Ada Lovelace, and his friend Adolphe Quetelet; Florence Nightingale and *her* collaborator, William Farr; Henry Buckle and the woman he loved and to whom he proposed marriage, who left the best account of the man and his work, the feminist pioneer, Emily Shirreff. Two of the speakers who made the most impression at the Industrial Remuneration

---

[61] Peter Laslett, 'Comments', in Stone, *Some British Empiricists*, 402.

[62] See p. 298 below.

[63] For an outline of some of the themes that explain historical continuity across this period as a whole, see Lawrence Goldman, 'Social Reform and the Pressure of 'Progress' on Parliament, 1660–1914' in Richard Huzzey (ed.), *Pressure and Parliament. From Civil War to Civil Society, Parliamentary History: Texts and Studies*, 13, 2018, 72–88.

Conference in 1885 were women with a close understanding of working-class life: Emma Paterson, the pioneer of women's trades' unionism in Britain, and Edith Simcox, political activist and anthropologist.

The focus on such notable Victorians is not difficult to explain: it is evidence in itself that statistics were of cultural and intellectual significance in this period and came to fascinate some of the most notable people of the age. As has been pointed out by several of the leading historians of the social sciences, these disciplines have 'two principal sources: the politico-social ideas or doctrines on the one hand; the administrative statistics, surveys and empirical investigations on the other'.[64] To get at the latter we need to examine 'the men, the teams, and the organizations which built systems of political arithmetic, looked for regularities to prove divine order or societal laws, wanted to know what industrial society really did to people'.[65] One historian of the social sciences paid tribute to the statisticians of this era who 'are allowed to dwell in oblivion' nonetheless.[66] This book is not entitled 'Victorians and Numbers' by accident because it is intended to do justice to the contribution of these otherwise forgotten figures. It has been possible to focus on them because, in the past generation, we have found out, consolidated, and made available so much more biographical information than before. We are better able to place, contextualize, and relate together different Victorian lives, and show how they combined—in this case, in the pursuit of social knowledge. Where earlier historians made admirable inroads into this subject by focusing on institutions in studies of the statistical societies and scientific organizations of the era, and more recent studies have concentrated on 'disciplinary formation'— the development of the subject known as statistics[67]—this book is better able to explain the history of Victorian statistics and society through the biographies of some of the major and minor figures of the age, and through prosopographies— collective biography—of those brought together in the name of statistics. The real nature of the Manchester Statistical Society can only be understood by intense focus on the collective religious and social identities of its founders. John Simon brought together a remarkable group of young medical researchers and practitioners who used statistics reflexively in their work. The identity of the artisans in the London Statistical Society is difficult to establish, but collectively and self-consciously, they stood and spoke for their class.

There are other ways of writing this history, undoubtedly. Mary Poovey, for example, has examined the 'discourse of statistics' to uncover the different

---

[64] Raymond Aron, *Main Currents in Sociological Thought* (2 vols) (Harmondsworth, 1967 edn.), II, 8.

[65] Paul F. Lazarsfeld, 'Toward a History of Empirical Sociology' in E. Privat (ed.) *Mélanges en l'honneur de Fernand Braudel. Methodologie de l'Histoire et des Sciences Humaines* (Toulouse, 1973), 290.

[66] Edward Shils, 'Tradition, Ecology and Institution in the History of Sociology', *Daedalus*, 99 (1970), 766.

[67] Schweber, *Disciplining Statistics*, et seq.

approaches taken by a variety of authors to the new profusion of social numbers. She emphasizes 'the tensions and contradictions inherent in statistical discourse as it was consolidated in Britain in the first decades of the nineteenth century'.[68] Statistics were hailed as the building-blocks for new social theories and also as the empirical antidote to inappropriate theorizing. They were used to emphasize national unity but also deployed to demonstrate the differences between regions. They could be marshalled to explain and excuse the poverty of the poor by those adopting an environmentalist position, or to blame the poor for their situation by those more attracted to moralism.[69] The approach taken in this book, however, is largely sociological and contextual rather than discursive. The different ways of thinking about, and using numbers, whether among the savants in Cambridge, the entrepreneurs in Manchester, the political class in Westminster, or the artisans in Clerkenwell—respectively academic, commercial, governmental, and radical—are explained by reference to the people involved in each different case, their intellectual outlooks, their social interests, the institutional constraints which they accepted or against which they struggled, and their relationship to government and to politics. Hopefully, by focusing on the identity and experiences of those who played roles in the statistical movement, whether individually or in groups, the history of statistics will come alive as a quest, undertaken by many different figures from varied backgrounds, for different forms of social knowledge and amelioration.[70]

It will be evident already that this is no sort of 'Whig history' of intellectual endeavour and scientific progress. Whig history starts from the present and casts back to find the precursors of current beliefs, ideas, and practices, thereby over-looking or obliterating the more complex path that the actual history will have followed. Whig history writes the failures, the missed opportunities, and the defeated out of the record. In the history of science and of intellectual life it is characterized by a chain of successful inquiries, each building on the last, and together taking knowledge forward in a linear fashion up to the present under-standing. The incorrect ideas, those ultimately disproved, and the false turns thereby taken by science, are ignored. But in this study of the impact of numbers in the nineteenth century it will be clear already that many of the projects included for discussion were heroic failures and that the overriding aim of finding pattern, order, and structure in society by examining social data was not achieved, often because the statistical techniques and technology required for sophisticated

---

[68] Mary Poovey, 'Figures of Arithmetic, Figures of Speech: The Discourse of Statistics in the 1830s', *Critical Inquiry*, vol. 19, 2, Winter 1993, 256–76. Quotation at 258.

[69] For further remarks on Poovey's approach to the history of statistics, see ch. 16, Conclusion, pp. 318–19 below.

[70] For some suggestions on the relationship of biography to the social sciences, see Lawrence Goldman, 'Foundations of British Sociology 1880–1930. Contexts and Biographies', *The Sociological Review*, 55, 3, August 2007, 431–40.

analysis did not then exist. Nor can this be what historians of science refer to as an 'internalist history', which is the history of the development of the key paradigms, concepts, and techniques within the subject itself. The historiography of statistics has benefited in recent decades from several very important additions to an internalistic account of its development, notably in the work of Lorraine Daston, Theodore Porter, Stephen Stigler, and Alain Desrosières.[71] But in general, and in early and mid-Victorian Britain specifically, in the field of social statistics there was limited internal development only.

Before Francis Galton, social statisticians 'often knew very little statistics'.[72] They 'spent little time reflecting on the hierarchy of knowledge, the boundaries between disciplines, and the status of vital statistics as a science'.[73] The statistical movement concentrated on aggregates and mean values. Numbers were collected, tabulated, averaged, and compared with growing confidence, for sure. Victorian social statisticians knew about, and made use of, the so-called error law, also known as the Gaussian or the normal distribution. In Quetelet's case it was crucial to his quest to determine 'the average man'.[74] William Farr devoted one of his famous 'Letters' in 1843 to the mathematics of constructing a life table to compute life expectancy.[75] But genuinely innovative mathematical operations on the data were uncommon. Most work was limited to 'descriptive statistics' so-called, which involved its classification and representation in simple line graphs, bar charts, histograms, pictograms, and the like, and the computation of sample means, variances, and ranges. Until relatively late in the nineteenth century 'the emphasis of statistics...remained upon the accumulation, organization, and careful presentation of numerical data'.[76] Farr 'made little attempt to use the most sophisticated mathematical methods and never became expert in mathematical theory'.[77] His equivalent in Berlin, Ernst Engel, was 'mathematically unsophisticated' as well.[78] According to one modern author, 'statistics as understood in early Victorian Britain...was a science of individual facts and observations that showed remarkable indifference to mathematics'.[79] Another author, in search of

[71] Stephen M. Stigler, *The History of Statistics: The Measurement of Uncertainty before 1900* (Cambridge, MA, 1986); Theodore M. Porter, *The Rise of Statistical Thinking, 1820–1900* (Princeton, NJ, 1986); Desrosières, *The Politics of Large Numbers*, passim.
[72] Daston, 'Introduction', in Krüger, Daston, and Heidelberger (eds), *The Probabilistic Revolution*, vol. 1, 3.
[73] Schweber, *Disciplining Statistics*, 94.
[74] Victor L. Hilts, 'Statistics and Social Science', in Ronald N. Giere and Richard S. Westfall (eds), *Foundations of Scientific Method. The Nineteenth Century* (Bloomington, Ind., 1973), 214. Theodore M. Porter, *The Rise of Statistical Thinking 1820–1900* (Princeton, NJ, 1986), 6.
[75] [William Farr], General Register Office *Fifth Annual Report of the Registrar General* (1843), xviii–xix.
[76] Joseph J. Spengler, 'On the Progress of Quantification in Economics', *Isis*, 52, 2, June 1961, 261.
[77] Bernard-Pierre Lécuyer, 'Probability in Vital and Social Statistics: Quetelet, Farr, and the Bertillons', in Krüger, Daston, and Heidelberger (eds), *The Probabilistic Revolution*, vol. 1, 327.
[78] Hacking, 'Prussian Numbers 1860–1882', 389.
[79] Lécuyer, 'Probability in Vital and Social Statistics', 323.

mathematics in the *Journal of the Statistical Society of London* during its first fifty years could find 'practically nothing of the sort'.[80] Another still has written of the Society's 'mathematically almost void first half-century'.[81] Before Galton there was no correlation or regression. With the exception of work for the growing insurance businesses of the Victorian era, in 'statistical writings within the reformist tradition…advanced mathematics was wholly absent'.[82] Quetelet had started his career in astronomy, a science in which 'probability-based statistical methods were a commonplace' by the mid-nineteenth century. But in the social sciences, to which he graduated in the 1830s, there were no such methodological advances.[83]

This book must take, therefore, what is known as an 'externalist' approach, one that explains the science in relation to its social, political and institutional contexts. No academic subject had, or has, more direct links with 'society' than statistics, and the people featured in this story played their parts in many other aspects of Victorian history, whether they were enthusiasts for statistics like Prince Albert and Florence Nightingale, or critics of their collection and use like Thomas Carlyle, John Ruskin, and Charles Dickens. Formed-up in societies that were, for example, deeply integrated into the culture of Manchester factory owners, metropolitan Whig reformers, and working-class radicals, only an 'externalist' approach could capture the role played by statistics in the many different institutions and causes of the era. The Manchester Statistical Society was, from the late 1830s until the mid 1840s, temporarily folded into the Anti-Corn Law League, the great Victorian pressure group for free trade. The Statistical Society of London was another notable partisan for the doctrines and practice of free trade and free exchange. On the other hand, the artisans who formed themselves into the London Statistical Society in 1825, a quite separate body, blamed free trade and laissez-faire, above all things, for the worsening condition of the poor, and argued for the economic regulation of society. I am interested in the use of numbers in different social contexts to defend key interests, make a political or economic case, address a current problem, right contemporary wrongs. This is not a history of the development and use of statistical concepts in the nineteenth century, important as that history must be to the intellectual history of mathematics. Nor is it a study of the development of the understanding of probability, the so-called 'probabilistic revolution' of the nineteenth century, though many aspects of the story to follow were shaped by attempts to understand risk and chance, and to

---

[80] I. D. Hill, 'Statistical Society of London – Royal Statistical Society. The First 100 years: 1834–1934', *JRSS*, series A, vol. 147, Pt. 2, 134.

[81] John Aldrich, 'Mathematics in the London/Royal Statistical Society 1834–1934', *Electronic Journal for History of Probability and Statistics*, 6, 1, June 2010, 2. http://www.jehps.net/indexang.html.

[82] Theodore Porter, 'The Mathematics of Society: Variation and Error in Quetelet's Statistics', *British Journal for the History of Science*, 18, 1, March 1985, 56. Lorraine J. Daston, 'The Domestication of Risk: Mathematical Probability and Insurance 1650–1830' in Krüger, Daston, and Heidelberger (eds), *The Probabilistic Revolution*, vol. 1, 237–60.

[83] Stigler, 'The Measurement of Uncertainty in Nineteenth-Century Social Science', 288–9.

remove their destabilizing effects from daily life.[84] Rather, it is a social and cultural history of the collection and manipulation of numbers, and of their relationship to political events, social developments, scientific methods, and cultural argument in Britain in the period between the 1820s and 1880s. It is because this story can only be told and explained in an 'externalist' manner, with frequent reference to wider political and social themes, that it can be written by a historian of Victorian politics and society rather than by a historian of mathematics or by a practising statistician.

For this reason also, no reader of this book requires any technical knowledge of statistics: this is a study about the collection and organization of numerical data and about early efforts to make the data unlock their secrets without the use of complex mathematical procedures. Often, the information didn't require sophisticated manipulation: high death rates and truncated life expectancies told their own story without the need for any further analysis or even commentary from the statisticians.[85] Merely to collect and publish the data was enough. But equally, no present-day statistician should neglect the early history of their discipline because it seems to have nothing to teach them about the practice of statistics. This is a history of sundry attempts to win social recognition, social authority, and political support for statistics, and to ensure wide dispersion of statistical results and understanding, all of them themes with contemporary relevance. It is also a cautionary tale on the matter of objectivity and the use and misuse of data: many of the groups and individuals in this story were animated by a cause, from the first statistical society founded in Manchester in 1833 through to Galton and his use of statistics to find evidence in support of eugenics. To contend that the emergence of the modern discipline of statistics has no relationship to the early Victorian statistical movement is to misunderstand much in the history of statistics, not least the vital roles of ideology and social context in the emergence of the modern discipline itself.[86]

This book attempts to contextualize each statistician and his or her specific contribution by recovering, as best as it can, the aims and intentions of each practitioner as an historical figure, the manner in which their statistical ventures were constructed, and by explaining these things in relation to wider intellectual and social trends of the era. Nothing less could be expected of research that began in Cambridge, where and when the history of political thought, and intellectual history in general, were understood as exercises in the recovery of intentions and

---

[84] Lorenz Krüger, 'Preface', in Krüger, Daston, and Heidelberger (eds), *The Probabilistic Revolution*, vol. 1, xv.

[85] Simon Szreter, *Fertility, Class and Gender in Britain 1860–1940* (Cambridge, 1996), 92.

[86] Mackenzie, *Statistics in Britain 1865–1930*, 8.

context.[87] With this as the aim, primary sources, often lodged in archives, have been used wherever possible so that the historical actors are read in their own words, whether in publications of the period or in the letters, journals, and memoranda they left behind. Yet ideas grow beyond the confines of a specific institution, document, or text; they have an afterlife once written down or published; they influence colleagues, readers, and followers; in the process of time they are changed. If contextualization is synchronic, analysing events or a text at a moment in time, this book must also be diachronic, moving through time in the attempt to construct a connected history and narrative of the development of statistics. Many studies of the history of statistics, and of social investigation more widely, tend to present a sequence of unconnected, lone endeavours, one thing after another. In this book, by noting the personal connections that linked together many of the projects and subjects, the role that key individuals played across the period as a whole, the institutional affiliations that led from one society and organization to the next, and by relating all these developments to an overarching ideological history which moved from liberal to authoritarian from the 1820s to the 1880s, there is a conscious attempt to unite different elements in a single story and to present the history as a unity.

This book also attempts, in its conclusion, to relate the past to the present, and as that might seem a contradiction of the claim already advanced that it does not take a 'whiggish' approach to intellectual history and the history of science, it must be explained carefully. When research incorporated here was first undertaken in the 1980s it was typed-up on an old, mechanical typewriter, whereas the typescript of this book has been prepared on a powerful laptop using the full resources of the internet for research, both of which are products of the many technical changes which are captured in the term 'digital revolution'. These changes to the availability, presentation, circulation, and analysis of data have a bearing on the way the history of *Victorians and Numbers* will be presented at the end of the book. If we go back to that room in Trinity College, Cambridge in June 1833, it was Malthus and Jones who first caught the eye and imagination. On first acquaintance, the story to be told seemed to be about the development of statistics as a critique of the then conventional and orthodox methods in political economy, the leading social discourse of the age.[88] This relatively new discipline was then dominated by deductive reasoning based on broad assumptions about the nature of human behaviour, notably its universality and predictability. What was true of economic behaviour in England was supposedly true of France, Spain,

---

[87] The *locus classicus* for this approach, associated with Cambridge's historians of political thought since the 1960s, is Quentin Skinner, 'Meaning and Understanding in the History of Ideas', *History and Theory*, vol. viii, no. 1, 1969, 3–53.

[88] Lawrence Goldman, 'The Origins of British "Social Science": Political Economy, Natural Science and Statistics, 1830–35', *The Historical Journal*, 26, 3 (1983), 587–616.

the United States, everywhere, and at all times. To Malthus and Jones, and Jones's close friend William Whewell, a better way to understand economic reality involved collecting and analysing data—statistics—on economic institutions and transactions in the real world, and noting how these differed according to time, place, culture, and context. This book examines in some detail the inductive economics they founded.

Today, however, the figure in that college room who compels attention and who links many of the themes in this story is Charles Babbage, who not only constructed a mechanical calculating engine, but devised and planned the first computer, the Analytical Engine, and who, it is suggested here, also foresaw the era of Big Data. At exactly the time that he was at his most creative in these fields, the early and mid-1830s, he also took the lead in establishing the institutions of the statistical movement in Cambridge and London. It would push the argument too far to suggest that the computer, whether mechanical or electronic, was the result of the 'information revolution' of the 1830s. But the argument *is* being made that it was out of this milieu, this emerging culture of numbers, that the concept of mechanized calculation took root and was nurtured. Babbage was located at the heart of the statistical movement at the very time that he was at his most creative in imagining machines that could process numbers and mathematical functions. With his Cambridge friends and colleagues, Babbage pursued 'Humboldtian Science', which mandated measurement. In Babbage's case specifically, this meant measurement and analysis of the new industrial civilization developing in Britain, which he undertook in researching and writing his pioneering volume of 1832, *On the Economy of Machinery and Manufactures*. Humboldtian Science would encourage the discovery and appreciation of otherwise hidden structures, patterns, and relations in the economy, as well as in society and nature, demonstrated numerically. Both Babbage and his expositor and friend, Ada Lovelace, could envisage a time in the future when the progress of science would depend on the processing of numbers at such scale and volume as to be too great for conventional human calculation. Science, as Babbage saw as early as 1822, would then require the assistance of machines if it were to advance.[89] In Babbage's biography, in other words, statistics and computing go together, mutually reinforcing each other. This relationship has only become clearer, and Babbage's role in the story to follow has only appeared more central, as events and developments in our own age have followed the premonitions of the man who, in his essay of 1832, discussed below, on 'The Constants of Nature and Art', imagined an international project to collect voluminous data about all the most important phenomena of the natural and social worlds which we might justifiably call today 'Big Data'. I trust that readers who have followed the

---

[89] Charles Babbage, 'On the Theoretical Principles of the Machinery for Calculating Tables. In a letter to Dr. Brewster', 6 Nov. 1822 ', *Edinburgh Philosophical Journal*, vol. 8, 1823, 128. See below, p. 123.

argument from the beginning of this book, through carefully-contextualized case studies, will not find the comparisons in the conclusion either fanciful or forced.

In this sense, what has happened in the present has encouraged a fresh understanding of the past, with the emphasis moving from one set of historical actors to another. In the reinterpretation that follows here, Malthus, Jones, and Whewell still have their place, and the critique they mounted in the 1830s of a purely deductive academic methodology is still an important part of the history presented. But in the light of more recent technical and social developments, from the manufacture of cheap and enormously powerful semi-conductors to the collection of huge data sets, it is the link between past and present through Charles Babbage that forms a pathway through this history. This book contextualizes, therefore, to make sense of key developments and people in their own words and time. It also narrates a story which closely follows ideological developments and changes across the nineteenth century. And it reinterprets that story in the new light that developments in the present cast on the past. Without destroying or simplifying the history, recent technological developments have encouraged a different focus and a changed emphasis. In linking past and present in this way, it may be possible to broaden public interest in what might seem to be, at first sight, an uninviting subject.

There are three related aims in this book, therefore. The first is to recover the intellectual and social projects which grew out of 'the first Data Revolution' of the early nineteenth century, recreating, thereby, the precise cultural milieu of the early and mid-Victorian statistical movement. The second is to explain why this intellectual quest lost traction and relevance from the 1860s onwards as new intellectual paradigms and techniques, alterations to the politics and the map of Europe, and a transformation in the ideological orientation of the era as a whole, changed not only social statistics but the liberal social values that had encouraged their emergence in the 1830s. The third is to acknowledge that this first Data Revolution has implications for the way we understand changes in forms of communication and the processing of information during our own lifetimes, and to recognize the links between the 1820s and 2020s. To do that does not compromise the history related here but enriches it with further meaning.

Prologue

Statistics at the Zenith. The International Statistical Congress, London 1860

The Statistical Movement in Victorian Britain reached its zenith at the fourth meeting of the International Statistical Congress which was held in London for a week in mid-July 1860. The ISC had been founded at the Great Exhibition in London in 1851, and had convened on three prior occasions in Brussels in 1853, Paris in 1855, and Vienna in 1857. Three years later, returning to London, the event drew together over 500 statisticians, civil servants, social reformers, and politicians for lengthy discussions on the collection, compilation, standardization, and comparability of social statistics. As the preeminent British statistician of the age, Dr. William Farr, told the Congress, 'The object has been to bring our plans together, and to endeavour to suggest the best method of carrying out statistics on the same plan in every part of the world.'[1] Comparability and uniformity were not ends in themselves but prerequisites for the use of social statistics in the improvement of health, education, and living standards. This was an era brimming with confidence in the potency of scientific method allied to numbers. As one periodical dedicated to social reform in Britain had expressed it two years earlier,

> Statistics are the foundation of our knowledge of all outward phenomena. They are the bones and sinews, the nerves and muscles of legislation. Reforms are impossible without their aid. No lasting improvements can be accomplished without their assistance.[2]

The Congress was held in Somerset House in the Strand and the adjoining King's College, London. Many of the museums, libraries and public buildings in the capital including the Reading Room of the British Museum, the Royal Geographic Society, the Geological Museum, the Zoological Gardens in Regent's Park, and

---

[1] *Report of the Proceedings of the Fourth Session of the International Statistical Congress Held in London July 16th, 1860, and the Five Following Days* (London, Her Majesty's Stationery Office, 1861), 210.
[2] *Meliora. A Quarterly Review of Social Science in its Ethical, Economical, Political, and Ameliorative Aspects* (London, 1858), vol. 1, no. 3, 201.

Victorians and Numbers: Statistics and Society in Nineteenth Century Britain. Lawrence Goldman, Oxford University Press.
© Lawrence Goldman 2022. DOI: 10.1093/oso/9780192847744.001.0002

several metropolitan hospitals were specially opened and made available to the members of the Congress.[3] More than 70 of the participants, many of them high government officials, had come from Europe and North America. The inclusion on equal terms of representatives from the Australian states, New Zealand, Canada, Ceylon, Jamaica, and Barbados was an innovation for the Congress which had not hitherto invited participation from European imperial territories.[4] It was, in fact, one of the first large-scale international academic gatherings ever held in Britain: as Farr put it at the end of the week's events, 'this was a combination which we had not attempted before in this country'.[5] The press, both national and provincial, covered each day's discussions. There were editorials in *The Times* and all the leading newspapers and journals. The foreign delegates were guests at a special reception at Buckingham Palace on the first day of the congress. The Liberal prime minister, Viscount Palmerston, hosted a reception for participants at his London home, Cambridge House. He addressed them also at a dinner in the Mansion House where they met more of Her Majesty's ministers, among them the foreign secretary, Lord John Russell; the home secretary, George Cornewall Lewis; Sir Charles Wood, secretary of state for India; and Edward Cardwell, Chief Secretary for Ireland. The government had pledged 'to afford every facility in their power for the deliberations of this Congress'[6] and the two ministers charged with assisting its organization, The President of the Board of Trade, Thomas Milner Gibson, and the First Commissioner of Works, W. F. Cowper, were also present. The Society of Actuaries and the Statistical Society of London threw a special dinner for the foreign representatives at the Freemason's Hall.

The Congress brought together 'delegates, statesmen, and men of literature and science from all parts of the world'.[7] Many of the foreign representatives were drawn from their respective national statistical bureaux—from the Prussian General Statistical Department (Dr Ernst Engel), the French Central Statistical Department (Alfred Legoyt), the Dutch Central Statistical Commission (Dr T. Ackersdyck), the Central Statistical Commission of the Russian Ministry of the Interior (Dr J. Wernadski), and so forth. The United States was officially represented by Judge Augustus Baldwin Longstreet, a southern lawyer, college president, and uncle to James Longstreet, the Confederate General during the Civil War. From the German state of Mecklenburg-Schwerin came one, the Baron Maltzahn, 'first president of the Patriotic Agricultural Society'.[8] The Congress ended eight days after it began with a 'grand fete at the Crystal Palace' itself,

---

[3] *The Times*, 16 July 1860, 5.
[4] Nico Randeraad, *States and Statistics in the Nineteenth Century. Europe by Numbers* (Manchester, 2010), 91.
[5] *Report of the Proceedings*, 210.
[6] Ibid., General Meeting: First Day, Somerset House (Mr. Milner Gibson), 1.
[7] 'The International Statistical Congress', *Illustrated London News*, 21 July 1860, 69.
[8] 'Official Delegates', *Report of the Proceedings*, ix–xi.

Sir Joseph Paxton's brilliant creation of plate glass and iron which had housed the Great Exhibition of 1851 and which still stood in Hyde Park in the centre of the city.[9] The creative force behind the Great Exhibition, Albert, the Prince Consort, gave the opening address on the first day of the Congress. Often considered the finest of all his speeches—'the Prince has made many good speeches, but none better than this'—it was his last public oration before his untimely death in the following year.[10] No gathering of social scientists in Britain will ever match the 1860 London Congress for pomp, grandeur, and national significance.

The key figure at the heart of proceedings in London in 1860 was William Farr, Superintendent of the Statistical Department in the Registrar-General's Office. With a hand in the design of the five decennial British censuses from 1841 to 1881, Farr was a member of the small executive committee that organized the Congress. Afterwards he edited and published its proceedings. He will play a key role in the story to come.[11] He was assisted by Dr William Guy, Professor of Forensic Medicine at King's College, London and another physician/statistician of note whose contribution to Victorian statistics will be discussed later.[12] The full Committee of Organization contained many of the great and good of Victorian Britain from the Governor of the Bank of England and Lord Mayor of London to the Astronomer Royal and Queen's Physician. It brought together 'persons belonging to all parties in politics, and every liberal pursuit in the country.'[13] Palmerston, Russell, and William Gladstone, the chancellor of the exchequer, were joined on it by the Leader of the Conservative Opposition in the House of Commons, Benjamin Disraeli, though none of them played a role in the Congress proper. Disraeli, however, attended the prime minister's reception and sent a characteristically vivid account of what he saw there to the wealthy widow who paid off his debts and who is buried beside him in the churchyard on his estate, Hughenden. Apparently, Lady Palmerston's

crowded salons at Cambridge House were fuller than usual, for she had invited all the deputies of the Statistical Congress, a body of men who, for their hideousness, the ladies declare, were never equalled. I confess myself to a strange gathering of men with bald heads, and all wearing spectacles. You associate these traits

[9] *Daily Telegraph*, 21 July 1860, 4.
[10] *The Examiner*, 21 July 1860, 1. See also Harriet H. Shoen, 'Prince Albert and the Application of Statistics to Problems of Government, *Osiris*, 5, 1938, 293: 'the Prince's greatest and best speech was also his last public speech'.
[11] John M. Eyler, *Victorian Social Medicine. The Ideas and Methods of William Farr* (Baltimore and London, 1979).
[12] 'William Augustus Guy, 1810–1885', *Oxford Dictionary of National Biography* (Oxford, 2004–), hereafter *ODNB*.
[13] *Report of the Proceedings*, 209 (Viscount Ebrington).

often with learning and profundity but when one sees one hundred bald heads and one hundred pairs of spectacles, the illusion, or effect is impaired.[14]

The Bishop of Oxford, Samuel Wilberforce, was a member of the committee; so too was Thomas Huxley, Charles Darwin's champion. Within a matter of weeks these two would debate the new theory of natural selection as explained by Darwin in *The Origin of Species*, published the year before, at the Oxford meeting of the British Association for the Advancement of Science. It was there and then that 'Soapy Sam' is said to have asked Huxley if he was descended from an ape on his mother's or his father's side of the family.[15] Two of the most important intellectual founders of the British statistical movement, William Whewell, Master of Trinity College, Cambridge, and the great mathematician, Charles Babbage, inventor and designer of the first computer, were on the committee. Babbage contributed a memoir on the history of international collaboration in statistics to the published *Report* of the congress which emphasized his own contributions to the movement.[16] Whewell and Babbage, friends from their youth, will feature greatly in the story to come. So also will Florence Nightingale, one of only two women to communicate publicly to the Congress, and also a member of the committee. The other woman was Martha Baines, wife of Edward Baines, the journalist, politician, and son of the founder of one of the great Victorian newspapers, the *Leeds Mercury*.[17] The first 'respectable' middle-class woman to speak in public in Britain, the social reformer, Mary Carpenter, had only done so as recently as 1857; an organized feminist movement was only starting to emerge in Britain at this time.[18] From Dublin came a Mr W. R. Wilde from the Census Office in the city, who made contributions at the Congress on the classification of deaths in the Irish census according to geographical location and occupation. Trained as a surgeon, William Wilde helped design three mid-Victorian Irish censuses and was knighted for his public service. He deserves to be better known for his own expertise than merely as the father of Oscar Wilde.[19]

The work of the Congress was divided into six sections, each with its complement of distinguished officers, mostly professional men drawn from occupations related to each section's subject. Judicial Statistics was presided over by Lord Brougham, a founder of the *Edinburgh Review*, revered by this time for his long-ago

[14] Disraeli to Mrs Brydges Williams, in J. Bonar and H. W. Macrosty, *Annals of the Royal Statistical Society 1834–1934* (London, 1934), 86.
[15] J. Vernon Jensen, '"Debate" with Bishop Wilberforce 1860' in *Thomas Henry Huxley: Communicating for Science* (Cranbury, NJ, 1991), 63–86.
[16] 'Letter from Charles Babbage, Esq., FRS etc. etc.', *Report of the Proceedings*, 505–7.
[17] Mrs. M. A. Baines, 'Suggestions for Obtaining Statistics of Wet Nursing'. The paper drew attention to the higher mortality rates among children put out to wet nurses rather than nursed by their mothers. *Report of the Proceedings*, 280.
[18] On Mary Carpenter's famous speech to the inaugural congress of the National Association for the Promotion of Social Science at Birmingham in 1857 see Lawrence Goldman, *Science, Reform and Politics: The Social Science Association 1857–1886* (Cambridge, 2002), 115–16.
[19] *Report of the Proceedings*, 264, 266, 269–70. 'Wilde, Sir William Robert Wills 1815–1876', ODNB.

exploits in the causes of anti-slavery, the passing of the Great Reform Act of 1832, and the reform of the law.[20] Sanitary Statistics, the section concerned with public health, was led by the earl of Shaftesbury, the famous champion of factory reform and street ('ragged') children in industrializing Britain.[21] The president of the Industrial section, concerned with the statistics of agriculture, mining and manufacturing, was Lord Stanley, the son of the fourteenth earl of Derby, the mid-Victorian Conservative prime minister, and himself a future foreign secretary.[22] Commercial Statistics were the responsibility of the political economist, lawyer, and public servant, Nassau Senior.[23] The Census had its own section, and was overseen by the earl Stanhope, until 1855 known as Lord Mahon, a moderate Conservative (a Peelite) in parliament, a notable historian of the eighteenth century, and a founding trustee of the National Portrait Gallery.[24] The president of the sixth section on Statistical Methods was Adolphe Quetelet, the Belgian mathematician and astronomer and the greatest statistician of the age, the author of the single most influential work of social statistics in this whole era, *Sur L'Homme et le Développement de ses Facultés: Physique Sociale* (1835). Quetelet knew all the leading British practitioners, had played a formative role in the foundation of the key British statistical societies and institutions in the early 1830s, and was the founder in 1851 of the International Statistical Congress itself, created at his instigation while he was in London for the Great Exhibition.[25]

Among these six were natural and social philosophers, as the age would call them, and also philosophical statesmen. Edward Henry, Lord Stanley, was a particularly interesting example of the latter, a model ameliorist, whose reading and interests provide insight into the intellectual foundations of social reform in this era and the role of statistics in guiding those reforms. He frequented meetings of the Statistical Society of London and was a central figure in the foundations of the Social Science Association between 1855 and 1857, another British forum for the promotion of social reforms. He was so much the dedicated reformer with a naturally bureaucratic mind that he was never comfortable as a Tory in the party led by his father, and he eventually crossed the floor of the House of Commons to join a Liberal government led by Gladstone. To read through his journals in the years before 1860 is an education in the psychology of public service, a record of government reports read, meetings attended, asylums inspected. The journals also preserve the record of a very serious project of social-scientific self-education. In the mid-1850s Stanley read and made copious notes on Herbert Spencer's

---

[20] Chester New, *The Life of Henry Brougham to 1831* (Oxford, 1961); 'Henry Peter, first Baron Brougham and Vaux 1778–1868', *ODNB*.

[21] E. Hodder, *The Life and Work of the seventh earl of Shaftesbury* (3 vols. London, 1887); G. B. A. M. Finlayson, *The Seventh Earl of Shaftesbury* (London, 1981).

[22] *Disraeli, Derby and the Conservative Party: Journals and Memoirs of Edward Henry, Lord Stanley 1849–1869* (ed. John Vincent) (Hassocks, Sussex, 1978).

[23] 'Senior, Nassau William 1790–1864', *ODNB*.

[24] 'Stanhope, Philip Henry, 5th earl Stanhope 1805–1875', *ODNB*.

[25] On Quetelet, see Chapter 7 below.

Social Statics, Malthus's *Essay on Population*, Mill's *Political Economy*, George Henry Lewes's *Comte's Philosophy of the Sciences*, De Tocqueville's *Democracy in America* and *The Ancien Regime and the French Revolution* and Quetelet's *Sur L'Homme*. He read also works of social investigation such as Dr. Southwood Smith's *Philosophy of Health* and Alexandre Parent-Duchâtelet's influential *De la prostitution dans la ville de Paris*.[26] Soon after its publication in 1857 Stanley read another work that made social statistics accessible, popular, and the talk of the town, H. T. Buckle's *History of Civilization in England*, 'a work which more than any other I have seen embodies the results of recent discovery, and the tendency of modern thought' and which this book will consider as well in a later chapter.[27] It would be difficult to assemble a more well-chosen assortment of the central texts of mid-Victorian social science. Nor from Stanley's notes could one envisage a more thoughtful reader. At the heart of his faith in the possibility of social reform was the discipline of statistics. As Stanley wrote to Brougham in August 1856, he was

> A firm believer in the efficacy of Statistics on almost all subjects of human action – it is only on masses, not on individuals, that the operation of social laws can be traced. In other words, man can only be effectively studied in the aggregate. Without such study, legislation is for the most part empirical.[28]

We can get a sense of the interest and excitement that the ISC generated from the letters and manuscript autobiography of one of the participants, Dr Edward Jarvis of Dorchester, Massachusetts. Jarvis was a leading American physician, an expert in the treatment of the insane, president of the American Statistical Association (which he formally represented at the ISC in London) between 1852 and 1884, a consultant to the federal census, and a founder in 1865 of the American Social Science Association.[29] He spent five months in Britain in 1860 and attended the London Congress in July which, as he told his friend Farr, was 'the crowning pleasure of [his] European visit'. There he delivered three papers on the registration of vital statistics in the United States; on the American decennial census; and on the different criminal behaviours of men and women.[30] At the Congress he joined with

---

[26] For all these, see the papers of Edward Henry, 15th earl of Derby, Liverpool Record Office, 920 (DER) 15 39/1–3 (3 vols), 'Notes 1851-6'; 15 46/1–2 (2 vols) 'Notes Taken during the Year 1857'.

[27] Ibid., 'Notes taken during the Year 1857', ii (unpaged).

[28] Papers of Henry, Lord Brougham and Vaux, University College, London, B.Mss. 24334, 22 Aug. 1856.

[29] Gerald N. Grob, *Edward Jarvis and the Medical World of Nineteenth Century America* (Knoxville, TN, 1978); Margo Anderson, 'The US Bureau of the Census in the Nineteenth Century', *Social History of Medicine*, 4, 3, Dec. 1991, 497, 505.

[30] Edward Jarvis, 'On the Laws and Practice of Registration in America'; 'On the Crimes of Males and Females'; 'On the further inquiry in the Census as to the Personal Health and Power of each person'. For all his contributions to the Congress see *Report of the Proceedings*, 51–5; 176; 264–7; 271–2; 277–83; 446–7; 497–9.

'scientific men, statisticians, philanthropists and political economists' in the discussion of questions of 'sickness, insanity, mortality, crime, pauperism, population, business, finance, agriculture—all the interests of humanity came under their review.'[31] He met Lord Ebrington, a prominent member of the Health of Towns Association of the 1840s, who invited him 'to a party at his house where was a large collection of sanitarians', and where he made the acquaintance of Charles Babbage.[32] At the 'grand party at Lord Palmerston's' where Jarvis saw 'nobles, ministers, men of high degree, ladies noble and great in their grandest attire', Shaftesbury introduced him to the prime minister.[33] At Buckingham Palace Jarvis discussed Anglo-American co-operation with a fervent Prince Albert who told him that 'your nation and ours hold the destinies of civilization in our hands more than all other nations. We must, then, walk together in harmony.'[34] Perhaps an even higher honour was to have breakfasted three times during his visit at the home of Florence Nightingale: as he recorded, 'Tuesday morning at 9.45, breakfasted at Miss Nightingale's with a similar company to that of Friday—the elite of the philanthropists, the men of higher culture, of England and of Europe.'[35]

The daily work of the Congress and its six sections was not as grand as this, of course. Participants sat through interminable reports on the state of national statistics in all participating countries. One delegate from New South Wales, with the directness that has always characterized his countrymen, referred to 'the digressive tendencies of most papers'.[36] But a theme common to all discussion was the need for agreed procedures in collecting and analysing social data. Lawyers wanted national registers of real property and maps of all landholding in their states, compiled to internationally-agreed standards. Physicians wanted a uniform system of sanitary statistics. Enthusiasts for the metric and decimal systems lobbied for their introduction and met stiff British resistance. There were lengthy debates on developing a uniform classification of crime. The fifth section set out the questions and subjects that should be included in every national census. Many of the discussions pitted advocates of uniformity against those concerned that new procedures would entail the loss of detail and conflict with national institutions and traditions. As one leading British statistician, Leone Levi, explained the dilemma: 'We must consult convenience in Statistics; we cannot always follow a strict philosophical classification.'[37] As a physician commented, 'if we enter into all the details, it will be impossible to obtain any results at all'.[38] So the Congress saw its work as the definition of general standards and procedures,

---

[31] Edward Jarvis, Mss Autobiography, Houghton Library, Harvard University, ff. 295–6.
[32] Ibid., f. 306.
[33] Edward Jarvis, 'European Letters, 1860', 3 vols, Concord Free Public Library, Concord, Mass. To his wife Almira, 23 July 1860, vol. 3, f. 203.
[34] Jarvis, Mss Autobiography, f. 298.
[35] Edward Jarvis to Almira Jarvis, 23 July 1860, 'European Letters, 1860', vol. 3, f. 209.
[36] *Report of the Proceedings*, 55 (Edward Hamilton).   [37] Ibid., 225.
[38] Ibid., 257 (Dr. James Bird).

leaving individual states and jurisdictions to interpret those generalities in their own fashion according to local needs. In Shaftesbury's words, 'our business here is to lay down great principles, and leave others to reduce them into operation and see how far they are practicable or not'.[39]

For press, public and participants alike, Prince Albert's speech, delivered in the hall of King's College, London, was the highlight of the Congress.[40] Albert met with William Farr beforehand to discuss the address, and Farr subsequently submitted to the Prince a set of 'detached notes' ('unhewn stones' as Farr described them) based on the Prince's ideas.[41] Farr commented later that 'his address was entirely his own'.[42] It bears careful examination.[43] The Prince began by making a point about the culture of Victorian Britain—that it was a public and participative culture rather than one defined by elites, officials, and experts. Though the Congress could have been 'a private meeting of the delegates of different Governments, discussing special questions of interest', because it was being held in Britain 'it had to assume a public and a national character, addressing itself to the public at large and inviting its co-operation'. This difference was to assume greater significance in the subsequent history of the ISC; it also forms a tension in the history of Victorian statistics in Britain specifically, where the very popularity of statistics militated against higher intellectual functions and the establishment of a social science founded upon them.

Albert went on to remind his audience that the Congress had its origins in 1851 at the Great Exhibition, his greatest personal triumph. He emphasized social improvement and welfare in 1860 as he had nine years before:

> It is the social condition of mankind, as exhibited by [those] facts, which forms the chief object of the study and investigation undertaken by this Congress; and it hopes that the results of its labours will afford to the statesman and legislator a sure guide in his endeavours to promote social development and happiness.[44]

Albert also endorsed international collaboration: these periodic Congresses not only brought together savants and officials from different countries 'but they pave the way to an agreement among different Governments and nations to follow up

---

[39] Ibid., 256–7.   [40] *Illustrated London News*, 21 July 1860, 22.

[41] William Farr to Sir Charles Phipps (Treasurer and Private Secretary to Prince Albert), 29 June 1860, Royal Archives, RA PPTO/PP/QV/MAIN/1860/4953. The meeting occurred on the previous day, 28 June 1860. For Albert's printed copy of his address, see Prince Albert's Official Papers, 'The Opening of the International Statistical Congress', RA Vic/MAIN/Z/271/34, https://albert.rct.uk/collections/royal-archives/prince-alberts-official-papers/the-opening-of-the-0.

[42] Theodore Martin, *The Life of His Royal Highness The Prince Consort* (5 vols, London, 1879), vol. 5, 138.

[43] Albert, Prince Consort, 'Inaugural Address', *Report of the Proceedings*, 2–7. The speech was republished in Albert, Prince Consort, *The Principal Speeches and Addresses of His Royal Highness the Prince Consort* (ed. Sir Arthur Helps) (London, 1866), 229–46.

[44] *Report of the Proceedings*, 5.

these common inquiries, in a common spirit, by a common method, and for a common end. (*Great Applause*)'.[45] Britain was not exempt from princely criticism in this respect: procedures here showed a lamentable variability and exceptionalism which Albert put down to the 'want of such a central authority or commission . . . to direct on a general plan all the great statistical operations'. This was not a universally popular view among British statisticians, many of whom prized their independence from the British state.

Displaying considerable intellectual bravery Albert ventured to discuss one of the popular themes of the age, whether the study of statistics encouraged philosophical determinism, or what the age sometimes termed 'fatalism'. The always similar annual frequencies of social phenomena like births, deaths, marriages, illnesses, and even suicides in large populations had been uncovered by statisticians long before. More recently, in 1857 the historian Henry Thomas Buckle had caused a literary sensation by claiming in his *History of Civilization in England*, the same work that had so impressed Lord Stanley, that these regularities made possible a scientific history based on predictive laws of human behaviour, with obvious implications for the role and capacity of the individual as an agent capable of exercising free will. As Albert explained, the subject of statistics was held in suspicion by some because it

> deprives man of his dignity, of his virtue and morality, as it would prove him to be a mere wheel in this machine, incapable of exercising a free choice of action, but predestined to fulfil a given task and to run a prescribed course, whether for good or for evil.[46]

Albert did not believe that the regularities of the natural world and of social life 'destroyed or diminished' the power of God, however. He pointed out that recurrent events were not ineluctable laws governing nature and behaviour, but probabilities only, thus leaving open the possibility of deviation from any pattern. The 'essence of Statistical Science', he argued, was 'that it only makes apparent general laws, but that these laws are inapplicable to any special case'. Thus, nature allowed for uncertainty; men and women could exercise will and volition; creation was not determined. It is unlikely that Albert solved this famous philosophical problem in his carefully constructed conclusion that granted men and women 'individual self-determination' in a world nevertheless established by God on the basis of laws 'conformable to His eternal nature'.[47] But in raising the subject he was nothing if not current, and his remarks open up another important theme in the development of Victorian statistics. This book will return to Buckle and his concept of history, and also consider other statisticians accused of determinism.

[45] Ibid.   [46] Ibid., 4.   [47] Ibid., 5.

The Prince Consort built to a climax in which he emphasized the mutual dependence of nations 'for their progress' and endorsed 'the maintenance of good will' between them. 'Let them still be rivals, but rivals in the noble race of social improvement', he averred. The 'long continued applause' that followed was endorsement of a speech that was liberal, improving, internationalist, universalist. As such, it was a statement of the aims of the statistical movement itself. This book will show how those aims developed and why they were undermined. Indeed, in the subsequent history of the International Statistical Congress there emerged elements the very reverse of Albert's hopes: nationalism, particularism, specialization, and divergence.

As interesting as the speech itself is the story of how Albert came to deliver it. By a chance of history, in the summer of 1836 Quetelet himself had tutored in mathematics the 17-year old Albert and his elder brother Ernest, then princes of Saxe-Coburg and Gotha awaiting entry to Bonn university. Quetelet's enduring friendship with his erstwhile pupils was the reason that the Prince Consort opened the Congress. Tutor and pupils stayed in close and affectionate contact, exchanging regular greetings.[48] The boys' interest in the application of mathematics to society, which had been encouraged by Quetelet's lessons in probability, was authentic. Ernest later recalled how Quetelet 'fixed our attention more on mathematics and statistics, in order to lay a preparatory foundation for further study of political science'.[49] He recalled Quetelet's profound influence over his brother, which is captured in a letter Albert wrote to his tutor in 1837:

> I cannot tell you, Sir, how much the subject of our correspondence interests me...I am curious to see the application of the calculus of probability to social and natural phenomena. These phenomena are always regarded too superficially, and even if they resolve themselves into a system, ordinarily they lack firm and incontestably true bases. The calculus, on the contrary, presents things in their universality, exactly as they occur in nature, without being altered by individual opinion.[50]

So interested was he in the applications of mathematics that in 1842 Albert and his Austrian uncle, Count Mensdorff (Emmanuel Graf von Mensdorff-Pouilly), who was also the uncle of Queen Victoria, visited the workshops of Charles Babbage to view the famous Difference Engine—often described as the first computer, though mechanical rather than electronic—that the great mathematician

---

[48] Shoen, 'Prince Albert and the Application of Statistics to Problems of Government', 276. Walter F. Willcox, 'Development of International Statistics', *Milbank Memorial Fund Quarterly*, 27, 2, 143–53.

[49] *Memoirs of Ernest II, Duke of Saxe-Coburg-Gotha* (4 vols, London, 1888–90), I, 72–3.

[50] Albert to Quetelet, 10 July 1837, quoted in Shoen, 'Prince Albert and the Application of Statistics to the Problems of Government', 281.

and engineer had designed and assembled. It is now in the Science Museum in London. As Babbage observed 'Albert was...sufficiently acquainted with the higher departments of mathematical science to appreciate the influence of such an instrument on its future progress.'[51] The Prince had been a patron of the Statistical Society of London since 1840 and he attended three of its meetings in 1843, 1849, and 1855.[52] On the second of those occasions he heard a paper on the relations between education, environment and crime.[53] It was almost automatic, therefore, that William Farr, in planning the Congress, should encourage his friend Quetelet to invite Albert to preside. Farr explained to Quetelet that

> We wish the Prince to give the congress his immediate countenance, to attend some of the meetings & to favour us with his opinions on certain points...If you will kindly point out to him the importance & utility of the Congress, looked at from a scientific and social point of view, it will be of use. I trust that he has not forgotten the excellent lessons you gave him...If the Prince Consort takes the thing up – he will naturally stimulate the present Government & there will be no delays.[54]

As Quetelet then wrote to the Prince,

> I know there has been a great desire to have Your Royal Highness for President...I should be happy to see the Prince whose first steps in statistics I followed, take today in sure hands the direction of a science which will be grateful to him for what he does in its behalf. Add to this the fact that this science is especially governmental.[55]

After the Congress, and Albert's triumph, Quetelet wrote with his congratulations: 'Nothing could have been more pleasing to me, in fact, than to see a science, essentially governmental, raised for the first time by an enlightened and generous prince to the level to which it belongs.'[56]

---

[51] Charles Babbage, *Passages from the Life of a Philosopher* (London, 1864), 170.

[52] 'Ninth Annual Report of the Council of the Statistical Society of London, 1842–3', *JSSL*, vol. 6, no. 2, 89; 'Twenty second Anniversary Meeting of the Statistical Society. Session 1855–56. Report of Council', *JSSL*, vol. 19, no. 2, 97.

[53] Joseph Fletcher, 'Moral and Educational Statistics of England and Wales', *JSSL*, xii, 1849, 151–76, 188–335.See below, pp. 75–6.

[54] William Farr to L. A. J. Quetelet, 7 Jan. 1859, Quetelet Papers, Correspondence with William Farr, file 990, Académie Royale de Belgique, Brussels.

[55] Quetelet to Albert, 30 Dec. 1858, in Schoen, 'Prince Albert and the Application of Statistics to Problems of Government', 316. This original invitation was to a London congress expected to take place in 1859. The Franco-Austrian War of that year, over Italian unification, led to the postponement of the Congress until 1860 and to fresh invitations to Albert at the end of 1859. See Schoen, ibid., 294.

[56] See M. Gossart, 'Adolphe Quetelet et le Prince Albert de Saxe-Cobourg 1836–1861', *Bulletins de la classe des lettres et des sciences morales et politique*, Académie Royale de Belgique (Brussels, 1919), 246.

Albert was not the only international celebrity at the International Statistical Congress, however. He shared pride of place with another ardent statistician of the age, Florence Nightingale. Four years after the end of the Crimean War, during which her exploits in nursing the wounded and dying had made her world famous, Nightingale was at the height of her influence as a sanitary reformer in both military and civilian life: she began with army medical statistics and then turned to hospital statistics. She was in professional allegiance and close correspondence with Dr Farr himself and could count on his support at the Congress. To this she never actually came, however. Instead, she invited leading participants to breakfast at her London home before the day's debates. They talked among themselves as she listened from the other side of a drawn curtain, neither seen nor heard except by a select few whom she received upstairs.[57] She contributed a paper that was much discussed at the Congress, and two letters that she wrote during the proceedings were read to the delegates by the earl of Shaftesbury. The first of these was on the statistics of mortality in civilian and military life;[58] the second, presented at the final session, was a long list of issues and items which the Congress was invited to consider and take into its competence in the future.[59]

Her major aim, and the principal business before the Sanitary Section, was her proposal for a 'uniform plan of hospital statistics' by which all hospitals would record patients and their illnesses under an agreed classification of diseases and on a single printed form. Comparisons might then be made between types of treatment and between different hospitals, and conclusions drawn about the incidence and virulence of disease in different locations and among different social groups.[60] Some of the London hospitals had trialled the forms already.[61] Though some delegates spoke in favour of the idea, it was altogether too neat, and in discussion the clinicians pointed to the existence of different classifications for the same diseases—which would they use? – and to the problem of classifying complex conditions that might be composed of several different illnesses and complaints. Nevertheless, Nightingale was feted throughout the Congress and the subject of a special vote of thanks at its concluding meeting. Her paper on 'Hospital Statistics' was summarized and published in the leading feminist periodical of that era, the *English Woman's Journal*.[62] Subsequently, Nightingale made further efforts to have her scheme adopted in Britain, but it ran into further opposition: a committee of the Royal College of Surgeons reported against the

---

[57] Her report to her father of these events was not quite accurate: 'They meet at my rooms a good deal for business (I of course not seeing them) under Dr. Farr's Presidency – and I am obliged to give them to eat.' Florence Nightingale to W. E. Nightingale, 12 July 1860, in M. Vicinus and B. Nergaard (eds), *Ever Yours. Florence Nightingale. Selected Letters* (London, 1989), 208.

[58] *Report of the Proceedings*, 177–8.   [59] *The Times*, 23 July, 8.

[60] Florence Nightingale, 'Proposals for an Uniform Plan of Hospital Statistics', *Report of the Proceedings*, 173–4.

[61] Nightingale to Farr, 31 Jan. 1860, Farr Papers, 5474/16, Wellcome Library, London.

[62] 'Notices of Books', *English Woman's Journal*, vol. lx, Aug. 1860, 421–3.

forms, and they were dropped.[63] She contributed a paper to the fourth Congress in Berlin in 1863, read for her there by Farr, and commented to him that 'railroads do more good than Sisters of Charity - & international congresses & exhibitions than philanthropy'.[64]

A third international celebrity at the Congress, whose very presence created controversy, came from an entirely different background and demonstrates another side to mid-century British liberalism. Martin Delany was a free African-American, born in Charles Town, Virginia in 1812 and raised in Pittsburgh, Pennsylvania. He collaborated there with another famous black abolitionist, Frederick Douglass, writing and publishing the leading abolitionist journal, the *North Star*. Like many free blacks and runaway slaves, he lived for a time across the border in Canada. Later, during the American Civil War, he joined the Union army, rising to the rank of Major. In the 1850s Delany contended that African-Americans could achieve no lasting equality in the United States and should depart for a country of their own. He is remembered, therefore, as the first black nationalist. In Pittsburgh he had worked as an assistant to a physician, remaining in the city to treat patients during the cholera epidemics of 1833 and 1854, though other doctors left. In 1850 Delany and two other African-American men were admitted to Harvard Medical School, but then ejected when their white fellow students objected to their presence.[65] With this background, it was natural for Delany to attend the sanitary section at the Congress where Shaftesbury invited him to speak about the treatment of cholera.[66]

Delany's participation was used by Lord Brougham to embarrass the United States, which, in 1860, was still a slaveholding nation, and to remind the Congress and wider society of his past credentials as one of the leading opponents of the slave trade and slavery itself. In a remarkable, and probably planned intervention, as soon as Albert had finished his address, in the presence of the Prince, Brougham rose to his feet and spoke directly to the American Minister to the Court of St. James's—the American ambassador, George M. Dallas of Philadelphia, the United States Vice-President between 1845 and 1849—who was in the audience:

> I hope my friend Mr Dallas will forgive me reminding him there is a negro present, a member of the Congress – (Loud laughter and vociferous cheering) – After the cheering had subsided Mr Dallas made no sign but the negro in question, who

[63] Florence Nightingale, 'Hospital Statistics and Hospital Plans', *Transactions of the National Association for the Promotion of Social Science*, 1861 (London, 1862), 554–60. On the opposition of the *Medical Times*, see Nightingale to Farr, 6 Sept. 1860, Farr Papers, 5474/25. Edward Cook, *The Life of Florence Nightingale* (2 vols, London, 1913), I, 430–4.

[64] Florence Nightingale to William Farr, 15 Aug. 1863, Farr Papers, Wellcome Library, London, 5474/62.

[65] 'Martin Delany, 1812–1885', Henry Louis Gates Jr. and Emmanuel Akyeampong, *Dictionary of African Biography*, vol. 6 (New York, 2012), 177–9.

[66] *Report of the Proceedings*, 285.

was understood to be Dr. Delany, rose amid loud cheers, and said: 'I pray your Royal Highness will allow me to thank his Lordship, who is always a most unflinching friend of the negro, for the observation he has made, and I assure your Royal Highness and his Lordship that I am a man'. This novel and unexpected incident elicited a round of cheering very extraordinary for an assembly of sedate statisticians.[67]

Dallas apparently 'maintained a grave silence'.[68] But such a gratuitous, direct, and undiplomatic assault had consequences. Judge Longstreet, the official American delegate, immediately withdrew from the Congress claiming insult to his nation. Brougham was rumoured to have received 'a very angry personal letter' from Dallas.[69] Both governments were involved.[70] It was evidently thought to require an apology, so some days later, Brougham clarified his remarks. Regretting that they had been taken as disrespectful, he merely doubled-down and turned the apology into further criticism by comparing the United States to Brazil and Spain, in other words to a country, and to the remnants of an empire in Latin America, where slavery was still legal.

When I called attention, in the presence of our friend, Mr Dallas, to the, in my opinion, important statistical fact that a most respectable coloured gentleman, from Canada, was a member of the Congress, I only called his attention to it just as I would the attention of our excellent friend the representative of the Brazils, who is here today; and, God knows, I do not entertain the slightest disrespect of the Brazils. I ought also to have called the attention of the Count de Ripaldi (the Spanish representative) to the same subject; they have colonies, and they have persons of various colours in their possessions. I call his attention to it hereby.[71]

This was no apology. Brougham had slithered out of his first insult by simply widening the circle of those insulted. These further remarks burnished his reputation as a champion of freedom and emancipation, but of greater significance was the reaction of the savants, officials, and representatives present, who had cheered in solidarity with Delany and the cause of anti-slavery. At the concluding meeting Delany thus rose to thank the Congress 'for the cordial manner in which I have been received...on terms of the most perfect equality'. But this was more than a personal compliment: 'I am not foolish enough to suppose that it was from any individual merit of mine, but it was that outburst of expression of sympathy for

---

[67] *The Times*, 17 July, 5.   [68] *Sheffield Independent*, 21 July, 9.   [69] Ibid.
[70] Grob, *Edward Jarvis and the Medical World of Nineteenth Century America*, 160. Frank A. Rollin, *Life and Public Services of Martin R. Delany* (Boston, 1868), 99–133; Victor Ullman, *Martin R. Delany: The Beginnings of Black Nationalism* (Boston, 1971), 238. See also Edward Jarvis to Almira Jarvis, 'Liverpool, England, July – 1860', Jarvis, 'European Letters', vol. 3, ff. 297–316.
[71] *The Times*, 21 July, 9.

my race whom I represent.'[72] The affair reinforced the liberal and reformist credentials of the statistical movement, though by this stage in British history the freedom of slaves everywhere was an article of national faith, subscribed to by most parties, factions, and groups.[73] We shall have reason to return to America and anti-slavery when we consider William Farr later in this book.

The press had a field-day, and not merely because of the cabaret that Brougham had provided. The scale of these events and the celebrity of those intimately involved could hardly be ignored. *The Times* was respectful but concluded that the British were a practical rather than a contemplative people and more likely to do something useful than merely analyse figures.[74] The commentary in the *Saturday Review*, the most critical and sardonic of the mid-Victorian periodicals, was predictable: 'probably our generation will have died out before an International Statistical Congress will be generally welcomed as an exciting inter-ruption of the monotony of life'. But even this writer could applaud Albert's efforts to grapple with the evidence that 'an irresistible fate appears to rule the average conduct of masses of individuals'.[75] *The Economist* was both insightful and appre-ciative of a meeting which it took to be a 'symptom of the present time'. Good government required an understanding of 'social forces':'men are gradually find-ing out that all attempts at making and administering laws which do not rest upon an accurate view of the social circumstances of the case, are neither more nor less than imposture'. Statistics could provide this accuracy: they were 'the allies of medicine, of police administrators, of sanitary authorities' which would unlock the evidence required for sound policy. The journal looked forward to the compilation of uniform sets of social statistics 'for each civilised state' which would make possible the comparison of death rates, life expectancy, criminality, wages, exports per capita and so forth. In this way 'we should have before us a chart of the social economy of the world almost as complete as the charts we already possess of its physical geography'. Here was prefigured the type of interna-tional organization we recognize today in bodies like the Organisation for Economic Cooperation and Development (OECD) which collect leading social and economic indicators from states in the developed world.[76]

Not all participants agreed with a utilitarian conception of statistics as merely providing the raw material for better governance, however. This issue was caught in exchanges in the sixth section, that on 'Statistical Methods', over the rival terms 'statist' and 'statistician'. The first of these implied expertise in statecraft and the skills of social administration. It presented practitioners as servants of the

---

[72] *Report of the Proceedings*, 207.
[73] Richard Huzzey, *Freedom Burning. Anti-Slavery and Empire in Victorian Britain* (Ithaca NY and London, 2012).
[74] *The Times*, 17 July 1860, 9.
[75] 'The Statistical Congress', *The Saturday Review*, 21 July 1860, 72.
[76] 'The International Statistical Congress. The Present Position of Statistical Inquiry', *The Economist*, 4 Aug. 1860, 841–2.

state, 'stat-ists', using numbers for social betterment. As Dr William Guy expressed this point of view,

> I contend that, according to the sense we attach to the word, we really mean statist, not as a ruler, but a scientific statesman; and I think the true meaning to be attached to 'statistics' is not every collection of figures, but figures collected for the sole purpose of applying the principles deduced from them to questions of importance to the state.[77]

Though Guy received influential support in this definition from Charles Babbage and from a young scholar named Henry Fawcett who would soon occupy the chair of political economy in Cambridge and a seat in the House of Commons,[78] Quetelet, the most technically competent practitioner among them, preferred the title 'statistician', as did several other foreign participants, because it implied the existence of an academic discipline independent of practical application, and the possession of mathematical and scientific skills, without which the study of statistics would be merely an exercise in collection and tabulation. These differences would surface at later meetings of the ISC, particularly in The Hague in 1869.[79]

The professional independence of statisticians was a related issue. One after another, the heads of foreign governmental statistical bureaux, employees of their states, stood up at the London congress to present reports on the development of social statistics in their nations. But there was no equivalent British department that was in overall control of national statistics. The report was made instead, therefore, by the statistician William Newmarch on behalf of the Statistical Society of London. As the Prince Consort had made clear in his address, the absence of a British statistical commission or co-ordinating committee was, for some, a national blemish. Reporting on the first International Statistical Congress in Brussels in 1853, Leone Levi had lamented his countrymen's failure to centralize the collection of social data under government: 'In most of the principal countries of Europe general statistical departments have been established, and in others, as in the United Kingdom, statistics are collected by each department of public administration.' The General Register Office kept details of vital statistics—births, marriages and deaths—and oversaw the census. In Levi's view the Statistical Department of the Board of Trade made merely haphazard attempts to report on the commerce of Great Britain.[80] This hardly compared with well-established

---

[77] Dr. William Guy, 'Statistical Methods and Signs', *Report of the Proceedings*, 380.

[78] Lawrence Goldman (ed.), *The Blind Victorian. Henry Fawcett and British Liberalism* (Cambridge, 1989).

[79] Libby Schweber, *Disciplining Statistics. Demography and Vital Statistics in France and England, 1830–1885* (Durham, N. C. and London, 2006), 122–3.

[80] Leone Levi, *Resume of the Statistical Congress, held at Brussels, September 11th, 1853, for the purpose of introducing unity in the Statistical Documents of all Countries. Read before the Statistical Society of London, 21st November 1853* (London, 1853).

statistical departments at the heart of European bureaucracies. The Prussian central statistical bureau dated from 1805, that in Bavaria from 1808, that in Austria from 1829, and the Statistique Générale de la France from 1833.[81]

Yet from the very beginning of the British statistical movement, its founders had shown a reluctance to trade their intellectual independence for state support. Speaking in Cambridge in 1833 to the small assembly who founded the Statistical Section of the British Association for the Advancement of Science, the political economist, Richard Jones, had admitted that although there were aspects 'of statistical enquiry which can never be satisfactorily dealt with without the aid of the general government', nevertheless, the 'first object must be not to demand such aid but to show by our own exertions that we are not unworthy of it should we hereafter seek it'.[82] In nineteenth century Britain 'self-help' was the default position, even for learned societies, and established independent endeavour was more likely to win public funding than direct appeals for state aid. In development of this position, in his remarks to the Congress, William Newmarch, who was a well-known economic individualist and a paid-up member of the Victorian awkward squad, revelled in the independence of British statisticians:

On the part of the Statistical Society, which is an entirely voluntary association of individuals, not connected with the State, and, I think I may say, not in the smallest degree desiring to be connected with the State, we pride ourselves upon our entire independence, we feel that if we are to maintain our ground in this country of free and open thought, and free and open competition, where every man, we firmly believe, obtains the reward which his merit deserves... we desire no sort of extraneous aid; that if our labours are so useful, and are so deserving of notice, as to obtain the support of the intelligent part of the public, and of those persons who take an interest in this important branch of knowledge, we shall obtain that support, and we shall obtain that support with sufficient constancy to keep us free from all Government trammels.[83]

William Farr seemed to agree: as he explained to the Congress, 'this country is not satisfied with the work of merely official people... the statists of other countries will, we hope, be induced to found similar independent societies'.[84] Scientific endeavour in nineteenth century Britain was often the preserve of gifted amateur gentlemen with the means to give them leisure for research and make them

[81] Lucy Brown, *The Board of Trade and the Free-Trade Movement 1830–42* (Oxford, 1958), 80. Alain Desrosières, 'Official Statistics and Medicine in Nineteenth-Century France: The SGF as a Case Study', *Social History of Medicine*, 4, 3, Dec. 1991, 515–37.
[82] Rev. Richard Jones, 'Sketch of the objects of the Section', 28 June 1833, 'Minutes of the Committee of the Statistical Section of the British Association, June 27, 1833', in Transcript from the note-book of Mr. J. E. Drinkwater, f. 3, Archives of The Royal Statistical Society, London.
[83] *Report of the Proceedings*, 116.
[84] *Report of the Proceedings*, 114–15.

independent of government and its demands.[85] Newmarch and Farr also reflected prevailing British attitudes in the favour they expressed for 'independence', conferring intellectual freedom, and in their implied criticism of European savants who were really civil servants and lacking such liberty. They were prescient also, because this issue would compromise the subsequent work of the ISC as states, notably the new German empire after 1871, refused to allow their delegates to share information and associate freely with other participants. Liberal Britain worked to a different pattern.

National differences were largely contained at the London congress, but never eradicated. Though the executive committee had decided that the use of both French and English was permissible in discussion, and the subsequent transactions published speeches and papers in both languages, there were determined efforts to limit the congress to French which more nearly 'approaches to a general language'.[86] Throughout the history of the Congress there were disagreements over whether participants should be limited to the official languages only. In line with the professed aim to achieve universal systems and procedures, many of the participants were keenly in favour of the adoption of the metric system for weights and measures and of the decimalization of all currencies. These were popular progressive causes of this era in many countries.[87] The Congress acted 'in unison' with the International Association for obtaining a Uniform Decimal System of Measures, Weights and Coins which was allowed to add a resumé of its work at the end of the transactions of the congress.[88] Of the two causes, decimalization was the more popular, even among British participants, several of whom were eager to decimalize the pound sterling, then divided into 20 shillings and 240 pence, a reform not achieved until 1971. They seemed to hope that in this way the pound would reach and retain the status of the dominant international currency.[89] But the coinage was the easier of the two projects. More speakers were against recommending the adoption of the metric system which, in the present state of British opinion, would be a 'waste of paper'.[90] Few went as far as General Sir Charles Pasley, a long-term opponent of the metric system, however: 'I would stand up to the last in defence of the national measures, weights, and money of this country; but let all other nations adopt the French if they please; it never will be

---

[85] J. Morrell and A. Thackray, *Gentlemen of Science: Early Years of the British Association for the Advancement of Science* (Oxford, 1981).
[86] *Report of the Proceedings*, lii, 15, 164.
[87] Martin H. Geyer, 'One Language for the World: The Metric System, International Coinage, Gold Standard, and the Rise of Internationalism, 1850–1900', in Martin H. Geyer and Johannes Paulmann, eds, *The Mechanics of Internationalism. Culture, Society, and Politics from the 1840s to the First World War* (Oxford, 2001), 55–92.
[88] *Report of the Proceedings*, 503–4.
[89] See the remarks by Samuel Brown of the Society of Actuaries, *Report of the Proceedings*, 195. See also ibid., 386–7.
[90] Theodore Rathbone, *Report of the Proceedings*, 387.

adopted in this country; it is an impossibility.'[91] This has a decidedly contemporary register, a recognizable tone and message more than a century later. The *Daily Telegraph*, then a Liberal newspaper, saw the humorous side in its report:

> In the advocacy of this plan, men after a while display a sort of fanaticism, as if the eternal interests of the human race depended upon sundry applications of the number ten. On the other hand, twelve has its advocates, who obviously persuade themselves that doomsday would be upon us at once if we exchanged it for a decimal number.[92]

There could be no doubting the official internationalist doctrines of the Congress, therefore. They showed themselves again in remarkable discussions in the fifth section on military statistics. It was one thing to request information on the sanitary condition of military forces across Europe and other developed nations, but quite another to ask for uniform returns from all nations on the number of their service personnel, the total expenditure broken down into many different categories of each national military, 'the ratios of those sums to the gross annual revenues of the state', and per capita expenditure on the armies and navies in each case.[93] The Accountant-General of the Royal Navy, attending the discussions on these matters, was commendably restrained in his observation that 'the Government would be very jealous of handing over to another person unconnected with the department any of the reports or documents in the archives, with a view to the extraction of any matter whatever from them'.[94] There was a kind of technocratic utopianism driving these requests for information. Bureaucrats were seeking perfect data despite the costs and burdens of finding out this information and without regard for the implications for national security of publishing it at all. But more than technocracy was in play. As explained by the key participant whose paper set these discussions in train, he had been asked to contribute by Farr himself and 'I can speak with confidence for my friend Dr Farr, that this subject was brought under the notice of the Congress in the interests of Peace.'[95] Mid-Victorian Britain gave birth to an organized peace movement and some of the Liberal leaders of the age, most notably the famous combination of Richard Cobden and John Bright, gave articulate and reasoned support to the cause of international peace.[96] In seeking to further this cause the international statistical movement was not out of step with some strains of opinion, therefore. But in their touchingly naïve efforts to further peace by circulating information on the costs of defence and of war—as if nations would willingly provide this information and allow its publication—William Farr and the ISC made further evident

[91] Ibid., 193.   [92] *Daily Telegraph*, 21 July, 4.   [93] *Report of the Proceedings*, 166–7.
[94] Ibid., 361 (Sir Richard Bromley).   [95] Ibid., 165 (W. Barwick Hodge, FSS, FIA).
[96] Donald Read, *Cobden and Bright. A Victorian Political Partnership* (London, 1967).

what may be called the ideologies of the statistical movement at its zenith: liberal, reformist, internationalist and also, in some guises, pacifist.

These themes were threaded through the week's events. Shaftesbury's public reflections on presiding over the sanitary section caught the mood:

> When I consider that there were so many foreign gentlemen come to England for the sole purpose of advancing these great principles, and that they were joined together with the British delegates in the advancement of the same object, I could not but thank God, in the belief that we were entering upon a new page in the history of humanity.[97]

Similar ideas were encapsulated in the closing remarks of William Cowper, MP, a conscientious workhorse in several mid-Victorian Whig and Liberal governments, the sponsor and author of many improving bills in parliament, and most notable personally for the Cowper–Temple clause he added to the famous 1870 Elementary Education Act.[98] Speaking on behalf of the British government, in words and a tone perhaps unfamiliar from a man of public business never famed for flights of rhetoric, there was a genuine sincerity and modesty in his ardent internationalism:

> We do desire to have stronger links and warmer sympathies with other countries. We feel that we have a great deal to learn upon all subjects of science and philanthropy from other nations, and we rejoice at such opportunities as this of knowing what is going on elsewhere, that we may learn something that will be a guidance for ourselves, that we may have the future satisfaction of explaining to others what we are doing, so that the advantage may be mutual between us and others. We feel that the purpose which has assembled us here is one which lies at the foundation of all efforts for the benefit of the human race.[99]

The *Daily Telegraph* might offer caution and scepticism in response, reminding its readers in one of its dispatches from the congress that 'all the pursuits of the philosopher are flung aside, and men take to mutual destruction' when the more usual situation of national rivalry supervened.[100] Yet Cowper went a stage further than even this idealistic internationalism to a kind of universalism, premised on the equality of races and peoples, that would provide the basis, beyond the co-operation of states and nations, for a genuinely comprehensive social science applicable to all:

---

[97] *Report of the Proceedings*, 179. See also Shaftesbury's similar sentiments at 293.
[98] 'Temple, William Francis Cowper-, Baron Mount-Temple (1811–1888)', *ODNB*. The Cowper-Temple clause mandated non-denominational elementary education in the rate-aided schools established under the 1870 Act. William Cowper was almost certainly the son of Palmerston.
[99] *Report of the Proceedings*, 206.   [100] *Daily Telegraph*, 21 July 1860, 4.

We are convinced that the human mind is substantially the same in all countries; that though there may be varieties, yet that man is substantially the same being, under whatever tribe or under whatever coloured skin he may be. And in order to study human nature, we cannot confine ourselves to the limit of any single kingdom, but we must endeavour, as far as we can, to extend our observations over the whole human race.[101]

Events in London in July 1860 thus illustrate what may be called 'the politics of statistics'. To practitioners and the public alike the collection, analysis, and application of statistics offered new insights into social structures and behaviours and a new method of social administration based on the possession of accurate social information. Statistics would provide unimpeachable and unquestionable authority for the rational decisions of government based on the best data. Governments would share knowledge and best practice between themselves and would be held accountable by their peoples through the collection and publication of standard measures of the efficacy of their social administration: 'league tables' of performance are not an invention of the present, be it noted. Statistics when applied at societal and community levels would provide for welfare, betterment, improvement. From shared knowledge and procedures would come international co-operation and amity. Statistics were co-extensive with 'progress' itself, a key component of mid-nineteenth century liberalism in Britain and abroad. This book will explain the liberal origins of social statistics which led up to the London Congress in 1860, but also consider how, in the later years of the century, the desire to collect, sort and classify social data led to the division and demise of the International Statistical Congress itself, and to the development of a quite different 'politics of statistics'.

---

[101] *Report of the Proceedings*, 206.

PART I
POLITICAL ARITHMETIC AND
STATISTICS 1660–1840

1

Before the Victorians

1.1 Political Arithmetic and 'Statistik'

On Saturday afternoon, 15 March 1834, in the rooms of the Horticultural Society in Regent Street, London, a distinguished group of politicians, civil servants, lawyers, doctors and savants, numbering more than three hundred, met together to inaugurate the Statistical Society of London.[1] The chairman on that occasion, and afterwards the Society's first president, was Henry Petty-Fitzmaurice, third marquess of Lansdowne, then a member of the Whig cabinet as the Lord President of the Council. Lansdowne was one of the leading parliamentarians of the first half of the nineteenth century, a member of many Whig and Liberal governments who could have been prime minister on occasions in the 1840s and 1850s had he wanted the highest office. Those who *did* want it had first to ensure that Lansdowne, who would have outranked them in status, experience, and perhaps ability also, did not. Against slavery, an unreformed parliament, and disabilities on Roman Catholics, Protestant Dissenters, and Jews, Lansdowne was a moderate Whig, naturally in the centre of British politics, and after the death of the Duke of Wellington in 1852, the unofficial adviser on constitutional questions to the Queen. If Lansdowne's presence was a measure of the importance accorded to statistics in Britain, there was something of even greater significance in his role on that day. He was a direct descendant of Sir William Petty, the late-seventeenth-century practitioner of 'political arithmetic', the synthesis of what we would now differentiate into the separate disciplines of economics, demography, and social statistics. By using Lansdowne to lead their venture, a new generation in British intellectual life and social administration showed its recognition and respect for what had gone before, and on which it intended to build. Indeed, the oldest of the founder members of the Statistical Society of London, Sir John Sinclair, was also a notable practitioner of political arithmetic, famed for his work as long ago as the 1790s. In short, before there were 'Victorians and Numbers', there were 'Georgians and Numbers', and before social statistics there was political arithmetic, defined by one of its exponents, Charles Davenant, as 'the art of reasoning by figures, upon things relating to government'.[2] According to another, John Arbuthnot, 'those

---

[1] 'Statistical Society', *The Times*, 17 March 1834, 3; 'Statistical Society', *The Gentleman's Magazine*, n.s., 1, April 1834, 422.
[2] Charles Davenant, *The Political and Commercial Works of that Celebrated Writer Charles D'Avenant* (ed. C. Whitworth) (5 vols, London, 1771), i, 128.

Victorians and Numbers: Statistics and Society in Nineteenth Century Britain. Lawrence Goldman, Oxford University Press.
© Lawrence Goldman 2022. DOI: 10.1093/oso/9780192847744.003.0001

that would judge or reason about the state of any nation must go that way to work, subjecting all...particulars to calculation. This is the true political knowledge.'[3]

Some would have traced a much more ancient history of statistics. William Farr began his account of its history with the *Domesday Book* of 1086, the great survey of the country taken after the Norman Conquest.[4] Many went back further still. The *Book of Numbers*, the fourth in the Old Testament, takes its name from the several censuses taken by the Israelites in their wanderings through the desert to Canaan. The plague visited on them after King David's census of Israel and Judah was another point of reference, a precedent employed by those who opposed the counting of the people.[5] Educated Victorians were immersed in the bible, and the classics also. Prometheus's claim that he had brought fire to mankind, and with it all the arts of civilization, 'and numbers, too, the subtlest science, I invented for them', would have been familiar.[6] It was frequently remarked that Plato, in *The Republic*, would have had the guardians, 'the principal men of our State...learn arithmetic' because 'those who have a natural talent for calculation are generally quick at every other kind of knowledge'. Beyond the needs of everyday statecraft, the study of abstract number elevated the intellect and the soul; it employed 'pure intelligence in the attainment of pure truth'.[7] That may have been the theory in classical utopias: we shall see just how close to the pure truth came some Victorian statisticians.

We are on firmer ground with Sir William Petty (1623–87) and his friend and collaborator John Graunt (1620–74), the acknowledged founders of political arithmetic in the 1660s and 1670s, the initiators of a tradition in Britain that stretched to the early decades of the nineteenth century.[8] Neither came close in type to Plato's philosopher kings, it must be admitted. Both were from humble backgrounds. Petty was the son of a clothier sent away to sea. His very worldliness, his capacity to take opportunities, to enrich himself with cash and land, and to become an intimate of successive English and Irish regimes, is as far removed from the Platonic ideal as could be.[9] Graunt was a haberdasher, a member of the

[3] John Arbuthnot, 'Essay on the Usefulness of Mathematical Learning', *Life and Works of John Arbuthnot* (ed. George A. Aitken) (Oxford, 1892), 421–2.

[4] William Farr, 'Report on the Programme of the Fourth Session of the Statistical Congress', *Programme of the Fourth Session of the International Statistical Congress to be held in London on July 16th and Five Following Days* (London, Her Majesty's Stationery Office), 12.

[5] 2 Samuel 24:1. See Viviana A. Rotman Zelizer, *Morals and Markets. The Development of Life Insurance in the United States* (New York, 1979, 2017 edn), 48–9. James H. Cassedy, *Demography in Early America* (Cambridge, MA., 1969), 69–70.

[6] Aeschylus, *Prometheus Bound*, line 455.

[7] *The Republic of Plato* (Tr. Benjamin Jowett) (3rd edn, Oxford, 1888). Jowett was Master of Balliol College, Oxford. For his support for statistics in association with Florence Nightingale, see below pp. 286–7.

[8] D. Glass, M. Ogborn, and I. Sutherland, 'John Graunt and His Natural and Political Observations [and Discussion]'. *Proceedings of the Royal Society of London. Series B, Biological Sciences*, 159 (1974), 2–37.

[9] E. C. Fitzmaurice, *The Life of Sir William Petty, 1623–1687* (London, 1895).

Drapers' Company, without much formal education, though he became a man of repute and influence in the City of London. He grasped how available data on mortality in the city, drawn from the published Bills of Mortality which, each week, listed those who had died and the causes of their deaths, could be used in the construction of the first ever life table based on reliable information. The data enabled him to calculate the likely duration of life at different ages for Londoners in the 1660s—what he termed 'Survivors to different birthdays and deaths between birthdays'—in his *Natural and Political Observations...made upon the Bills of Mortality*.[10] It was an early example of time-series analysis.[11] Data on deaths and also on the number of christenings in the city made it possible for Graunt to calculate the population of London in three different ways, each confirming a total of approximately 380,000. This was much less than the grossly inflated guesses of contemporaries, but close to our estimates today.[12]

They were followed by Charles Davenant (1656–1714) and Gregory King (1648–1712), public officials and office-holders, who wrote in the 1690s and the first years of the eighteenth century. John Arbuthnot (1667–1735), a physician, satirist, and intimate of Pope, Swift, and Gay, and William Derham (1657–1735), a clergyman, both used numbers in the early eighteenth century to demonstrate the God-given order and the providential design of the world. Arthur Young (1741–1820) used political arithmetic in surveys of agriculture and its improvement which he published in travelogues of his extensive tours through farming districts in Britain and France.[13] Sir John Sinclair (1754–1835) and Sir Frederick Morton Eden (1766–1809) conducted large social surveys at the end of the eighteenth century to produce, respectively, the *Statistical Account of Scotland*, still the most cited of all Scottish historical sources, and *The State of the Poor* (1797), an account of the history and condition of the labouring classes in England.[14] William Playfair (1759–1823) made brilliant methodological breakthroughs in the graphing and depiction of social statistics—he was the inventor of the time-series graph and pie chart in publications in 1787 and 1801 respectively—but received scant

[10] John Graunt, *Natural and Political Observations mentioned in a following Index, and made upon the Bills of Mortality, by John Graunt, Citizen of London. With reference to the Government, Religion, Trade, Growth, Ayre, Diseases, and several Changes of the said City* (London, 1662).

[11] Judy L. Klein, *Statistical Visions in Time: A History of Time Series Analysis 1662–1938* (Cambridge, 1997), 25–53.

[12] Richard Stone, *Some British Empiricists in the Social Sciences 1650–1900* (Cambridge, 1997), 210–13.

[13] For a fine survey of the political arithmeticians, see Joanna Innes, 'Power and Happiness: Empirical Social Enquiry in Britain, from "Political Arithmetic'" to "Moral Statistics"', in Innes, *Inferior Politics. Social Problems and Social Policies in Eighteenth-Century Britain* (Oxford, 2009), 109–75.

[14] Sir John Sinclair, *The Statistical Account of Scotland* (21 vols, 1791–99). Frederick Morton Eden, *The State of the Poor, or, An history of the labouring classes in England from the conquest to the present period; in which are particularly considered their domestic economy with respect to diet, dress, fuel and habitation; and the various plans which, from time to time, have been proposed and adopted for the relief of the poor etc.* (3 vols, London, 1797).

recognition until the twentieth century.[15] John Rickman held various official positions in the early nineteenth century, but is remembered for the design of the first British census of 1801, and for the increasing sophistication of the decennial censuses subsequently conducted down to 1841.[16]

Many more names could be added to the list. These were only some of the political arithmeticians who, during what is sometimes called 'the long eighteenth century' from the Restoration to the Regency, tried to make sense of social structure, demographic trends, economic development, the state of trade, the distribution of wealth, and God's purposes for man in this world. Some, like Sinclair and Eden, were gentlemen; others held administrative office in government, or through their publications sought such offices. Their talents were required for public administration, and official experience gave them access to material, some of it statistical, on which to base their published work. The eighteenth-century 'military-fiscal state' in Britain was designed to raise taxes which were largely spent in equipping a navy and funding armies—frequently armies raised by other states, but useful in the pursuit of British ends. Compared with European state-structures of this period, the British state was small in size and also efficient. Its civil servants developed among themselves a 'culture of numbers' that equipped them for running national institutions and which provided experience in manipulating data.[17]

Political Arithmetic is best understood as the analytical study of key demographic, social, and political variables. Beyond the mere compilation of numerical data and description, it implied the application of some mathematical procedure or analysis to uncover more complex relationships and results. It was dynamic, often seeking to understand a trend over time, such as birth or death rates. It is customary to distinguish between political arithmetic as developed by the British, and the tradition of 'statistik' which emerged in the eighteenth-century German states.[18] According to William Farr 'the new science was called, at its origin, political arithmetic in England. It obtained its present name in the universities of Germany, where the first courses of lectures in statistics were delivered.'[19] As its etymology suggests, the discipline of statistik was related to the creation and administration of the state. It amounted to the compilation of information on the population, finance, industry, armed forces, political structures, and trade of states, frequently set out in tabular form. It was usually static and descriptive: a statement of national resources and organization at a moment in time, and a way of making simple international comparisons. German 'Statistik' was also known

[15] Joseph J. Spengler, 'On the Progress of Quantification in Economics', *Isis*, 52, 2, June 1961, 261n. See William Playfair's *Commercial and Political Atlas* (1787); *Lineal Arithmetic* (1798); *Statistical Breviary* (1801); and *Decline and Fall of Powerful and Wealthy Nations* (1805).

[16] O. Williams, *Lamb's Friend the Census-taker: Life and Letters of John Rickman* (London, 1912); David Eastwood, 'John Rickman 1771–1840', *ODNB*.

[17] John Brewer, *The Sinews of Power. War, Money and the English State 1688–1783* (London, 1989), 65–6.

[18] Anthony Oberschall, *Empirical Social Research in Germany 1848–1914* (The Hague, 1965), 4.

[19] Farr, 'Report on the Programme of the Fourth Session of the Statistical Congress', 3.

as 'Staatswissenschaft', the science of the state, or 'Staatenkunde', the comparative study of states. Modern authors sometimes refer to 'state-istics' to better define its meaning and qualities.[20]

Its practitioners included Hermann Conring (1606–1681), Gottfried Achenwall (1719–1772) and August Ludwig von Schlözer (1735–1809). Achenwall, the most famous of the trio, was trained in philosophy and law: he first used the term 'statistik' as a noun in 1749.[21] Conring is credited with first lecturing on 'Statistik' much earlier in 1660. Von Schlözer was a prolific and influential historian: he described statistics as 'stationary history' in 1804.[22] All three were professors in German universities.[23] It makes an interesting and instructive contrast with Britain where no eighteenth-century political arithmetician had any sort of formal academic career. Most were drawn from the 'middling sort', gained skills and experience in commerce or government as young men, and generally found a place in public administration. Nor did any of the leading practitioners of Victorian statistics have a base as a statistician in a British university: they were businessmen, officials, doctors, writers, or like Francis Galton, gentlemen of independent means.

Michel Foucault, the influential social theorist of the late twentieth century, suggested that it was the very instability and insecurity of the disunited German states of the seventeenth and eighteenth centuries that promoted the study of 'statistik'. Their conflicts, vulnerabilities, and misfortunes 'obliged them to weigh and compare themselves against others' and encouraged the development of the first modern bureaucratic state in Prussia, before all other states, precisely to compensate for this underdevelopment.[24] An older tradition in British historiography, however, encompassing the so-called 'Tudor revolution in government' in the sixteenth century, the growth of a civil service under both monarchy and republic in the seventeenth century, and the development of the 'military-fiscal state' in the eighteenth century, runs counter to this argument. The origins of the modern state, and of political arithmetic within it, were the products in this interpretation of national unification and assertion, political stabilization, the centralization of political authority, and growing wealth in early modern Britain.[25]

---

[20] Martin Shaw and Ian Miles, 'The Social Roots of Statistical Knowledge', in *Demystifying Social Statistics* (eds John Irvine, Ian Miles, and Jeff Evans) (London, 1979), 31.

[21] Mary Poovey, *The History of the Modern Fact. Problems of Knowledge in the Sciences of Wealth and Society* (Chicago and London, 1998), 308.

[22] David F. Lindenfeld, *The Practical Imagination. The German Sciences of State in the Nineteenth Century* (Chicago, 1997), 18–20. August Ludwig von Schlözer, *Theorie der Statistik: nebst Ideen über der Politik überhaupt. Erstes Heft. Einleitung* (Göttingen, 1804), 86.

[23] Alain Desrosières, *The Politics of Large Numbers. A History of Statistical Reasoning* (1993) (Cambridge, MA, and London, 1998), 19, 326.

[24] Michel Foucault, 'The Birth of Social Medicine' (1974) in Foucault, *Power. Essential Works of Foucault 1954–1984: Volume 3* (Paris, 1994) (2000 edn, London) (Colin Gordon, ed.), 138–9.

[25] Geoffrey Elton, *The Tudor Revolution in Government. Administrative Changes in the Reign of Henry VIII* (Cambridge, 1953); Gerald Aylmer, *The King's Servants: The Civil Service of Charles I, 1625–42* (London, 1961); Gerald Aylmer, *The State's Servants: The Civil Service of the English Republic,*

We may conclude that whether the state was small or large, successful or challenged, wealthy or immiserated, accurate social and economic data were required for good government.

The first English uses of the word 'statistics' followed German meanings. In 1770 'statistics' denoted the science which 'teaches us what is the political arrangement of all the modern states of the known world'.[26] In 1797 'statistics' was defined by the *Encyclopaedia Britannica* as a 'word lately introduced to express a view or survey of any kingdom, country or parish'.[27] A similar definition was used by the cartographer and compiler Benjamin Capper in 1808:

> Statistics are that comprehensive Part of municipal Philosophy, which states and defines the Situation, Strength, and Resources of a Nation, and is a Kind of political Abstract, by which the Statesman may be enabled to calculate his Finances, as well as guide the OEconomy of his Government.[28]

When Sir John Sinclair explained his understanding of statistics he added another element, however, differentiating between German 'Statistik' and his own approach, which was focused less on the description of the state and more on the welfare of its people, on their 'quantum of happiness'.[29]

> Many people were at first surprised at my using the words "statistical" and "statistics", as it was supposed that some term in our own language might have expressed the same meaning. But in the course of a very extensive tour through the northern parts of Europe, which I happened to take in 1786, I found that in Germany they were engaged in a species of political enquiry to which they had given the name "statistics," and though I apply a different meaning to that word—for by "statistical" is meant in Germany an inquiry for the purposes of ascertaining the political strength of a country or questions respecting matters of state—whereas the idea I annex to the term is an inquiry into the state of a country, for the purpose of ascertaining the quantum of happiness enjoyed by its inhabitants, and the means of its future improvement; but as I thought that a new word might attract more public attention, I resolved on adopting it, and I hope it is now completely naturalised and incorporated with our language.[30]

1649–1660 (London 1973); Gerald Aylmer, *The Crown's Servants: Government and Civil Service under Charles II, 1660–1685* (Oxford, 2002); Brewer, *The Sinews of Power*, passim.

[26] J. F. von Bielfeld, *The Elements of Universal Erudition* (trans. W. Hooper) (3 vols, London, 1770), iii, 269.

[27] *Encyclopaedia Britannica* (3rd edn, Edinburgh, 1797), vol. XII, 731.

[28] B. P. Capper, *A Statistical Account of the Population and Cultivation, Produce and Consumption of England and Wales* (London, 1801), vii. Capper also compiled and published *A Topographic Dictionary of the United Kingdom* in 1808.

[29] Maxine Berg, *The Machinery Question and the Making of Political Economy 1815–48* (Cambridge, 1980), 299.

[30] Sinclair, *Statistical Account of Scotland*, vol. 20, viii.

This recalls Graunt's remarks in 1662 when he asked of his own calculations, 'to what purpose tends all this laborious buzzling and groping?' He answered that 'the art of governing and true politiques is how to preserve the subject in peace and plenty'.[31]

Are sharp distinctions between different meanings of 'statistics', and between statistics and political arithmetic, worth upholding as some historians have contended?[32] Perhaps they shouldn't be pushed too far. Many British authors worked both statically and dynamically with the numbers at their disposal, and there were domestic genres and traditions which bear resemblance to 'statistik' but actually predate German descriptive statistics. There are overlaps and alternatives to the conventional binary division into two different national types and styles of analysis, in other words, which make for a more complex and more interesting intellectual history. For example, there was a pre-existing, English antiquarian tradition of historical, political, social and geographical description, to which many authors contributed, from William Camden's *Britannia*, published in 1586, onwards.[33] Works in this genre were general collections of historical, natural, useful, and diverting information about people, places, social and intellectual institutions, physical features, natural wonders, and man-made landscapes. It was common to include information not only on the historic inhabitants, but the most worthy contemporaries as well, and thus garner sponsorship towards the costs of publication. They include late-seventeenth-century 'county histories' such as Robert Plot's *Natural History of Oxfordshire, being an Essay toward the Natural History of England* of 1676, and John Aubrey's *The Natural History and Antiquities of the County of Surrey,* published posthumously.[34] Aubrey also compiled a 'Naturall Historie of Wiltshire', over several decades in the second half of the seventeenth century, though it was never published.[35] The genre probably reached its zenith a century later in a controversial work still widely read and discussed today, Thomas Jefferson's *Notes on the State of Virginia*. Completed in1781 and published four years later, it was a survey of the geography, agriculture, institutions, people, and races in that American state at the time of the Revolution.[36]

Rigid differentiation between statistik and political arithmetic also fails because many political arithmeticians in the English tradition were also authors of descriptive works of statistics. Petty wrote a *Political Anatomy of Ireland* in 1672

[31] Graunt, *Observations on the Bills of Mortality* quoted in Stone, *Some British Empiricists*, 223.
[32] See, for example Alain Desrosières, *The Politics of Large Numbers,* 16, 18, 22, 24; Libby Schweber, *Disciplining Statistics. Demography and Vital Statistics in France and England, 1830–1885* (Durham N.C., and London, 2006), 4.
[33] Wyman H. Herendeen, 'William Camden 1551–1623', *ODNB*.
[34] Roy Porter, *The Making of Geology. Earth Science in Britain 1660–1815* (Cambridge, 1977), 32–61.
[35] A. J. Turner, 'Robert Plot (1640–1696)', *ODNB*. Adam Fox, 'John Aubrey (1626–1697)', *ODNB*. John Aubrey, *The Natural History and Antiquity of the County of Surrey* (ed. Richard Rawlinson) (5 vols, 1718–19).
[36] Lawrence Goldman, 'Virtual Lives: History and Biography in an Electronic Age', *Australian Book Review*, June 2007, 37–44.

(published in 1691) and published the first ever atlas of Irish counties in his *Hiberniae Delineatio* of 1685: he is famous for both the dynamic study of population *and* the descriptive presentation of human geography and political institutions. From the *Political Anatomy* it was possible to trace the shift of land ownership from Catholic to Protestant in mid-seventeenth century Ireland, the distribution of wealth by the size and quality of house, and the occupational structure in Ireland. As the epitaph erected by his second wife explains, Gregory King was many things: 'a Skilful herald, a good Accomptant Surveyor, and Mathematician, a Curious pen Man, and Well Cerit in Political arithmetick'.[37] He was interested in how population changed over time, certainly, but he is most famous for the social structure he devised at a single moment in time for his *Natural and Politicall Observations Upon the State and Condition of England*, published in 1696, which broke the nation down into 26 social groups.[38] As King explained in the preface to the work, 'the true state and condition of a nation, especially in the two main articles of its people and wealth' is that 'piece of political knowledge, of all others, and at all times, the most useful, and necessary'.[39] In consequence, he concluded the *Observations* with an analysis of the rival economies of Britain, France, and the Netherlands, then engaged in the War of the League of Augsburg, through a simple comparison of national income per head.[40] In the late seventeenth century, political arithmetic, like statistik, was also largely focused on such calculations of power, wealth and population as between Britain and other states.[41]

Sinclair's *Statistical Account of Scotland* covered the geography, history, economy, and society of all Scottish parishes, nearly a thousand in total. The information was gathered by questionnaires sent to the local clergy, and, when these failed to elicit a response, visitations from his editorial assistants.[42] Sir Frederick Eden surveyed the poor, but he also calculated the population of Britain and Ireland at 16.5 million, remarkably close to the likely figure of 15.9 million.[43] Arthur Young published a work in 1774 that nicely conflates both of these

---

[37] Julian Hoppitt, 'Gregory King 1648–1712', *ODNB*. The epitaph is in The Church of St. Benet Paul's Wharf, London.

[38] 'A Scheme of the Income & Expence of the Several Families of England Calculated for the Year 1688', in Gregory King, *Natural and Politicall Observations Upon the State and Condition of England* (1696) (ed. G. E. Barnett, Baltimore, 1936), 31. G. S. Holmes, 'Gregory King and the Social Structure of pre-Industrial England', *Transactions of the Royal Historical Society*, 5th ser., 27 (1977), 41–68. P. H. Lindert and J. G. Williamson, 'Revising England's Social Tables, 1688–1812' and 'Reinterpreting Britain's Social Tables 1688–1913', *Explorations in Economic History*, 19, 1982, 358–408 and 20, 1983, 94–109.

[39] Gregory King, 'Preface', *Natural and Politicall Observations*.

[40] 'The General Account of England, France & Holland for the Years 1688 & 1695', in King, *Natural and Politicall Observations*, 55. The relative amounts per head in 1695 were £7.80 for England, £5.49 for France and £8.15 for Holland. See also Stone, *Some British Empiricists*, 102.

[41] E. A. Wrigley, 'Comment' in Stone, *Some British Empiricists*, 426.

[42] Rosalind Mitchison, *Agricultural Sir John: The Life of Sir John Sinclair of Ulbster* (London, 1962).

[43] Frederick Morton Eden, *An Estimate of the Number of the Inhabitants in Great Britain and Ireland, 1800. Written while the Census Bill was before Parliament; partly extracted from The State of the Poor* (London, 1800). See Stone, *Some British Empiricists*, 298–301.

numerical traditions and makes our point: *Political Arithmetic: Containing Observations on the Present State of Great Britain and the Principles of her Policy in Encouragement of Agriculture.* It is surely of some relevance that Gottfried Achenwall and August Von Schlözer taught at the preeminent university for the study of statistik in Germany, Göttingen in Hanover. George II, British king and elector of Hanover both, had founded the university in 1737, and there was extensive contact between its faculty, the British court, and British savants for the rest of the century. In this way 'statistik', 'staatenkunde', and 'political arithmetic' came together with a common purpose to describe, explain and anatomize the different elements of a state and its people.

Drawing precise and tight definitions of the different ways in which statistics were used in the eighteenth century also makes it more difficult to appreciate the continuities linking the Georgian with the Victorian period. For example, data was collected and deployed in criticism of government as well as in establishing the strength and resources of the state. Political arithmetic was cultivated in England in the second half of the eighteenth century by some religious dissenters and radicals, such as the Rev. Richard Price, the non-conformist minister, philosopher, and demographer, precisely because it enabled a more searching critique of the state, its church, and its finances.[44] Such a tradition of 'critical statistics' only grew more potent in the nineteenth century as the examples to follow, from the artisan statisticians of the 1820s to the working-class activists of the 1880s, substantiate.[45] Victorian statisticians uncovered predictable regularities in certain social behaviours that set off a long-running debate in the 1840s and 1850s about determinism and the scope for free will in human affairs. More than a century earlier these same regularities had been used by political arithmeticians as evidence to prove the role of providential design in nature.[46] Arbuthnot published *An Argument for Divine Providence, taken from the constant Regularity observ'd in the Birth of both Sexes* in 1710 which was based on the analysis of the London Bills of Mortality since 1629. Observing that more males were born than females in a ratio of 18 to 17, and that, on average, more men than women died young before marriage because they led more hazardous lives, he reasoned that the resulting balance of numbers between the sexes at marriageable age could not be ascribed to chance but to design, divine planning.[47]

[44]   Peter Buck, 'People Who Counted: Political Arithmetic in the Eighteenth Century', *Isis*, 73, 1, March 1982, 28–45. Richard Price, *An Essay on the Population of England from the Revolution to the Present Time* (London, 1780); Richard Price, *Observations on Reversionary Payments: on Schemes for Providing Annuities for Widows, and for Persons in Old Age; on the Method of Calculating the Values of Assurances on Lives; and on the National Debt* (London, 1783).

[45]   See Chapters 4 and 15 below.

[46]   Lorraine Daston, *Classical Probability in the Enlightenment* (Princeton, N.J., 1988), 131–2.

[47]   Daniel R. Headrick, *When Information Came of Age. Technologies of Knowledge in the Age of Reason and Revolution, 1700–1850* (Oxford, 2000), 64; Theodore Porter, *The Rise of Statistical Thinking 1820–1900* (Princeton, N. J., 1986), 50.

Among innumerable footsteps of Divine Providence to be found in the Works of Nature, there is a very remarkable one to be observed in the exact Ballance that is maintained, between the Numbers of Men and Women; for by this means it is provided that the Species may never fail, nor perish, for by this means it is provided since every Male may have its Female, and of a proportionable Age. This Equality of Males and Females is not the effect of Chance but Divine Providence, working for a good End, which I thus demonstrate.[48]

The argument was made again, and with greater impact across Europe, a generation later by Johann Peter Süssmilch in *Die göttliche Ordnung*.[49] Directly after Arbuthnot, William Derham delivered the Boyle Lectures in London which were published under the title *Physico-Theology, or, a Demonstration of the Being and the Attributes of God from his Works of Creation*. This also was a work of natural theology, finding evidence of divine existence and beneficent intent in the physical world around us.[50] It reached its twelfth edition as soon as 1754. As we shall see, the Cambridge coterie responsible for institutionalizing statistics in the 1830s were schooled in eighteenth-century natural theology and strongly influenced by it. Florence Nightingale, one of the keenest of Victorian statisticians, also saw the natural and social regularities revealed by statistics as evidence of God's existence and His divine order.

At the end of the Victorian period Charles Booth, another of the subjects of this study, used the human intelligence of School Board Visitors whose job it was to ensure that all households in London in the 1880s sent their children to school, to build a picture of living standards in the city, street by street, for his famous survey, *The Life and Labour of the People of London*, published between 1889 and 1903. A century earlier, Eden had employed a paid investigator and a number of clergymen to visit parishes throughout England to report on the state of the poor.[51] And in the 1830s, the very first projects attempted by both of the new statistical societies founded in Manchester and London, were surveys of the urban poor living close by, based on data collected by visiting inquirers. Booth may have worked a century after Sinclair and Eden, but like them, he financed himself: from the proceeds of his businesses he paid for the most influential social survey ever undertaken in Britain. It was the outcome of a national controversy in the

    [48] John Arbuthnot, 'An Argument for Divine Providence, taken from the constant Regularity observ'd in the Birth of both Sexes', *Philosophical Transactions* (Royal Society), vol. 27, 1710–12, 186–90.
    [49] Johann Peter Süssmilch, *Die Göttliche Ordnung in den Veränderungen des Menschlichen Geschlechts, aus der Geburt, dem Tode und der Fortpflanzung Desselben Erwiesen* (1741); Porter, *The Rise of Statistical Thinking*, 21–3.
    [50] Innes, 'Power and Happiness: Empirical Social Enquiry in Britain', 126.
    [51] Richard Stone, 'Frederick Morton Eden and the Poor of England' in Stone, *Some British Empiricists in the Social Sciences*, 277–301.

mid-1880s over the extent of poverty in Britain.[52] Like many Victorian surveys and statistical inquiries it was the response to a perceived social concern, therefore, the dimensions of which it sought to ascertain. But this, also, was nothing new: many late-eighteenth century projects in social research were driven by social problems, by the desire or need to know more about 'poverty, vagrancy, crime, and imprisonment for debt' in order to reform social institutions and preserve public order.[53] The cost of administering the old Elizabethan poor laws as pauperism grew from the 1790s was a constant focus of national debate and statistical investigation before they were replaced by the New Poor Law of 1834.[54] As we shall see, the artisan statisticians of Clerkenwell who were at work in the 1820s, were also focused on social problems, but their analysis of the causes of immiseration and pauperism, and their recommended solutions, were very different from those advanced by the Manchester bourgeoisie or by the Poor Law Inquiry Commission in London.

A fascination with numbers was as much a part of eighteenth-century domestic and commercial life as it was characteristic of later Victorian family enterprise. Francis Galton described his grandfather, Samuel John Galton, the founder of the family's wealth in the 1770s and 1780s, and a member of the famous Lunar Society of Birmingham, as 'a scientific and statistical man of business' who loved 'to arrange all kinds of data in parallel lines of corresponding lengths, and frequently used colour for distinction'. He passed on this 'statistical bent' to his children 'in a greater or lesser degree'.[55] In his grandson, the habit of keeping detailed business and domestic accounts became an all-round academic fascination with numbers, counting and measurement. 'Whenever you can, count' was Francis Galton's motto.[56] But it could have served just as well for his grandfather who 'harvested facts and figures about horses, canals, building materials and anything else that aroused his interest'.[57] Statistical interests, vocations, and methods were shared across the eighteenth and nineteenth centuries.

Of all the acknowledged practitioners of political arithmetic, the first, Sir William Petty, may be the most representative and also the most interesting.[58] Not all the political arithmeticians played the political game so successfully. Charles Davenant, for example, was a strong Tory, too close to James II, and he lost out in the Glorious Revolution, only returning to some sort of favour in the

[52] See below, pp. 307–8.
[53] Innes, 'Power and Happiness: Empirical Social Enquiry in Britain', 142.   [54] Ibid., 158–9.
[55] Francis Galton, *Memories of My Life* (London, 1908), 3.
[56] Karl Pearson, *The Life, Letters and Labours of Francis Galton* (4 vols, Cambridge, 1914–30), II, 340.
[57] Martin Brookes, *Extreme Measures. The Dark Visions and Bright Ideas of Francis Galton* (London, 2004), 4.
[58] Fitzmaurice, *The Life of Sir William Petty*, passim.

reign of Queen Anne.[59] He was also the 'foremost propagandist' for the Royal African Company, which transported and sold more slaves from Africa than any other British institution of this or any other era.[60] Petty, on the other hand, was highly successful, upwardly mobile, affable, humorous and able to offer useful service to successive regimes of the mid- and late-seventeenth century. According to Samuel Pepys he was 'one of the most rational men that I ever heard speak with a tongue, having all his notions the most distinct and clear.'[61] Petty was at various times a sailor, a scholar, a physician, a professor, a surveyor, a founding member of the Royal Society, an MP, and an admired author. According to John Evelyn, 'If I were a prince, I should make him my second counsellor at least.'[62] Mathematically inclined as a child, Petty had the good fortune to be injured and then abandoned in France where he remained to be educated by Jesuits. In Paris he spent a period in the company of mathematicians and natural philosophers gathered around Marin ('Father') Mersenne. On his return to England in the 1640s he studied medicine in Oxford, rising very fast to take the chair of anatomy in the university and to enjoy a private practise of his own. Then the revolutionary state had use for him in the early 1650s, first as physician-general to the army in Ireland and then overseeing the expropriation of Irish estates belonging to those landholders considered disloyal to the Protestant Cromwellian regime. The land was granted to members of the invading army and those who had provided financial support for the campaign to subjugate Ireland. Petty conducted the so-called Down Survey of Irish lands and their division among the victors: it was the sort of technical and arithmetical exercise at which he excelled. In the process he was hugely enriched. It led him to conduct and eventually publish, at his own expense, the first accurate mapping of Ireland, *Hiberniae Delineato*. It led also to controversy when an investigation into his alleged corruption was begun in 1658, though this collapsed at the Restoration two years later. Petty has been censured in our own time, as well: he was a willing servant and architect of the English and Protestant ascendancies in Ireland in which land was forcibly transferred to the ownership of invaders, and the indigenous culture rendered subservient.[63]

Petty had the confidence of Henry Cromwell, son of the Protector and Lord-Lieutenant of Ireland in the 1650s; he amused Charles II, always sound policy, who knighted him in 1662; he was on good terms with his successor, James II. This very worldliness has led to the underestimation of Petty as a savant in his own right. From the Restoration until his death in 1687, he sought an opportunity—though

[59] D. A. G. Waddell, 'Charles Davenant 1656–1714 – a Biographical Sketch', *Economic History Review*, 2nd series, 11, 1958–9, 279–88.

[60] William A. Pettigrew, *Freedom's Debt. The Royal African Company and the Politics of the Atlantic Slave Trade, 1672–1752* (Chapel Hill, NC, 2013), 49.

[61] Samuel Pepys, *Diary* (London, 1893 edn), vol. iv, 23–4.

[62] John Evelyn, *Diary* (London, 1906 edn), vol. ii, 307–8.

[63] Poovey, *The History of the Modern Fact*, 121, 122, 136–7.

it never came to him—to overhaul and reform government, on what might be called statistical lines, by making it more accurate, efficient and precise in collecting taxes and safeguarding public welfare.[64] He wanted to be the 'King's Accomptant' and left notes and memoranda on the information he would have collected in that role.[65] Often called a political economist, the general term for a writer on economics throughout the nineteenth century, his interests as evidenced in his publications were very wide: economic policy, national income and accounting, money and its circulation, taxation, demography, surveying, medicine—each of them a technical subject requiring at least a modicum of mathematics. Among other works he was the author of A Treatise of Taxes and Contributions (1662); Verbum Sapienti (1665, published in 1691) on the distribution of taxes, a work that includes the first 'set of national accounts ever to have been made';[66] Quantulumconque ('Something, be it ever so small') Concerning Money (1682, published 1695), and an Essay Concerning the Multiplication of Mankind (1682). He had no statistical training or acumen beyond the calculation of averages. He was a statistician only in the sense that he used quantitative data extensively and reflexively, but that would be true of many whom we call statisticians up to Francis Galton at the very end of this story. Yet in the range of his interests and his use of them in the service of the state, Petty was a genuine founding figure.

Petty used the term 'Political Arithmetic' in the titles to several of his publications. One concerned 'the Growth of the City of London: with the Periods, Causes and Consequences thereof' (1683); another was about 'the People, Housing, Hospitals etc. of London and Paris...tending to prove that London hath more people than Paris and Rouen put together' (1686); a third, perhaps begun in 1671, completed in 1676 or 1677, and published posthumously, was a study of the British state entitled Political Arithmetick, or a Discourse concerning the extent and value of Lands, People, Buildings; Husbandry, Manufacture, Commerce, Fishery, Artizans, Seamen, Soldiers; Public Revenues, Interest, Taxes. Charles Davenant admired Petty's work, but pointed out that he had depended on inexact and limited sources, essentially the returns for customs, excise, and the hearth tax, 'so that the very grounds upon which he built his calculations being probably wrong he must, in many instances, be mistaken in his superstructure'.[67] Whether accurate or not, what distinguished and united this body of work and set an example that was both celebrated and followed, was Petty's commitment to work from data, numbers, statistics, and details contextualized over time. He was an

[64] Leopold von Ranke, History of England, Principally in the Sixteenth and Seventeenth Centuries (1859–67) (Eng. tr., 6 vols, Oxford, 1885), iii, 586.
[65] Stone, Some British Empiricists, 46. Poovey, The History of the Modern Fact, 13, 93, 131–2.
[66] Ibid., 31.
[67] Charles Davenant, Discourses on the Publick Revenues, and on the Trade of England (1698) quoted in Stone, Some British Empiricists, 52.

inductivist, drawn to explain economic behaviour from the evidence of actual wealth creation, distribution, and exchange expressed numerically. He tried to apply analytical techniques to the statistics he had collected. Petty explained himself thus:

> The method I take is not yet very usual; for instead of using only comparative and superlative words, and intellectual arguments, I have taken the course (as a specimen of political arithmetic I have long aimed at) to express myself in terms of *number*, *weight*, or *measure*; to use only arguments of sense, and to consider only such causes as have visible foundations in nature, leaving those that depend upon the unstable minds, opinions, appetites, and passions of particular men, to the consideration of others.[68]

Petty attended the very earliest meetings in Oxford in the late 1640s from which the Royal Society emerged and he was among its first members. The Society was to be an 'invisible college for the promoting of physico-mathematical experimental learning'. Its early motto, originally used by Francis Bacon, was '*nullius in verba*', 'take no man's word for it', or as sometimes rendered, 'nothing in words'.[69] After decades of political and religious dispute, expressed in millions of words of tracts and pamphlets which had fed civil strife in the 1620s and 1630s and led to the Civil War itself in the 1640s, it was time enough for a different intellectual approach, one that replaced contention and opinion expressed in rhetoric and speech with unimpeachable fact and the certainty of calculation. After the Restoration, natural philosophy, and cultural life more generally, were to be reformulated using the new language of figures, or so it was hoped. According to Thomas Sprat, the first historian of the Royal Society, its members were resolved 'to reject all the amplifications, digressions, and swellings of style...bringing all things as near the Mathematical plainness as they can'.[70] Petty, and his Political Arithmetic, embodied this.[71] It was written of him that he 'made it appear that Mathematical Reasoning, is not only applicable to Lines and Numbers, but affords the best means of Judging in all concerns of human life'.[72]

A hundred and fifty years later, or thereabouts, the newly-founded Statistical Society of London also wished to avoid controversy and disputation and to deal only in attestable fact which was best expressed in numerical form. It took for itself a similar motto, 'Aliis Exterendum', which means 'to be threshed out by

---

[68] Sir William Petty, *Several Essays in Political Arithmetic* (1691) (4th edn, London, 1755), 98.

[69] https://royalsociety.org/about-us/history/.

[70] Thomas Sprat, *The History of the Royal Society of London, for the Improving of Natural Knowledge* (London, 1667), 113.

[71] Derek Hirst, *Authority and Conflict: England 1603-1658* (London, 1986), 359-63.

[72] *Philosophical Transactions of the Royal Society*, 16, 1686, 152, quoted in Peter Buck, 'Seventeenth-Century Political Arithmetic: Civil Strife and Vital Statistics', *Isis*, 68, 1977, 81.

others': its original seal showed a wheatsheaf with the motto running around the bundle. According to its initial prospectus in 1834, the society 'will consider it to be the first and most essential rule of its conduct to exclude carefully all Opinions from its transactions and publications – to confine its attention rigorously to facts'.[73] The statisticians would mine the archives and the ledgers, count the poor and enumerate the sick, but would leave to others the messy business of interpreting and acting upon the numbers they had thereby uncovered or generated.[74] It is doubtful if the worldly men who founded statistical societies and statistical departments in the 1830s believed this was possible or even welcome if statistics were to play their full role at the heart of the state. But 'aliis exterendum' was a profession of neutrality that everyone could respect: it might restrain some from inciting controversy and divert others into different channels. We shall see how difficult it proved to live up to, not least in the very first statistical society, founded in Manchester in 1833, which was organized by men devoted to a cause.

1.2 The Statistical Moment: 'Reform' in Britain, 1828–36

This leads to a question: if it is the case that by the early 1830s, 'the machinery of government in Britain was indissolubly tied to the collection of numerical information', why had there been this palpable acceleration of interest in the discipline of social statistics, leading to its institutionalization?[75] Systematic collection of data by the state before this was rare: the empirical social inquiries undertaken by parliamentary select committees on pauperism and popular education of 1812–15 and 1817 respectively, were exceptional. Up to this time political arithmeticians had largely worked alone, without the benefit of learned societies, journals, intellectual exchange, and public recognition. Political arithmetic was the pastime of mathematically-minded officials and the preserve of concerned gentlemen. If traditions of numerical analysis can be traced as far back as the 1660s and have a more or less continuous history through the eighteenth century, what was special about the 1830s that led to the formation and recognition of a 'statistical movement'? The answer lies at the intersection of three related developments: the growth of the state since the 1790s and its need for expertise with numbers; the political crisis of the late 1820s when Tory government collapsed and the Whigs took their place; and the reform of Britain's political and governmental

---

[73] Prospectus of the Statistical Society of London, 1834, quoted in I. D. Hill, 'Statistical Society of London – Royal Statistical Society. The First 100 Years: 1834–1934', *Journal of the Royal Statistical Society*, ser. A, 147 (1984), 133.

[74] Victor L. Hilts, 'Aliis Exterendum, or, the Origins of the Statistical Society of London', *Isis*, 1978, 69, 21–43.

[75] Poovey, *The History of the Modern Fact*, 317; David Eastwood, "Amplifying the Province of the Legislature": The Flow of Information and the English State in the Early Nineteenth Century', *Historical Research*, 62, 149, October 1989, 276–94.

institutions to bring them into line with the economic, social, and religious changes that had been underway since the mid-eighteenth century.

As taxes soared and the national debt ballooned, the generation that fought the Napoleonic Wars depended on improved and efficient national administration at home to collect revenue, equip the navy, fund armies in the field, and galvanize all levels and sectors of British society.[76] After the wars, there was a host of technical social and economic questions to be addressed concerning the stabilization of the currency, protective tariffs on imports of raw materials and foodstuffs known as the 'corn laws', the sliding scale that modified those tariffs from 1828, and the cost and organization of pauperism, which each invited the help of the statistically-minded. But Waterloo in 1815 begat Peterloo in 1819: the wars were followed by difficult years of readjustment marked by high unemployment, political unrest, and the repression of radicalism and democracy as at St. Peter's Field in Manchester in August 1819 when a very large meeting called to press for manhood suffrage was broken-up violently by the local yeomanry. After 1822, as the economy stabilized, the opportunity to liberalize and reform national institutions presented itself. However, the degree to which the state should recognize and accommodate itself to new social forces and groups outside a landed, Anglican establishment—such as industrialists and industrial workers in the new cities of the north, and Protestant nonconformists and Irish Roman Catholics chafing at religious and social exclusion—split the governing Tory party in the 1820s. So-called 'ultras' opposed any concessions to pluralism; liberal Tories argued that without concessions there was a risk of inciting serious unrest, not short of revolution. Their Whig opponents, meanwhile, though without serious power for two generations since the 1780s, argued that through the granting of liberal concessions it would be possible to stabilize society and construct a new and broader, if more diverse, class of propertied men, who would be naturally supportive of the state once their interests had been recognized by the concession of the suffrage and other institutional reforms.

In this battle between different strategies for political and social stabilization, it was the Whig version that won out. Tory cabinets were constructed and collapsed between 1827 and 1830. Concessions that equalized the rights of different Christian denominations, the repeal of the Test and Corporation Acts and Catholic Emancipation, were wrung out of weak governments in 1828 and 1829 respectively. At the end of 1830 the Whigs came to power and delivered on their promises. The events between 1827 and 1832—between the resignation of the long-serving Tory prime minister, Lord Liverpool, and the passage of the Great Reform Act—have been interpreted as the end of the *ancien regime* in Britain.[77] That certainly undervalues both the effect of long-running changes in British society, which,

---

[76] Roger Knight, *Britain Against Napoleon. The Organization of Victory 1793–1815* (London, 2014).
[77] J. C. D. Clark, *English Society 1660–1832. Religion, Society and Politics During the Ancien Regime* (Cambridge, 1985), 501–63.

since the 1760s, had been challenging and altering the status-quo, and also the degree of flexibility shown by the British state in response to these challenges.[78] But it does capture the sense of an accelerating trend of change and modernization which overwhelmed the 'protestant eighteenth-century constitution' by which Britain was still being governed. It was in this context, the context of 'Reform', indeed a veritable host of reforms in the first half of the 1830s, that the statistical movement was born. There was a palpable acceleration in the collection and analysis of social data in the 1830s. During these years the state had need of numbers and reformers recognized that social statistics could provide them with potent justifications for change.

In the year of his death, 1832, Jeremy Bentham, the social thinker famed for developing utilitarianism as a philosophy of government, supported the organization of a statistical society by some of his followers to collect information to provide justification and guidance for the wholesale reform of British institutions.[79] Utilitarianism has many meanings and implications and it is common to accuse its devotees in the early nineteenth century of forms of dehumanization. In the wrong hands, or applied in the wrong way, utilitarian approaches to social issues could be used to control, incarcerate, and divide. But in the context of the early 1830s, utilitarianism implied the application of the test of utility to institutions which had lost their rationale or efficacy and were no longer fit for their purpose. Did they possess genuine utility or were they obstacles to the best working of society and its members? Utilitarianism was the creed of the reformer, the antidote to the privileges and corruption of the *ancien regime*, and many of Bentham's followers found their way into government in order to apply the tests of utility to political institutions, departments of state, the Church, the law, and legal institutions.[80] Albany Fonblanque, for example, one of the most ardent and visible of the 'philosophic radicals' of the 1830s, as the Benthamites were known, was made the Statistical Secretary of the Board of Trade in 1847, a post he held until shortly before his death in 1872.[81] Those tests depended upon the use of numbers to count the unenfranchised men in industrial towns and cities, the number of paupers claiming 'outdoor relief', the monetary value of commuted tithes under the Tithe Commutation Act of 1836, the average length of a case before the Chancery courts, the death rates in the worst districts of industrial cities like Manchester

---

[78] Joanna Innes, 'Jonathan Clark, Social History and England's "Ancien Regime"', *Past & Present*, 115, 1, 1987, 165–200.

[79] Mary P. Mack, *Jeremy Bentham. An Odyssey of Ideas* (London, 1962), 115–16, 235–40; J. R. Poynter, *Society and Pauperism: English Ideas on Poor Relief, 1795–1834* (London, 1969), 129–30.

[80] A. V. Dicey, *Lectures on the Relation Between Law and Public Opinion in England during the Nineteenth Century* (London, 1905); S. E. Finer, 'The Transmission of Benthamite Ideas 1820–50', in G. Sutherland (ed.), *Studies in the Growth of Nineteenth-Century Government* (London, 1972), 11–32; Lawrence Goldman, *Science, Reform and Politics in Nineteenth Century Britain* (Cambridge, 2002), 272–8.

[81] James A Davies, 'Albany William Fonblanque (1793–1872)', *ODNB*.

and Leeds. Statistics were the chosen weapon of the reformer in that they uncovered outmoded, failed, or plain wrong ways of organizing society.

All of this is caught in the comment of a vice-president of the Statistical Society of London, the physician Dr Frederic Mouat, looking back from 1885 at the time of the society's jubilee:

> Young as we are in the history of the world, we made our appearance with kindred institutions of similar character only when the age was ripe for the more exact observation of phenomena and facts of all kinds and classes in physical science, as well as in the moral and social relations of man, in the mixed and complex conditions of modern civilization and progress.[82]

Those 'mixed and complex conditions' included Catholic Emancipation in 1829; the drafting of successive Reform bills in 1831–2 for the reform of parliamentary representation, and the passage of the Reform Act itself in June 1832; in 1833, slave emancipation, the first Factory Act regulating the hours worked by children and youths in textile mills, and the first grants of central funds for elementary education; the Poor Law Amendment Act of 1834; the Municipal Corporations Act of 1835 which brought accountable and representative government to British towns and cities; and the Civil Registration Act of 1836.

All these measures required investigation, analysis, the writing of reports, parliamentary scrutiny and public debate. In almost every case, they involved counting and thus the help of a new class of public servant. Counting was made easier because it was not just the quantity of statistics that was increasing: their quality was improving, as well. Where political arithmetic, the discipline of lone scholars, offered just snapshots of social and economic issues, or analyses based on limited data, by the 1830s it was possible to spot national trends over time and justify actions more confidently. Since the 1790s discussion of the Poor Laws at national level, and the minutiae of their daily administration at parish level, had been altered by the development of 'a new language of numbers...in contrast to an earlier language of custom and moral right.'[83] The Royal Commission on the Poor Laws of 1834 had available 'continuous figures for poor relief expenditure since 1813' which made it all the easier to achieve its pre-determined outcome, a system that was supposed to reduce the costs dramatically, though there is evidence that the humble recipients of relief were also empowered by the proliferation of data.[84] As the economy grew and diversified and population increased, new

[82] Frederic J. Mouat, 'History of the Statistical Society of London', *Journal of the Royal Statistical Society*, Jubilee Volume, June 1885, 49.

[83] Steven King, '"In These You May Trust". Numerical Information, Accounting Practices, and the Poor Law, c. 1790 to 1840', in Tom Crook and Glen O'Hara (eds), *Statistics and the Public Sphere. Numbers and the People in Modern Britain, c. 1800–2000* (New York and Abingdon, 2011), 52.

[84] Ibid., 58–62; Eastwood, '"Amplifying the Province of the Legislature,"' 288.

sources of data became available, thus meeting the conditions which Davenant had laid down for success more than a century before:

> He who will pretend to compute, must draw his conclusions from many premises; he must not argue from single instances, but from a thorough view of many particulars; and that body of political arithmetic, which is to frame schemes reducible to practice, must be composed of a great variety of members...for as in common arithmetic, one operation proves another; so in this art, variety of speculations are helpful and confirming to each other.[85]

The economist J. R. McCulloch complained about 'The State and Defect of British Statistics' in an article of that title in the *Edinburgh Review* in April 1835. Casting backwards, he lamented the failure to collect information systematically:

> If we had possessed circumstantial, and at the same time really accurate accounts of the various changes, however minute, in the wages, habits, accommodations, and conditions of the population since the Peace of Paris in 1763, we should now have been able to try principles and doctrines by the test of experience; and to appreciate, with considerable accuracy, the influence of particular systems and measures. But we have no such information.[86]

Yet even as he wrote, through new or improved central agencies, such as Royal Commissions of Inquiry, the state was freeing itself from dependence on local government and the local gentry for information. It was better able to collect the data it needed itself 'in partnership with experts',[87] using 'the charmed circle of the Whig-Liberal intelligentsia who were to dominate the parliamentary enquiries and statistical investigations of the 1830s'.[88] The alliance between aristocratic ministers and expert advisors had become even clearer by the 1850s and 1860s when a body like the Social Science Association, founded in 1857, brought both groups together before the public to debate social improvement, and the very word 'expertise' entered the language.[89]

The centralization of social knowledge proceeded with, and made possible, the centralization of policy-making in the 1830s. Whig reforms depended on statistics, and the Statistical Society of London, headed by Lord Lansdowne, was, as we

---

[85] Charles Davenant, *Discourses on the Publick Revenues, and on the Trade of England* (1698) quoted in Stone, *Some British Empiricists*, 54.

[86] J. R. McCulloch, 'The State and Defects of British Statistics', *Edinburgh Review*, 1835, 176–7.

[87] Eastwood, '"Amplifying the Province of the Legislature"', 293.

[88] M. J. Cullen, *The Statistical Movement in Early Victorian Britain. The Foundations of Empirical Social Research* (Hassocks, Sussex, 1975), 21.

[89] Lawrence Goldman, 'Experts, Investigators and The State in 1860: British Social Scientists through American Eyes', in M. Lacey and M. O. Furner (eds), *The State and Social Investigation in Britain and the United States* (Cambridge, 1993), 99.

shall see, almost an arm of Whig government, especially close to the Board of Trade. It was not by accident that the Society's first Prospectus, dated 23 April 1834, considered it 'desirable that the Society should as soon as possible endeavour to open a communication with the statistical department established by government at the Board of Trade'.[90] Nor was it an accident that the first paper published by the new Society on 'the collation, concentration & diffusion of statistical knowledge regarding the state of the United Kingdom' was written by William Jacob, the Comptroller of Corn Returns in the Board of Trade since 1822.[91] Jacob's two reports on agricultural conditions in northern and eastern Europe of 1826 and 1828, published by the House of Commons, and arguing that free trade would not result in a destabilizing influx into the British market of cheap foreign corn, were crucial documents in the developing case for the repeal of the corn laws.[92] Their author was especially welcome at the Statistical Society of London where the majority of members supported repeal.

Examples of the government's requirement for data, and for the expertise to understand and use it, abound amidst the reforms of the 1830s, especially in relation to the first and foremost of the changes, parliamentary reform. The redistribution of parliamentary seats from the over-represented south and west of England to the under-represented north had been calculated initially in 1830–31 with reference to the most recent census of 1821. But its information was now out of date, and many parliamentarians were inveterately hostile to the implication in the use of census data that the new representative system in a post-Reform parliament would forsake the representation of property in favour of population. That smacked of American-style democracy.[93] With the defeat of this first plan, the Whig cabinet changed its approach and required a team of parliamentary boundary commissioners to reconstruct the electoral map in its entirety on a broader basis, including consideration of the number of houses and the amount of taxes paid in each borough, as well as population.[94] They comprised 18 commissioners

[90] James Bonar and Henry W. Macrosty, *Annals of the Royal Statistical Society 1834–1934* (London, 1935), 24.

[91] William Jacob, 'Observations and Suggestions Respecting the Collation, Concentration, and Diffusion of Statistical Knowledge Regarding the State of the United Kingdom', *Transactions of the Statistical Society of London*, vol. I, pt. i, 1–2. The paper was written in 1831–2, circulated within the Board of Trade, and delivered and discussed in December 1834 and January 1835. The *Transactions* predate the *Journal of the Statistical Society of London* and reside only in the library of the Royal Statistical Society. See Cullen, *The Statistical Movement*, 19–20, 157n.

[92] The two reports were entitled 'Report on the Trade in Corn and the Agriculture of Northern Europe' (1826) and 'Report on Agriculture and Trade in Corn in some of the Continental States of Northern Europe' (1828). See William Jacob, 'Report on the Agriculture and the Trade in Corn in Some Continental States of Northern Europe', *The Pamphleteer*, 29, no. lviii, 361–456. Susan Fairlie, 'The 19th Century Corn Laws Reconsidered', *Economic History Review*, 2nd ser., 18, 1965, 562–75.

[93] S. J. Thompson, '"Population Combined with Wealth and Taxation". Statistics, Representation and the Making of the 1832 Reform Act', in Crook and O'Hara (eds.), *Statistics and the Public Sphere*, 205–23.

[94] Martin Spychal, 'Constructing England's Electoral Map. Parliamentary Boundaries and the 1832 Reform Act', unpublished PhD thesis, 2017, Institute of Historical Research, University of London.

and 30 surveyors and draftsmen. Henry Brougham, the leading Whig of the 1810s and 1820s who had linked together the party in parliament with its supporters in the country, was now Lord Chancellor with oversight of parliamentary reform, and he looked out for men with experience as surveyors and with statistical acumen. Much earlier in his career, as the leading exponent of anti-slavery on economic grounds, Brougham himself had been the first to use statistical material in his essays and speeches against the slave trade and slavery itself.[95]

Under the estimable Captain Thomas Drummond of the Royal Engineers, the commissioners and their assistants set about the task with a determination to do it impartially and 'scientifically'. They calculated the size of boroughs and constituencies, estimated population by counting households, and calculated tax assessments and wealth to determine who would now qualify for the vote. For this, they were 'cheered by liberals and opposed by traditionalists'.[96] They were part of a 'burgeoning culture of scientists, geographers, statisticians and political economists…who wanted to create a science of government'.[97] The boundary commission set the tone and deployed the techniques that would be used by many more investigations in the coming years. It was hoped that the very impartiality of their methods and numerical calculations, and the transparency and openness of their procedures, would win public trust. As William Jacob had put it in that first paper published by the Statistical Society of London, which had been written during the Reform crisis of 1831–2,

The best mode of allaying disquietude and of diffusing contentment on the subject of public affairs is an open and clear disclosure of their conditions and management…A more general diffusion of accurate knowledge regarding the state of public affairs would tend to check that excitement and party spirit which has often been created by misrepresentation or exaggeration, and which has produced an annoyance to the government and at least a temporary disaffection to the public mind.[98]

In 1833, the Emancipation Act and the distribution under it of £20 million in compensation to former slaveholders, as determined by the Commissioners of Slave Compensation, required a census of all slave-holdings in the British empire

[95] John Morewood, 'Henry Peter Brougham and Anti-Slavery 1802–1843', Unpublished PhD dissertation, 2021, Institute of Historical Research, University of London, 64–5.
[96] Stanley H. Palmer, 'Thomas Drummond (1797–1840)', ODNB.
[97] 'Crown and Country: Research Profiles in the IHR', Martin Spychal in conversation with Lawrence Goldman, Past and Future: The Magazine of the Institute of Historical Research, Issue 21, Spring/Summer 2017, 12–13.
[98] Jacob, 'Observations and Suggestions', 1. See fn. 91 above. On the theme of public trust, see Tom Crook, 'Suspect Figures. Statistics and Public Trust in Victorian England', in Crook and O'Hara (eds.), Statistics and the Public Sphere, 165–84.

and the reckoning of monies due. The amount set aside was remarkable in itself, equal to about 40 percent of the Treasury's annual income.[99] Half of the funds were paid out to owners living in the Caribbean and Africa; the remainder to absentee owners living in Britain, many hundreds of British families, in fact.[100] In some cases a single slave was part-owned by several owners. At the other extreme, Henry Lascelles, second earl of Harewood, the acknowledged leader of the 'West Indian interest' in parliament at the time of emancipation, owned 1277 slaves on six plantations and received £26,309 in compensation.[101] Whatever contemporaries might think of this remarkably generous settlement, and there were many critics of it in the anti-slavery movement, it all took calculation.

The Factory Act and the Poor Law Amendment Act both depended on the prior research of two Royal Commissions of Inquiry, or so it was thought. The 1830s saw the majestic rise of Royal Commissions as a new arm of the state, and fifty-three of them were initiated and reported between 1830 and 1842. Assistant commissioners would fan out across the land asking questions, collecting data, and writing reports for the Commissioners to study and use in making their recommendations.[102] In this manner government aspired to reach Plato's 'pure truth'. But if we judge from the most famous of the Royal Commissions of this era, the Poor Law Commission of 1832–4, accurate data might have only the most marginal influence on recommendations and legislation. In this case, the chairman, Edwin Chadwick, one of the great and controversial bureaucrats of the nineteenth century, had decided the conclusions before the Commission was even brought into being and never intended to test his theory against the information. Indeed, the Royal Commission deliberately set out to influence public opinion in its favour in advance, by circulating thousands of copies of *Extracts of Information* in 1833 to garner support for its preferred solutions.[103]

The state had need of numbers, but the development of statistics within early Victorian government was haphazard rather than planned, accidental rather than foreordained. The Civil Registration Act of 1836 in its establishment in the following year of the General Register Office (now the Office for National Statistics) to register births, marriages, and deaths, gave an institutional home to statistics in government and would seem, at first sight, to fit neatly into this pattern of governmental development, data collection, and the growing use of professional

---

[99] *Parliamentary Papers*, 1837–8, 215, vol. 48.

[100] Nicholas Draper, *The Price of Emancipation. Slave-Ownership, Compensation and British Society at the end of Slavery* (Cambridge, 2010); Richard Huzzey, *Freedom Burning: Anti-Slavery and Empire in Victorian Britain* (Ithaca, NY, 2012). For full information on the history of British slave compensation, see the *Legacies of British Slavery* website, https://www.ucl.ac.uk/lbs/.

[101] On Harewood, see John Morewood, 'Henry Brougham and Slave Emancipation 1803–1843', 123, 133.

[102] H. M. Clokie and J. W. Robinson, *Royal Commissions of Inquiry* (Stanford University Press, 1937).

[103] S. E. Finer, *The Life and Times of Sir Edwin Chadwick* (London and New York, 1952), 48–9.

expertise. Among those who wanted to reform and rationalize the registration system, one of the most potent arguments they could deploy was that, in the collection and analysis of demographic data, Britain lagged well behind many other European states. It still depended on the local, parochial registration of baptisms, marriages and burials, dating from the Tudor period. An industrial and commercial society like Britain in the 1830s had obvious need of reliable demographic data, especially in a period marked by the first steps towards systematic social policies sponsored by the state. The GRO became a centre for the collection of statistics in government, and the institutional home of Dr. William Farr for more than four decades. Farr ensured that the data was used openly and effectively in a consistent campaign to improve public health and well-being.[104]

Yet the origin of the GRO is more complicated than this, and also more 'political'.[105] It is a caution against any argument that the development of Victorian statistics can be easily understood as if the pure product of the rational growth of government to meet new national requirements. Civil Registration was not simply an administrative measure designed to improve the quality of demographic data available to government in an era of more active social intervention. In a long-standing interpretation, Civil Registration was part of a response to the inequalities and grievances experienced by the fast-growing communities of religious nonconformists—those Christians organized in sects outside the established church—in Britain.[106] In a more recent interpretation, the measure was also required so that the registration, descent and inheritance of property could be made more accurate and secure.[107]

The registers of births, marriages and deaths maintained by dissenting communities were not admissible as evidence in court. Nor, with some exceptions, were dissenters' marriages in their own chapels legally recognized.[108] This had long been controversial and seemed not only discriminatory but positively antediluvian by the early nineteenth century. As Edward Higgs has explained, the reforms to registration were effected in two distinct pieces of legislation which contemporaries and some historians have confused.[109] Alongside the 1836 Civil Registration Act, the Marriage Act of that parliamentary session licensed all places of worship to perform legal and admissible marriages. It was this piece of

---

[104] Simon Szreter, *Fertility, Class and Gender in Britain 1860–1940* (Cambridge, 1996), 86–7.

[105] Edward Higgs, *Life, Death and Statistics: Civil Registration, Censuses and the Work of the General Register Office 1837–1952* (Hatfield, 2004), 1–89.

[106] M. J. Cullen, 'The Making of the Civil Registration Act of 1836', *Journal of Ecclesiastical History*, 25, 1974, 39–59.

[107] Edward Higgs, 'A Cuckoo in the Nest? The Origins of Civil Registration and State Medical Statistics in England and Wales', *Continuity and Change*, 11, 1, 1996, 115–34.

[108] The exceptions were Quakers and Jews.

[109] Higgs is right to include the present author among those hitherto confused. He is grateful to be put right. See Lawrence Goldman, 'Statistics and the Science of Society in Early Victorian Britain: An Intellectual Context for the General Register Office', *Social History of Medicine*, 4, 1991, 418.

legislation that dealt directly with the religious problem.[110] In 1836 the Whigs were redeeming a debt they owed to Dissenters for their political support, and in expanding religious liberty they were signifying their fealty to a key component of their own political tradition. The government was not especially interested in the bureaucratic rationale of the measure, or in the collection of demographic data as a virtue in itself; the Marriage Act was essentially a political matter.

By recording the cause of death in each individual case, civil registration provided information that was of great value to medical research and public health reform. Posterity has therefore associated the GRO with campaigns for the improvement of sanitary conditions, above all. However, in establishing a national system for the registration of vital life events under the Civil Registration Act, a key motive was to ensure that better records existed for proof of legal identity, a pre-requisite for establishing ownership, bequest, and inheritance of property in an age of growing population and of growing transmissible wealth, much of it now in the form of personal as opposed to real property, i.e. land. Many of the witnesses who gave evidence to the 1833 Select Committee on Parochial Registries called for accurate and accessible demographic information as a basis for secure property transactions.[111] In the mid-1850s a committee of the Treasury placed the GRO's legal responsibilities before its medical functions. It was 'to furnish the means of tracing the descent of property, of calculating the expectation of life and the laws of mortality, and of ascertaining the state of disease and the operation of moral and physical causes on the health of the people and the progress of the population'.[112] Notwithstanding the opinion of Lord Ellenborough that civil registration was being introduced 'just to gratify the statistical fancies of some few philosophers, in order that they might know how many persons died, and how many were born in a year', as he explained to the House of Lords in July 1836, outright opposition to the principle of the Civil Registration bill, as opposed to its details, came only from a group of Tory Anglican diehards.[113] We shall see how another group of 'statisticians', gathered together at exactly this time in the Manchester Statistical Society, were also animated by religious motives and a religious cause.

The Statistical Department of the Board of Trade, with which the Statistical Society of London was so anxious to liaise, was established in 1832 under the direction of George Richardson Porter, another well-connected Whig functionary and strong free-trader. He was the brother-in-law of the great economist, David Ricardo, and a founder member of the Statistical Society. The Statistical Department's name was misleading, however, and may have misled subsequent historians. This was not a department that would be focused on the statistics of

---

[110] Ibid., 120–1.   [111] Ibid.

[112] *Report on the General Register Office, London*, 20 July 1855, 1, quoted in D. V. Glass, *Numbering the People: The Eighteenth Century Population Controversy and the Development of Census and Vital Statistics in Britain* (Farnborough, Hants, 1973), 142–3, n. 70.

[113] *Hansard*, third series, House of Lords, xxxv, 87.

trade only, but was seen at its creation as an agency that would collect data for use across the whole of government. As a subsequent report into the work of the Board of Trade, written by Stafford Northcote and Charles Trevelyan, explained in 1853, 'The Statistical Department...was originally intended to serve the purpose of collecting and digesting statistical information of all kinds.'[114] Rather than undertaking separate projects of research to meet the specific requests of different parliamentary committees and MPs, and in line with the wider commitment to generate and make data accessible, at its foundation the Statistical Department set about trying to provide annual series of data across a host of different subjects.[115] But collecting the information with a small staff, let alone systematizing it, proved very difficult, and the department did not make the contribution to policy-formation that was expected. George Porter tried to extract economic information from chambers of commerce, known individuals, even his own friends, but the returns were thin and unsystematic.[116] Some of the department's statistical reports were published too late; some were not adapted 'to the requirements of the present day'; and overall, 'the very voluminous and miscellaneous character of the returns' rendered them unusable in the forms presented.[117]

Northcote and Trevelyan had a choice: either to enhance the Statistical Department so as to make it equal to the many responsibilities expected of it, and construct something much closer to a central statistical bureau as in one of the leading countries of Europe, or to recommend that the collection of statistics should be devolved throughout government. They chose the latter course. They made the point that since 1832 'several separate departments have been established for the particular management of some of the chief branches of our social economy' including the Poor Law Board, the General Register Office, the Emigration Office and the Committee of Council on Education. They recommended, therefore, that with some coordination to make the different returns mutually conformable, there should be wholesale devolution:

We consider that the several Departments of Government should collect the statistical information which is connected with their own business; and each should publish it separately, but in a form harmonising with that adopted by others. Thus the Finance Accounts should be published by the Treasury; the Criminal Statistics by the Home Office; the Emigration Returns by the Colonial

[114] 'Board of Trade', *Report of Committees of Inquiry into Public Offices and Papers Connected Therewith*, Parliamentary Papers, 1854, vol. 87, 129–55. Dated 23 March 1853. Quotation at 147. The 1853 report was written by Charles Trevelyan and Stafford Northcote, who, before their famous 1854 'Report' on the reform of the civil service and its future, were collaborating in reviews of various government departments. They were assisted by James Booth, who was secretary to the Board of Trade from 1850 to 1865.
[115] Lucy Brown, *The Board of Trade and the Free Trade Movement 1830–42* (Oxford, 1958), 76–9. Simon Szreter, 'The GRO and the Historians', *Social History of Medicine*, 4, 3, Dec 1991, 408.
[116] Brown, *The Board of Trade*, 82–3.
[117] 'Board of Trade', *Report of Committees of Inquiry into Public Offices*, 148.

Office; and other matters should fall to the other Offices; leaving to the Board of Trade, the Statistics of Navigation, of Agriculture, and of Railways.[118]

This institutionalized the difference between the collection and publication of official statistics in Britain and other leading countries, and explains why, at the London meeting of the International Statistical Congress in 1860, it was the Statistical Society of London that had to 'speak for England'. It also exemplifies the nature of the growth of government in nineteenth century Britain which was empirical, organic, a process of learning from mistakes, and very largely dictated by the pervasive anti-centralism of the early and mid-Victorian periods.[119]

Whatever the problems of the Statistical Department, outside his official duties, George Porter himself produced a famous digest of national information, *The Progress of the Nation*, in 1837, which went through several editions, gradually becoming more comprehensive and reliable as a guide to the economy and society of Victorian Britain. That a senior civil servant could publish in his own right, and achieve success as an editor and author while never quite achieving the expected outcome in his day-job, as the phrase goes, tell us something important about the nature of British bureaucracy at this stage. Leading civil servants enjoyed remarkable freedom to mix and associate outside the confines of government and to undertake private projects that drew on their knowledge and the expertise within their departments. The boundaries of government service were essentially permeable. Many of the statisticians to follow in this account would use that to advantage.

In these contexts it is hardly surprising that the process of embedding statistics in government, even with the cry of 'reform' on every street corner and in every political meeting, was haphazard and far from automatic. Government in the 1830s was frequently unprofessional, unsystematic, decentralized, unstable, experimental. Yet that also offered advantages to the experts and statisticians required in the departments: they served at a time when the conventions, codes of conduct, and boundaries of the civil service were still to be written and defined. William Farr worked in the General Register Office, initially as Superintendent of Abstracts; he was also an active member and later President of the Statistical Society of London, a founder of the Social Science Association in July 1857, the organizing secretary of the London meeting of the International Statistical Congress, and the ally of Florence Nightingale in many a public sanitary campaign. Or take Joseph Fletcher, a barrister, the secretary of the Statistical Society of London, and the editor of its journal from 1842 until his death ten years later. He

---

[118] Ibid., 148.
[119] The classic statement of this view of government growth is Oliver MacDonagh, 'The Nineteenth Century Revolution in Government: A Reappraisal', *Historical Journal*, 1, 1958, 52–67. For the classic riposte see Jennifer Hart, 'Nineteenth Century Social Reform: A Tory Interpretation of History', *Past & Present*, 31, 1965, 39–61.

served as secretary to the Royal Commissions into the handloom weavers (1837-41), on the Health of Towns (1840) and on children's employment (1843) and as an inspector of schools with a strong personal interest in the relationship of education and crime on which he delivered several papers to the Statistical Society in the 1840s.

Men like these had freedom to express their opinions, present their data, combine with others outside of government, and then take to their departments or commissions the ideas generated elsewhere. Farr turned the annual reports of the Registrar General, nominally his superior, into 'the vehicle for the expression of passionately held personal views, for propaganda directed against the opponents of public health reform, and for agitation for state intervention in a new field to a degree' that, as one historian has put it, 'would send cold shivers down the spine of a modern civil servant'.[120] Later in the period, after the reforms of the civil service laid down in the famous Northcote–Trevelyan Report of 1854, a more tightly-drawn bureaucracy under effective political control developed, and it was one that was staffed by genuine officials, first-class men from Oxford and Cambridge who obeyed orders, rather than by experts—often mavericks—who had gained direct knowledge of social issues through experience outside government.

Senior civil servants in the 1830s and 1840s did not simply seize opportunities as they arose. Given the vicissitudes of public opinion, the fact that formal party politics had yet to concern itself with social policy (a development which occurred later in the Victorian period) and the resistance to any extension of state control which sometimes came from vested interests, sometimes from the ideology of *laissez-faire*, and often from the sheer incompetence of local officials, then, in a sense, the responsibility for initiative and innovation could only come from the expert in government. Such people became, whether by design or by default, so-called 'statesmen in disguise'.[121] They used the freedom they enjoyed to pursue reforms and, in several cases, such as that of Edwin Chadwick, to build empires.[122] During this transitional phase, the patronage system was used to bring talent into government, and this elite group of officials used their influence in turn to bring into subordinate positions like-minded men—physicians, statisticians, sanitarians, Benthamites. The statistical movement provided many of these people for government: middle class, professional men with a talent for numbers, employed on some practical project, who turned their experience into a learned paper delivered at one of the statistical societies and thereby came to the notice of a Whig grandee. But before the development of a modern and codified bureaucracy, and

[120] M. W. Flinn, 'Introduction' to Edwin Chadwick, *Report of the Sanitary Condition of the Labouring Population of Great Britain* (1842) (Edinburgh, 1965), 28.
[121] G. Kitson Clark, '"Statesmen in Disguise": Reflexions on the History of the Neutrality of the Civil Service', *Historical Journal*, 2, 1959, 19–39.
[122] Finer, *Life and Times of Sir Edwin Chadwick*, passim; Anthony Brundage, *England's 'Prussian Minister'. Edwin Chadwick and the Politics of Government Growth, 1832–1854* (London, 1988).

in a milieu where they depended on patronage, talent was never enough to secure these men permanency and continued influence.

Many of the key figures in this story—men like William Farr, Sir John Simon, Sir James Kay-Shuttleworth, and Chadwick himself—ended their careers early, and in states of despair, as the political context changed and they were ejected from their positions. Lauded in the professional worlds from which they came and in the learned societies to which they belonged, the loss of office by an aristocratic patron or a sudden change in parliamentary opinion was enough to depose or undermine them, or in Simon's case, to merge their department with another. Aristocratic government had need of their expertise, but political power and influence before 1900 still depended on wealth, birth, and title, rather than brains.

Charles Babbage, the most brilliant and also the most politically radical of all the figures in the statistical movement, was a persistent public critic of aristocratic involvement in science, a stance the may explain why successive British governments in the 1830s and 1840s were not wholly sympathetic to his appeals for financial assistance from the state. Encouraged by his friends, Babbage had shown interest in becoming the first Registrar-General at the newly created GRO but was dismayed to discover that the post was reserved for T. H. Lister, the beneficiary of Whig nepotism. He was the brother-in-law of both Lord John Russell, then Home Secretary, and Lord Clarendon. The author of a romantic novel, *Granby*, published in 1826, Lister was dismissed by Chadwick as 'a very good novelist who cared nothing for the subject'.[123] Babbage was equally discouraged, when, on Lister's death a few years later, he learnt that his successor was already chosen: Major Graham, the brother-in-law of another Secretary of State.[124] Unsurprisingly, in 1855 Babbage was an early subscriber to the new Administrative Reform Association, which supported the wholesale reform of aristocratic government and its patronage system in response to the military debacle of the Crimean War. Two years later he supported the idea of an 'Educational Franchise' to weight voting for parliament in favour of graduates and professional men and thus counteract the influence of aristocratic incompetence and amateurism in government.[125] But neither campaign was successful. Meanwhile, the commercial and industrial bourgeoisie looked with suspicion on the 'service middle class' who worked for government, as likely to regulate their enterprises, factories and workshops, and raise their taxes to pay for social reforms. As we shall see, there was quite some sociological and ideological difference between statistically-minded officials in London and the members of the Manchester Statistical Society. In short, the haphazard nature of government in the early and mid-nineteenth century was littered with both opportunities and threats for the statisticians.

[123] Edwin Chadwick to D. T. Laycock, 13 April 1844, Chadwick Papers, University College, London.
[124] Charles Babbage, *Passages from the Life of a Philosopher* (London, 1864), 478; Maboth Moseley, *Irascible Genius. A Life of Charles Babbage, Inventor* (London, 1964), 130.
[125] Goldman, 'Experts, Investigators, and the State', 118–19.

PART II

THE ORIGINS OF THE STATISTICAL MOVEMENT 1825–1835

2

Cambridge and London

The Cambridge Network and the Origins of the
Statistical Society of London, 1833–34

The origins of the Statistical Society of London, from 1887 the Royal Statistical
Society, and the most important institution for the study of Victorian statistics,
can be traced to June 1833. At the end of that month the British Association for
the Advancement of Science held its third meeting in Cambridge, after having
convened in York in 1831 and Oxford in 1832. This so-called 'parliament of science', constructed while the reform of Parliament itself was the great issue before
the nation, was designed to bring together different scientific communities,
whether in the metropolis, provinces, or universities, in the promotion of science
to a wider public.[1] It was still at a plastic stage, capable of being bent and shaped
to the will of its membership, when, in the rooms being used by the Rev. Richard
Jones in Trinity College, an unofficial 'statistical section' was convened. 'At a
meeting of Gentlemen desirous of forming a Statistical Section of the British
Association at Cambridge, Thursday Morning June 27th 1833' where 'the prospects
and most desirable methods of establishing the section were discussed', the
following were present: 'Professor Malthus (Chairman), Mr. Aug. Quetelet,
Rev. Dr. D'Oyley (sic); Rev. Richard Jones; Professor Babbage; Lt. Col Sykes;
Dr. Somerville and Mr Drinkwater (secretary)'.[2]

Malthus, Professor of Political Economy at the East India College, Haileybury,
was the famous author of the *Essay on the Principle of Population*, published in
1798 and one of the greatest and most controversial works in the history of economics. D'Oyly was a Cambridge-educated clergyman and theologian, treasurer
of the Society for Promoting Christian Knowledge, and also a principal figure in
the foundation of King's College, London in the early 1830s.[3] Colonel William
Henry Sykes was a long-serving army officer with the East India Company,
recently retired and returned to Britain, who had been responsible for a detailed

---

[1] Jack Morrell and Arnold Thackray, *Gentlemen of Science. Early Years of the British Association for
the Advancement of Science* (Oxford, 1981).
[2] 'Minutes of the Committee of the Statistical Section of the British Association, June 27, 1833', in
Transcript from the note-book of Mr. J. E. Drinkwater, f. 1, Archives of the Royal Statistical
Society, London.
[3] L. C. Sanders, 'George D'Oyly 1778–1846', *ODNB*.

Victorians and Numbers: Statistics and Society in Nineteenth Century Britain. Lawrence Goldman, Oxford University Press.
© Lawrence Goldman 2022. DOI: 10.1093/oso/9780192847744.003.0002

census of the population of the Deccan, and for other statistical reports in India in the 1820s.[4] Somerville had been a military surgeon through the Napoleonic Wars and from 1819 was surgeon to the Royal Hospital, Chelsea. 'A man of liberal outlook', his wife, whom he always encouraged, was the famous mathematician and scientist, Mary Somerville.[5] Assisted by two key figures in the history of the statistical movement, Babbage and his close friend, the astronomer John Herschel, Mary Somerville had published to acclaim in 1831 *The Mechanism of the Heavens*. In 1834 she would publish *On the Connection of the Physical Sciences*, another very popular work that, in its synthesis of different scientific subjects from astronomy to geography, shared a common intellectual origin with the statistical movement itself.[6] John Elliot Drinkwater, from 1836 known as John Elliot Bethune, was a Cambridge mathematician who, as Counsel to the Home Office in the 1830s and 1840s, drafted some of the most important legislation of the era, including the Municipal Corporations Act of 1835 and the Tithes Commutation Act of 1836. Later, he was a notably liberal member of the supreme council of India. In 1833 he had published lives of Galileo and Kepler for the Society for the Diffusion of Useful Knowledge.[7]

Some of these figures would play relatively insignificant roles in the subsequent statistical movement. Somerville went abroad when his health declined at the end of the 1830s; Drinkwater was in India in the 1840s; D'Oyly was in essence a local clergyman, the rector of Lambeth. The wide range of their interests and backgrounds is indicative of the kinds of professional people who would be attracted to the statistical movement in the coming decades: physicians, military officers, civil servants, clergymen. Of far greater importance to the foundations of statistics in Britain, however, were four other figures in Cambridge that week, all of them of the highest intellectual significance: Richard Jones, Charles Babbage, Adolphe Quetelet, and William Whewell, though Whewell was not present at the irregular first meeting of the statistical section, probably because he had wider duties as the overall secretary of the Cambridge meeting. These four saw in statistics not just a way of representing social conditions and directing social change, but a method for studying all the sciences in a new manner.

Malthus, who chaired the first two meetings, was a special case rather than a fifth member of the key group. From an earlier generation, he was evidently an inspiration to the founders and in sympathy with their intellectual project to construct an empirical economics and social science based on the measured (rather than the assumed) behaviour of people in society. The first edition of his

---

[4]   B. B. Woodward, 'William Henry Sykes' (1790–1872), *ODNB*.

[5]   E. M. Clerke, 'William Somerville' (1771–1860), *ODNB*.

[6]   Mary R. S. Creese, 'Mary Somerville' (1780–1872), *ODNB*.

[7]   Katherine Prior, 'John Elliott Drinkwater' (1801–51), *ODNB*. His father's change of surname led him to follow suit. See Francis Espinasse, 'John Drinkwater (later John Drinkwater Bethune)' (1762–1844), *ODNB*.

infamous *Essay on Population* had been 'written on the impulse of the occasion, and from the few materials which were within my reach in a country situation'.[8] At that stage it was, in essence, a taut, deductive analysis of the consequences of the imbalances Malthus observed between rates of population and resource growth. He then set out to find evidence to support his thesis, and to the second and subsequent editions he added copious historical, demographic, and statistical detail in illustration of his arguments, including material on population growth, food prices, agricultural returns, and mortality. Malthus acknowledged only four sources in his first edition: the second edition was nearly four times as long and contained a bibliography of nearly 200 items. It is not surprising, therefore, that he should have given his blessing and support to a concerted attempt to remedy the over-reliance on theory, untested by data.[9] As he wrote,

> This branch of statistical knowledge...may promise...a clearer insight into the internal structure of human society from the progress of these inquiries. But the science may be said yet to be in its infancy, and many of the objects, on which it would be desirable to have information, have as yet been either omitted or not stated with sufficient accuracy.[10]

His support for the Cambridge group was not in doubt, therefore. As Whewell had written of Jones to Herschel in 1832, 'I have just been with him to Hayleybury [sic], where Malthus and he had divers palavers of no common length'.[11] But Malthus died in the following year, 1834: he was the Statistical Movement's first great patron.

Jones had graduated from Cambridge in 1816 and had taken the parish of Brasted in Kent in 1823 after his marriage. In 1833 he was elected professor of political economy at King's College, London. After Malthus's death Jones succeeded him at Haileybury.[12] Babbage, already involved in the famous though abortive construction of his 'calculating engine', dependent on a physical mechanism rather than electronic impulses, was Lucasian professor of mathematics in Cambridge between 1828 and 1839.[13] Whewell, a fellow of Trinity College from

---

[8] T. R. Malthus, *An Essay on the Principle of Population* (1798), Preface to the 2nd edn (London, 1803), ix. See *The Works of Thomas Robert Malthus* (eds E. A. Wrigley and David Soudan) (8 vols. London, 1986), vol. 2, iii–iv.

[9] Mary Poovey, *A History of the Modern Fact. Problems of Knowledge in the Sciences of Wealth and Society* (Chicago and London, 1998), 290.

[10] Malthus, *An Essay on the Principle of Population* (3rd edn, London, 1806), Book 1, ch. ii, 25.

[11] William Whewell to John Herschel, 18 Sept. 1832, John Herschel Papers, Royal Society, London, 18.183.

[12] William Whewell D. D. (ed.), 'Prefatory Notice', *Literary Remains Consisting of Lectures and Tracts of the late Rev. Richard Jones* (London, 1859). L. G. Johnson, *Richard Jones Reconsidered* (London, 1955).

[13] M. Moseley, *Irascible Genius. The Life of Charles Babbage, Inventor* (London, 1964); Anthony Hyman, *Charles Babbage. Pioneer of the Computer* (Oxford, 1982).

1817, and twice a professor in the university (remarkably, of mineralogy and then of moral philosophy), was Master of Trinity from 1841 until his death in 1866. He cut a leading figure in academic life and natural science for half a century and if his precise scientific contributions are now forgotten, his coining of the term 'scientist' in 1833 makes him an immortal.[14] Jones, Whewell, and Babbage were united in their opposition to the then predominant method in the study of economics which, following the example set by David Ricardo in his *Principles of Political Economy and Taxation* (1817), was rigidly deductive in approach, constructing a superstructure for the science of political economy on a set of questionable assumptions about human behaviour in all places at all times. Finally Quetelet, who trained as an academic mathematician and was a noted astronomer at the Belgian Royal Observatory, became president of the Belgian Commission Centrale de Statistique in 1841, and from 1853 was founding president of the International Statistical Congress.[15] Whewell and Babbage had both invited him to the second meeting of the BAAS in Oxford in 1832 without success. He came to Cambridge as the official delegate of the Belgian government in 1833 after further invitations from Whewell and the Scottish scientist, James David Forbes.[16]

The founders of the Statistical Section were of long acquaintance. Whewell, Babbage, and Jones had known each other as undergraduates at Cambridge, where John Herschel had been part of their circle, as well.[17] They remained in close contact for the rest of their lives, with Whewell, who never really left Trinity College, Cambridge, as their fixed point and anchor. Babbage 'never forgot the early friendships and valuable connections which bound him to the University'.[18] They had often discussed scientific procedure, in particular their commitment to scientific induction—to measuring, collecting, tabulating, and calculating, and then to reasoning on the basis of the data collected. When Babbage read Jones's most important published work, the first and only part of his *Essay on the Distribution of Wealth and on the Sources of Taxation,* published in 1831, he apparently 'declare[d] that he recognises the fruit...of the undergraduate confabulations of the good old set on every page'.[19] As Jones wrote to Herschel in the same year,

---

[14] Laura J. Snyder, *The Philosophical Breakfast Club. Four Remarkable Friends who Transformed Science and Changed the World* (New York, 2011), 3. Isaac Todhunter, *William Whewell D.D.: An account of his writings with selections from his library and scientific correspondence* (2 vols, London, 1876).

[15] Kevin Donnelly, *Adolphe Quetelet, Social Physics and the Average Men of Science 1796–1874* (London, 2015); F. H. Hankins, 'Adolphe Quetelet as Statistician', *Columbia University Studies in History, Economics and Public Law*, xxxi, 1908; D. Landau and Paul F. Lazarsfeld, 'Quetelet', *International Encyclopedia of the Social Sciences*, xiii (New York, 1968), 247–56.

[16] Morrell and Thackray, *Gentlemen of Science*, 374.

[17] Johnson, *Richard Jones Reconsidered*, 6–7.

[18] H. W. Buxton, *Memoir of the Life and Labours of the Late Charles Babbage* (1872–80) (ed. A. Hyman, Cambridge, Mass., 1988), 17.

[19] Whewell papers, Trinity College, Cambridge, Add Mss c. 52/27, 9 March 1831.

Do you remember sitting one night at St. John's [College], with feet on your fender, and the aiming and wishing that we might, three of us [Jones, Herschel and Whewell] meet some day in mature or even declining life with a comfortable feeling that each had done his part to leave the map of knowledge – the chariot wheels of knowledge I think was your expression – a little in advance of where we found them?[20]

The group considered founding an academic review to publicize their approach and style: as Whewell was to write to Jones, 'taking such a line of moral philosophy, political economy and science, as I suppose we should, we might partly form a school which would be considerable in influence of the best kind.'[21] They were all admirers of Quetelet's work, and both Babbage and Whewell had met Quetelet before, Babbage in 1826 in Paris at a dinner given by the astronomer Bouvard, and Whewell in 1829 in Heidelberg at the Gessellschaft Deutscher Naturforscher und Ärtze, a national gathering of German savants, and a model for the subsequent BAAS.[22] Quetelet visited Babbage in England in 1830.[23] All three were in regular correspondence, as well.

Herschel left England for his observatory at the Cape of Good Hope in November 1833 from where he mapped the southern night skies, so was not party to the events leading up to the institutionalization of statistics in British life in 1833–4. But he combined all the key elements of the group in his person: a dedication to observation, measurement, and empirical science in general; a determination to conduct science by the most appropriate and rigorous methods; and the sense that these procedures might then be applied to the study of society. All of this was combined in Herschel's manual of scientific method, the *Preliminary Discourse on the Study of Natural Philosophy*, published in 1830 and influential on all intellectuals at this time, from William Whewell to John Stuart Mill and Charles Darwin. 'Now recognized as a classic in the empiricist tradition of the philosophy of science, it was also a major statement of the important place of natural science in culture and society.'[24] Herschel tentatively extended science to society:

The successful results of our experiments and reasonings in natural philosophy, and the incalculable advantages which experience, systematically consulted and dispassionately reasoned on, has conferred in matters purely physical,

---

[20] Jones to Herschel, 10 Jan. 1831, Herschel Papers, 10.350.
[21] Whewell to Jones, 24 April 1831, Whewell papers, Add MSS. c. 51/104. For Jones's suggestion, see Add MSS c 52/34, 23 April 1831.
[22] For Babbage's first meeting with Quetelet, see L. A. J. Quetelet, 'Extracts from a notice of Charles Babbage', *Annual Report of the Board of Regents of the Smithsonian Institution* (1873), 184. For Whewell's meeting with Quetelet, see Todhunter, *William Whewell*, I, 41.
[23] Edouard Mailly, 'Essai sur le vie et les ouvrages de Quetelet', Annuaire de L'Académie Royale des Sciences, des Lettres et des Beaux-Arts de Belgique, 46 (1875), 201.
[24] Michael J. Crowe, 'Sir John Frederick William Herschel' (1792–1871), *ODNB*.

tend of necessity to impress something of the well-weighed and progressive character of science on the more complicated conduct of our social and moral relations. It is thus that legislation and politics become gradually regarded as experimental sciences.[25]

Herschel was a more cautious and equable person than either Whewell or Babbage and his language was not as confident and direct as theirs when he considered the benefits that science might bring to social improvement. Yet he agreed nonetheless that 'reason which has enabled us to subdue all nature to our purposes should...achieve a far more difficult conquest' in 'enabling of the collective wisdom of mankind' for the improvement of man's estate.[26]

The group of friends can be placed, therefore, at the centre of the so-called 'Cambridge Network' which Susan Faye Cannon distinguished in the 1970s as the 'progressive centre of English thought of the period, with respect to which other intellectuals can objectively be classified as narrow or cosmopolitan, old-fashioned or modern, radical or reactionary'. The network was associated with liberal Anglican and Broad Church promotion of moderate political and religious reforms. It supported 'the modernisation and internationalisation of English science, scholarship and religious thought.' It comprised 'a loose convergence of scientists, historians, dons and other scholars with a common acceptance of accuracy, intelligence and novelty'.[27] It was centred on Trinity College, and led, if any single figurehead can be found, by Whewell himself.[28] The mathematicians and astronomers George Peacock, G. B. Airy, and George Boole were all within the Cambridge network, as well.

At the first irregular meeting of what became Section F, the Statistical Section of the British Association, held on the morning of 27 June 1833, Quetelet 'communicated to the meeting some of the results of his inquiries into the proportion of crimes at different ages and in different parts of France and Belgium'.[29] These would soon be incorporated in the book that would make Quetelet's name across Europe and North America, and serve as the leading inspiration for the statistical movement, *Sur L'Homme et le Développement de ses Facultés: Physique Sociale*, published in 1835. Babbage presented some similar work of his own 'in illustration of Mr. Quetelet's previous remarks'. This included 'curves he had constructed of the number of persons committed for drunkenness in the metropolis in the several months of the year' and 'curves of mortality, marriage, and fecundity in Sweden and Finland'. From the very outset, the Statistical Section and the

[25] J. F. W. Herschel, *Preliminary Discourse on the Study of Natural Philosophy* (1830) (London, 1851 edn), 72–3.
[26] Ibid., 74.
[27] S. F. Cannon, *Science in Culture. The Early Victorian Period* (New York, 1978), 30, 63.
[28] Morrell and Thackray, *Gentlemen of Science*, 21.
[29] 'Minutes of the Committee of the Statistical Section of the British Association, June 27, 1833', in Transcript from the note-book of Mr. J. E. Drinkwater, f. 1.

subsequent Statistical Society, were concerned with so-called 'vital statistics', the statistics of life and death, and also with 'statistique morale', the statistics of human behaviour, notably of crime and suicide.

Babbage also addressed the group on something more technical: he 'exhibited a set of curves drawn with a view to shew the effect of averages in producing a perceptible law in phenomena apparently of the most arbitrary character when considered separately'.[30] This was likely what statisticians call today the Central Limit Theorem, a key component of probability theory. According to this, the mean values of a set of independent and randomly generated variables will rapidly converge with a normal distribution. Whatever the distribution of data, with the proviso that the samples are identical, as the sample size increases, the distribution of the mean values of the sample will approximate to the bell-shaped normal curve. The more samples that are taken, the more the graphed results take that familiar shape. The French mathematicians Laplace and Poisson had, respectively, proved and refined the theorem earlier in the nineteenth century. Quetelet had met them both during a sojourn in Paris in 1823.[31] Babbage, sometimes with Herschel as his companion, had also befriended both men on trips to Paris, and was here, in Cambridge in 1833, discussing their ideas with his select audience.[32]

To the emerging statistical movement the applicability to almost any distribution of the Central Limit Theorem, and of other mathematical concepts like it, for example Bernoulli's Law of Large Numbers, which we shall encounter later, instilled confidence in the scientific endeavour just beginning.[33] Numbers could apparently bring order out of chaos. They demonstrated that social processes were regular; that these processes displayed an overarching pattern or principle at work; that they were therefore explicable and perhaps controllable, as well. The study of statistics might thus disclose fundamental, universal, and usable truths about humans and the world they inhabited. Quetelet, who tried harder than anyone to discover and expose those supposedly immutable principles, is famous for the first use of the normal distribution—often known as the error curve—for something other than the analysis of observational or sample error. In his anthropometric measurements and research, such as into the chest measurements of 5738 Scottish soldiers drawn from different regiments, he demonstrated later that many human physical characteristics are distributed normally.[34] As for Babbage,

---

[30] Ibid. I am grateful to Professor Geoff Nicholls of St. Peter's College, Oxford, for identifying the Central Limit Theorem from these lines in Drinkwater's notebook.

[31] Donnelly, *Adolphe Quetelet*, 88–95; Theodore M. Porter, *The Rise of Statistical Thinking 1820–1900* (Princeton, N.J., 1986), 7.

[32] Charles Babbage, *Passages from the Life of a Philosopher* (London, 1864), 196–7.

[33] See below, Chapter 7, 149.

[34] A. Quetelet, *Lettres à S.A.R. le Duc Regnant de Saxe-Cobourg et Gotha, sur la Théorie des Probabilités, Appliquée aux Sciences Morales et Politiques* (Brussels, 1846), lettre XX, 136–8, 400. Porter, *The Rise of Statistical Thinking*, 100.

the interest displayed in both social statistics and statistical theory by the man who first envisaged and then constructed a computer, gives this meeting, and the subsequent history of the statistical movement, more than merely local or even just national significance.

Later, in the evening of the same day, at a second meeting again chaired by Malthus, a larger group of 'noblemen and gentlemen intimated their desire of attaching themselves to the section'. They included the third earl Fitzwilliam (1786–1857), known by the courtesy title of Viscount Milton until 1833, and up to that time the radical MP for the counties of Yorkshire and then Northamptonshire. Fitzwilliam had presided over the first meeting of the BAAS in York in 1831 and would be the president of the Statistical Society of London on three occasions in the future.[35] Also present were Henry Hallam, the historian; George Wood, MP for South Lancashire, a leading figure in business and culture in Manchester; and John Minter Morgan, a notable and wealthy promoter of Owenism, the co-operative socialist philosophy of the cotton spinner and radical, Robert Owen. Minter Morgan was evidence in his very person of the wide range of groups interested in social science in this period. As Richard Jones told this meeting, 'care must be taken not to circumscribe its utility by narrowing the field of inquiry'. He called for 'an accurate classification of the several heads of inquiry, and enumeration of the wants of the science'. Babbage, who had just published a pioneering study of the methods and organization of industry in Britain, *On the Economy of Machinery and Manufactures*, added the suggestion 'that much information could be derived from the proprietors of large manufactures'.[36]

When the adjourned meeting re-convened on the following morning, Jones presented 'a sketch of the objects of the section'. He made it clear that it was to be more than just an assistant to the study of political economy—that it would not be limited to those subjects that 'bear directly or indirectly upon the production or distribution of public wealth'.

> It is with wider views that such an Association at the present would approach the subject. It may be presumed that they would think foreign to the objects of their inquiries no classes of facts relating to communities of men which promise when sufficiently multiplied to indicate general laws.

It was a felicitous formulation that was repeated several times in the early history of the statistical movement. Jones went on to caution against 'premature speculation' and to suggest, in prudence, that 'they limit as far as possible their reception of such matter to facts capable of being expressed by numbers'. These would fall

[35] G. B. Smith, 'Charles William Wentworth Fitzwilliam, third Earl Fitzwilliam', *ODNB*.
[36] 'Minutes of the Committee of the Statistical Section of the British Association, June 27, 1833', in Transcript from the note-book of Mr. J. E. Drinkwater, f. 2.

naturally, he continued, into 'four distinct compartments': the 'economic, medical, political, and moral and intellectual statistics of nations'. It was the first of many such taxonomies of social knowledge—subdivisions of the type and extent of social statistics—drawn up by members of the statistical movement.[37] Jones also suggested the foundation of a permanent committee in London, with Babbage as its president, to take the subject further. That committee, he suggested, might be 'converted into a Statistical Society' that would be financially and intellectually independent of the British Association. As the meeting ended Quetelet presented some of his publications, among them his pamphlet entitled *Sur la possibilité de mésurer l'influence des causes qui modifient les Élémens Sociaux*.[38]

The British Association was the careful construct of natural philosophers and savants who sought to unite in a single organization the different factions and groups within British science. Its founders and managers were conscious of its potential fragility and of the need to retain consensus and agreement as the organization developed. This may explain why the president of the Cambridge meeting, Adam Sedgwick, Woodwardian Professor of Geology in the university after whom the Sedgwick Museum in Cambridge is named, showed such sensitivity and procedural fastidiousness when faced with the 'irregular' birth of a sixth section without the sanction of the Association.[39] At the General Meeting in the Cambridge Senate House on the evening of Thursday, 27 June, he called on Babbage to set out 'the reasons and circumstances of this proceeding'. Babbage did so, explaining that 'they had been assisted by a distinguished Foreigner, possessing a budget of most valuable information' and he asked the meeting for 'a bill of indemnity, for having broken the laws'.[40] That seemed to be forthcoming and the meeting evidently dispersed in the general understanding that the statistical section had been accepted and incorporated. But by the following day's General Meeting, Sedgwick was evincing public doubt over the procedural legalities and the very subject-matter of the new section. His announcement yesterday, he now explained, 'was altogether out of order'. He had been taken by surprise and had been over-awed by 'the great names' of those involved. Now he feared that 'he should establish a very bad precedent, and risk the integrity of the constitution' of the British Association which had just been drafted. The matter was being sent back to the General Committee, therefore.[41]

[37] On these taxonomies, see Chapter 10 below, pp. 192–6.
[38] 'Transcript of the Notebook of Mr. J. E. Drinkwater', ff. 2–4. See L. A. J. Quetelet, *Sur la possibilité de mésurer l'influence des causes qui modifient les Élémens Sociaux*, [On the Possibility of Measuring the Influence of the Causes which change Social Structures] (Brussels, 1832).
[39] J. A. Secord, 'Adam Sedgwick' (1785–1873), *ODNB*.
[40] *Lithographed Signatures of the Members of the British Association for the Advancement of Science, who met at Cambridge, June MDCCCXXXIII with a Report of the Proceedings at the Public Meetings During the Week; and an Alphabetical List of the Members* (Cambridge, 1833), 82.
[41] Ibid., 90.

By the time that the weeklong Meeting ended and Sedgwick gave his concluding address, the matter had been settled and Section F had been incorporated into the British Association by proper and agreed means. Yet concerned by these events, however trivial they may now seem, Sedgwick felt it necessary to discuss their implications in remarks at the very outset of the statistical movement which remained relevant to the rest of its history. Given the need to justify the new section, Sedgwick spoke on 'what I understand by science'. His definition was in line with what was then current scientific thinking, and was in no manner hostile to the statisticians: 'all subjects, whether of a pure or mixed nature, capable of being reduced to measurement and calculation'. Hence, he continued, there would be no difficulty in bringing statistical inquiries into the organization:

> so far as they have to do with matters of fact, with mere abstractions, and with numerical results. Considered in that light they gave what may be called the raw material to political economy and political philosophy; and by their help the lasting foundations of those sciences may be perhaps ultimately laid.

But such studies were 'most intimately connected with moral phenomena and economical speculations' and so touched 'the mainsprings of passion and feeling'. Against 'these higher generalisations' he issued a warning:

> if we transgress our proper boundaries, go into provinces not belonging to us, and open a door of communication with the dreary world of politics, that instant will the foul Daemon of discord find his way into our Eden of Philosophy.[42]

These might be dismissed as the personal concerns of an elderly and conservative don, but points like Sedgwick's were made frequently in subsequent decades, not least in the 1870s when there was a formal attempt to evict Section F from the British Association. Quite soon the British Association learnt that to reduce political tensions and minimize controversy, Section F was best assigned to small rooms and the papers delivered there were best represented in published form by their numbers only, shorn of their more provocative commentary.[43] At the end of the concluding meeting, the Association's General Committee approved the formation of a Statistical Section. In accordance with Sedgwick's definition of a science, 'It was resolved that the inquiries of this Section should be restricted to those classes of facts relating to communities of men which are capable of

---

[42] *Report of the Third Meeting of the British Association for the Advancement of Science; Held at Cambridge in 1833* (London, 1834), xxvii–ix.
[43] J. Morrell and A. Thackray, *Gentlemen of Science. Early Years of the British Association for the Advancement of Science* (Oxford, 1982), 296.

being expressed by numbers, and which promise, when sufficiently multiplied, to indicate general laws'.[44]

Quetelet produced a summary of these events in Cambridge:

Les recherches de statistique ont également occupé M. Babbage, et comme cette science n'était pas comprise au nombre de celles dont les comités avaient à s'occuper à Cambridge, nous nous réunîmes d'abord pour en parler avec MM. Malthus et Jones, dont j'avais eu l'honneur de faire la connaissance. Quelques personnes témoignèrent le désir de prendre part à ces conférences toutes particulières, qui reçurent bientôt une extension telle que l'association reconnut, en séance générale, un sixième comité pour la statistique, mais en resserrant cette science dans sa partie purement numérique.[45]

The General Committee also voted a 'sum not exceeding £100…towards a plan proposed by Professor Babbage, for collecting and arranging the Constants of Nature and Art', a vast and characteristic project to collect the most important and significant numerical data in heaven and earth that will be explained later.[46] Babbage had certainly played a central role in the creation of Section F as a prime mover and chief exponent: it forms a largely overlooked 'passage in the life of a philosopher' as his autobiography was entitled, though it is highly suggestive of Babbage's convictions and aims. Indeed, these events evidently meant so much to Babbage that he left four different accounts of them, three of them published in 1851, 1860, and 1864 respectively,[47] and one of them written for Quetelet when Babbage was in Brussels in 1853 at the first International Statistical Congress.[48]

---

[44] *Report of the Third Meeting*, xxxvii.

[45] 'Statistical research had occupied Babbage as well, and as this science was not included among the sections at the Cambridge meeting, we came together beforehand to discuss it with Malthus and Jones whose acquaintance I had already had the honour of making. Several people expressed the desire to take part in these discussions which, happily, were recognized by the Association as a sixth, statistical section, at its general meeting. But this science was narrowed to the purely numerical'. L. A. J. Quetelet, 'Notes Extraites d'un Voyage en Angleterre, aux mois de Juin et Juillet 1833', *Correspondance mathematique et physique*, III, i (Brussels, 1835), 14. Much later, in 1869, Quetelet produced a second, inaccurate, account of these events. See 'Congrès international de statistique des délégués des different pays', *Extrait des Bulletins*, 2 seie, tome XXVII, no. 5, 1869, International Statistical Congress, The Hague, 1869, held in the Quetelet Papers, Bibliotheque Royale de Belgique, Brussels, file 3551.

[46] *Report of the Third Meeting*, xxxvii, 490. See below, Chapter 5, pp. 107–9.

[47] Charles Babbage, *The Exposition of 1851; or, Views of the Industry, the Science and the Government of England* (London, 1851), 16–18; 'Letter from Charles Babbage, Esq., FRS', *Report of the Proceedings of the Fourth Section of the International Statistical Congress. Held in London, July 16, 1860 and the Five Following Days* (London, 1861), 505–7; Charles Babbage, *Passages from the Life of a Philosopher* (London, 1864), 432–5.

[48] 'Note sur l'origine de la société de statistique de Londres par M. Babbage', Quetelet Papers, Bibliotheque Royale de Belgique. See 'Babbage's Note Respecting the Origin of the Statistical Society, Brussels, Sept. 1853' in *JRSS*, vol. 124, no. 4, 1961, 546: 'Charles Babbage's Note and a French translation by Quetelet were photographed and presented by Dr. Leopold Martin, President of the Biometric Society, to the Royal Statistical Society in March 1961'.

Each differs slightly, but in them all he strove to write himself into the centre of events. He recognized before 1833 that the British Association had too narrow a base, requiring 'other sciences besides the physical' and a section 'to interest the landed proprietors or those members of their families who sat in either house of parliament'. Nor, he added 'was there much to attract the manufacturer or the retail dealer'. Thus he turned the 'accidental circumstance' of Quetelet's presence in Cambridge to advantage:

> At the Third Meeting of the British Association at Cambridge in 1833, I happened, one afternoon, to call on my old and valued friend the Rev. Richard Jones, Professor of Political Economy at Haileybury, who was then residing in apartments at Trinity College. He informed me that he had just had a long conversation with our mutual friend M. Quetelet, who had been sent officially by the Belgian Government to attend the meeting of the British Association. That M. Quetelet had brought with him a budget of statistical facts, and that as there was no place for it in any section, he (Professor Jones) had asked M. Quetelet to come to him that evening, and had invited Sir Charles Lemon, Professor Malthus, Mr. Drinkwater (afterwards Mr Bethune), and one or two others interested in the subject, to meet him, at the same time requesting me to join the party. I gladly accepted this invitation, and departed. I had not, however, reached the gate of Trinity College before it occurred to me that there was now an opportunity of doing good service to the British Association. I returned to the apartment of my friend, explained to him my views, in which he fully coincided, and I suggested the formation of a Statistical Section.[49]

It is doubtful if it happened quite like this, however. Nor is it clear that Quetelet's presence was simply accidental. The move had been planned, or at least suggested, some months in advance, and not by Babbage. In his inaugural lecture at the new King's College, London on 27 February 1833, four months before Section F was established under his guidance, with Whewell, Babbage, and Herschel in the audience, Jones had called for the establishment of an academic society to develop knowledge of 'mankind and his concerns'.[50] And it was Whewell who had written to Jones on 24 March 1833 that 'I want to talk with you about getting statistical information, if the British Association is to be made subservient to that, and

[49] 'Letter to Dr. Farr, On the Origin of the International Statistical Congresses', *Proceedings of the Fourth International Statistical Congress* (London, 1860), 505–7. See also Babbage, *Works*, vol. 11, 321–7. According to Babbage, these events took place on 26 June 1833.

[50] Richard Jones, 'An Introductory Lecture on Political Economy, Delivered at King's College, London, February 27th, 1833', in Whewell (ed.), *Literary Remains*, 571. Snyder, *The Philosophical Breakfast Club*, 149–50.

which I think would be well.'[51] Nevertheless, when the British Association meeting was over, it was Babbage who had the responsibility of taking the Cambridge initiatives further. Jones encouraged him to 'keep the section alive, at least for Chalmers to dry nurse it if you can do no more—perhaps Drinkwater & you may do more if you are so minded.'[52]

The scene then moved from Cambridge to London. As Jones reported to Whewell, at a dinner at Babbage's house, 1 Dorset Street, Marylebone, in February 1834, with Malthus, Drinkwater and the historian Hallam present as well, 'I prevailed on Babbage (who was not reluctant) to call a general meeting of the Committee of the [British] Association for next week and get an authority for them to set about forming a society as the best means of carrying the spirit of the Cambridge Instructions into effect.'[53] Three days later on 21 February, the statistical committee set up after the foundation of Section F (as suggested by Jones) met again at Babbage's house 'to establish a Statistical Society in London, the object of which shall be the collection and classification of all facts illustrative of the present condition and prospects of society'.[54] Babbage, Jones, Malthus, Drinkwater, Sykes, and George Wood, MP, from Manchester, were all present. The nascent society began to prospect for members, and names were collected, many of them Whig MPs, political economists, physicians and lawyers. On the premise that 'accurate knowledge of the actual condition and prospects of society is an object of great national importance, not to be attained without careful collection and classification of statistical facts', the Statistical Society of London was brought into being at its inaugural meeting in the following month. As we have seen, Henry Petty-Fitzmaurice, third marquess of Lansdowne, was in the chair on the occasion. On a motion proposed by Babbage, Quetelet was made its first foreign member.[55]

When Lansdowne spoke he referred to the 'painful difficulties which lay in the way of the literary and scientific man, and of the political economist, arising from the absence of any combination of those facts for practical purposes'.[56] But he also reinforced the national and governmental rationales for the society: 'His lordship informed the meeting that the Government would be glad to avail itself of the labours of such an institution; which, in return, should have the assistance of Government when it was necessary.'[57] He was followed by Henry Goulburn, a

---

[51] Whewell to Jones, 24 March 1833, Whewell Papers, Add MSS c. 51/154.

[52] Richard Jones to Charles Babbage, 3 July 1833, Babbage Papers, British Library, London, Add Mss 37188, f. 4. The reference is to the planned BAAS meeting in Edinburgh in the next year, 1834, where Thomas Chalmers, the Scottish Presbyterian minister and political economist, would, Jones hoped, look after the new section F.

[53] Jones to Whewell, 18 Feb. 1834, Whewell Papers, Add. Mss c. 52/60.

[54] 'Meeting of the Committee at Babbage's House, 1 Dorset Street, 21 Feb. 1834', Transcript from the note-book of Mr. J. E. Drinkwater, 5–6. Ibid.

[55] 'Statistical Society', The Times, 17 March 1834, 3.   [56] Ibid.

[57] 'Statistical Society', The Gentleman's Magazine, n.s., 1, April 1834, 422.

Tory (though, it must be said, a liberal Tory) to give political balance to the proceedings. He had been chancellor of the exchequer when Wellington was prime minister between 1828 and 1830 and would hold the same office again when Peel was prime minister between 1841 and 1846. In seeking the financial support of government for the construction of his calculating engines Babbage had dealings with Goulburn personally during both of these periods.[58] Goulburn remarked 'that one of the greatest difficulties he had experienced when in office, was the want of completeness or arrangement, in the statistical returns to which he required to refer'.[59] Evidence was misrepresented by 'persons who only adduced those facts in support of some favourite theories'. Meanwhile, many 'an aspiring mind' was discouraged from important work by the sheer impossibility 'of one individual being able to collect together the vast body of authenticated facts' which were necessary to support a valid theory. He was 'decidedly favourable to the appointment of such a society', therefore.

To give a national character to the proceedings, the third speaker was the Lord Advocate, the chief legal officer of the crown in Scotland. It was no ordinary lawyer who rose to speak, however, but Francis Jeffrey, one of the founders of the *Edinburgh Review* in 1802, and the great periodical's first full-time editor, a post he held for 27 years before turning back to the law. At the time of this meeting, in March 1833, he was also Whig MP for Edinburgh. Jeffrey embodied intellectual whiggery, the journey this outlook had made in the early nineteenth century from Edinburgh to London, and the impact it had on Whig politics in general.[60] Like Goulburn, Jeffrey was concerned by the misrepresentation and manipulation of facts: 'how difficult it would be to find one unbiased man to whom so important a duty could be intrusted', he remarked, warning against the social influences 'arising from prejudice, or from sinister interests'. But he added a new theme to the discussion, as well, as befitted the editor of a journal that had published such distinguished historical reviews: the study of the past.

> It was by a comparison of the present with the past, that the future could with safety and probability be predicted. An accurate statement of the present condition only of a system perpetually undergoing change would not be a sufficient ground for prediction and arrangement of the future; and he had no doubt, although the collection of past facts was not referred to in the resolution, yet that it was one of the objects of the society to include them.

[58] Maboth Moseley, *Irascible Genius. A Life of Charles Babbage, Inventor* (London, 1964), 101–2, 145–6; Babbage, *Passages from the Life of a Philosopher* (London, 1864), 75, 93–6. In 1829 Goulburn accompanied Wellington himself on a visit to Babbage's workshop and the application for further funds was successful. Later, in November 1842, Goulburn was party to a joint decision with Peel to reject further support for Babbage's engines.
[59] 'Statistical Society', *The Gentleman's Magazine*, n.s., 1, April 1834, 422.
[60] Biancamaria Fontana, *Rethinking the Politics of Commercial Society: The Edinburgh Review 1802–1832* (Cambridge, 1985).

Babbage then proposed the 'appointment of a society for the collection and classification of facts to be called the Statistical Society of London'. Richard Jones seconded him, adumbrating the principal branches of their subject: political statistics concerning public, legal and economic matters; medical statistics, including vital statistics and 'other matters closely connected with the principles of population'; and moral and intellectual statistics concerning 'the state of literature, of crime, and of moral and religious instruction'. A provisional committee was left in charge of the new society's business. It comprised Babbage, Jones, Hallam, and Drinkwater.[61] Henry Hallam was a significant contributor to the Victorian construction of the 'whig interpretation of history' with important published works on Britain and Europe in the early modern period.[62] He had been an early writer for the *Edinburgh Review* and was a member of the Bowood Circle of Whig savants that Lansdowne had created at his Wiltshire estate. Through his friendships, historical views, and political opinions he fitted perfectly into the Whig world of the new Statistical Society and was its treasurer for its first six years until 1840.[63] The Council of the Society in its early years has rightly been described as 'a subcommittee of a Whig cabinet'.[64]

These events were summarized by Drinkwater, acting as secretary for the new Society, in a letter to Quetelet on 18 March 1834:

I am happy to be able to tell you that at last we have triumphed over all our difficulties, and have established 'The Statistical Society of London' in the most satisfactory manner. The Committee at the foundation of which you assisted did nothing until about a month ago, when we all suddenly waked up, and decided on endeavouring to set on foot an independent society. Accordingly, we called a public meeting for that purpose, which was very well attended. The Marquess of Lansdowne took the chair, & some of the most distinguished men in the country brought forward & supported resolutions. Mr Babbage, Mr Jones, Mr Hallam & myself have been appointed a Provisional Committee to draw up Regulations for the Society, & when that is done, which will be in about a month, we shall go regularly to work. In consideration of the great influence which your presence at Cambridge had upon the formation of the section, out of which this society has arisen, you were elected by acclamation the first Foreign Member of the Society, & as soon as we have our machinery in order, you will receive your diploma.[65]

[61] 'Statistical Society', *The Times*, 17 March 1834, 3.
[62] Henry Hallam, *The Constitutional History of England from the Accession of Henry VII to the Death of George II* (2 vols, London, 1827); *Introduction to the Literature of Europe in the Fifteenth, Sixteenth, and Seventeenth Centuries* (4 vols, London, 1837–9).
[63] Timothy Lang, 'Henry Hallam 1777–1859', *ODNB*.
[64] Philip Abrams, *The Origins of British Sociology 1834–1914* (Chicago and London, 1968), 15.
[65] J. E. Drinkwater to L. A. J. Quetelet, 18 March 1834, Quetelet Papers, Académie Royale de Belgique, Brussels, File 898.

It is important to note the change of focus and subject between Cambridge in the summer of 1833 and London in the spring of 1834. In Cambridge, Section F of the British Association had been established by natural scientists, mathematicians, and political economists who saw a place for the inductive study of society by statistics inside a new organization dedicated to the unity and popularization of the sciences in general. There was no talk of government in Cambridge: far from it, as the concern was precisely that the new section might introduce partisanship into a body dedicated to a concept of science as pure and unaligned. But in the translation to London and the establishment of a single-discipline society with a broad and influential membership, the raison d'être of the Statistical Society of London had altered. This was suspected from the start by its Cambridge founders. Setting out his views on its academic remit and the wide range of its interests, in the very week of the Society's creation in February 1834, Jones wrote to Whewell that 'There will be some trouble in getting Londoners to lay the whole broad foundation here sketched, but we must try.'[66] The Society rapidly became an association, indeed a social project, for men of affairs, those who governed. Most of these men, moreover, were of a single political persuasion: Whig, the party then in power. They needed data for policy and public administration, and they evidently feared the falsification of the facts. Indeed, in their speeches, they referred far more frequently to facts, the stuff of public debate and ministerial affairs, than to numbers and statistics, the focus of interest for the Cambridge savants who had proposed the establishment of the society. This transformation may have followed from the organizational requirement to garner the patronage of leading public figures and politicians, and Lansdowne and Goulburn were indeed leaders, as were other early members of the Statistical Society of London. With this sort of backing the society would succeed. But the transition from the pursuit of social knowledge for intellectual purposes to the collection of accurate and verifiable information to assist government illustrates the different conceptualizations of the purposes of statistics at this formative stage, a variety that becomes yet more diverse when the simultaneous project in Manchester of generating statistics to promote the city and assist radical political and religious causes is added to the mix.

The Statistical Society of London grew and thrived after its foundation, but not in the way that its founders in Cambridge had hoped and projected. The *Westminster Review* argued that the injunction 'to exclude carefully all opinions from its transactions and publications' had deprived it of intellectual focus and purpose: 'No mere record and arrangement of facts can constitute a science.'[67]

---

[66] Jones to Whewell, 18 Feb. 1834, Whewell Papers, Add. Mss. c. 52/60. For Jones's enumeration of those aims, see Chapter 10, pp. 193–4.

[67] [G. Robertson] 'Transactions of the Statistical Society of London', [vol. 1, pt. 1, 1837]. *The London and Westminster Review*, xxxi, i (April–Aug 1838), 68–9.

The opening pages of the first volume of the society's journal in 1838 did indeed state baldly that the science of statistics 'proceeds wholly by the accumulation and comparison of facts, and does not admit of any kind of speculation'.[68] But it is just as likely that its swelling membership did not appreciate the imperatives of its founders and impressed on the Society their own concerns and interests. As most were drawn from the interlocking worlds of politics and public administration, the Society was quickly drawn into policy making, government, and the provision of data rather than its analysis. Its intellectual mission to construct a statistical social science was part of its rhetoric in the first few years, but was never pursued actively. It was never an academic forum. Nor was it ever impartial.

One explanation for this lies with the originators. Malthus died in 1834 and Quetelet was domiciled in Brussels. Whewell and Jones were initially involved in the practical organization of the society, sitting on its 'Committee on Moral and Intellectual Statistics' in 1834–5. On 15 May 1834 'Mr Whewell's enumeration and classification of facts bearing upon the subject allotted to this department was discussed' and, suitably amended, presented to the society's council.[69] But to add to his notorious dilatoriness, Jones was appointed to Malthus's professorial chair, drew up the parliamentary bill for the commutation of tithes in 1836 (in which task he collaborated with another of the founders of Section F, Drinkwater) and was then made one of the three tithe commissioners, thereby setting a pattern: many a subsequent Victorian statistician was also drawn into the work of government. As Whewell wrote of him to their mutual friend Herschel, 'he employs himself rather with other matters (drawing tithe bills and the like) than with his theories, which I hold it his duty under any circumstance to complete and publish'.[70] But as Whewell also explained to Herschel, Jones was paid a salary of £1500 a year as a tithe commissioner, three times his remuneration as a professor (in which position he continued).[71] To Whewell's lasting credit, he published all Jones's essays and fragments after Jones's death in a book of *Literary Remains* in 1859. As for Whewell himself, as he explained to Jones 'I am myself much inclined to thrust my fingers into your economical statistics, but I have so much else to do', which was something of an understatement from a man whose energy in so many scientific fields was legendary.[72] Of them all, Babbage retained the greatest interest in, and strongest lifelong connections to, the Statistical Society of London, though he was no leader. He had neither the taste nor patience for

[68] *Journal of the Statistical Society of London (JSSL)*, i, 3.
[69] Minute Book 7, 'Committee on Moral and Intellectual Statistics, 1834–5', Archives of the Royal Statistical Society.
[70] Whewell to J. Herschel, 6 April 1836, Herschel Papers, Royal Society, London, 18.186.
[71] Whewell to J. Herschel, 4 Dec. 1836, in Todhunter (ed.), *William Whewell D.D.*, II, 249.
[72] Whewell to Jones, ('1833'), Whewell Papers, Add MSS c. 51/161.

committee work, but he never lost his fascination for statistics, nor pride in his foundational role in 1833–4.

But there is also evidence of an intellectual disaffection with the society. Jones was very soon reported as 'somewhat disturbed at some of the vagaries which it appears likely to take',[73] and Whewell, from the first month of its existence, was urging the need for 'work of classification and arrangement of materials with reference to *results* as well as of collection'.[74] In August 1834 he wrote to Quetelet that 'our Committee has had several meetings, but we are still somewhat embarrassed by the extent of our subject'.[75] Six months later he was critical of the emphasis in the new organization on collecting information rather than analysing it: 'I confess my interest in it is strong only when the facts appear to furnish some indication of a principle'.[76] There was a need for some overarching theory as well as data:

> They will, I have no doubt, obtain a great deal of information; but my opinion at present is that they would go on better if they had some zealous theorists among them. I am afraid you will think me heterodox; but I believe that without this, there will be no zeal in their labour and no connexion in their results. Theories are not very dangerous, even when they are false; (except when they are applied to for action) for the facts collected and expressed in the language of a bad theory may be translated into the language of a better; but unconnected facts are of comparatively small value.[77]

His comments recalled his speech to the British Association in June 1833 when he had observed that 'facts can only become portions of knowledge as they become classed and connected...they can only constitute truth when they are included in general propositions'.[78] Whewell always maintained that 'there could be no rigid separation of theory and fact'.[79] He was hostile to the discussion of social policy, as well. It infringed the objectivity of science and was in no sense the subject, the true science of society, he had envisaged. As he wrote of Section F to the geologist, Roderick Impey Murchison,

> Who would venture to propose...an ambulatory body, composed partly of men of reputation and partly of a miscellaneous crowd, to go round year by year,

[73] Whewell to Jones, 22 April 1834, Whewell Papers, Add MSS c 51/165.
[74] Whewell to Jones, 23 March 1834, Whewell Papers, Add MSS c 51/164.
[75] Whewell to Quetelet, 4 Aug. 1834, Quetelet Papers, File 2644.
[76] Whewell to Quetelet, 3 Feb. 1835, ibid.   [77] Whewell to Quetelet, 2 Oct. 1835, ibid.
[78] William Whewell, 'Address', *Report of the Third Meeting of the British Association for the Advancement of Science, Held at Cambridge, 1833* (London, 1834), xxi.
[79] Richard Yeo, 'William Whewell, Natural Theology and the Philosophy of Science in Mid Nineteenth Century Britain', *Annals of Science*, 36, 5, 1979, 501.

from town to town and at each place to discuss the most inflammatory and agitating questions of the day?[80]

At the British Association in 1836 Whewell referred to the Manchester Statistical Society, which was already focusing it attentions on national educational reform, as 'a menace'.[81] In 1837, just before the annual meeting, he told the Association's General Council 'that the statistical section ought never to have been admitted, and went much too far'.[82]

The problem was as much sociological as intellectual: it lay with the people who joined the society and also with the clashing conception of the society's role which was held by Babbage. For in Babbage a refined interest in empirical science, inductive political economy, and in statistical compilation and theory, was joined to a tenacious concern to link science with society. The Cambridge mathematician who had spent months touring the industrial districts of Britain to write *On the Economy of Machinery and Manufactures* was involved in Whig-radical politics, unsuccessfully seeking a parliamentary seat in the early 1830s.[83] He believed that 'the greatest benefit will accrue to science', following the foundation of the British Association, in 'the intercourse' which its meetings could not 'fail to produce between the different classes of society. The man of science will derive practical information from the great manufacturers.'[84] At the second meeting in Oxford in 1832 he had asserted that 'attention should be paid to the object of bringing theoretical science in contact with that practical knowledge on which the wealth of the country depends'. He therefore wanted the third meeting held in Manchester, the great centre of manufacturing, rather than Cambridge.[85] At that meeting in Cambridge he had put in a word for 'the proprietors of large manufacturies' and in his various accounts of the foundation of Section F he had always made his motivation plain: to interest landholders, parliamentarians, and businessmen in the British Association so as to bring science and its representative institutions broad social support. As he put it in his second version of these events, 'the principle of extending the basis of the Association so as to unite the various classes, was steadily and unremittingly pursued'.[86]

[80] Whewell to Sir R. I. Murchison, 2 Oct. 1840, in Todhunter (ed.), *William Whewell DD*, II, 289.
[81] *The Athanaeum*, 1836, 657.
[82] Sir R. I. Murchison to W. V. Harcourt, 18 Sept. 1837, *The Harcourt Papers* (ed. E. W. Harcourt) (1880–1905, Oxford), xiii, 362.
[83] M. J. Cullen, *The Statistical Movement in Early Victorian Britain: The Foundation of Empirical Social Research* (Hassocks, Sussex, 1975), 80.
[84] Charles Babbage, *On the Economy of Machinery and Manufactures* (London, 1832), 311–12.
[85] 'Proceedings of the Second Meeting of the British Association for the Advancement of Science', *Transactions of the BAAS*, 1 (1831–2), 107. J. Morrell and Arnold Thackray, *Gentlemen of Science. Early Years of the British Association for the Advancement of Science* (Oxford, 1981), 166–7.
[86] Charles Babbage, *The Exposition of 1851; or, Views of the Industry, the Science and the Government of England* (London, 1851), 18.

The 313 'original members' who had joined before the inauguration of the Statistical Society of London, and the 98 subsequently admitted in 1834, included Edward Adolphus Seymour, the 11th duke of Somerset, a genuine scholar and mathematician and a close friend of Babbage;[87] Lord John Russell, later a Liberal prime minister; Sir George Grey, the stalwart member of many Whig cabinets to come; the radical MP Sir William Molesworth; and even the young William Gladstone, not yet a Liberal. As Whewell informed Quetelet in August 1834, 'You will find that the statistical section, which sprang up under your auspices at Cambridge, is grown into a Statistical Society in London with many of our noble-men and members of parliament for its members.'[88] The society's members were predominantly parliamentarians and notable figures drawn from the commercial and professional classes with little conception of the scientific aims and ambitions of the founders. And there was no bar to the membership of orthodox political economists, or those who shared their views, either. Among the initial members were Nassau Senior and J. R. McCulloch. Yet in 1831 Jones had derided Senior's view, derived from Ricardo, that 'the foundations of political economy [were] a few general propositions – deduced from observations from consciousness and generally admitted as soon as stated' as 'nonsense.'[89] As for McCulloch, he had written a hostile review of Jones's *Essay on Rent* which had prompted Whewell to look forward to a time 'when Ricardists and McCullochites have become like Aristotleians and Cartesians now are.'[90] McCulloch was an acknowledged disciple and popularizer of Ricardo in their eyes.[91] He was also recognized as a leading 'statistical author' of the period. But his conception of the purpose of statistics was entirely different from the holistic science of society that Jones and Whewell hoped to see created on the basis of the collection and analysis of social data. McCulloch held to a rigid division of labour between fact-gathering and theory. In 1824, he had written that

Besides being confounded with Politics, Political Economy has sometimes been confounded with Statistics; but they are still more easily separated and distin-guished. The object of the statistician is to describe the condition of a particular country at a particular period; while the object of the political economist is to discover the causes which have brought it into that condition, and the reasons by which its wealth and riches may be indefinitely increased. He is to the statisti-cian what the physical astronomer is to the mere observer. He takes the facts

---

[87] Hyman, *Charles Babbage*, 55.
[88] Whewell to Quetelet, 4 Aug. 1834, Quetelet Papers, File 2644.
[89] Jones to Whewell, 24 Feb. 1831, Whewell Papers, Add MSS c 52/20.
[90] Whewell to Jones, 'Dec. 1831', Whewell Papers, Add MSS c 51/122.
[91] For a revision of this traditional view, see D. P. O'Brien, *J. R. McCulloch: A Study in Classical Economics* (London, 1970), 16, 121–3, 403–5. O'Brien places McCulloch in the tradition of Smith and Hume into which 'he attempted to incorporate Ricardian elements' (p. 121).

furnished by the researches of the statistician, and after comparing them with those furnished by historians and travellers, he applies himself to discover their relation.[92]

This was the very antithesis of the synthetic and inductive science of society, incorporating as one element within it the study of economic interactions, that Jones, Whewell, and Babbage projected. But it undoubtedly exerted influence over the Statistical Society of London. According to the very first page of the Society's new *Journal*, published in 1838,

> The Science of Statistics differs from Political Economy, because, although it has the same end in view, it does not discuss causes, nor reason upon probable effects; it seeks only to collect, arrange, and compare, that class of facts which alone can form the basis of correct conclusions with respect to social and political government.[93]

This division of labour and self-denying ordinance on the part of statisticians, in combination with the political character of its membership, enforced on the Society a practical, policy-orientation. Data was collected for public purposes and as an end in itself, but not as the material of a science of society. In 1838 and 1839, the Society's annual reports, bearing a Jonesian imprint, had spoken explicitly of creating a 'social science'. But by 1840, as practical and utilitarian functions began to dominate, the Sixth Annual Report was something of a lament:

> It was not to perfect the mere art of 'tabulating' that [the society] was embodied;- it was not just to make us hewers and drawers to those engaged in any edifice of physical science:- but it was that we should be ourselves the architects of a science or sciences, the perfectors of some definite branch or branches of knowledge.[94]

Once it was decided to extract this new social science from the medium of intellectual discourse in lectures, books and reviews and put it into the public arena, creating sections and societies, it was required to attract influential members to provide funds and prestige. As Jones wrote in excitement to Whewell just before the inaugural meeting of the society,

---

[92] J. R. McCulloch, *A Discourse on the Rise, Progress, Peculiar Objects and Importance of Political Economy: Containing an Outline of a Course of Lectures on the Principles and Doctrines of that Science* (Edinburgh, 1824), 75.

[93] *Journal of the Statistical Society of London*, May 1838, 'Introduction', 1.

[94] 'Sixth Annual Report', *JSSL*, iii (1840), 1–2.

We expect Whigs and Lords in abundance…We want tories and look too whiggish and ministerial at present – I am to ask Goulburn to ask Peel etc. – Catch us some tories if you can, ditto radicals…I dine with the Bishop of London tomorrow and hope to get him to come and take part in its formation.[95]

Such a membership, drawn from the early-Victorian governing class, brought with it practical notions of the use of social data and a marked readiness to defer to orthodox economic thought: the Society was unlikely to have the time, inclination, or wit, to grasp the innovatory intentions of its founders in Cambridge. Moreover, this was the age when 'economists as backbenchers had an impact on legislation by voting, by speeches, by sponsoring of legislation, and by activities on committees, that is unmatched by any other time or in any other country'.[96] Trying to teach the governing classes a new way of thinking about society was always going to be difficult, in other words.

British natural science, meanwhile, was seeking widespread social recognition and support in the early 1830s, and popularization set up 'a contradiction between the advancement of science and its promotion' and inculcated 'an empirical and utilitarian image of science which neglected its theoretical dimension'. Scientific populism and ostentation 'encouraged a perception of natural knowledge as a collection of interesting and useful facts, and this view militated against a proper appreciation of science'.[97] It made it possible for hundreds of Victorians to join statistical societies, but it also undercut intellectual endeavour and attainment. Babbage had written to Whewell in 1820 that 'All sorts of plans, speculations and schemes are afloat, and all sorts of people, proper and improper, are penetrated with the desire of wielding the sceptre of science'.[98] Babbage himself would play a most prominent part in these struggles as the main protagonist in the 'Decline of Science' debate in the early 1830s, at the time of the election of the Duke of Sussex as the President of the Royal Society.[99] Babbage, a radical, took the Duke's nomination as a slight against genuine 'gentlemen of science' and persuaded Herschel to stand against him. Herschel lost by only a few votes. At a time of flux and change in the institutionalization of British science, when the nature of the scientific community was a subject for debate, the Cambridge network were drawn to adopt popular but indiscriminate styles of organization. They were academics engaged in debates over economic and scientific theory who chose, as

---

[95] Jones to Whewell, 18 Feb. 1834, Whewell Papers, Add MSS c 52/60.

[96] F. W. Fetter, 'The Influence of Economists in Parliament on British Legislation from Ricardo to John Stuart Mill', *Journal of Political Economy*, lxxxiii, 5, 1975, 1060.

[97] Richard Yeo, 'Scientific Method and the Image of Science', in R. Macleod and P. Collins (eds), *The Parliament of Science: The British Association for the Advancement of Science 1831–1981* (London, 1981), 73, 77.

[98] Babbage to Whewell, 15 May 1820, Whewell Papers, Add MSS a. 200/192.

[99] Snyder, *The Philosophical Breakfast Club*, 130–2.

it transpired, an inappropriate means through which to make their case. They might have been better served by one of their original ideas, which was to found a journal or a review that would set forth and continually reinforce their ideas.

Without doubt, the echo of their academic project to use the patterns and regularities in statistics to understand social life can be heard for the rest of the century in figures as varied as William Farr, Florence Nightingale, Prince Albert, and the historian H. T. Buckle. But it was usually subordinate to more obviously useful and practical rationales for the collection of social data. In the 1840s the Statistical Society of London drew in large numbers of civil servants, professional men, and those employed as actuaries in the insurance business, and these formed the core of the membership through the mid-Victorian decades.[100] For these groups, statistics denoted 'a science of government' and the statistician 'was to be the scientific expert in matters of policy'.[101] According to the society, statistics was 'the science of the arts of civil life' and it was to 'statists' that 'the statesman and the legislator must resort for the principles on which to legislate and govern'.[102] As McCulloch put it in his review of British statistics in 1835,

> The accumulation of minute and detailed information from all parts of the country would, at length, enable politicians and legislators to come to a correct conclusion as to many highly interesting practical questions that have hitherto been involved in the greatest doubt and uncertainty.[103]

Because of this practical and utilitarian rationale, the society showed little inclination to develop statistics as a branch of mathematics for the first half-century of its existence. Its members collected and arranged social data; they did not strive to find new ways of analysing them. It wasn't until 1912, when Francis Edgeworth became president of the then Royal Statistical Society, that it could boast a Professor at its head. Edgeworth remained the only president of the Society to have held a chair in its first century.[104]

In the years to follow its foundation in the 1830s, the 'governmentality' of the statistical movement would dominate: statistics would be understood in most contexts as the social knowledge required for political and civic life to function efficiently. But in Cambridge, among a select group, statistics were to play a quite different role, providing a new way of understanding and analysing human society that, in its methodology, was comparable to the physical sciences. Yet even here, among Jones, Whewell, Babbage, Quetelet, and their collaborators, there

---

[100] John M. Eyler, *Victorian Social Medicine. The Ideas and Methods of William Farr* (London and Baltimore, 1979), 14.
[101] Ibid., 16.   [102] 'Sixth Report of the Council', *JSSL*, 2.
[103] J. R. McCulloch, 'State and Defects of British Statistics', *Edinburgh Review*, vol. 61, April 1835, 178.
[104] 'Francis Ysidro Edgeworth, Statistician', *JRSS*, series A, 141 (1978), 291.

were differences of outlook, each of which requires analysis and explanation. Social and economic statistics, evidence of the real economy, would provide Jones with the data he required to mount an assault on the method and doctrines of orthodox political economy as then understood. Whewell joined him in that, but looked also to the numbering of the natural world to provide insight into the fundamental processes of the earth and heavens. For Babbage, his involvement in the statistical movement was an extension of his fascination with numbers and encouraged him in the development of a mechanical engine that could process them. For Quetelet and others, social data would offer deeper insight into vital statistics and anthropometry—into key life events and the physical qualities of humans—and also into their behaviour *en masse*, and would thereby provide the basis for a predictive social science. The advent of statistics in the early 1830s was an event in intellectual history, in other words, though a complex event with several different strands running through it which must be untangled.

3

Manchester: The Manchester Statistical Society

Industry, Sectarianism, and Education

The literary and cultural critic Raymond Williams found it 'very striking' that it was during the 1830s that 'the classic technique devised in response to the impossibility of understanding contemporary society from experience, the statistical mode of analysis, had its precise origins'.

> Without the combination of statistical theory, which in a sense was already mathematically present, and arrangements for collection of statistical data, symbolised by the foundation of the Manchester Statistical Society, the society that was emerging out of the industrial revolution was literally unknowable…After the industrial revolution the possibility of understanding an experience in terms of the available articulation of concepts and language was qualitatively altered… New forms had to be devised to penetrate what was rightly perceived to be to a large extent obscure.[1]

Charles Babbage, who was not only at the heart of the statistical movement but was also a profound student of the organization of industrial production, found it necessary in building his Difference Engine, to devise an entirely new 'mechanical notation'. This was based on a system of signs he had devised himself to overcome 'the imperfection of all known modes of explaining and demonstrating the construction of machinery'. In this way even 'the most complicated machine' could be understood 'almost without the aid of words'.[2] To be understood, and to harness the productive powers and technologies now available, the industrial society of the nineteenth century required new languages and forms of representation. Statistics provided a new grammar and vocabulary; they were a new means by which to capture and analyse a changed reality. They offered a way of anatomizing more precisely than before a new set of social and technological procedures which depended on precision. We shall see later how critics of both

---

[1] Raymond Williams, *Politics and Letters* (London, 1979), 170–71.
[2] Charles Babbage, *Passages from the Life of a Philosopher* (London, 1864) 142.

Victorians and Numbers: Statistics and Society in Nineteenth Century Britain. Lawrence Goldman, Oxford University Press.
© Lawrence Goldman 2022. DOI: 10.1093/oso/9780192847744.003.0003

industrialism and precision felt and wrote about statistics as the idiom of the new industrial society.

In these contexts the origins of the Manchester Statistical Society, which met for the first time on 2 September 1833, might seem both obvious and appropriate. That in the very centre of the Industrial Revolution, in the great new city of mechanized production, where the steam-powered factories were employing tens of thousands of workers, there should arise an organization for the study of the new processes and commerce of the age, using this new language, might hardly seem to require much explanation. The counting of cotton threads, spindles, horsepower, workers ('hands'), and yards of cloth in the local factories gave rise naturally, it might be thought, to a learned society where the statistics of business could be collected, tabulated and set forth for analysis and discussion.

The existence of other statistical societies in other provincial cities also undergoing rapid social and economic change in the early nineteenth century only adds to the sense that the Manchester Statistical Society's existence was predictable.[3] The Statistical Society of Glasgow was founded in February 1836, though it had folded by 1838, to be replaced in 1840 by the Glasgow and Clydesdale Statistical Society. Bristol's statistical society dates from November 1836. Three statistical societies were founded in the first months of 1838 in Liverpool, Leeds, and Belfast (the Ulster Statistical Society).[4] Like several of these societies—Bristol, Liverpool, and the second Glasgow society follow the same pattern—a society in Newcastle was encouraged into existence in 1838 by the visit to the city in that year of the British Association for the Advancement of Science. The Birmingham Educational Statistical Society (also known as the Birmingham Statistical Society for the Improvement of Education) surfaced in 1838 as well, almost certainly in response to the battles over elementary education which consumed the Manchester Statistical Society and which are the subject of this chapter. In Leeds, the Statistical Committee of the town council predated the formation of the local society. In Sheffield there was never a society but the council's statistical committee was the forum in which Dr George Calvert Holland conducted important research on the welfare of the woollen workers in Yorkshire. Unsurprisingly, this was critical of the factory system, and of mechanization itself, for their effects on health.[5] One of the two secretaries of the Liverpool Society, Dr William Henry Duncan, was a

---

[3] Michael J. Cullen, *The Statistical Movement in Early Victorian Britain. The Foundations of Empirical Social Research* (Hassocks, Sussex, 1975), 119–33. See also Walter F. Willcox, 'Note on the Chronology of Statistical Societies', *Journal of the American Statistical Association*, 29, 188, Dec. 1934, 418–20. This includes a long, international list of societies founded in the nineteenth century and beyond.

[4] Christopher O'Brien, 'The Origins and Originators of Early Statistical Societies: A Comparison of Liverpool and Manchester', *JRSS*, series A, 174, 1, Jan. 2011, 51–62.

[5] Ibid., 131. George Calvert Holland, *Inquiry into the Moral, Social and Intellectual Condition of the Industrious Classes of Sheffield* (London, 1839) and *The Vital Statistics of Sheffield* (London, 1843). Cullen, *The Statistical Movement*, 131–2; Maxine Berg, *The Machinery Question and the Making of Political Economy 1815–48* (Cambridge, 1980), 310–12.

leading sanitarian, from 1847 the city's first medical officer of health and, as such, the first local medical officer of health in England.[6] In Bristol the society organized two important surveys of the condition of the local working class and on 'statistics of education' in the city.[7] Further societies in other provincial towns and cities were mooted and may have met, but have no traceable institutional existence. With the exception of that in Manchester, all of the provincial statistical societies faded away in the course of the 1840s. That founded in Liverpool lasted barely a year and had probably ceased to function by 1839.[8] As the *Journal of the Statistical Society of London* lamented in June 1846, the month in which the Corn Laws were repealed, 'how ephemeral is the existence, or transitory are the exertions, of the local Statistical Societies generally.'[9] Even the Manchester Statistical Society temporarily disappeared from sight during that decade, though for a reason which will become clear in due course and which was related to 'repeal'.

Manchester's urban story was so remarkable that the possession of a statistical society to analyse and explain it seems altogether unremarkable. With a population of perhaps 20,000 in 1770 the city had grown to more than 140,000 by 1831. James Phillips Kay, a local physician and one of the founding members of the Manchester Statistical Society, was the author famous for his social survey of the city in the following year, published as *The Moral and Physical Condition of the Working Classes Employed in the Cotton Manufacture in Manchester*. As he expressed it, 'Visiting Manchester, the metropolis of the commercial system, a stranger regards with wonder the ingenuity and comprehensive capacity, which, in the short space of half a century, have here established the staple manufacture of this kingdom.'[10] Until 1832, however, under the old electoral map of the unreformed House of Commons, it went unrepresented and returned no MPs to parliament. Nor until 1838 was it a municipal borough with an elected city corporation. Yet from the earliest years of the nineteenth century it was home to the largest concentration of cotton spinning mills in the world. In 1830, the Liverpool to Manchester line opened, the world's first steam passenger railway, which also allowed for the fast transportation of raw cotton imports in one direction and finished cotton textile exports in the other.

Alexis de Tocqueville, the great French liberal theorist, visited the city in the summer of 1835.[11] He met Kay, who accompanied him around Manchester, and

---

[6] W. M. Frazer, *Duncan of Liverpool. Being an Account of the Work of Dr. W. H. Duncan, Medical Officer of Health of Liverpool 1847–63* (London, 1947).

[7] 'Condition of the Working Class in the City of Bristol', *JSSL*, 2, Oct. 1839, 369; 'Statistics of Education in Bristol', *JSSL*, 4, Oct. 1841, 250–1.

[8] O'Brien, 'The Origins and Originators of Early Statistical Societies', 54.

[9] 'Twelfth Annual Report of the Statistical Society of London, 1845-6', June 1846, vol. 9, *JSSL*, 9, June 1846, 99.

[10] James Phillips Kay, *The Moral and Physical Condition of the Working Classes Employed in the Cotton Manufacture in Manchester* (London, 1832), 46.

[11] Michael Drolet, 'Tocqueville's Interest in the Social: Or How Statistics Informed His "New Science of Politics"', *History of European Ideas*, 31, 2005, 464–5.

they went together to a meeting of the Manchester Statistical Society (hereafter MSS) on the evening of 1 July 1835 where Tocqueville met other leading members of the Society. In his journal Tocqueville famously described the city as wreathed in black smoke, experiencing only semi-daylight, its population moving and working ceaselessly amidst a never-ending clatter of noise, their lives lived in squalor.[12] But Manchester was also a city of culture and reasoned enquiry, the home to a Literary and Philosophical Society, founded in 1781, a Board of Health founded in 1796, a Natural History Society founded in 1821, the Manchester Institution, an art gallery, opened in 1823, a Mechanics' Institution for working-class education dating from 1824, and the District Provident Society to assist the poor, founded in 1833, which Tocqueville also visited. Disraeli's observation in his novel *Coningsby*, published in 1844, that 'a Lancashire village has expanded into a mighty region of factories and warehouses. Yet rightly understood, Manchester is as great a human exploit as Athens', was entirely different from Tocqueville's, but perhaps just as accurate.[13] The city was a magnet to which many were drawn to gain an insight into the future, and it encouraged a sort of 'factory tourism'. Visitors came to see for themselves the *'Chaos vom Fabriken'*, the title of an unpublished memoir of a visit to Manchester in 1840 to purchase machinery by a young Viennese, Joseph Mohr, the son of a cotton manufacturer. His astonishment, the more authentic because not written for publication, sits somewhere between Tocqueville's revulsion and Disraeli's admiration.[14] He was thus one of the more moderate tourists: according to one historian, 'German visitors from pleasant country towns thought of Manchester as a kind of inferno'.[15]

The story of the Manchester Statistical Society, which encompasses many of these themes—urban growth, cultural development, entrepreneurship, personal enrichment, technical advance, factory labour, exploitation, squalor, and social reform—is more complex and less intuitive than might be thought.[16] It depends very decidedly on the identity of the Society's membership, their collective social and religious consciousness, the politics of the city between the 1820s and 1850s, and the issue of education rather than industrialization. Raymond Williams was right to explore the broad and universal relationship between numbers and modern production, but the story of the Manchester Statistical Society is specific to a set of men with a particular outlook and with a cause—the education of all

[12] *Alexis de Tocqueville, Journeys to England and Ireland* (New Haven, Ct., 1958), 2 July 1835, 106–8.
[13] Benjamin Disraeli, *Coningsby: Or the New Generation* (1844), vol. 2, book 4, ch. 1, 2.
[14] Rebecca Ryan, '"Chaos vom Fabriken": A German insight into the Industrial Revolution in Britain in 1840', unpublished BA dissertation, Oxford History Schools, 2010 (St. Peter's College). On factory tourism, see Steven Marcus, *Engels, Manchester, and the Working Class* (New York, 1974) (1985 edn), 30–66; Francis D. Klingender, *Art and the Industrial Revolution* (London, 1972), 109–17.
[15] John Pickstone, 'Manchester's History and Manchester's Medicine', *British Medical Journal*, 295, 19–26 Dec. 1987, 1604–08. Quotation at 1604.
[16] The standard institutional account is T. S. Ashton, *Economic and Social Investigation in Manchester, 1833–1933. A Centenary History of the Manchester Statistical Society* (London, 1934).

children, irrespective of religion—through which they wished to attack the social and religious establishment of the day. The best way to write that story is through prosopography, the collective biographies of the founders and members of the MSS: we cannot call them statisticians, as will become clear.

The Society has been written about in two different ways. The first, coming out of the history of sociology, has wanted to ask why the first British statistical society, predating that in London and the smaller and more ephemeral statistical societies in other British cities, failed to fully institutionalize itself and build its initial social survey work into academic structures—university chairs, journals, learned societies, public lectures, and so forth—for the scientific study of society.[17] Under the influence of a slightly earlier debate about the supposed absence of sociology in Britain, which failed to establish itself as a sanctioned and respected academic discipline in this country at any time before the 1960s, and perhaps not even then, these studies tried to show how the Manchester Statistical Society failed to develop the attributes of a learned body of savants dedicated to social research and theorizing.[18] The objection that this was never the aim of the founders, who were very far from being academics or theorists, was not considered. As will be argued here, the Society is better understood as an outgrowth of Manchester's business elite at its most politically influential than as a sociological forum. The Society was the product of the sociability and political concerns of the Manchester 'millocracy', closer in nature, therefore, to a body like the Lunar Society of Birmingham which had united entrepreneurs, technologists, and savants in the Midlands in the 1770s and 1780s, than to an academic organization or learned society.[19]

A second approach was much more concerned with the social class, and the class interests, of these men, but placed too much emphasis on these factors alone and failed, thereby, to take seriously the particular issues that animated the founders. Their class position as bourgeois factory owners, bankers, lawyers, and ancillary professionals dominated this interpretation at the expense of considering their religious identity and their resulting consciousness of social exclusion. So it was suggested 'that the need to provide a coherent justification for the factory system, at least in its more humane forms, was a factor in the founding of

[17] David Elesh, 'The Manchester Statistical Society: A Case Study of Discontinuity in the History of Empirical Social Research', and Stephen Cole, 'Continuity and institutionalization in science: A Case Study of Failure' in Anthony Oberschall (ed.), *The Establishment of Empirical Sociology: Studies in Continuity, Discontinuity and Institutionalization* (New York, 1972), 31–72, 73–129.

[18] Perry Anderson, 'Components of the National Culture', *New Left Review*, 50 (1968), 1–57. Philip Abrams, *The Origins of British Sociology 1834–1914* (Chicago and London, 1968). For a more recent discussion of this subject, see Matthew Grimley, 'You Got an Ology? The Backlash against Sociology in Britain, c. 1945–90', in Lawrence Goldman (ed.), *Welfare and Social Policy in Britain Since 1870* (Oxford, 2019).

[19] Jenny Uglow, *The Lunar Men: Five Friends Whose Curiosity Changed the World* (London, 2002).

the Manchester Society'.[20] It was argued that a new organization dominated by successful cotton capitalists, many of whom were among the leading businessmen in Manchester and its environs, was keen to de-emphasize, if not ignore totally, the relationship between poverty, low wages, long hours, and the physical break-down of workers. Its member sought to keep social investigators out of their factories for that would be the first stage on the road to state regulation of hours and conditions.[21] Agitation for this, in the shape of the Ten Hours Movement to limit the working day in factories, was already underway by the early 1830s, with Parliament now taking an acute interest in the conditions of factory labour. M. T. Sadler's famous select committee inquiry of 1831, and a subsequent Royal Commission, had led to the first significant Factory Act in 1833, which fixed the minimum age for textile factory employment at nine, and regulated hours worked up to the age of 18.[22] In this view, the focus of the MSS on public health reform and the provision of elementary education was a deliberate strategy to avoid official scrutiny of the factory system itself and a consequent assault on its members' interests. Capitalists placed their emphasis, by design, on general urban sanitary and social ills, removed from the real source of Manchester's problems, which was located in the factories they owned.

There is evidence to support this contention, without question. It is hardly coincidental that one of the first reports presented to the Manchester society by the brothers Samuel and William Rathbone Greg, mill owners both, was a review of the evidence taken by the 1833 Royal Commission of Inquiry into factory conditions. They contended that the large majority of the testimony collected was favourable to the factory system and they reached an extreme conclusion:

That the health and morals of the people employed in cotton mills are at least equal to that of those engaged in other occupations in the towns in which they are situated; that the long hours of labour do *not* over-fatigue the children, or injure their health and constitutions; that the general charges of cruelty and ill-treatment, which have been so repeatedly alleged, are entirely groundless; that the education of the factory children, as compared with others, is more carefully attended to; that the poor rates are lower in Lancashire than in any other county

---

[20] Cullen, *The Statistical Movement*, 108. For a similar argument see also Martin Shaw and Ian Miles, 'The Social Roots of Statistical Knowledge', in John Irvine, Ian Miles, and Jeff Evans (eds), *Demystifying Social Statistics* (London, 1979), 33: 'The gentlemen researchers who belonged to these societies sought knowledge that would establish the kinds of reform necessary to secure a stable hierarchical society, and would be evidence in favour of them.'

[21] Berg, *The Machinery Question*, 301–14.

[22] R. W. Cooke-Taylor, *The Factory System and the Factory Acts* (London, 1894), 64–76.

in the kingdom; and that the wages of labour are such as, in the agricultural districts, would be regarded as positive opulence.[23]

When Charles Poulett Thomson, the MP for Manchester 1832–9, and the then Vice-President of the Board of Trade, met with James Phillips Kay and other founders of the Society in January 1834 to advise on its future work, he encouraged surveys of employment and wages among the working class as well as of their membership of provident societies and cultural organizations.[24] Yet when these surveys began they were predominantly concerned with working-class religion and education. Without doubt the early members of the Society sought to keep inquirers away from their factories and tried to argue that poverty and its attendant ills had their origins in general city conditions, rather than the specific working environment. But coming from a particular sectarian and social background, for these men the critical battles and targets were religio-political—laws and conventions designed to exclude them from local society and national power because of their identity—and they shaped the Manchester Statistical Society accordingly. When James Heywood led a house-to-house survey 'into the state of 176 families in Miles Platting, within the Borough of Manchester in 1837' it contained some information on wages, certainly, but it concentrated on religion, church attendance, bible ownership, literacy and education.[25] This was because Heywood was primarily involved in building a case against the exclusivity of the national church, and the state's folly in failing to educate its people, two issues which were intimately related. It has been argued that there was a 'contradiction in the ideas of men like Kay and Greg' between 'their generally laissez-faire approach to some social issues (such as the poor law and factory acts) and their highly interventionist beliefs in others (such as education and sanitary reform)'.[26] But this would not have seemed at all inconsistent to the founders of the Manchester Statistical Society who looked upon these issues as fundamentally dissimilar in nature, and whose primary aim was to end the related ills of ignorance and exclusion on religious grounds.

---

[23] Samuel Greg and W. R. Greg, *Analysis of the Evidence Taken Before the Factory Commissioners as far as it relates to the Population of Manchester and the Vicinity Engaged in the Cotton Trade. Read Before the Statistical Society of Manchester, March 1834* (Manchester, 1834). Ashton, *Economic and Social Investigation*, 17–18.

[24] J. P. Kay, 'Results of a Conference with Mr Thomson concerning Objects towards which the Society should direct its attention', Appendix to the Minutes of the Statistical Society, Manchester, 1833', f. 4, GB127.BR MS f 310.6 M5, Manchester Central Library; Ashton, *Economic and Social Investigation*, 16–17; Cullen, *The Statistical Movement*, 26, 110–11.

[25] James Heywood, 'Report of an Enquiry Conducted from House to House into the State of 176 families in Miles Platting, within the Borough of Manchester in 1837', *JSSL*, 1, 34–6.

[26] Cullen, *The Statistical Movement*, 110.

Historians have protested that 'Door-to-door surveys of working people inquired into children's education and the presence of books in the home – but not into family income or hours of work.'[27] But in treating the concentration on popular education in Manchester as some sort of gambit, device, deviation, or deliberate side-show, and ignoring its centrality to the first members of the MSS, the society's true origins and its members' intentions have been obscured. These can be appreciated far better and more accurately in the context of Vic Gatrell's argument that questions of class and class relations were not the dominant considerations in the politics and consciousness of the Manchester bourgeoisie in the 1820s and 1830s. 'The political issues that counted for them were not centrally concerned with the demands of the poor. They were more to do with who among themselves should wield local power and the status attached to it.'[28] Their Liberalism led them to struggle to replace 'Tory and High Church authority' in the city, which was stronger and better organized than is commonly imagined, and which was an affront to a group very largely drawn from the ranks of Protestant nonconformity.[29] Given that it was these Liberals who founded the Manchester Statistical Society, in its first years the organization is better conceptualized as a sectarian pressure group than as an academic society or a cabal of local capitalists. This was almost admitted in the society's very early self-description as an association of 'gentlemen accustomed for the most part to meet in private society, and whose habits and opinions are not uncongenial.'[30] A new member needed the support of 12 existing members, and initially it was agreed to limit the membership to 50 only. By 1835 there were 40 members.[31]

Nor were the society's origins purely local: it had several links with the events in Cambridge and London in 1833–4 that we have examined already. Benjamin Heywood, the central figure in the origins of the Manchester society, and his brother James, attended the Cambridge Meeting of the British Association for the Advancement of Science in June 1833.[32] Among those present at the informal meetings in Trinity College, Cambridge at which the Statistical Section of the British Association was founded during that week, were the Rev. Edward Stanley, Rector of Alderley in Cheshire, later Bishop of Norwich; the then MP for South Lancashire, George William Wood; and John Kennedy.[33] Stanley was a keen

[27] Shaw and Miles, 'The Social Roots of Statistical Knowledge', 33.
[28] V. A. C. Gatrell, 'Incorporation and the Pursuit of Liberal Hegemony in Manchester 1790–1839', in Derek Fraser (ed.), *Municipal Reform and the Industrial City* (Leicester and New York, 1982), 17.
[29] Ibid., 18, 23.   [30] Ashton, *Economic and Social Investigation*, 13.
[31] Christopher O'Brien, 'The Manchester Statistical Society. Civic Engagement in the 19th Century', *Transactions of the Manchester Statistical Society*, 2011–12, 98.
[32] Ibid., 97.
[33] They are listed by name at the second, evening 'meeting of Gentlemen' on 27 June 1833, having 'intimated their desire of attaching themselves to the Section'. Transcript from the Note-Book of Mr J. E. Drinkwater. Minutes of the Committee of the Statistical Section of the British Association, Archives of the Royal Statistical Society, London.

educationist and one of the 14 acknowledged founders of the MSS. Kennedy, also a member of this founding group, was a leading cotton spinner in the city from the 1790s to the 1830s.[34] In partnership with another founder of the Manchester Statistical Society, James M'Connel, Kennedy owned the Sedgewick Mill in Ancoats, then the largest cast-iron-framed building in the world at eight stories high. A member of the 1825 Liverpool-Manchester railway committee, he was also one of three judges at the famous Rainhill Trials of 1829 at which George Stephenson convinced the railway's projectors to use steam locomotives rather than move carriages using fixed steam engines and cables.[35] George Wood was the son of a Unitarian minister, a prominent Manchester businessman, an officer for two decades of the Manchester Literary and Philosophical Society, a founder of other local civic organizations, and a fellow of both the Linnaean and Geological Societies in London. He sat for Lancashire South from 1832 to 1835 and then for the borough of Kendal between 1837 and 1843. Wood was also a member of the committee of Section F, the Statistical Section, formed in Cambridge at the end of the British Association meeting there. He was present at the meetings in Babbage's London house on 19 and 21 February 1834 where, along with Malthus, Richard Jones, and others, it was decided to press ahead with the foundation of a statistical society in London.[36] His son, William Wood, was an early member of the MSS, as well.[37]

George Wood was also among the delegation of five MPs who attended the third meeting of the MSS in Manchester on 27 November 1833 to see the new society for themselves, evidence of the wide interest in the Manchester initiative.[38] He was accompanied on that occasion by two other local MPs, Viscount Molyneux (1796–1855), later the third earl of Sefton, who was then the other member for Lancashire South, and Mark Philips, a Unitarian and an entrepreneur, and one of the first two MPs to represent Manchester in the Parliament of 1832, a position in which he continued for another 15 years. The fourth MP was the remarkable Sir George de Lacy Evans (1787–1870) an army officer, a veteran of the Peninsular campaign, of the Battles of New Orleans and Waterloo, a later participant in the Crimean War (from which he returned as a national hero), and also the radical Liberal MP for Westminster from 1833 until 1865.[39] The fifth MP

[34] C. H. Lee, 'John Kennedy 1769–1855', ODNB; W. Fairbairn, 'A brief memoir of the late John Kennedy, esq.', Memoirs of the Literary and Philosophical Society of Manchester, 3rd ser., 1 (1862), 147–57.
[35] Christian Wolmar, Fire and Steam (London, 2007), 36. https://www.steamindex.com/people/kennedy.htm.
[36] Transcript from the note-book of Mr. J. E. Drinkwater, f. 5.
[37] O'Brien, 'The Origins and Originators of Early Statistical Societies', 55.
[38] Ashton, Economic and Social Investigation, 14.
[39] Edward M. Spiers, 'Sir George de Lacy Evans 1787–1870', ODNB.

was inscribed as the 'earl of Kerry'. This had been the courtesy title of the 3rd Marquess of Lansdowne, the inaugurating president of the Statistical Society of London, before he took his seat in the House of Lords in 1810. The title had passed to his son, William Thomas Petty-Fitzmaurice, in 1818, and he was MP for Calne in Wiltshire from 1832 until his early death in 1836. The latter was a more likely visitor to Manchester in 1833 than his father. Both Lansdowne and Kerry, father and son, had been included as members of the embryonic Statistical Society of London in the list prepared by John Drinkwater on 21 February 1834.[40] Notable also on that list of founding members of the London society was the 'Statistical Society of Manchester' itself, a member in its corporate capacity. The brothers, Benjamin and James Heywood, were also members of the Statistical Society of London from 1834.[41]

All this goes to show that events in Manchester were linked to those in Cambridge in the summer of 1833, and to those in London in the following year: the Manchester Statistical Society was not purely the product of independent, provincial endeavour. One conjecture is that the concept of a statistical society was carried back from Cambridge to Manchester in the summer of 1833 by the Heywoods, Wood, Stanley, and Kennedy. Or they may have gone to Cambridge to meet with other founding statisticians because of a pre-existing commitment on their part to the establishment of a statistical society in their own city. Citizens in Manchester required local information for their own purposes, for sure, but the means by which this was to be collected may have been influenced by models already under discussion at the British Association.

The key text for understanding the origins of the MSS is the inaugural address of the Society's then president, Thomas Read Wilkinson, delivered on 17 November 1875. As it was not composed until four decades later and was based on the recollections of 'those members still living who took an active interest in its foundation', it may be unreliable.[42] Quite possibly, it omits the names of some founders while including those who played only minor roles. But in its description of 'a nucleus of active supporters' in 1833, numbering 14 in all, it includes both plausible and representative figures whose identities help us to understand the society's purpose.[43] The founders comprised two bankers. Benjamin Heywood was MP for Lancashire in 1831–2 and President of the Manchester District Provident Society which was also established in 1833, a few months before the Manchester Statistical Society. William Langton, who was Wilkinson's primary source, was chief cashier for Heywood's Bank, and one of two honorary secretaries

[40] Transcript from the Note-Book of Mr. J. E. Drinkwater, f. 6.
[41] O'Brien, 'The Origins and Originators of Early Statistical Societies', 56.
[42] Wilkinson, 'On the Origin and History of the Manchester Statistical Society', 9.
[43] Ibid., 12–13.

of the District Provident Society.[44] The cottonocracy—both mill owners and cotton merchants—were in the clear majority, including the brothers Samuel Greg and William Rathbone Greg, John Kennedy, William M'Connel, Henry Newbery, James Aspinall Turner, Samuel Robinson, and James Murray. Samuel Dukinfield Darbishire was a solicitor.[45] Lieutenant-Colonel James Shaw Kennedy was in command of the Manchester garrison from 1826 to 1835. James Phillips Kay, later and better known as the civil servant James Kay-Shuttleworth, was senior physician to the Ardwick and Ancoats dispensary and the other secretary of the District Provident Society.[46] It is entirely plausible that the work of the Provident Society in sending out visitors to assess those who sought its support, and in the keeping of records associated with this activity, encouraged the foundations of the Manchester Statistical Society.[47] Another of its probable sources was the organization of the city in the previous year, 1832, to monitor and control the spread of cholera during the first of the nineteenth century's European cholera pandemics.

These men were overwhelmingly religious dissenters—Protestant nonconformists, chapel-goers rather than Churchmen. Eleven were Unitarians at some stage in their lives, or were drawn from historically Unitarian families: Langton, Heywood, the Greg brothers, Kennedy, M'Connel, Darbishire, Turner, Robinson, Murray, and James Phillips Kay.[48] They were members, therefore, of a Christian denomination that, emerging from the religious schisms of the seventeenth century, rejected the trinity and took a notably intellectual, rational, and high-minded approach to faith. Three of these 11—Heywood, Darbishire, and Turner—were trustees of the Cross Street Chapel, where the wealthiest and most fashionable Unitarians worshipped, whose minister from 1828 was one, William Gaskell, husband of the novelist Elizabeth Gaskell.[49] A new generation of Unitarians, including the Gregs, the younger Philipses, and G. W. Wood, worshipped at the Moseley Street Chapel.[50] The one founder who came from an entirely different background might seem to be the Rev. Edward Stanley, an Anglican educated in

---

[44] James Bevis, 'Kay, Heywood & Langton: 19th Century Statisticians and Social Reformers', *Transactions of the Manchester Statistical Society*, 2010–11, 65–77.

[45] Cullen lists Darbishire as a cotton master at page 109, whereas Thomas, in his account of the Cross Street chapel community, has him as a solicitor. He may have been both. See Thomas Baker, *Memorials of a Dissenting Chapel . . . Being a Sketch of the Rise of Nonconformity in Manchester* (London, 1884), 113–14.

[46] His change of name was part of the settlement made on his marriage in 1842 to Janet Shuttleworth, heiress to the Gawthorpe estate near Padiham in Lancashire. The Shuttleworth family did not approve of an alliance with a mere doctor and civil servant and sought to dignify it by adding the family name.

[47] O'Brien, 'The Manchester Statistical Society', (2011–12), 96.

[48] The exceptions were Stanley (Anglican), Shaw Kennedy (as an army officer, likely to be Anglican), and Henry Newbery whose denomination cannot be established.

[49] Thomas Baker, *Memorials of a Dissenting Chapel*; E. L. H. Thomas, *Illustrations of Cross Street Chapel* (Manchester, 1917); H. McLachlan, *The Unitarian Movement in the Religious Life of England* (London, 1934).

[50] Gatrell, 'Incorporation and the Pursuit of Liberal Hegemony', 25.

Cambridge. But he was the exception that proved the rule. Stanley was a Liberal in politics, active in promoting better relations between Anglicans and dissenters, keen to meet with nonconformists in person and to assist their educational aims. Against general opinion in the Church of England he supported the Whig government's educational grants and schemes in 1839, and the endowment of Maynooth College, a Roman Catholic seminary in Ireland, in 1842. He was just the sort of clergyman to make common cause with these Unitarians.[51]

Most of these men had gone straight into business, but of those who had enjoyed a higher education, three had attended Edinburgh University—Kay and the Greg brothers—while Benjamin Heywood had gone to Glasgow. None (with the exception of Stanley) had attended an English university.[52] While it was possible for a dissenter to study at Cambridge, nevertheless to take a degree there entailed formal assent to the 39 articles of the Church of England. In Oxford, that assent had to be given on arrival, at matriculation. These prohibitions were not rectified until the passage of the Oxford and Cambridge University Acts of 1854 and 1856 respectively. There were no Tories among these 14: they comprised various shades of Liberalism from radical to so-called 'Palmerstonian whig'. Interestingly, James Turner was the latter, and when he stood as such in Manchester in the general election of 1857, and defeated both John Bright and Thomas Milner Gibson, he thereby ended the age of the 'Manchester School', a generation in which liberals and radicals from nonconformist backgrounds, espousing the doctrines of free trade and laissez-faire, controlled Manchester's political and cultural life.[53] There was some intermarriage between the families of the founders as might be expected of a provincial elite: two of William Langton's daughters married sons of Benjamin Heywood; two of John Kennedy's daughters married a Heywood and a Robinson respectively. Mark Philips's sister married R. H. Greg.[54] There were business partnerships among them also, notably that between John Kennedy and the M'Connel family, which began in the 1790s and lasted for three decades.[55] There was notable wealth, as well: the firm of A. & G. Murray, from which James Murray derived, was the largest spinner of cotton yarn in the world in the first years of the nineteenth century.[56] But there was also learning and literary accomplishment: the Greg brothers were important

---

[51] R. E. Prothero, 'Edward Stanley 1779–1849', *ODNB*. Stanley was the father of Rev. A. P. Stanley, the famous Dean of Westminster.

[52] It should be noted that James Heywood, who was not one of the core founders of the MSS, studied at Trinity College, Cambridge.

[53] Norman McCord, 'Cobden and Bright in Politics, 1846–1857', in R. Robson (ed.), *Ideas and Institutions in Victorian Britain* (London, 1867), 87–114.

[54] Gatrell, 'Incorporation and the Pursuit of Liberal Hegemony', 25.

[55] C. H. Lee, *A Cotton Enterprise. A History of M'Connel and Kennedy, Fine Cotton Spinners* (Manchester, 1972).

[56] Ian Miller and Chris Wild, *A & G Murray and the Cotton Mills of Ancoats* (Lancaster, 2007).

contemporary essayists and writers, and Samuel Robinson has a place in the *Dictionary of National Biography* as a scholar of Persian.[57]

The characteristics shared among the fourteen founders have been found in other prosopographies of the early membership, as well. It has been calculated that '15 of the first 28 members were Unitarians, and 17 of the first 40'.[58] Another study has focused on 51 out of the 60 members who had joined the Manchester Statistical Society by 1840 and about whom details can be traced.[59] Out of these 51, some 25 were definitely nonconformists, and six only were Anglicans. Whigs and Liberals numbered 23 and there was just a single Tory; the politics of the rest are unknown. There were nine past, present and future MPs in the group as a whole, two Mayors of Manchester, six city aldermen and two local newspaper editors. Out of the 51, 23 were related to at least one other member. Eleven were members of the Manchester Chamber of Commerce. Many were active supporters of the Anti-Corn Law League and of its forerunner also, the Manchester Anti-Corn Law Association. They included James Phillips Kay, James Heywood, R. H. and W. R. Greg, the brothers and cotton masters Henry and Edmund Ashworth, Samuel Darbishire, James Turner, W. Raynor Wood, and Richard Cobden himself, the most famous and influential of all 'repealers' and one of the great liberal thinkers and politicians of the century. Cobden had joined the MSS in November 1835.[60] Equally significant, 19 of the 51 belonged to one or more society that pressed for the adoption of a national system of elementary education, at this stage a key nonconformist demand emanating from the city. As this larger study confirms, the Manchester Statistical Society was an alliance of the commercial elite of the city, brought together by business, religion, politics, and a civic culture that they had done much themselves to create: the Literary and Philosophical Society had 29 members among these 51, and it was a group of the most prominent Unitarian families—the Gregs, Heywoods, Henrys, McConnels, Philipses, and Robinsons—which imparted to the 'Lit and Phil' its activism and tone.[61] With an annual subscription of two guineas, the Manchester Statistical Society could only have been meant for the wealthy.

Yet despite their wealth, local status and position, the Manchester Unitarians lacked genuine equality: laws affecting political participation, education, and religion discriminated against them, and they suffered also from informal social stigma for taking a position outside the establishment. It was not until as

---

[57] Stanley Lane-Poole, 'Samuel Robinson 1794–1884', *ODNB*.

[58] O'Brien, 'The Origins and Originators of Early Statistical Societies', 59.

[59] This is drawn from the excellent biographical research accomplished by David Elesh. See Elesh, 'The Manchester Statistical Society: A Case Study of Discontinuity', 38–39.

[60] Ashton, *Economic and Social Investigation*, 12. O'Brien found that out of the first 40 members of the MSS, 10 'were members of the Manchester Anti-Corn Law Association Council in 1839–40'. See O'Brien, 'The Origins and Originators of Early Statistical Societies', 60.

[61] Arnold Thackray, 'Natural Knowledge in Cultural Context: The Manchester Mode', *American Historical Review*, vol 79, 3, June 1974, 698, 705.

late as 1813 that the clauses of the Toleration Act of 1689, which had excluded anti-trinitarians from the terms of that measure, were repealed. Though Unitarians had been to the fore in the foundation of the Manchester Athenaeum in 1835, an attempt was made three years later to exclude them from holding office in the institution. In the 1839 Manchester election W. R. Greg was the Liberal candidate and his Unitarianism was made an issue in the campaign by his Tory opponent. For two decades from 1825 the Cross Street chapel was enmeshed in legal proceedings stemming from the bequest of property 'to their trinitarian ancestors in the seventeenth century'.[62] Understandably, the 'self-appointed mission' of the Manchester Unitarians, 'was to destroy the High Church and Tory hegemony in the town which denied them, it seemed, the proper status which their wealth and education merited'.[63] This struggle spilled over and into the Manchester Statistical Society and its campaigns for local and national education.

Over a period of three years, 1834–7, the MSS surveyed working-class life in Manchester and adjoining towns. What began as an inquiry conducted by James Phillips Kay in 1834 into local handloom weavers grew into a wider survey of more than four thousand households in the police divisions of St. Michaels and New Cross, conducted by an agent going door-to-door. Out of approximately 12,000 children in the districts, only 252 were attending day school, with another 4680 at Sunday Schools.[64] This study, in turn, was widened to consider other parts of Manchester and Salford, as well as the towns of Bury, Ashton, Stalybridge, and Dukinfield. Over a 17-month period, four agents collected information on household members, occupations, county of birth, religion of the head of the household, the size and amenities of the home and the length of residence there, the state of the water supply, the number of books in the home, the number of children at school, and the cost of that schooling.[65] The results were read to Section F of the British Association at Liverpool in 1837 and published in 1838. A separate inquiry into education in Manchester in 1834-5 was undertaken by a paid agent, James Riddell Wood, written up by Samuel Greg, and presented by James Phillips Kay to a committee of the House of Commons.[66] This was the first of a series of reports undertaken by Wood on behalf of the Society into the state of education in Salford, Bury, Bolton, and Liverpool between 1835 and 1837.[67] There were also inquiries into the state of education in York in 1836–7 and Hull in 1840 where it

---

[62] Gatrell, 'Incorporation and the Pursuit of Liberal Hegemony', 28.   [63] Ibid., 23.

[64] 'On 4,102 Families of Working Men in Manchester' (1834) Mss. report, Manchester Statistical Society, Manchester Central Library.

[65] *Report of a Committee on the Condition of the Working Classes in an Extensive Manufacturing District in 1834, 1835, and 1836* (Manchester, 1838). See Wilkinson, 'On the Origin and History of the Manchester Statistical Society', 13.

[66] *On the State of Education in Manchester in 1834* (1835). See *Report from the Select Committee on Education in England and Wales: Together with the Minutes of Evidence, Appendix, and Index* (House of Commons, 1835).

[67] Wilkinson, 'On the Origin and History of the Manchester Statistical Society', 14.

was found that both towns had better day-school provision than Manchester (as was probably the intention in choosing them). The eight major investigations into schooling conducted up to 1840 'were, for the period, a model of careful research'. They were also 'totally condemnatory of the situation then existing'.[68]

These educational reports and findings were communicated annually to Section F of the British Association in stormy and controversial sessions, to the consternation of the Association's managers who no doubt remembered Adam Sedgwick's stern warnings against the politicization of science when the Statistical Section had supplicated for admission at Cambridge in 1833. William Langton presented the results of the 1835 educational survey of his city at the Manchester meeting of the British Association in the same year. In the following year at the Bristol Meeting, Langton, William Greg, and Henry Romilly, a Manchester merchant, 'continued their polemics on behalf of the Manchester Society' and called for the establishment of a national Board of Public Instruction.[69] In 1837, at Liverpool, delegations from the two cities engaged in 'municipal brawling' over the state of education in their respective cities and the Manchester men, setting forth the results of their surveys, repeated the call for state education overseen by a national Board:

> For the attainment, therefore, of this object, of which everyone in the present day will admit the paramount importance, what recourse is left but in the active agency of the Government? An agency which surely might be so conducted as in no degree to interfere with the spirit of British institutions. The task is certainly one of great magnitude and cannot fail to meet with both honest and interested opposition. But the country ought not, on this account, to shrink from it; and we feel persuaded that the establishment of a Board of Public Instruction would be hailed by all who have seen the glaring deficiencies of the present state of education as the first step in the performance of a duty which is imperative with every enlightened Government.[70]

Langton and Greg were admonished for making such a direct political appeal.[71] Without doubt 'the Mancunians were quite transparently using the Association, as well as Government committees, to further their social and political purposes'.[72] Following the 1837 Meeting, in October of that year, the Manchester Society for Promoting National Education was established with, at its core, Cobden and Mark

[68] P. H. Butterfield, 'The Educational Researches of the Manchester Statistical Society, 1830–1840', *British Journal of Educational Studies*, 22, 3, Oct. 1974, 340–59. Quotations at 344, 354.
[69] Jack Morrell and Arnold Thackray, *Gentlemen of Science. Early Years of the British Association for the Advancement of Science* (Oxford, 1981), 293–4.
[70] *Report of the Manchester Statistical Society on the State of Education in the Borough of Liverpool* (London, 1836), 42. Wilkinson, 'On the Origin and History of the Manchester Statistical Society', 15.
[71] Ibid., 15.   [72] Morrell and Thackray, *Gentlemen of Science*, 293.

Philips, who were founders, and Samuel Darbishire, Benjamin Heywood, and R. H. Greg who were life members, among others.[73] It was affiliated to the Central Society of Education, a national body founded in the previous year, which drew its model of non-religious schools from the United States. Released in June 1837, *The First Publication of the Central Society of Education* included a summary of the reports of the MSS on education.[74]

Out of 23 surveys conducted by the Manchester Statistical Society between 1834 and 1841, 11 were explicitly on some aspect of education and schooling.[75] Other surveys in addition to these included educational data. Only four surveys were 'on the condition of the working classes' in specific districts and towns, though the 1838 survey on the working class 'in an Extensive Manufacturing District' was an extensive piece of work for a small society to undertake, for sure.[76] Why was education the deliberate focus of the Manchester Statistical Society? First, because the neglect of children's schooling was a national scandal in itself; second, because its members believed that education would assist social stability, cohesion and right moral conduct in a city that was always changing, always in flux, and subject to severe cycles of boom and bust. An educated labour force, at least in the basics, would be a technically competent, hard-working, and more compliant labour force.[77] Through education James Phillips Kay looked forward to 'the rearing of hardy and intelligent workingmen, whose character and habits shall afford the largest amount of security to the property and order of the community'.[78] But the religious complexities of educational provision in Britain at this time were especially galling, whether to radical Anglicans like Cobden or to the ranks of Unitarians at the apex of Manchester's civic life, and they form a third and compelling reason for the Society's interest.

[73] S. E. Maltby, *Manchester and the Movement for National Elementary Education* (Manchester, 1918); Gatrell, 'Incorporation and the Pursuit of Liberal Hegemony', 45.

[74] 'Analysis of the Reports of the Committee of the Manchester Statistical Society on the State of Education in the boroughs of Manchester, Liverpool, and Salford and Bury' in *The First Publication of the Central Society of Education* (London, 1837).

[75] The educational surveys whether published or in manuscript were, in order: On the Number of Sunday Schools in Manchester and Salford (1834); On the State of Education in Manchester in 1834 (1835); On the State of Education in Bury (1835); On the Means of Religious Instruction in Manchester and Salford (1836); On the State of Education in Salford in 1835 (1836); On the State of Education in Liverpool in 1835–36 (1836); On the State of Education in Bolton (1837); On the State of Education in York, 1836–7 (1837); On the State of Education in Pendleton, 1838 (1839); On the State of Education in the County of Rutland in the year 1838 (1839); On the State of Education in Kingston-upon-Hull (1840). See 'Appendix C. List of Reports and Papers', Ashton, *Economic and Social Investigation*, 141.

[76] See ibid.: On 4,102 Families of Working Men in Manchester (1834); On the Condition of the Working Classes in an Extensive Manufacturing District in 1834, 1835, and 1836 (1838); On the Condition of the Population in Three Parishes in Rutlandshire (1839); On the Condition of the Working Classes of Kingston-upon-Hull (1841).

[77] Brian Simon, *Studies in the History of Education 1780–1870* (London, 1960) (1974 edn), 169.

[78] James Phillips Kay, 'On the Establishment of County or District Schools for the training of the Pauper Children Maintained in Union Workhouses, Pt.1', *JSSL*, 1, May 1838, 23.

For self-made men and their sons whose wealth came from industry and who were not members of the established church, the use of Anglican endowments and funds for the education of children in Church schools only, was an affront. The first grants of state funds for education, paid annually as from 1833 to two educational societies, one Anglican, the National Society, and the other nonconformist, the British and Foreign School Society, was a step in the right direction. But the available sums were small, English elementary education remained embarrassingly backward when compared to its provision in other nations of Europe and North America, and beyond the actualities, the *principle* of a nonsectarian system available to all children was the rational and equitable solution to a longstanding national disgrace. That the current situation discriminated against the interests of children from dissenting communities and districts was especially galling. On the other side, however, the Anglican establishment was jealous of its privileges and independence, generally unwilling (and sometimes legally unable) to share its educational endowments outside the Church, and for various reasons in the 1830s and beyond, hostile to the idea of state intervention in its affairs. The inquiries in that decade of Church Commissioners and Tithe Commissioners in related efforts to modernize the Church of England and make it more acceptable to a religiously-plural country had led many Anglicans to suspect the motives of Whig governments.

The Manchester Statistical Society was never likely to launch a campaign to improve working conditions and wages in the city's factories which were largely owned by its membership. Yet those same members, from their specific background, found a righteous cause, as they saw it, in breaking down the sectarianism that prevented the development of a genuine *system* of education. As rational, calculating and successful capitalists they believed in doing things systematically. The formation of the Manchester Society for Promoting National Education, emerging from the surveys of the Manchester Statistical Society, was just the start of a campaign that lasted until the Education Act of 1870 created some sort of national system of elementary education. Later, in 1848, the Lancashire Public Schools Association was formed in the city to campaign for free public education. Its Anglican opponents, conscious of its Unitarian roots, referred to its committee as 'the godless men of Cross Street' in reference to the chapel there.[79] Two years later, as it widened the campaign, it became the National Public Schools Association.

For decades, during the early and mid-Victorian periods, the mutual suspicion of these confessional communities, and their concerns that intervention by the state might favour one side over the other, prevented significant educational change and improvement. Even after 1870 the newly established structures still divided between denominational and secular schools funded and governed in

---

[79] *Manchester Guardian*, 19 Feb. 1848, 5, quoted in Butterfield, 'The Educational Researches of the Manchester Statistical Society', 355.

different ways. Though the MSS failed in its early years to establish a state-aided system of unsectarian elementary education, its programme was later adopted by a much more famous and ultimately more effective organization, the Birmingham Education League of the 1860s, which became the National Education League in 1869. In Joseph Chamberlain, it, too, was led by a Unitarian of great ability, from a provincial industrial background, who resented the privileges of the establishment and who ultimately became one of the dominant statesmen of the late-Victorian generation. According to a later, highly distinguished educationist, Michael Sadler (the great-great-nephew of the aforementioned factory reformer M. T. Sadler) who was Professor of Education at the University of Manchester during the Edwardian period, 'the work of the educational reformers in Manchester between 1840 and 1865, though it seemed to be a failure, was in fact the foundation on which the new system of English public elementary education was based.'[80]

Explaining why the Manchester Statistical Society went into temporary decline and effectively disappeared for a decade at the end of the 1830s, issuing no more annual reports after 1841, is much easier once its nature as a campaigning organization is evident. It has been argued that the Society was weakened by the election to Parliament of some of its leading members, the translation to the civil service of others, notably James Phillips Kay, and the engagement of many in the local and national campaigns of the Anti-Corn Law League in the years leading up to the repeal of tariffs on agricultural imports in 1846.[81] However, if we appreciate the nature of the Manchester Statistical Society correctly, these should not be taken as causes of decline or as examples of institutional failure, but the reverse. The MSS had been used by leading local figures to campaign for educational and social reform. As the issues changed and the campaigners moved on, so the institutional embodiment of those campaigns changed as well, and key figures devoted time to other organizations better adapted to immediate political struggles—from 1840 or thereabouts, to the Anti-Corn Law League. By the end of the 1840s the Manchester Statistical Society was almost moribund, a 'sinking patient', kept alive mainly by the exertions of Dr John Roberton, a Scottish physician who, in three decades, 1834–65, presented 27 papers to the Society.[82] Its dissolution was actually discussed at its annual meeting in 1849.[83] But the feeling was against winding-up, and the MSS was revived in the early 1850s: from 1853 it began to publish its own *Transactions* regularly. By this time the Manchester School had achieved some of its key objectives and was beginning to lose its influence over the nation,

---

[80] Michael Sadler, 'The Story of Education in Manchester', in W. H. Brindley (ed.), *The Soul of Manchester* (Manchester, 1929), 51. Roy Lowe, 'Sir Michael Ernest Sadler', 1861–1943, *ODNB*. On Sadler, see also Lawrence Goldman, *Dons and Workers. Oxford and Adult Education since 1850* (Oxford, 1995), 61–6, 78–81.

[81] Elesh, 'The Manchester Statistical Society: A Case Study of Discontinuity', 57–65.

[82] O'Brien, 'The Manchester Statistical Society. Civic Engagement in the 19th Century', 99–100.

[83] Wilkinson, 'On the Origin and History of the Manchester Statistical Society', 16–17.

a process made manifest when John Bright, with Cobden the other great leader of the Anti-Corn Law League and of the cause of 'repeal', lost his seat for the city in 1857 and found a new political home in Birmingham. Then it was timely for the MSS to revert to the model of a learned society, albeit one focused on the issues of most interest to its membership such as the statistics of banking and investment, credit and trade cycles, and other technical economic and financial subjects.[84] One third of the papers between 1853 and 1875 were on sanitary subjects as well, largely contributed by physicians.[85]

According to the first of its annual reports, issued in July 1834, 'The Manchester Statistical Society owes its origin to a strong desire felt by its projectors to assist in promoting the progress of social improvement in the manufacturing population by which they are surrounded.'[86] Some of the Society's later critics have expected 'social improvement' to include, almost exclusively, the reform of working conditions in the projectors' factories. It is an entirely understandable and laudable expectation, but it was only met over a long period of time because of successive interventions by the state in the form of the nineteenth-century Factory Acts. It did not come from the voluntary and willing actions of the membership of the Society. To them, however, it would appear that the term 'social improvement' implied something different: literacy, bible reading, arithmetic, as well as sanitary reform. To educate the children, moreover, required that the Society agitate for religious equality in law, in practice, and in the everyday conventions that governed the interaction of different communities. This, too, was a meaning of 'social improvement' in the 1830s.

There is no reason to doubt the sincerity of the commitment of the Manchester Society to popular education: it was shared across the early statistical movement as a whole among Whig, liberal and radical members of all shades who were united in the desire to use the state to provide decent elementary schooling. This explains the hostility of many of the leading figures to the conclusions of a study by the Parisian lawyer, André-Michel Guerry, published in 1833 under the title *Essai sur la Statistique Morale de la France*.[87] In an argument that threatened to undermine the case made by English educational reformers, Guerry contended that in France education and crime were in a direct rather than an inverse relationship. In districts where educational provision was most advanced, crime was highest. It did not help that Guerry deployed an array of statistical diagrams and tables that, technically, were in advance of anything yet published by the infant statistical movement in Britain.[88] From Manchester, W. R. Greg took up the

[84] Ibid., 20–1.   [85] Ibid., 19.
[86] Quoted in Ashton, *Economic and Social Investigation*, 13.
[87] Cullen, *The Statistical Movement*, 139–44.
[88] Anthony Oberschall, 'The Two Empirical Roots of Social Theory and the Probability Revolution' in Lorenz Krüger, Gerd Gigerenzer and Mary S. Morgan (eds), *The Probabilistic Revolution. Vol. 2: Ideas in Society* (Cambridge, Mass., 1987), 108–9, 113.

challenge. In a work on the social statistics of the Netherlands, read first to the British Association in Dublin in 1835, he contended that the high crime rates in areas of advanced education represented crimes of property against wealthier communities. But in these very districts, serious crimes of violence were less frequent than in educationally-deprived regions.[89] Then, in a paper read to the Statistical Society of London in December 1835, G. R. Porter tried to demonstrate Guerry's over-reliance on criminal statistics for a single year, 1831.[90] He also deployed the argument, which was echoed by other leading London statisticians like Rawson Rawson and Joseph Fletcher, that if Guerry were correct, French experience simply showed the requirement for high-quality elementary education: the crimes he had logged had been committed by the half-educated, not the well-educated. Poor education might not have deterred crime, but good education would: the argument with Guerry could be turned to the advantage of educational reformers by emphasizing the need for the best education possible.[91] The issue evidently bothered British statisticians for most of two decades, and it generated a large literature, almost all of which was designed to refute Guerry's conclusions. Manchester's commitment to education was shared across the movement. They were all good environmentalists.

The Society's legacy as an early sponsor of local surveys has won it a deserved place in the history of social investigation, even if, as is argued here, those surveys were used for a specific political purpose: to build the case for the related causes of religious and educational reform.[92] The MSS also nurtured one of the most famous social investigators of the era in James Phillips Kay, and Kay's work in turn was an influence on, and won the approval of, the man who left the most important of all studies of Manchester in this period, Friedrich Engels, author of *The Condition of the Working Class in England in 1844*. Engels corrected Kay for conflating 'the working-class in general with the factory workers', but *The Moral and Physical Condition of the Working Classes* was 'otherwise an excellent pamphlet' which he used for his own account of Manchester.[93] Kay was central to the statistical movement and the arc of his career, which took him from provincial medical practice into government, is an early example of a pattern that recurs throughout this book and this story. Engels, on the other hand, was an outsider

---

[89] W. R. Greg, *Social Statistics of the Netherlands* (Manchester, 1835).

[90] G. R. Porter, 'On the Connexion between Crime and Ignorance, as exhibited in Criminal Calendars', *Transactions of the Statistical Society of London*, I, 97–103.

[91] Rawson W. Rawson, 'An Enquiry into the Condition of Criminal Offenders in England and Wales, with respect to Education; or, Statistics of Education among the Criminal and General Population of England and other countries', *JSSL*, III, 1841, 331–52. Joseph Fletcher, 'Moral and Educational Statistics of England and Wales', *JSSL*, xii, 1849, 151–76, 188–335.

[92] Richard W. Selleck, 'The Manchester Statistical Society and the Foundation of Social Science Research', *Australian Educational Researcher*, 16, 1, 1989, 1–14.

[93] Friedrich Engels, *The Condition of the Working Class in England in 1844* (1845) (London, 1952 edn), 49.

who went to Manchester from his family home in Barmen, Westphalia, at the end of 1842 to work for his father's textile firm, Ermen and Engels.[94] Both were drawn to explain the effects of the industrial revolution in the city most associated with its birth. In Kay's case, and in light of the argument being advanced here, *The Moral and Physical Condition of the Working Classes* reads as if a guide to, and manifesto for, the views of the Manchester Statistical Society.

It was published in 1832 in the wake of that year's cholera epidemic in the city, and it was based on research generated in the attempt to understand and control the spread of the disease. A Board of Health had been established in each of the city's 14 police districts, each with two or more inspectors, who, armed with 'tabular queries' – a set of standard questions—visited every street and home to build a picture of the extent of the cholera and of local conditions generally.[95] Using these returns, as well as public documents and 'personal observation', Kay, who was secretary to the Board in 1832, in addition to his role as physician to the Ancoats and Ardwick dispensary, provided an account of Manchester which, in light of the foregoing analysis of the Manchester Statistical Society, reads like a guide to its ideas and campaigns rather than any sort of objective and unbiased study of the local working class. Kay evinced a mixture of sympathy, revulsion, and contempt for the labouring class, their habits, and their behaviours. He accepted that the long hours of repetitive labour depressed and weakened workers, but contended that they were paid a reasonable living wage.[96] He was critical of the choices they made: the worker 'lives in squalid wretchedness, on meagre food, and expends his superfluous gains on debauchery...the artisan has neither moral dignity nor intellectual nor organic strength to resist the seductions of appetite...he sinks into sensual sloth, or revels in more degrading licentiousness.'[97] Kay was especially and repeatedly harsh in his attitude to the Irish migrants who had come to work in Manchester's factories and whom he blamed for having 'demoralized' the local people.[98] This aspect of the pamphlet has generated the most criticism from modern historians.[99] Kay also lamented 'the absence of religious feeling' and religious observance among the factory workers and their fondness for drink.[100] But he would not blame their degraded life on liberal capitalism itself:

[94] W. O. Henderson, *The Life of Friedrich Engels* (2 vols, London, 1976).
[95] Kay, *The Moral and Physical Condition of the Working Classes*, 5–6.
[96] Ibid., 26–7.   [97] Ibid., 8–11.   [98] Ibid., 6–7, 50, 54.
[99] John Walton, *Lancashire. A Social History 1558–1939* (Manchester, 1987), 285; Frank Emmett, 'Classic Text Revisited. The Moral and Physical Condition of the Working Classes (1832), *Transactions of the Historic Society of Lancashire and Cheshire*, 162 (2013), 185–219; M. A. Busteed and R. I. Hodgson, 'Irish Migrant Responses to Urban Life in Early Nineteenth-Century Manchester', *The Geographical Journal*, 162 (1996), 139–53.
[100] Kay, *The Moral and Physical Condition of the Working Classes*, 34–9.

A system, which promotes the advance of civilization, and diffuses it over the world – which promises to maintain the peace of nations, by establishing a permanent international law, founded on the benefits of commercial association, cannot be inconsistent with the happiness of the *great mass of the people.*

The fault lay elsewhere with the distortion of free markets and free trade by protectionism: low wages and unemployment were the fault of tariffs which limited reciprocal trade. If 'the extent of the market for manufactures be diminished, the demand for labour will be confined within the same limits.'[101] The condition of the working classes could not improve 'until the burdens and restrictions of the commercial system are abolished'.[102] But there were mitigations to be made nonetheless, and the final section of the work set them out: improvements to the sanitary state of the city including better drainage and the repair of homes; 'provident associations and libraries' for the workers; the provision of services to the workforce on the model of the factory paternalism of Thomas Ashton of Hyde, who employed 1200 workers.[103] Kay also wanted to reduce the social distance between master and man and encouraged 'cordial association of the higher and lower orders' by meetings and visits to working-class homes to advise on domestic improvements.[104] But the most potent force for good was education: 'Above all [the master] should provide instruction for the children of his workpeople: he should stimulate the appetite for useful knowledge and supply it with appropriate food.'[105] If education were neglected and 'the higher classes [were] unwilling to diffuse intelligence among the lower' they would leave it free for others to fill the void 'who are ever ready to take advantage of their ignorance'.[106] Education was a civil, religious, and class responsibility, and also a national duty:

Ere the moral and physical condition of the operative can be much elevated, a general system of education must be introduced, not confined to the mere elementary rudiments of knowledge. He should be instructed in the nature of his domestic and social relations, of his political position in society, and of the moral and religious duties appropriate to it.[107]

Written a year before the MSS first met, by a man who was central to its establishment and to its first years of existence, *The Moral and Physical Condition of the Working Classes* is something of a blueprint for the Society's views and policies, and its objectivity, and hence its status as a classic of social investigation, have

[101] Ibid., 55.   [102] Ibid., 59.   [103] Ibid., 63–71.
[104] On this aspect on inter-class relations see Martin Hewitt, *Making Social Knowledge in the Victorian City. The Visiting Mode in Manchester 1832–1914* (Abingdon, Oxfordshire, 2020).
[105] Kay, *Moral and Physical Condition of the Working Classes*, 64.
[106] Ibid., 72.Trygve R. Tholfsen, *Sir James Kay-Shuttleworth on Popular Education* (London, 1974), 9.
[107] Kay, *The Moral and Physical Condition of the Working Classes*, 61.

been questioned.[108] The causes of immiseration are many; they are not simply the result of the factory system itself or of low wages; and they cannot be solved by the regulation of the workplace and other restrictive legislation, for that will only put the owners of businesses, and their workers also, out of work. There were certainly paternalistic duties which capital should perform for the good of labour, but the most potent solution to moral and physical decay, after free trade to expand the market overall, was the provision of education to civilize and moderate behaviour and to lay the basis for religion. When Kay called for 'a general system of education' he was foreshadowing the louder and more insistent calls of the Manchester Statistical Society a few years later.

If most of the active membership of the MSS was consumed from the end of the 1830s in the struggle for free trade, Kay, who had by then left Manchester, was enabled to pursue the educational reforms he and the Society had sketched out from the start. Thanks to Whig patronage, in 1839 he became assistant secretary, the leading official, of the newly established Education Committee of the Committee of the Privy Council. Kay's boss there, the Lord President of the Council, was none other than Lord Lansdowne, the leading Whig minister, the first president of the Statistical Society of London, and the father of the earl of Kerry who had visited the MSS in late 1833. Kay's job was to oversee the central expenditure on schools begun by the Whig administration in 1833 and he turned it into an opportunity to expand and improve education in general, to make it more nearly into that 'system' which he and his colleagues spoke and wrote about from the inception of the Manchester Statistical Society (and no doubt before that as well). In a decade at the Privy Council from 1839 to 1849 he established a model teacher-training college at Battersea; began the regular inspection of schools and improved their design; established a central education bureaucracy; and against the rigidities, rote-learning, and corporal punishment of early nineteenth-century pedagogy, he encouraged a more sympathetic, pupil-centred learning. Above all, though he himself had become an Anglican by this time, he braved the opposition of the established church (and sometimes that of the nonconformists, also) to his many attempts to sustain and expand the right of the state to control education and make policy for the good of all children. He set out his views in his 1847 pamphlet *The School, in its Relation to the State, the Church and the Congregation*.[109] As he wrote to his supporter and protector in this role, the prime minister, Russell, at the end of 1848,

Every hour from 1839 to the present period the Education Department has been subjected to the skirmishes of controversy, and the parliamentary struggles of 1839, 1842 and 1846 will be memorable in the annals of party conflict. Your

---

[108] Emmett, 'Classic Text Revisited', 207.
[109] R. J. W. Selleck, *James Kay-Shuttleworth: Journey of an Outsider* (London, 1995).

lordship is aware what part I have borne in the daily labours of the office in the preparation of parliamentary schemes, in the constant vigilance required for the direction of the public administration, and in the violent party struggles which have occurred.[110]

According to Matthew Arnold, another great educationist, though Kay was disliked and distrusted by Church and sects alike, 'Statesmen like Lord Lansdowne and Lord Russell appreciated him justly; they followed his suggestions, and founded upon them the public education of the people of this country.'[111] When Thomas Read Wilkinson presented his history of the Manchester Statistical Society he described Kay as 'foremost in the fight with ancient prejudices and to whom, more than to any single individual is due the success of the Committee of Council on Education.'[112] By the end of the 1840s the Society might have seemed moribund, but its two founding ideas, free trade and the extension of public education, had been very greatly advanced. If we see the Society for what it was in its first years, a pressure group to extend and secure the social and religious interests of Manchester's dissenting middle class, it was a very successful one indeed. Then, all passion spent, in the 1850s it could revert to what a statistical society might be expected to do and count bales of cotton, yards of cloth, and the savings in the local banks.

We have seen, therefore, the great differences between rival and sometimes clashing conceptions of statistics in different locations. In Cambridge, statistics amounted to a natural scientific methodology that would change the way in which science and economics were studied. In London, statistics were the basis for enhanced social administration, a procedure and a discipline to make government more expert. In Manchester, statistics denoted a sectarian and class-based agenda of issues and ideas which was delivered successfully in the course of the 1840s by a mass campaign for free trade on the one hand, and, on the other, by the placing of one of the founders of the Society in a position of influence over national education at the heart of government. But there was also a statistical society in Clerkenwell, London, some distance, both geographically and metaphorically, from Westminster and Whitehall. Who had established it, and why?

---

[110] James Phillips Kay to Lord John Russell, 20 Dec. 1848, quoted in Frank Smith, *The Life and Work of Sir James Kay-Shuttleworth* (1923) (1974 edn, Trowbridge, Wilts), 216–17.
[111] Matthew Arnold, 'Schools', in T. Humphry Ward (ed.), *The Reign of Queen Victoria: A Survey of Fifty Years of Progress* (2 vols, London, 1887), ii, 240.
[112] Wilkinson, 'On the Origin and History of the Manchester Statistical Society', 18.

4

Clerkenwell

The London Statistical Society and Artisan Statisticians, 1825–30

In 1825, a small volume was published in London entitled *Statistical Illustrations of the Territorial Extent and Population, Commerce, Taxation, Consumption, Insolvency, Pauperism and Crime of the British Empire. Compiled for and Published by Order of the London Statistical Society*. By 1827 it had gone through two further editions. This chapter is an attempt to track down and identify the compilers of the *Statistical Illustrations* and relate their work to the wider statistical movement. That a body calling itself the London Statistical Society predated by almost a decade the formation of the much more famous Statistical Society of London, founded in 1834, may be of interest. That it was organized not by academics, savants, entrepreneurs, civil servants, and politicians, but by artisans and working men in order to collect data to demonstrate that the working classes had recently endured social and economic immiseration, makes it worthy of research and historical reflection. The *Statistical Illustrations* was one of a series of publications between 1819 and 1827 that can be linked to a national network of artisans, organized in 'district committees', who were protesting against the fiscal and political arrangements of the British state in the post-Napoleonic period.[1]

The slim volume, originally published in both quarto and pocket-sized editions, was noted in the centenary history of the Royal Statistical Society, published in 1934, but as a curiosity only: the authors did not think it worth finding out more about this earlier statistical organization.[2] They did comment, however, that 'it seems to have been the private adventure of men of small means'.[3] One more

---

[1] *Select Committee of Artisans, 1823* (London, 1824), 5.
[2] They noted that 'it was compiled by a Committee of Artisans, probably of Owenite leanings, and the tables seem quite honestly handled, though they betray a tendency to select and accentuate the gloomiest facts and figures of that depressed time'. J. Bonar and H. W. Macrosty, *Annals of the Royal Statistical Society, 1834–1934* (London, 1934), 3. In the 1980s I discovered the copy of the work they had probably encountered themselves in a cupboard in the Royal Statistical Society in London which then held the society's archives.
[3] Ibid.

Victorians and Numbers: Statistics and Society in Nineteenth Century Britain. Lawrence Goldman, Oxford University Press.
© Lawrence Goldman 2022. DOI: 10.1093/oso/9780192847744.003.0004

recent historian has dismissed the London Statistical Society as 'non-existent'.[4] Another contended that there was 'no evidence that any such society actually existed' and relegated it to a footnote.[5] But that is to take a literal approach to the definition of a 'society' and to overlook thereby the existence in this period of a network of artisan researchers, a statistical counter-culture in fact, which was in critical dialogue with the dominant forms of knowledge, and which must also be found its place in the history of the origins of the statistical movement.

It is not easy to get at the history of this artisan network and of its relationship to other working-class societies and publications, however, and a degree of caution about the level of organization behind the various interconnected publications is certainly required.[6] Nevertheless, in the words of the preface to the first edition of the *Statistical Illustrations*,

> The privation and misery endured by the productive classes of society in Great Britain in 1816 and 1817, led to the formation of an Association in London for the purpose of investigating the nature and extent of that misery; and of ascertaining, if possible, how far it resulted from avoidable or from unavoidable causes; and how far repetitions of similar ills were likely or not to occur.[7]

The network seems to have established its own so-called 'Select Committee' 'to inquire into the causes which have led to the extensive Depreciation, or Reduction in the Remuneration of Labour', which reported in 1824. The report, first presented at a public meeting in Bolton, Lancashire, was based on the research of 'District Committees' of working men who collected data on wages and social conditions which they forwarded to the Select Committee as the raw data for its Report.[8] The Select Committee called for the formation of a 'permanent committee...in

---

[4] M. J. Cullen, *The Statistical Movement in Early Victorian Britain. The Foundations of Empirical Social Research* (Hassocks, Sussex, 1975), 21. Cullen, as well as some university libraries, have ascribed the *Statistical Illustrations* to John Marshall, another statistical writer of this period. Marshall's name does not appear in the work and he is nowhere referenced by the artisans, whereas each edition in turn makes it clear that it is the work of a committee of artisans.

[5] According to Victor Hilts, 'The third edition of a volume published in 1827 and entitled *Statistical Illustrations....of the British Empire* is ascribed to a Statistical Society of London [sic], but there is no evidence that any such society actually existed.' In fact, it is ascribed to 'the London Statistical Society'. Victor Hilts, '*Aliis exterendum*, or, the Origins of the Statistical Society of London', *Isis*, 1978, vol. 69, 30n.

[6] In correspondence with the historian who knew most about this radical artisan milieu, Iowerth Prothero of the University of Manchester suggested that the London Statistical Society might have been linked to the Society for the Encouragement of Industry and Reduction of the Poor Rate which met at the King's Head, Poultry, in the 1820s. The latter society was connected with the General Association attended by Gast and Powell in 1827. Iowerth Prothero, letter to the author, 7 April 1981.

[7] *Statistical Illustrations* (1825 edn), iii.

[8] 'That the foregoing Resolutions be printed, and a copy of them transmitted to each of the District Committees of Artisans throughout the Kingdom'. See *Select Committee of Artisans* (1824), Resolution 16, p. 4. As the *Select Committee* explained, 'The Artisans of Great Britain, having deputed a Select Committee to inquire into the causes which have led to the extensive Depreciation, or Reduction in the Remuneration for Labour, and the extreme privation and calamitous distress consequent thereupon,

London, as the focus of information and intelligence from all parts of the Kingdom'.[9] It is possible that the body called the London Statistical Society was this 'permanent committee', several years before anyone else, whether in the capital or in provincial Britain, thought of founding statistical societies. Whatever the institutional basis of what must remain a shadowy movement, the motive was caught clearly in the observation in 1826 by the network's most significant known figure, John Powell of Clerkenwell, London, that 'there are many working and intelligent men among the productive classes, who reflect much, but do not know how to come to the truth... for want of conclusive facts to guide them in their inquiry'.[10]

It was believed that 'a collection of naked facts'[11] would speak for themselves: 'they can only repeat that they have been actuated by no motive or desire but that of the exhibition of facts'.[12] As they wrote in 1825, 'the object has been to exhibit facts, and such authentic data, as could be obtained, for the purpose of enabling the enquirer in Social Economy to exercise his own judgment and draw his own conclusions'.[13] The facts were gleaned from the 'examination of every authentic document which had been presented to the British Parliament during the last forty years'.[14] More up-to-date information—for example, about cotton consumption and corn importation—was added to each of the successive editions of the *Statistical Illustrations* in 1826 and 1827 respectively. An 'Appendix' was also published in 1826.[15] Each version began with an explanatory preface which remained essentially the same. This gave little information about the network responsible for the research and publication, but provides some insight into its purpose and views. For example, the authors took issue with the claim that because parochial assessments to pay for poor relief had declined since 1817, this was evidence of the growing prosperity of the people. It was rather, they argued, evidence of declining contributions made to paupers under the poor law. Despite the evidence that in the 1820s 'the means of comfort and of social order have multiplied around them', nevertheless 'Pauperism, the consequence of Poverty, and its

the Select Committee have at length concluded their labours, and drawn up a Report, which was read at a public meeting, held at Bolton, on Tuesday the 13th of January, 1824...'

[9] Ibid., Resolution 14.

[10] John Powell, *An Analytical Exposition of the Erroneous Principles and Ruinous Consequences of the Financial and Commercial Systems of Great Britain. Illustrative of their Influence on the Physical, Social and Moral Condition of the People. Founded on the Statistical Illustrations of the British Empire* (London, 1826), 8.

[11] *Statistical Illustrations* (1825 edn), unpaginated advertisement at end. Note that there are slight differences between the different editions of the *Statistical Illustrations*, and also between copies of the same editions held in the British Library, the Cambridge University Library, the Goldsmith's Library of the University of London, and the Royal Statistical Society, where this research was conducted. Some of the most revealing commentary occurs in advertisements and prefaces which appear in some editions and not in others.

[12] *Statistical Illustrations* (1825 edn), iv.   [13] Ibid., xiv.   [14] Ibid., iii.

[15] *Appendix to the First Edition of the Statistical Illustrations* (London, 1826).

inseparable concomitant Crime, with all its train of disorder, has increased in a corresponding ratio.'[16]

Beyond the preface, however, the *Statistical Illustrations* offered little running commentary on the state of the nation, few interpretive asides or explanatory footnotes: it was essentially a compilation of statistical data across page after page, tables of figures which apparently told in themselves, without discussion, a story of increasing immiseration. One can sense the unspoken desire to correlate rising crime with falling incomes, and rising pauperism with increased taxation. But before the advent of techniques for correlation developed in the late nineteenth century, the artisans could do no more than demonstrate apparently related trends in the public and social accounts. They lacked the means to build connections between different variables and to undertake multivariate analysis. Decades later, from the 1890s, correlation would 'be applied routinely…to a variety of social problems, such as the relations among crime, poverty, ignorance, alcohol, and poor health'.[17] Nevertheless, merely presenting the data in tabular form would, it was hoped, be enough. Among those workers able 'to think, to decide, and to act for themselves' it was believed to be

> unnecessary to offer any further remarks, either of illustration or of proof, the extensive range of facts in themselves will suffice to enable them to draw their own conclusions, whilst to the pretending and time-serving politicians of today, all efforts to instruct and to convince are a mere waste of time.[18]

Some years before this, the banker Samuel Tertius Galton had also made an attempt at multivariate analysis, though in his case the indicators were financial rather than economic and social. His pamphlet, *A chart exhibiting the relation between the amount of Bank of England notes in circulation, the rate of Foreign Exchanges, and the Prices of Gold and Silver Bullion and of Wheat, accompanied with explanatory observations*, dealt with the great fiscal and currency questions at the end of the Napoleonic wars. It demonstrated in its very title the urgency of the quest for a means of calculating precisely, rather than merely charting, the association between variables.[19] The quest would end with the invention of the correlation coefficient by Samuel Tertius's son, Francis Galton, in the 1880s.

In explaining the London Statistical Society and its relationship to the history of British social statistics more generally, the place to begin is with a man called John Powell, a watchmaker who lived in Rosoman Street, Clerkenwell. This was

[16] *Statistical Illustrations* (1825 edn), xv.
[17] Theodore Porter, *The Rise of Statistical Thinking 1820–1900* (Princeton N.J., 1986), 296.
[18] *Statistical Illustrations*, xviii.
[19] Samuel Tertius Galton, *A chart exhibiting the relation between the amount of Bank of England notes in circulation, the rate of Foreign Exchanges, and the Prices of Gold and Silver Bullion and of Wheat, accompanied with explanatory observations* (London, 1813).

close to Spa Fields where, at the end of 1816, two famous meetings of artisans and their political leaders to petition the government for economic relief, had begun several years of radical agitation and repression in Britain.[20] Powell was more involved with these various publications than anyone else and has been credited as leader of the network and the editor of the *Statistical Illustrations*.[21] It had proved impossible to unionize the watchmakers, who suffered 'destitution' at the end of the Napoleonic Wars, and who experienced declining incomes and craft status during the period as a whole, and Powell became one of their advocates.[22] He was active also among the Spitalfield silk weavers in the 1820s and was one of the key figures behind the *Trades' Newspaper and Mechanics' Weekly Journal*, the world's first trade union paper. This had emerged from the campaign in the mid-1820s to prevent the repeal of legislation passed in 1824 to legalize trade unions in Britain. Writing in the *Trades' Newspaper*, Powell 'developed in a striking way' a critique of orthodox political economy and of the policies built upon it.[23] He condemned free trade and the reduction of society 'to a system of purely monetary relationships', what Thomas Carlyle would later term 'the cash nexus'. This, Powell argued, had led to stark social divisions between rich and poor, and to a decline in the standard of living of the majority.[24] In a critical essay published in 1826 and specifically 'founded on the *Statistical Illustrations of the British Empire*', Powell condemned:

> the incontrovertible tendency of the existing order of society, and of the doctrine called political economy, to produce a convergence of money influence, or despotism, which is the worst of all despotisms, as it weakens the bonds of society, by reducing the affections of human nature to a money value…increasing wealth in few hands, and extending privation among greater numbers, the unerring indicators of national decay according to all experience and all history.[25]

In the context of the statistical movement as a whole, Powell's vindication of the 'immutable laws of social economy' against the doctrines of political economy is of more than casual significance. He is last traced delivering a speech on

---

[20] I. J. Prothero, *Artisans and Politics in Early Nineteenth Century London. John Gast and his Times* (London, 1979), 224.

[21] Stanley H. Palmer, *Economic Arithmetic. A Guide to the Statistical Sources of English Commerce, Industry, and Finance 1700–1850* (New York, 1977), 45–6.

[22] Prothero, *Artisans and Politics*, 65.

[23] Prothero, *Artisans and Politics*, 224–5. See the following signed articles by Powell in the *Trades' Newspaper* in 1826: 5 Feb., 475; 6 Aug, 25; 27 Aug, 49; 3 Sept., 57; 17 Dec., 182. See also, Powell, *A Letter addressed to Edward Ellice esq., M.P.*, et seq.

[24] Prothero, *Artisans and Politics*, 224–5.

[25] Powell, *An Analytical Exposition*, quoted in Prothero, *Artisans and Politics*, 225.

parliamentary reform to the Clerkenwell Reform Union in September 1831.[26] He was also a member at this time of the National Political Union, a radical organization campaigning for parliamentary reform.[27]

Another way of situating and understanding John Powell is to see him as the close associate in the 1820s of John Gast, the Thames shipwright, who, as Iowerth Prothero demonstrated in 1979, was at the heart of successive artisans' movements from the 1790s to the 1830s and became the dominant figure among organized workers in London in this era.[28] To the historian Edward Thompson, Gast was 'one of the...truly impressive trade union leaders who emerged in these early years'.[29] According to Peter Linebaugh, he was 'the most indefatigable trade unionist and artisanal organizer during the stormy first third of the nineteenth century'.[30] Gast was born in Bristol in 1772 and was apprenticed as a shipwright in the city. As he wrote later, 'I am a working shipwright and one who has contributed towards erecting those best bulwarks of the nation – her wooden walls – which is the principal safeguard of our country'.[31] Moving to London, he organized a famous strike of Thames shipwrights in 1802. He spent 28 years in one Deptford yard on the Thames, in charge of a group of 16 or so shipwrights. Gast was a leading official of the Hearts of Oak benefit society, formed after the 1802 strike, and the first secretary of the Thames Shipwrights' Provident Union, established in 1824. He was a Deist and lay preacher who would commonly berate working men from the pulpit for their fondness for 'the pot'. Instead, he preached the virtues of 'skill, hard work, sobriety and thrift'. He was influenced by Paine's ideas and was a supporter of Henry 'Orator' Hunt, and thus of universal manhood suffrage.[32] In 1818 he led the formation of the Philanthropic Hercules, an attempt to construct a general union of all trades.[33] Unifying the trades to increase the leverage of working men was one of his strongest instincts and the key to much of his political career. As Gast explained somewhat later,

> He wanted to see, not merely isolated unions in several trades, but a general union made up out of the whole. In 1818, he proposed a society which he had

[26] John Powell, *Plain Reasons for Parliamentary Reform, in familiar letters to a friend; Being the substance of a speech delivered at a meeting of the Clerkenwell Reform Union, September 21, 1831* (London, 1831).
[27] John Powell should not be confused with Thomas Powell, an associate at this time of another leader of London radicalism, Francis Place.
[28] For a different view of Gast, see Peter Linebaugh, 'Labour History without the Labour Process: A Note on John Gast and His Times', *Social History*, 7, 3, Oct. 1982, 319–28.
[29] E. P. Thompson, *The Making of the English Working Class 1780–1830* (1963), 774.
[30] Linebaugh, 'Labour History without the Labour Process', 319.
[31] *Independent Whig*, 18 Sept. 1808, 725, quoted in Prothero, *Artisans and Politics*, 18.
[32] Prothero, *Artisans and Politics*, 331–2.
[33] J. L. Hammond and Barbara Hammond, *The Town Labourer 1760–1832* (London and New York, 1917), 311; G. D. H. Cole, 'A Study in British Trade Union History. Attempts at a "General Union" 1829–1834', *International Review for Social History*, 4, 1938, 359–462.

called the Philanthropic Hercules; it being, in fact, a Parliament of working men. He proposed that each trade should elect delegates, in proportion to its extent; and that the delegates thus elected should constitute a working man's Parliament.[34]

He was one of the leaders of a new Thames shipwrights' union which organized another strike in 1825, though this one failed, and he became the manager of the *Trades' Newspaper*. Gast composed the founding document for the paper, a call to intellectual arms, for only by education, information, and reason would the workers prevail:

A new and important era has commenced in the history of our class of society. We have become universally weary of the state of ignorance in which we have been so long sunk; we are grown sensible of the inefficacy of all attempts to better our condition, either extensively or permanently, until we better ourselves; we have begun to call up the power of our *minds* to assist in the improvement of our humble lot, and to cultivate and improve these by all possible means; and now it may be safely predicted that the day is close at hand when by Reason alone, we shall assert, successfully, our rights and interests at the bar of Public Opinion.[35]

Gast was the founder also of the General Association in February 1827, another short-lived attempt at a union of all trades.[36] He lived long enough to take a hand in 1836, a year before he died, in the formation of the London Working Men's Association which drafted the Six Points of the People's Charter, the foundational text of the Chartist movement.[37]

The aim of the London Statistical Society was to raise political awareness by providing the information from which a new consciousness might develop:

[to] awaken the attention of the public to the calamity that awaits them, so as to induce them to seek for the maintenance of social order in the only course that can possibly supply the desired desideratum, the examination of facts, and thereby be led to influence the direction of a system of social order, founded on permanency and certainty, instead of expediency and speculation...[38]

Powell joked that 'the conductors of the press' if they deigned to consult his compilation, would 'undeceive themselves'. '*Statistical Illustrations* are fatal to mere

---

[34] *The Constitutional*, 25 Nov. 1836, p. 7, quoted in Prothero, *Artisans and Politics*, 102.

[35] 'Laws and Regulations of the Trades' Newspaper Association', *Trades' Newspaper*, 17 July 1825, p. 1, cited in Prothero, *Artisans and Politics*, 186.

[36] *Trades' Newspaper*, 25 Feb. 1827, vol. ii, no. 85, 259; Prothero, *Artisans and Politics*, 216, 225.

[37] Prothero, *Artisans and Politics*, passim. Iowerth Prothero, 'Gast, John (c. 1772–1837), trade unionist and radical', *ODNB*. See John Gast's autobiographical article in the *Trades' Newspaper*, 31 July 1825.

[38] *Statistical Illustrations* (3rd edn, 1827), 'Introduction to the Third Edition', vi.

speechification' and destroyed 'the jargon of opinions'.[39] When the General Association was founded by Gast and his associates, it placed great emphasis on collecting and analysing information: 'the Committee offer an earnest invitation to the industrious classes, to study the questions which concern their most important interests'[40] Powell prefaced one of his essays with Hosea's lamentation: 'the people are destroyed by lack of knowledge'.[41]

The various publications were not, in fact, based on original and fresh research but used data compiled from other published sources, especially parliamentary returns. These included the reports of the Finance Committee of the House of Commons, created in 1797 to investigate the public accounts, and then, from 1800, the annual accounts presented to Parliament. There was much use of the 1821 census, as well. The artisans complained of the difficulty of finding reliable information: 'such is the prevalence of pretension and speculation, and utter disregard to matters of fact, that no person can imagine the difficulty of obtaining worthy information but by experience of the task'.[42] This same rationale was used by their social superiors in justifying the foundation of the Statistical Society of London at its initial meeting in 1834, nine years later. In a section on demography added to the third edition of the *Statistical Illustrations*, the compilers called for a system of civil registration fully a decade before it was introduced:

As correct information of the numbers and relations of a community is essentially necessary to its complete social organisation, it is much to be regretted that some measure is not devised and put into practice, to render certain, a correct annual account of the Marriages, Births and Burials throughout the whole British territories. No stronger proof need be adduced of the vain pretension to legislative attainment of the English People than their imperfection in, and indifference to accurate information on this elemental principle of social organisation.[43]

Given the inadequacies of published data, the artisans consulted and used the library of M. Cesar Moreau, the French vice-consul in London, and the compiler of works in the 1820s on the commerce of Britain since the founding of the Bank of England in the 1690s; on Indian statistics; on the history of the British silk trade; and on the history of British and Irish manufactures and exports.[44] The

[39] Powell, *An Analytical Exposition*, 40.

[40] *A Narrative and Exposition of the Origin, Progress, Principles, Objects etc. of the General Association, Established in London, for the Purpose of Bettering the Condition of the Manufacturing and Agricultural Labourers* (London, 1827), 20.

[41] Hosea, iv, 6.   [42] *Statistical Illustrations* (1825 edn), 69.

[43] *Statistical Illustrations* (3rd edn 1827), 163n.

[44] *Statistical Illustrations* (3rd edn 1827), 'Introduction to the Third Edition', vii. See also *The Trades' Newspaper*, 5 Feb. 1826, 473n: 'We have to thank M. Moreau for his valuable communications'. On Cesar Moreau, see [G. Vapereau], *Dictionnaire Universel des Contemporains* (Paris, 1858), 1246–7. He is not to be confused with Alexandre Moreau de Jonnès, director of the newly-established Statistique Générale de la France, the French central statistical bureau, between 1833 and 1852.

third edition of the *Statistical Illustrations* is dedicated to Moreau who, on his return to France, founded both the Sociètè Française de Statistique Universelle and L'Académie de L'Industrie, Agricole, Manufactière et Commerciale.[45] The title of the latter organisation advertises the fact that Moreau was a Saint-Simonian, and the Saint-Simonians were to influence members of the Cambridge network a short time later in the early 1830s.[46] Because Moreau lived and worked in the official, rather than the counter-culture, he is one of the few figures involved with the *Statistical Illustrations* of whom anything is known in reliable detail. In London he was a member of the Royal Institution, the Royal Asiatic Society, and the Society of Arts, as well as similar societies in France. He alone seems to have offered the artisans help and was the 'one exception' to an otherwise dispiriting experience. The Society had

> not only met with nothing in the shape of patronage, either pecuniary or cheeri-ness of encouragement, but on the contrary, almost every effort to ascertain the accuracy of positions, have been thwarted by innumerable obstacles; and, in many cases, coldness and reserve have been experienced, where approbation and encouragement were most expected, and most likely to have been met with, merely because truth and demonstration exposed the fallacy of preconceived notions and long entertained opinions.[47]

What did the artisans believe in? What was the programme of this avowedly campaigning organization? Like so much of the radical economics of this era, it started with a version of the labour theory of value: 'all accumulation resolves itself, not into any accession of National Wealth, but into an abstraction from the fair and just reward due to the labour and skill applied in production, at the sacri-fice of privation of consumption.'[48] The transfer from workers to other groups was largely achieved through ruinous and unfair taxation. In the *Select Committee of Artisans* of 1824,

> it is shewn, that, as Taxation progressively increased, the remuneration of Labour as progressively decreased, and that the aggregate increase of Taxation, corre-sponds exactly with the aggregate decrease in the Remuneration of Labour, and proves to Demonstration, that not only direct Taxation, but that all subsistence,

[45] Cesar Moreau to L. A. J. Quetelet, 15 July 1832, 29 May 1834, Quetelet papers, Academie Royale de Belgique, Brussels, file 1843. Cesar Moreau to Henry, Lord Brougham, inviting Brougham to become an honorary member of the Sociètè Française de Statistique Universelle, Brougham papers, University College, London, 2 Nov. 1834, B. Mss 41336.

[46] See below, Ch. 6, pp. 133–5.

[47] *Statistical Illustrations*, 3rd edn, 1827, 'Introduction to the Third Edition', vi.

[48] Ibid., 30n.

and all Income acquired by any other means than productive occupation, resolve themselves into an abstraction from the fair reward due to productive labour.[49]

This was a familiar theme in working-class protest, evident in the works of John Wade, Henry Hunt, and William Cobbett among many. It was commonly complained that the tax structure which had evolved since the Napoleonic wars had transferred vast sums from the people to the parasites. As Powell put it in 1824, 'Taxation, monopoly, corn laws, machinery, commercial avarice and money jobbing, prey upon labour with raven appetite.'[50] Like many radical groups, they objected strongly to the expansion of credit under the paper regime used to fund the Napoleonic Wars and then its deliberate contraction at the Wars' end in the decision to cease cash payments and go back to the gold standard: 'the whole range of history furnishes no example of distress and misery so extensive, general and severe, as that experienced by the people of the United Kingdom of Great Britain and Ireland with but little intermission from the period of 1816 to 1822.'[51]

The exploitation of the workers was apparently proven in the statistics of British trade. One table in the *Statistical Illustrations* tried to demonstrate that the quantity of exports since 1797 had doubled, but that of imports had remained stable, while wages had fallen. Where was the reward for labour in this?[52] According to the *Statistical Illustrations*,

> While the external distribution of the products of British industry have more than doubled, notwithstanding an increase of more than one third in the number of consumers, the commodities received in exchange, calculated to add to the comforts of the people, and to reward them for the increase of toil, to which they have been subjected (after allowing for the increase of raw materials of manufacture worked up for re-exportation) have rather decreased than otherwise.[53]

In other words, labourers' wages had been driven down to make British exports affordable overseas, and workers could not themselves afford to purchase the

---

[49] Ibid., 29.

[50] John Powell, Watchmaker: *A Letter Addressed to Weavers, Shopkeepers, and Publicans, on the Great Value of the Principle of the Spitalfields Acts: in Opposition to the Absurd and Mischievous Doctrines of the Advocates for their Repeal* (London, 1824), 6.

[51] *Statistical Illustrations* (1825 edn), xvii.

[52] 'Table (P) showing the Increased Quantity of British Produce and Manufactures Exported in each year since 1797, and their Depreciation in Value since 1807, at the Expense of the Artizan and Labourer, in the Reduction of their Wages without any corresponding Equivalent; and the stationary Quantity of Imports as a consequence of the bulk of the People being precluded from consuming them', *Statistical Illustrations*, 36. This table was also published in *The Select Committee of Artisans, 1823*, at page 16, showing the degree of overlap between the various organisations and publications in the artisan network behind the London Statistical Society.

[53] Ibid., xv. It was later asserted that 'the quantity of colonial and foreign productions imported and retained for home consumption, in the four years 1819–22 has actually been less than it was in the four years 1798–1801', *Statistical Illustrations* (1825 edn), 36.

imports coming back the other way. In what we might term an 'export drive', 'the furious spirit of mistaken predilection for foreign trading has triumphed over, and subverted all regard for Domestic and Social Comfort'.[54] According to John Powell,

> Should the merchant buy foreign productions with the money for which he sells his goods and bring them home, he cannot find a market for them. He left so little to be distributed among the producers of the commodities he took out, that they have nothing left to buy what he brings in return, so that to effect a market abroad, he destroys his customers at home.[55]

In one of his few extant letters, John Gast wrote to Francis Place along these lines:

> Since the Peace trade and Commerce have fallen off, and additional population have been thrown into the market of Labour; foreigners meet Englishmen in the field of competition, and heavy taxation forces down the price of Labour, to enable Tradesmen and the manufacturer to meet the foreign markets, all sorts of schemes is had recourse to, to cheapen the Articles for exportation, and the working man [h]as been the greatest sufferer.[56]

This was demonstrated in the *Statistical Illustrations* in a table designed to show 'the vast depreciation' in the value of goods exported in 1823 and 1824 compared with their value a decade before. The consequence of the destruction of domestic demand was a rise in pauperism, a key theme running through all these publications. If, in these years, the gentry petitioned government for help with growing numbers of paupers, and with higher and higher poor rates, so the artisans complained about the increasing degradation of so many. As the *Statistical Illustrations* averred,

> If in addition to the vast increase of the Money Amount of Pauper Relief since the middle of the last Century, the increasing privation of the relieved, as a consequence of the diminished quantum of relief, be taken into account, the above Statement may be considered as exhibiting a picture of degradation without parallel in the history of society.[57]

Another of the trends that attracted the opprobrium of the artisans was rising crime, a consequence of growing misery, 'the committals for crime having [been] quadrupled in less than twenty years, the number being in 1805, 5600; and in 1823, 22,099'.[58] One of the tables compiled for the *Statistical Illustrations* put

---

[54] Ibid., 60.   [55] Powell, *An Analytical Exposition*, 20.
[56] John Gast to Francis Place, 3 July 1834, Place Papers, British Library, Add Ms. 27,829, f. 19.
[57] *Statistical Illustrations*, 81n.   [58] Powell, *An Analytical Exposition*, 29.

together the annual disbursements for poor relief for each county in England and
Wales with annual average committals for crime in each county over two periods,
1805–15 and 1816–25. A calculation of the number of annual committals out of
every 10,000 of the total population, broken down by county, was added for good
measure. The aim was to show a positive correlation between crime and pauperism and hence between rising crime and declining social conditions. In the event,
Middlesex and Lancashire showed the highest number of criminals committed,
and the largest amount of poor relief spent, though there are many ways of
explaining this: any relationship between crime and pauperism was bound to be
complex and subject to a wide range of local factors. In the absence of techniques
for calculating and demonstrating correlations between different social variables,
no proof of a direct relationship could be adduced. But by tabulating together the
data for crime and pauperism, the compilers of the *Statistical Illustrations* were
intending to make a clear inference.[59]

As the third edition of the *Statistical Illustrations* expressed it, theirs was
'a System founded in pretension, expediency and Speculation, the inherent and
indubitable consequences of which are Collusion, Chicane, Insolvency, Pauperism
and Crime, or in one sentence "social disorganisation".'[60] The expectation that society
be both organized and systematic suggests, in the use of an Owenite vocabulary at
this point, the underlying assumption that the 'derangement and perversion of the
several interests of the country' as it was referred to in 1824, was remediable.[61]
Powell had cited Robert Owen openly, approvingly, and at length, in previous
works, as well.[62] As E. P. Thompson explained, the artisans learned from Owenism
'to see capitalism, not as a collection of discrete events, but as a system'.[63]

Taken as a whole, the *Statistical Illustrations* and the literature associated with
it focused in many different ways on the gross contradictions of national life,
especially the transformation and massive expansion of productive capacity in
Britain but also the maldistribution of the proceeds of that transformation. Centred
as the artisans were on London, they nevertheless recognized in the cotton industry
the evidence of this contradiction at its most egregious. The industry's remarkable
development 'exhibits features without any parallel in the History of human
affairs', but its social effects were 'as discreditable to the talent and integrity of the
National Administration, and to the House of Commons, as it is lamentable on
the score of political and social derangement, and moral degradation'.[64]

---

[59] *Appendix to the First Edition of the Statistical Illustrations* (London, 1826), 46.
[60] *Statistical Illustrations* (pocket edn, 1825), 170.
[61] The phrase is drawn from the *Select Committee of Artisans*, proposition 10.
[62] *A Letter Addressed to Edward Ellice, esq., M.P,* 7–9. 'Though some of the theories of Mr. Owen, of
New Lanark, have been deemed objectionable, he has evidently laboured with great earnestness and
sincerity for the public good...'
[63] Thompson, *The Making of the English Working Class*, 806.
[64] *Statistical Illustrations* (pocket edn, 1825) 79n.

Their solution was a paternalistic and regulatory state. Powell published an open letter in 1819 to Edward Ellice, the Radical-Whig MP for Coventry, on the then ills of the apprenticeship system, drawn largely from his experience as a watchmaker and from the state of that trade in London and Coventry. It included, as well, 'remarks on the prevailing theories on freedom of trade and the justice and policy of Regulation'. He argued that

> the opinion that unlimited freedom of action is necessary in trade is obviously fallacious; as no man can look into the world or his family without being convinced that *some* regulation is necessary. It is by means of the regulation of parts and of the whole, that a people are progressively advanced from a state of barbarism to a state of civilisation.[65]

It was unjust that the state regulated the price of grain to the advantage of the land owners, argued Powell, but would not intervene in areas of vital interest to labouring men.[66] Such a policy was not merely immoral but 'by leaving the working classes unprotected and proportionately unprovided for, it causes immense parish burthens...while it also increases the number of offenders beyond all parallel in any age or country'.[67] Again we see the link made between low wages, unemployment, pauperism, and crime.[68]

Powell and his collaborators called for the legal enforcement of regulated wages; the regulation of the apprenticeship system to ensure controlled entry into skilled trades of men of proven ability; and an end to the exploitation of apprentices. They were improperly trained, therefore produced inferior goods, which undermined the reputation of British craftsmanship abroad, and they were paid a pittance. Undertaking the work of journeymen in 'large establishments of apprentices' was to the detriment of qualified workers and exploited apprentices, both.[69]

> The conflict for labour has led to the abrogation of the exclusive right of indentured apprentices in the arts of mechanism, which has inundated the trades with an excess of insufficient workmen, deteriorating our manufactured goods, destroying the confidence heretofore reposed in their superiority, and consequently the stimulus for demand.[70]

The remedy was 'that no man take an apprentice that is not himself a workman, and that no man have more than two at one time'.[71] In 1812, the Artisans' General Committee, also known as the United Artisans' Committee, had been formed as a

---

[65] John Powell, *Letter Addressed to Edward Ellice*, 3–4.   [66] Ibid., 25.   [67] Ibid., 4.
[68] Later in the *Letter Addressed to Edward Ellice*, at page 27, Powell reverts to the increase in crime caused by 'the immoral tendency of manufactures'.
[69] Ibid., 13.   [70] Ibid., 26.   [71] Ibid., 28.

national network to defend traditional apprenticeships. Though there is no obvious link between it and the network responsible for the *Statistical Illustrations*, both are examples of attempts in this period to bring together workers in crafts suffering from the erosion of customary rights and their position in the market.[72]

The Spitalfields silk weavers, beset by competition from abroad and from new silk businesses established in northern towns where workers were paid less, were a special concern of Powell's.[73] They could work fourteen or sixteen hours a day and yet earn barely sixteen shillings a week.[74] Powell defended the Spitalfields Acts of 1773, 1792 and 1811, under which magistrates legally endorsed and made binding the agreements reached between masters and men in the silk weaving trade, against 'the absurd and mischievous doctrines of the advocates for their repeal'.[75] These included not only masters who sought to force down wages but 'the pretenders to political economy' as well, dogmatists for laissez-faire.[76] Nevertheless, the Acts were repealed in 1824. Opposition to the 'unrestrained freedom of trade', and the call to legally enforce wage agreements, were consistent themes throughout this body of literature down to the formation of John Gast's General Association in 1827. 'The first principle of the Association [was] regulation' in contrast to the 'first principle of Ricardo, individual competition; which is subversive of order, tends to corrupt, brutalize, impoverish and enslave'.[77] Here, and in many other pamphlets and articles, the London artisans 'took Ricardo as the reference point' for their opposition to orthodox economic opinion.[78] Against him, the General Association advocated legislative regulation of wages for 'the whole body of industrious people'.[79] Powell, speaking in Spitalfields in 1828, a year after a failed strike there, mocked 'the miserable pretensions that were abroad about wages being left to find their own level'.[80] These artisans even opposed the free importation of foreign corn because it would lead immediately to a reduction in returns on the land and a consequent reduction in the wages of labourers. They called for 'that just principle of regulation that shall produce a social and general good, without any of that derangement so obviously likely to result from an importation, under any of the regulations hitherto proposed'.[81]

---

[72] Prothero, *Artisans and Politics*, 52.

[73] From 1825 the prohibition on the importation of foreign silks was lifted, to be replaced by a 30% duty. But as foreign silk weavers were paid so much less, this amounted to serious undercutting of the Spitalfields' workers.

[74] Prothero, *Artisans and Politics*, 210.   [75] Ibid., 39.

[76] Powell, *A Letter Addressed to Weavers, Shopkeepers, and Publicans*, 2.

[77] *A Narrative and Exposition of the...General Association*, 'Summary' of the argument of the Introductory Chapter.

[78] Linebaugh, 'Labour History without the Labour Process', 325.

[79] *A Narrative and Exposition of the...General Association*, 29.

[80] 'General Meeting of the Journeymen Silk Weavers of Spitalfields', *The Trades' Free Press*, 23 Feb. 1828, 247. Prothero, *Artisans and Politics*, 212.

[81] *Appendix to the First Edition of the Statistical Illustrations* (London, 1826), xviii.

The analysis of the ills of British society developed by the artisan groups can be linked with other and more familiar working-class movements and genres of protest in this era. Many of the tables of data in the *Statistical Illustrations* tell not only of immiseration but also of the maldistribution of wealth and the advancing corruption of the state, a theme associated with William Cobbett above all at this time. Like Cobbett as well, the third edition of the *Statistical Illustrations* presents London—Cobbett's 'Great Wen'—as a place of speculation and parasitism, making nothing while consuming the products and labour of others.

Taking the metropolis for example, with all its seeming advancement, activity, reciprocity of interchange, and liberal expenditure, the sum and substance of the subsistence of the whole does but resolve itself into a tax on the products of the labour of the people in the country: the direct form and pressure of the tax being rendered imperceptible to the ordinary understandings of man, by the peculiarly insidious intervention and working of money...the metropolis, in a general sense, produces nothing, and therefore can give nothing.[82]

There are obvious similarities between the *Statistical Illustrations* and the several compilations made by John Wade at this time, notably his *Extraordinary Black Book...presenting a complete view of the expenditure, patronage, influence, and abuses of the Government, in Church, State, Law and Representation*, which was first published in instalments in 1819; then as book in 1820; and then in various revised editions in the 1830s. Wade, originally a wool sorter, was, besides Francis Place, 'the most impressive fact-finder among the Radicals'.[83] His work was also an exposé of elite parasitism, bringing together compendious information and statistics on the widest range of abuses including church pluralism, government sinecures, aristocratic pensions, wasted expenditure, and public corruption. Its final section, more than 80 pages long, was a vast 'List of Placemen, Pensioners, etc.' with the amounts they were currently receiving from public funds.[84] The work was dedicated to the people as 'a record of the abuses from which they have long suffered, and of the means by which they may be alleviated'.[85] They were 'the tax-ridden, law-ridden, priest-ridden, deluded people of England'.[86] Like Wade and other radicals, the *Statistical Illustrations* differentiated between 'the proportion of the population productive, and the proportion unproductive'.[87] But *The Black Book*, for all its virtues and influence, was highly personal and idiosyncratic, a

---

[82] *Statistical Illustrations* (3rd edn, 1827), 'Introduction to the Third Edition', iii.
[83] Thompson, *The Making of the English Working Class*, 770.
[84] John Wade, *The Extraordinary Black Book...presenting a complete view of the expenditure, patronage, influence, and abuses of the Government, in Church, State, Law and Representation* (London, 1831 edn), 416–501.
[85] Ibid., v.   [86] Ibid., 302.
[87] *Statistical Illustrations*, 1826 (3rd edn, 1827), 'Introduction to the Third Edition', iii.

compendium of Wade's opinions as much as national facts, whereas the *Statistical Illustrations* were presented as definitive, requiring no commentary or further explanation, an unimpeachable and wholly objective statistical source on which the artisans and factory hands could build the strongest case.

Intriguingly, Powell and his fellows rejected orthodox political economy because of its deductivism, just as the Cambridge inductivists, those far more illustrious savants like Richard Jones, William Whewell, and Charles Babbage, rejected the methods of Ricardo, and tried to embody that rejection in the foundation of both Section F, the statistical section of the British Association for the Advancement of Science, and the Statistical Society of London. According to John Powell in 1826, 'Men too, of great pretension and overbearing presumption, are pressing on public attention certain dogmas that have no foundation in truth, and are totally inapplicable to British society, under the imposing and dignified title of Political Economy.'[88] According to the *Trades' Newspaper* in the following year, 'The great evil with men who have written on what they term political economy is, that they never take into consideration the habits and customs, and all the natural passions and propensities belonging to human nature.'[89] The Central Committee of the General Association ascribed 'a very large portion of the calamities that afflict the people of Great Britain to the influence which the erroneous doctrines of Ricardo and his disciples have acquired over public opinion.' These doctrines were 'contrary to the evidence of facts.'[90] Powell was more precise still: 'Nothing has hitherto been avowed or developed, that is calculated to lead to the conclusion, that either statesmen, or writers on political economy, have taken a sufficiently comprehensive view, either of the operations of the existing system of commerce, or of those principles which are calculated to obviate the evils of that system.'[91]

These comments could have been made by Richard Jones writing letters to William Whewell about the weakness of Ricardianism in the early 1830s; or by Charles Babbage complaining in *The Economy of Manufactures* that political economists had paid insufficient attention to the processes of industrial production. The Cambridge network also wrote and spoke of the narrowness of orthodox economics and the requirement to theorize on the basis of observation, empirical inquiry, and multiple inductions. According to John Powell in his account of Gast's General Association:

> The great misfortune of this country is, that the Legislators, instead of having formed their opinions on a comprehensive reference to facts and experience, the only true basis of national policy – have very generally, given preference to the doctrines of theorists, who on superficial grounds have presumed to lay

[88] Powell, *An Analytical Exposition*, 4.   [89] *Trades' Newspaper*, 12 Feb. 1826, 487.
[90] *A Narrative of…the General Association*, 4–5.   [91] Powell, *An Analytical Exposition*, 13.

down laws, and then bend all the operations of society to establish these pre-conceived opinions.[92]

To both groups—the artisans comprising the plebeian information networks of the 1820s, and the Cambridge 'gentlemen of science'—a far better term than *political* economy, was '*social* economy', denoting economics as it should be by indicating the breadth required to adequately explain social behaviour. In 1831 and 1832 Whewell referred to Jones as a 'social economist' and to his style of analysis as 'social economy'.[93] Some years earlier the *Statistical Illustrations* had been compiled

[so] that the succeeding generation should have the benefit of a comprehensive extent of accurate information in all the great practical pursuits of social econ-omy…the object has been to exhibit facts, and such authentic data, as could be obtained, for the purpose of enabling the enquirer in social economy to exercise his own judgment, and to draw his own conclusions.[94]

In a celebrated essay begun in 1831, published in 1836, and entitled 'On the Definition of Political Economy', John Stuart Mill established the context, mean-ing, and import of the term 'social economy'.[95] He explicitly differentiated between the restricted scope and purpose of political economy and the wider, synthetic intent of social economy. The latter denoted the science of the 'laws of human nature in a social state'. It showed 'by what principles of his nature man is induced to enter into a state of society…what are the various relations which establish themselves among men as the ordinary consequences of social union'. It 'embraced every part of man's nature'. But in comparison 'the words "political economy" have long ceased to have this extensive meaning'. According to Mill, political economy 'does not treat of the whole conduct of man in society. It is concerned with him solely as a being who desires to possess wealth…It makes entire abstraction of every other human passion or motive'.[96] John Powell and his artisan statisticians would have agreed. In fact, they were saying and writing the same thing a decade before Mill's essay was published. Given Mill's radical contacts in the 1820s, including his public discussions with Owenites at the

---

[92] Powell, *Narrative and Exposition…of the General Association*, 38–9.
[93] William Whewell to Richard Jones, 8 Nov. 1831 and 11 Oct. 1832, Whewell Papers, Trinity College, Cambridge, Add. Mss c. 51 (118) and c. 51 (142).
[94] *Statistical Illustrations* (1825 edn), 'Advertisement', iv; Preface, xiv.
[95] J. S. Mill to J. P. Nichol, Jan. 1834, *Earlier Letters of John Stuart Mill, 1812–1848*, ed. F. E. Mineka (Toronto, 1963), 211.
[96] J. S. Mill, 'On the definition of political economy; and on the method of philosophical investigation in that science', *London and Westminster Review*, iv and xxvi (October 1836), 10–12. This paragraph is drawn from Lawrence Goldman, 'The Origins of British "Social Science". Political Economy, Natural Science and Statistics, *The Historical Journal*, 26, 3, 1983, 605.

Co-operation Society which met weekly for debates in Chancery Lane, it is certainly possible that he had heard the term 'social economy' used by this counter-culture and that he brought it into use from that source. Mill described 'a *lutte corps à corps* between the Owenites and Political Economists' in these debates in 1825. He placed himself on the side of political economy.[97]

If the workers were using the same terminology as the savants and for the same reason—to make the emerging economics of the 1820s and 1830s more accurate and applicable, less dogmatic and theoretical; and if they had reached this advanced intellectual position before, rather than after the intellectual radicals in Cambridge, the history of the London Statistical Society, in whatever form it existed, calls into question a Foucauldian interpretation of statistics in the nineteenth century. In this view, now widely diffused through society as well as academe, the development of social statistics should be understood as an aspect of the social control and suppression of the working-class movements of this era.[98] To count, then to categorize, and then to divide and differentiate, is to control and to impose a dominant set of social values on people who have a different social experience. Yet among Powell, Gast, and their collaborators, we see workers using social statistics to interrogate and condemn the state for its corruption and expropriation; to accuse aristocratic and bourgeois interests of siphoning off the people's wealth and of undermining their living standards; and to draw attention to the lives of the poor.

This reversal of the usual relations between rich and poor, the investigator and the investigated, the powerful and the weak in the history of social enquiry, mandates that we write the history of the London Statistical Society in a different way. In general, historians present the poor as passive victims of middle-class designs, control, do-gooding, and also of their hostile interventions. The poor are practised upon, subordinated, victimized, perhaps pitied. This is another example of what E. P. Thompson famously termed 'the enormous condescension of posterity'.[99] Yet workers could and did play other and more active historical roles. In this case, they used the most advanced of contemporary methods, the collection and analysis of social statistics, to promote their own view of society. The poor counted the rich and the weak interrogated the strong to form a view of their own predicament, its causes, and its remedies.

The 1820s may not have experienced any very notable radical uprisings and events in comparison with the years immediately after the wars, and with Chartism in the 1830s and 1840s. But it was the formative period during which the basic political and economic positions that would guide the working-class

[97] J. S. Mill, *Autobiography* (London, 1873) 123–4. Jose Harris, 'John Stuart Mill 1806–1873', *ODNB*.

[98] For further discussion of Michel Foucault's ideas about social statistics, see the Conclusion, below, pp. 315–18.

[99] Thompson, *The Making of the English Working Class*, 13.

movement for the next generation were laid down.[100] Thompson's account of the 1820s in *The Making of the English Working Class* focused on several different radical traditions and tendencies which he traced through the unstamped press, the theatres and taverns, as well as in the factories, workshops, and shipyards.[101] First there was William Cobbett, who assaulted the ruling class for their profligacy and parasitism, and excoriated the British state for its exploitation of the people through unfair taxation and waste. This was 'Old Corruption' and Cobbett became the hero of artisans, shopkeepers, smallholders, and consumers when he cursed it. The rights of man, the old tradition of Tom Paine, was upheld now by Richard Carlile. There were working-class utilitarians, applying the tests of utility and efficiency to British institutions in the manner of Bentham and the Philosophic Radicals, among whom Thompson places John Wade.[102] There were the trade unionists, led by John Gast; and there were the Owenites, followers of Owen but branching out to explore the many different meanings and destinations of co-operation. Thompson described popular radicalism in these years as 'an intellectual culture', one in which 'the impulse of rational enlightenment', hitherto 'confined to the radical intelligentsia was now seized upon by the artisans and some of the skilled workers'.[103] This was the world of John Powell and the London Statistical Society, though in their case, rather than taking lessons from the radical intelligentsia, they were in advance of it. They developed a radical intellectual assault on the status-quo, built up by the painstaking collection of data, in an attempt to construct an unanswerable case against the policies of the state and the intellectual foundations on which they rested, orthodox political economy.

It may be presumptuous, but perhaps we should add another radical tradition, another genre of protest, to those identified by Thompson more than half a century ago. The artisan statisticians were trying to defeat dogma by data, and ideology by numbers. Through the collection of social statistics they hoped to better understand the forces that had reshaped their lives and communities in the preceding two generations. In this, remarkably, they were some years in advance of bourgeois and aristocratic groups who used statistics in constructing a case for moderate social reforms, and of those savants, equally frustrated by the narrowness of Ricardian economics, who hoped to use statistics in the construction of a genuine social science. Statistics were in vogue, and all groups and classes sought to employ them to advance their interests and their cause.

The history of these artisan statisticians adds an important new element to the understanding of the emergence of social statistics circa 1830 and shows how complex a movement it really was. Perhaps we should think not of a statistical movement in the singular, but of statistical movements in the plural, doing

---

[100] Prothero, *Artisans and Politics*, 267.
[101] Thompson, *The Making of the English Working Class*, Ch. XVI, 'Class Consciousness', 711–832.
[102] Ibid., 770.   [103] Ibid., 711, 726.

different things with the profusion of social numbers that had become available, sometimes converging, as in the shared critique of political economy, but often at odds with each other. Social statistics were a new and potent aspect of the age, employed by many different groups for political, economic and intellectual ends. Neutral in themselves, they could be bent to conflicting purposes but were amenable to use by all, even humble weavers, clockmakers, and shipwrights.

The avalanche of numbers from the mid-1820s stimulated many different social and scientific endeavours. Reform movements, academic societies, policy institutes, business forums, and radical agitations all took their cue from the new availability of social data that promised different things to different groups. There was not one statistical movement but several different statistical movements in Manchester, Cambridge, Westminster, Clerkenwell, and elsewhere. Nor was there a single conception of the meaning of social statistics in the early nineteenth century but several interlocked: raw data for an inductivism common to all sciences, social information as a basis for legislation and reform, the numbers required for the critique of the ruling class, and also the components of an innovatory social science. Many groups, suddenly made aware of new information, believed that they could use it to further their aims, whether they were political or strictly intellectual. It was as if a new and revolutionary technology had been discovered and distributed widely, allowing different groups to promote their interests as never before.

PART III
INTELLECTUAL INFLUENCES

5

Charles Babbage and Ada Lovelace

Statistics and the Computer

Of all the founders and contributors to the Statistical Movement, Charles Babbage is the most interesting and his legacy the most enduring and influential. Indeed, the man whose image adorns a page of the current British passport is one of the acknowledged makers of the modern age. Famous in his own day as a mathematician, engineer, and technologist, he comes down to us as the 'inventor of the computer'. When, in 1946, the magazine *Nature* included an article about one of the first large relay computers, it was entitled 'Babbage's Dream Comes True'. The Automatic Sequence Controlled Calculator, recently installed in Harvard University, was 'a realisation of Babbage's project in principle, although its physical form has the benefit of twentieth century engineering and mass-production methods'.[1] Babbage attempted to build mechanical rather than electronic computers between the early 1820s and mid-1840s, a project fraught with difficulty, expense, and ultimate frustration. But in the process he envisaged the basic architecture of the modern computer and foretold the range of functions that a 'calculating engine' might fulfil for later generations. Because his personal history and his projects have been left largely to historians of technology and computing, his role in the statistical movement and his fascination with numbers, as opposed to machines that might crunch those numbers, has been largely overlooked.

Working backwards from the present, most authors who treat of him 'tend to ignore the relation of Babbage's innovations to the thought or technology of his age'.[2] Yet Babbage was at the heart of the institutionalization of statistics in the early 1830s, and his mathematical imagination was more active and creative within the statistical movement than that of anyone else. He not only advocated the collection and analysis of numerical social data on a grand scale as the way to do social science, but he tried for two decades to construct a mechanical computer to process that data. Ada Lovelace, Babbage's pupil and expositor, admired, according to her mother, Lady Byron, the wife of the poet, the 'universality of his

---

[1] L. J. Comrie, 'Babbage's Dream Comes True', *Nature*, 26 Oct., 1946, 567–8. For some sceptical remarks on the framing of Babbage as the inventor of the computer see Doron Swade, '"It will not slice a pineapple": Babbage, Miracles and Machines', in F. Spufford and J. Uglow (eds), *Cultural Babbage. Technology, Time and Invention* (London, 1996), 37–41.

[2] I. Bernard Cohen, 'Foreword', in *The Works of Charles Babbage* (ed. Martin Campbell-Kelly), 11 vols (London, 1989), vol. 1, 11.

Victorians and Numbers: Statistics and Society in Nineteenth Century Britain. Lawrence Goldman, Oxford University Press. © Lawrence Goldman 2022. DOI: 10.1093/oso/9780192847744.003.0005

ideas', the way in which he tried to link together mathematics, mechanism, manufacturing, science, statistics and society.[3] In this he was at one with the aims of the statistical movement itself, though he did it better than anyone else. After Lady Byron had been to view Babbage's Difference Engine, she wrote that 'there was a sublimity in the views thus opened up of the ultimate results of intellectual power'.[4]

Babbage was polymathic: a brilliant mathematician and holder of the Lucasian Chair in Cambridge for more than a decade, 1828–39, though he never delivered a single lecture; an economist interested in what we would now call 'systems analysis' and the structure of industry; a technologist who tried to apply the latest techniques in mechanical engineering to his own and other projects; a controversialist who took aim at the aristocratic leadership of British science; and a political liberal to boot, 'a reformer by conviction', who stood for parliament in the early 1830s.[5] He lost on both occasions.[6] Born in 1791, the son of a banker, and independently wealthy throughout his life, he was educated in Cambridge between 1810 and 1814 where he was a member of the famous Analytical Society, a well-organized student group, composed of the most talented undergraduate mathematicians, which argued and lobbied successfully for the introduction into domestic mathematics of the continental, 'Leibnizian' notation. The members, too, were radical liberals in politics.[7] Babbage first thought of the calculation of logarithms 'by machinery' as early as 1812 or 1813, while still a student, when 'sitting in the rooms of the Analytical Society at Cambridge, my head leaning forward on the table, in a kind of dreamy mood, with a table of logarithms lying open before me'.[8] The idea of computing arithmetical tables mechanically recurred to him in 1819 and Babbage began to 'sketch out arrangements for accomplishing the several partial processes which were required'.[9] Two years later he and John Herschel were engaged by the Astronomical Society in the computation and publication of astronomical and navigational tables, notoriously prone to human error, to be

[3] Betty Alexandra Toole, 'Ada Augusta Byron (Ada King, countess of Lovelace) 1815–52', *ODNB*; See also Miranda Seymour, *In Byron's Wake. The Turbulent Lives of Lord Byron's Wife and Daughter: Annabella Milbanke and Ada Lovelace* (London, 2018), 259–79.

[4] Lovelace Papers, 21 June 1833, quoted in Doris Langley Moore, *Ada, Countess of Lovelace. Byron's Legitimate Daughter* (London, 1977), 44.

[5] Doron Swade, 'Charles Babbage 1791–1871', *ODNB*. H. W. Buxton, *Memoir of the Life and Labours of the Late Charles Babbage* (1872–80) (ed. Anthony Hyman) (Cambridge, Mass., 1988), 313–14.

[6] Maboth Moseley, *Irascible Genius. A Life of Charles Babbage, Inventor* (London, 1964), 120–21.

[7] Charles Babbage, *Passages from the Life of a Philosopher* (London, 1864), 29, 38–40; Buxton, *Memoir of the Life and Labours of the Late Charles Babbage*, 29–30; Anthony Hyman, *Charles Babbage. Pioneer of the Computer* (Oxford, 1982), 24; Moseley, *Irascible Genius*, 49. William J. Ashworth, 'Analytical Society', *ODNB*.

[8] Babbage, *Passages from the Life of a Philosopher*, 42. See also B. H. Babbage, *Babbage's Calculating Machine or Difference Engine* (London, 1872) quoted in Babbage, *Works*, vol. 2, 225; Buxton, *Memoir of the Life and Labours of the Late Charles Babbage*, 19, 45, 67n.

[9] Babbage, *Passages from the Life of a Philosopher*, 43.

included in the improved *Nautical Almanac*.[10] 'The intolerable labour and fatiguing monotony of a continued repetition of similar arithmetical calculations, first excited the desire, and afterwards suggested the idea, of a machine, which... should become a substitute for one of the lowest operations of the human intellect.'[11]

In an apocryphal tale of the sort common in the history of science, Babbage is supposed to have expostulated to John Herschel in frustration with their joint labours, 'I wish to God these calculations had been executed by steam!'[12] He did, in fact, write to his friend on 20 December 1821,

Can you come to me in the evening as early as you like I want to explain my Arithmetical engine and to open to you sundry vast schemes which promise to reach the third and fourth generation... Do let me see you for I cannot rest until I have communicated to you a world of new thought.[13]

An unpublished personal recollection by Babbage provides a more detailed account:

Being engaged in conjunction with my friend Mr Herschel about the conclusion of the last year in arranging and superintending some calculations of considerable extent, which were distributed amongst several computers, the delays and errors which are inseparable from the nature of such undertakings soon became sufficiently sensible... In the course of our conversations on this subject it was suggested by one of us, in a manner which certainly at the time was not altogether serious, that it would be extremely convenient if a steam-engine could be contrived to execute calculations for us; to which it was replied that such a thing was quite possible, a sentiment in which we both entirely concurred and here the conversation terminated. During the next two days the possibility of calculating by machinery (which I should never for a moment have doubted had I ever proposed it to myself as a question) recurred several times to my imagination... Finding myself at leisure the next evening and feeling confident not only

[10] Bruce Collier and James MacLauchlan, *Charles Babbage and the Engines of Perfection* (Oxford, 1998), 27.

[11] *On the Application of Machinery to the Purpose of Calculating and Printing Mathematical Tables. A Letter to Sir Humphry Davy* (London, 1822), in *Works of Charles Babbage*, vol. 2, *The Difference Engine and Table Making*, 6–14. Quotation at 6. See also from 1822 Charles Babbage, 'A Note Respecting the Application of Machinery to the Calculation of Astronomical Tables', Memoirs of the Astronomical Society, vol. 1, 1822, 309: 'I have been engaged during the last few months in the contrivance of machinery, which by the application of a moving force may calculate any tables that may be required../..and the arrangements are of such a nature that, if executed, there shall not exist the possibility of error in any printed copy of tables computed by this engine'.

[12] Doron Swade, *The Cogwheel Brain. Charles Babbage and the Quest to Build the First Computer* (London, 2000), 9–10, 17.

[13] Babbage to Herschel, 20 Dec. 1821, John Herschel Papers, Royal Society, London, 2. 169.

that it was possible to contrive such a machine, but that it would not be attended with any extraordinary difficulties, I commenced the task.[14]

Elsewhere he referred to this project at the time as 'machinery adapted to numbers'.[15] A later commentator, the journalist and popularizer of science Dionysus Lardner, wrote of 'reduc[ing] arithmetic to the dominion of mechanism'.[16] If Babbage initially saw the aim as removing the burden from calculation, doing it faster and more accurately, over time he developed a wider conception of the possible roles of the engine, and of the science of numbers itself.

Amongst his many identities, Babbage's commitment to the statistical movement has been largely overlooked and underplayed. In 1824 he was recruited to be chief actuary of the projected Protector Life Assurance Society, though in the event the business was never commenced. He read up on the whole subject of vital statistics and the calculation of probabilities on which the insurance business depends—which is one of the key sources out of which modern social statistics have emerged—and sought the help of friends and scientific colleagues. 'I am much obliged by your kindness in procuring for me information relative to the assurance of lives', he wrote to Quetelet. 'We must find some mode of communicating scientific intelligence and papers readily between our respective countries.'[17] Having committed time to actuarial studies, Babbage subsequently took the view that the public could benefit from a work that made life insurance accessible and comprehensible: 'I could not help observing how very imperfectly the merits of the numerous and complicated institutions for this purpose were understood . . . [and] to this circumstance may be attributed the publication of a work which has few claims either to novelty or originality.'[18] Babbage therefore published in 1826 *A Comparative View of the Various Institutions for the Assurance of Lives*, a largely neglected primer that explained the work of mutual assurance societies for policy-holders and potential customers alike, designed to reassure readers that this was a well-founded business based on careful calculation. A German edition of the book encouraged several German companies to adopt his tables of mortality.[19] As Babbage remarked after listening to a paper on the vital statistics of an Irish

---

[14] 'The Science of Number Reduced to Mechanism', Nov. 1822 (Mss in the Buxton Collection, Museum of the History of Science, Oxford, donated by Babbage to his friend H. W. Buxton).

[15] 'On the Theoretical Principles of the Machinery for Calculating Tables. In a letter to Dr. Brewster', *Edinburgh Philosophical Journal*, vol. 8, 1823, 122–8, quotation at 127.

[16] Dionysus Lardner, 'Babbage's Calculating Engine', *Edinburgh Review*, 59, 1834, 263–327 in Babbage, *Works*, vol. 2, 119.

[17] Babbage to L. A. J. Quetelet, 13 Nov. 1826, Quetelet Papers, File 267, Acadèmie Royale de Belgique, Brussels. Subsequently Babbage helped Quetelet procure a telescope; see Babbage to Quetelet, 15 Dec. 1827.

[18] Charles Babbage, *A Comparative View of the Various Institutions for the Assurance of Lives*(London, 1826), ix.

[19] 'Introduction', *Charles Babbage and his Calculating Engines. Selected Writings by Charles Babbage and Others* (eds. Philip Morrison and Emily Morrison) (New York, 1961), xxii.

parish at a meeting of the British Association, 'to discover those principles which will enable the greatest number of people by their combined exertions to exist in a state of physical comfort and of moral and intellectual happiness is the legitimate object of statistical science'.[20]

In the same year that he published his study of life insurance, he also published privately a different numerical study, a 'list of those facts relating to mammalia, which can be expressed by numbers'. Approximately two hundred copies were printed and all are now lost. But this curious paper became in 1832 the essay 'On the advantage of a collection of numbers to be entitled the Constants of Nature and Art', published in the *Edinburgh Journal of Science*.[21] It was republished in 1853 in the proceedings of the first International Statistical Congress held in Brussels with a foreword that referred back to the 1826 version,[22] and then again in 1857 by the recently founded Smithsonian Institution, which set aside funds to pursue at least part of Babbage's programme as there outlined.[23] Throughout his life Babbage sought to make the natural and social realms numerical: to put a number on whatever could be expressed in figures so as to make analysis possible by turning the descriptive into the quantitative. 'He remained a symbol of a new way to think about nature and our works: numerically.'[24] It was a common idea shared by others in the Statistical Movement, though no one went as far as Babbage in indicating what might be done and achieved in this manner.

In this 1832 essay he described a collaborative project to be undertaken by the leading scientific academies of Britain, France, and Germany to collect 'all those facts which can be expressed by numbers in the various sciences and arts'.[25] Babbage set out 19 different classes of facts in this precise order: 'About the solar system'; the elements and their chemistry; metals and their attributes; light; the 'number of known species of mammalia, birds, reptiles, fishes, mollusca, worms, crustacea, insects, zoophytes'; the anthropometry of mammalian species; the anthropometry of man; 'power of man and animals'; plants; the geographic distribution of all species; atmospheric phenomena; 'materials'; 'velocities'; rivers;

[20] Ibid., xxii.

[21] 'On the advantage of a collection of numbers to be entitled the Constants of Nature and Art', *Edinburgh Journal of Science*, n.s. xii (1832), 334–40. See Charles Babbage, *Works*, vol. 5, 138–50 from which quotations are taken. Lawrence Goldman, 'The Origins of British Social Science: Political Economy, Natural Science and Statistics, 1830–1835', *The Historical Journal*, 26, 3, 1983, 603; Ian Hacking, *The Taming of Chance* (Cambridge, 1990), ch. 7, 'The Granary of Science', 55–63.

[22] 'Sur les constantes de la nature – classe des mammieres', *Compte Rendu des travaux du congres general de statistique* (Bruxelles, 1853), 222–30. 'The following list of those facts relating to mammalia, which can be expressed by numbers, was first printed in 1826. It was intended as an example of one chapter in a great collection of facts which the author suggested under the title of the 'Constants of Nature and Art'...

[23] 'On Tables of the Constants of Nature and Art', *Annual Report of the Board of Regents of the Smithsonian Institution for 1856* (Washington D.C., 1857), 289–302. In this version there is a further section to the paper listing dozens of possible measurements of mammals, fish and birds in that order.

[24] Hacking, *The Taming of Chance*, 60.

[25] Babbage, 'On the advantage of a collection of numbers', 138.

population; buildings; 'weights, measures etc.'; 'the frequency of occurrence of the various letters of the alphabet in different languages'; and educational statistics including the 'number of books in great public libraries...number of students at various universities'. Babbage was

> confident that many a weary hour, now wasted in the search for existing knowledge, will be devoted to the creation of new, and that it will thus call into action a permanent cause of advancement toward truth, continually leading to the more accurate determination of established facts, and to the discovery and measurement of new ones.[26]

He presented his plan a year later at the Cambridge meeting of the British Association for the Advancement of Science. An abstract of his ideas was published by the Association, which, as we have seen, made him a grant of £100 for the work to begin. To Babbage, the scheme seemed 'absolutely necessary at the present time, and...would be of the greatest advantage to all classes of the scientific world'. The collection and periodic revision of these constants 'would be a work fraught with advantages to knowledge, by continually leading to the more accurate determination of established facts, and to the discovery and measurement of new ones'. Evidently a work in progress, the constants soon comprised twenty categories. These were the first five as published in 1834:

1. All the constant qualities belonging to our system; - as, distance of each planet, - period of revolution, - inclination of orbit etc., - proportion of light received from the sun, - force of gravity on the surface of each, etc.
2. The atomic weight of bodies.
3. List of the metals, with columns for specific gravity, - electricity, - tenacity, - specific heat, - conducting power for heat, - conducting power for electricity, - melting point, - refractive power, - proportion of rays reflected out of 1000, - at an incidence of 90°.
4. Specific gravities of all bodies.
5. List of mammalia, with columns for height, - length, - weight, - weight of skeleton, - weight of each bone, - its greatest length, - its smallest circumference, - its specific gravity, - number of young at birth, - number of pulsations per minute, - number of inspirations per minute, - period of blindness after birth, - of sucking, - of maturity, - temperature, - average duration of life, - proportion of males to females produced, etc. etc.[27]

[26] Babbage, 'On the advantage of a collection of numbers', 144.
[27] 'Constants of Nature and Art', *Report of the Third Meeting of the British Association for the Advancement of Science; Held at Cambridge in 1833* (London, 1834), 490–1.

In a tribute after Babbage's death, Quetelet described the great project as 'a register of everything capable of being measured'.[28]

The choice of the categories, and the divisions between them seem arbitrary and without rationale, a collection of random themes with no structure or logic, as bodies of knowledge, animate and inanimate, natural and man-made, social and physical, the easily collected and the utterly unknown, follow each other without an obvious structure or key. Some were physical constants, invariable features of the universe. In other cases, such as the measurement of anatomical features, the measurements might provide a stock of knowledge with which to compare species and their development. Some of the readings would never change; other measurements, *en masse*, might allow for the deeper understanding of the structures and processes of the natural world. Historians have tried to make sense of the curious list, though without much success. It is possible that Babbage was here issuing a challenge to see how far numbers could replace words; to see if it might really be possible to build bodies of numerical knowledge that could be better analysed because each fact or measurement had its unique figure or value. He remained committed to his project, certainly, always envisaging a collective endeavour that might take decades to fulfil. From the outset, he had thought it desirable

> to insert the heads of many columns, although not a single number could be placed in them; for they would thus point out many an unreaped field within our reach, which requires but the arm of the labourer to gather its produce into the granary of science.[29]

In 1853, on having read the transactions of the first International Statistical Congress in Brussels, he wrote to Quetelet that 'the Constants of Nature and Art are within your limits and I am more than ever convinced that even an attempt to collect a portion of them would contribute largely to the advancement of knowledge and to the economy of the time of its cultivators'.[30]

If this is correct, some of Babbage's other published papers fall into place as examples of the challenge accepted, experiments to see what might be gleaned from this sort of analysis, and attempts to establish its limits. One of his first, from 1821, was a probabilistic study of games of chance.[31] In 1829 he had published a study of the number of male and female children born in five different countries

---

[28] L. A. J. Quetelet, 'Extracts from a notice of Charles Babbage', *Annual Report of the Board of Regents of the Smithsonian Institution*, 1873, 184. The 'notice' Quetelet refers to here was his 'Notes Extraites d'un Voyage en Angleterre, aux moins de Juin et Juillet, 1833', *Correspondance mathematique et physique de l'Observatoire de Bruxelles*, Tome Dieuxième, (Brussels, 1835), 13.

[29] 'Constants of Nature and Art', *Report of the Third Meeting of the British Association*, 491.

[30] Babbage to Quetelet, [n.d., 1853], Quetelet Papers.

[31] Charles Babbage, 'An examination of some questions connected with games of chance', *Transactions of the Royal Society of Edinburgh*, 9, 1821, 153–77.

and regions—France, Naples, Prussia, Westphalia, and Montpellier—paying attention also to climate, physical geography and ethnicity. Babbage noted that among legitimate births, boys considerably outnumbered girls, but among illegitimate births the proportions were close to parity, though no conclusion was offered for this striking finding.[32] Babbage had a lifelong interest in ciphers and cryptography and at this time became interested in applying simple statistical analysis to words as a way of deciphering code. He published an essay in 1831, deriving from a letter he had sent to Quetelet, on the frequency of certain letters in certain languages—English, French, German, Italian, and Latin. Specifically, it was on the arcane subject of 'how many times in a sample of 10,000 words a letter is doubled in the middle of the word'. Babbage suggested further studies, also; for example, how many times a two-letter or a three-letter word occurred in a sample of 10,000 words.[33] Not long after this, Quetelet reported receiving from Babbage

> some curious documents, containing a list of all the drunken persons who have been arrested by the London police in the year 1832, and who were immediately released, because no charge was brought against them...If we possessed an extensive series of similar documents, we should find in them the most precious memorials of the manners of the English people, and in particular, all which relates to changes in the condition of the population.[34]

It was this strange return on London drunks that Babbage communicated to the little group gathered in Trinity College, Cambridge at the end of June 1833 at the inauguration of Section F, the statistical section of the British Association.

Stranger still, perhaps, was his calculation of the likelihood of the Resurrection itself. In his unofficial *Ninth Bridgewater Treatise*, published in 1837, he argued against Hume's famous refutation of miracles, calculating at great length that 'the miracle of a dead man being restored' was theoretically possible, though at odds of 200 billion to one.[35] When Tennyson wrote 'Every moment dies a man, Every moment one is born' in his poem *The Vision of Sin*, Babbage wrote to him in 1851 with a correction, suggesting that the line be changed to read "Every moment $1^{1/16th}$ is born".[36] His essay on 'The Statistics of Lighthouses', delivered to the 1853

[32] Charles Babbage, 'On the proportionate number of births of the two sexes under different circumstances', *Edinburgh Journal of Science*, n.s., vol. 1 (1829), 85–104.

[33] Charles Babbage, 'Sur L'emploi plus ou moins fréquent des mêmes lettres dans les différentes langues', Correspondence Mathématique et Physique, vol. 7, 1831, 135–7.

[34] L. A. J. Quetelet, *A Treatise on Man and the Development of His Faculties* (Edinburgh, 1842) (Tr. R. Knox), (Gainesville, Florida, 1969 edn), 72. The data was broken down into months of the year 1832, and the numbers of men and women released without charge in each of them.

[35] Charles Babbage, *The Ninth Bridgewater Treatise. A Fragment* (London, 1837), Ch. X, 'On Hume's Argument against Miracles' and Appendix E, 'Note to Chapter X on Hume's Argument against Miracles', 120–31, 192–203.

[36] Spufford and Uglow (eds.), *Cultural Babbage*, ix.

International Statistical Congress, was a descriptive account of the roughly 1500 lighthouses located across the world, of which approximately 680 were maintained by Britain, 380 by the United States and 200 by France.[37] Then there was his study of 'the relative frequency of occurrence of the causes of breaking of plate glass windows' published in *The Mechanics Magazine* in 1857.[38] Later, at the 4th International Statistical Congress in London in 1860 Babbage made a series of speeches on statistical methods, the presentation of data, and perhaps drawing on his analysis of languages and their differences, the 'immense quantity of time and talent...lost by the diversity of languages'.[39]

There was also Babbage's 'Analysis of the Statistics of the Clearing House During the Year 1839' delivered at the Statistical Society of London and published in its journal in 1856.[40] It was a study of the average traffic of cheques and money orders through the Clearing House located in Lombard Street in the City of London, the financial centre, on different days in different weeks of the year. At the Clearing House, the business between banks in London was tracked, calculated and assessed each day. By writing off what was owed to, against what was owed from, each bank to every other bank, 'a very small amount of bank-notes pass from hand to hand' and an otherwise highly complex and time-consuming set of payments was rationalized and simplified.[41] The sums flowing through the Clearing House fluctuated between £2 million and £6 million daily. Full of terms like 'inland bills', 'English and foreign settlements', and 'Saturday clearings', the essay was written for a highly specialized readership and audience. Babbage presented his paper as a taster for a bigger research project that would track the flows of money through the clearing house in the past, and also weekly, monthly, and annually in the present. This, he thought, would yield a deeper understanding of the banking and finance systems, though he didn't specify what he expected would be learnt thereby.

> In the [equally] complicated phenomena of currency and exchanges, we have little hope of fully unravelling their laws, except through patient observations, continually made and published, of those monetary transactions which are constantly fluctuating. The weekly returns of the Bank of England may perhaps be

---

[37] Charles Babbage, 'On the Statistics of Lighthouses', *Compte Rendu des travaux du congrès général de statistique* (Bruxelles, 1853), 230–7.

[38] Charles Babbage, 'Table of the relative frequency of occurrence of the causes of breaking of plate glass windows', *The Mechanics' Magazine*, vol. 66, Jan.–June 1857, 82.

[39] Babbage, 'Contributions to the Discussions at the Fourth International Statistical Congress, London, 1860', in Babbage, *Works*, vol. 5, 169–175. See *Proceedings of the Fourth International Statistical Congress* (London, 1860), 380, 381, 383, 394.

[40] Charles Babbage, 'Analysis of the Statistics of the Clearing House During the Year 1839', *JSSL*, 19, 1856, 28–48; Babbage, *Works*, vol. 5, 94–132.

[41] The description of the work of the London Clearing House in the 1856 article is taken from Charles Babbage, *The Economy of Machinery and Manufactures* (1832) (4th edn, London, 1835), 172–4.

considered as our economical barometer, and other commercial publications be likened to other instruments; but I believe, few will be found of greater importance than the publication of returns from the clearing house.[42]

It was not a random application of numbers, therefore, but a study he believed would highlight patterns and regularities and hence provide insight. As such, it was an initial exercise in proving the usefulness of statistics, even when unravelling the most complex flows. In consequence of Babbage's paper, the Statistical Society of London 'was enabled to induce the bankers of London to publish the weekly accounts of the Clearing House'.[43]

Whether taken together or singly, these papers might seem to suggest an almost unbalanced mental state in which Babbage was counting for its own sake and exhibiting the mathematical equivalent of an obsessive compulsive disorder. But the attempt to apply numbers to all phenomena—money, lighthouses, infants, drunks, letters of the alphabet—was a way of bringing order and system to society and nature, a way of taming the unknown and proving that it was not unknowable. It was amusing and diverting as well, no doubt, and Babbage was a man who took delight in such diversions. As Simon Schaffer has demonstrated, one surprising context from which Babbage's 'thinking machine' emerged was the popular delight in Regency England for automata. These were apparently self-directed machines, displayed to a paying public as mass entertainment by their none-too-scrupulous nor honest inventors, and from his boyhood Babbage was fascinated by them.[44] He would 'concentrate all the energies of his mind upon the construction of some infantile toy, or other trivial mechanical contrivance, intended to amuse his youthful friends'.[45] But Babbage was more than merely playing in his various statistical papers; each of these curious escapades was an experiment in the application of numbers in circumstances where numbers might not otherwise have gone, in order to determine where and how they might be used with advantage. Thus one commentator has praised 'his pioneering ability to inspire mathematical order in wide areas of a non-mathematical universe'.[46]

Babbage was also a central figure in the institutionalization of statistics in Britain. William Farr, then the most respected social statistician in Victorian Britain, delivered Babbage's obituary to the Statistical Society in November 1871 and lauded his contribution: 'He was attached to the Society: and was, in reality,

[42] Babbage, 'Analysis of the Statistics of the Clearing House During the Year 1839', Works, vol. 5, 106.

[43] William Farr, 'Inaugural Address as President of the Statistical Society of London, delivered 21st Nov. 1871', JSSL, vol. 34, 4, 413.

[44] Simon Schaffer, 'Babbage's Dancer and the Impresarios of Mechanism', in Spufford and Uglow (eds), Cultural Babbage, 53–80.

[45] Buxton, Memoir of the Life . . . of Charles Babbage, 346.

[46] John M. Dubbey, 'Mathematical Papers' in 'Foreword', The Works of Charles Babbage (ed. Martin Campbell-Kelly), (11 vols, London, 1989), vol. 1, 22.

more than any other man its founder.'[47] Farr recalled the roles played by Babbage in the origins of Section F in 1833 and the foundation of the Statistical Society itself in the following year.[48] Babbage was a trustee of the Society for the rest of his life, 'attended at intervals for thirty-seven years', and occasionally delivered papers.[49] Farr recalled that at the public meeting on 15 March 1834 at which the Society was founded 'Mr. Babbage moved two resolutions, one to the effect that the Statistical Society of London should be founded, and one relative to M. Quetelet' who was made its first Foreign Member.[50] His friendship with Quetelet was also lifelong and it was in a letter to him, dated 27 April 1835, that Babbage provided information not only on the assembly and delay of his famous Difference Engine but of 'another Engine of far greater power', the Analytical Engine, that he was now designing. The Difference Engine was limited in its operations. It was designed to cope with the most basic mathematical functions, using the method of differences, which requires only addition (eliminating the need for multiplication or division) so as to simplify the mechanism.[51] It was based on the movement of mechanical cogs, more than 25,000 moving parts in total. The answer was then printed out. But the Analytical Engine was more sophisticated and the letter to Quetelet was the first written communication concerning what is generally understood to have been the design of the first computer, making it a foundational document in the history of computing.[52]

The Calc.Eng. you saw which was made for the Govt. has not advanced *one step* during two years. The whole time has been lost in annoying discussions…and it seems probable that the Engine will not be resumed[,] at least by the English government. In the meantime, during the last six months I have been contriving another [*Analytical*] Engine of far greater power. I have given up all other pursuits and am making drawings of it and advance rapidly but it is not probable that it will be executed here. I am myself surprised at the powers I have given it which I should not have thought possible twelve months since.[53]

[47] William Farr, 'Inaugural Address as President of the Statistical Society', 411.
[48] Babbage, 'Letter to Dr. Farr', 505; Farr, 'Inaugural Address', 412; Hyman, *Charles Babbage*, 151.
[49] Farr, 'Inaugural Address', 411.   [50] Ibid., 412.
[51] Collier and MacLachlan, *Charles Babbage and the Engines of Perfection*, 40–3.
[52] 'Une Lettre à M. Quetelet de M. Ch. Babbage relativement à la machine à calculer', Académie Royale de Belgique, Bruxelles, *Bulletins*, 2, 1835, 123–6. This was a rough French translation of a letter in English sent by Babbage to Quetelet on 27 April 1835. The original letter is preserved in file 267 of the Quetelet Papers. It is the English *translation* of the published French version of this seminal document in the history of computing that has been quoted from extensively since the nineteenth century, while the original letter, which is used in the text here, has been overlooked. Alfred W. Van Sinderen in his published study of the letter used his own translation from the French, believing incorrectly that 'the exact date of the letter is not clear, and the original is not known to exist'. But it does. See Alfred W. Van Sinderen, 'Babbage's Letter to Quetelet', *Annals of the History of Computing*, 5, 3, July 1983, 263–7.
[53] Babbage to Quetelet, 27 April 1835, Quetelet Papers, Académie Royale de Belgique, Bruxelles, File 267.

The blueprints of the Analytical Engine show what Babbage called the 'mill', equivalent to the central processor of modern digital computers, and the 'store', equivalent to the memory.[54] The 'logical structure of his Analytical Engine remains visible in today's big electronic computers.'[55] It was to have been pro-grammed using punch cards, as we would now call them, just as the computers of the post-Second World War era were programmed. Babbage took the idea from the mechanism invented to control the warp threads on an eighteenth-century French Jacquard textile-weaving loom.[56] As Ada Lovelace put it, 'The Analytical Engine weaves algebraic patterns just as the Jacquard loom weaves flowers and leaves.'[57] It could repeat operations as programmed so to do. It would choose which possible action to take next, depending on the outcome of calculations already achieved. It would also print out the results. The Difference Engine pro-cessed only those numbers entered into it: it was limited to specific functions. The Analytical Engine was designed to work automatically and to be of general use, as required.[58] By the autumn of 1834, which was in many ways Babbage's *annus mirabilis* during which '*a principle of an entirely new kind* occurred to him', he had grasped the full potential of his second engine. By 1837 he had finished its design.[59] But only a small and basic section of the whole Analytical Engine was built during Babbage's lifetime, though he continued to draft, calculate, and tin-ker with its architecture in thousands of documents for the rest of his life.[60]

Babbage attended the first International Statistical Congress in Brussels in 1853 and published an account of its foundation.[61] He was a member of the orga-nizing committee of the fourth congress in London in 1860 where he spoke in a discussion on 'statistical methods and signs'. These institutional initiatives and affiliations were but the tip of the iceberg, public manifestations of his enduring interest in numbers and their social potential. Babbage was interested in the micro-economics of the individual firm and the organization of production within it. He was a founder of what is known today as operational research. Like Jones and Whewell, and like the compilers of the *Statistical Illustrations*, he was an instinctive inductivist, sharing their suspicions of the deductive orthodoxies of classical economics and desirous that social and economic principles be derived from observation and information. 'Babbage's whole approach to economic

[54] Babbage, *Passages from the Life of a Philosopher*, 117.
[55] 'Introduction', *Charles Babbage and his Calculating Engines,* xxi.
[56] Collier and MacLachlan, *Charles Babbage and the Engines of Perfection*, 66–8, 83.
[57] Ada Lovelace, 'Sketch of the Analytical Engine Invented by Charles Babbage by L. F. Menebrea. With Notes upon the Memoir by the Translator, Ada Augusta, Countess of Lovelace', *Taylor's Scientific Memoirs*, 3, 1843, 666–731, Note A. See also www.fourmilab.ch/babbage/sketch.html from where extracts have been taken.
[58] Babbage, *Passages from the Life of a Philosopher*, 49.   [59] Ibid., 62–3, 83 (Babbage's italics).
[60] Swade, *Charles Babbage and his Calculating Engines*, 12.
[61] 'Letter to Dr. Farr, On the Origin of the International Statistical Congresses', *Report of the Proceedings of the Fourth Session of the International Statistical Congress*, 505–7.

theory and practice was always based on the collection of carefully chosen statistical information.'[62] As he wrote in 1832 in his most important work of economics, *On the Economy of Machinery and Manufactures*, a book described as 'empirical and unashamedly inductive',[63]

> Political economists have been reproached with too small a use of facts, and not large an employment of theory. If facts are wanting, let it be remembered that the closet philosopher is unfortunately too little acquainted with the admirable arrangements of the factory; and that no class of persons can supply so read-ily…the data on which all the reasonings of political economists are founded, as the merchant and manufacturer; and, unquestionably, to no class are the deduc-tions to which they give rise so important. Nor let it be feared that erroneous deductions may be made from such recorded facts: the errors which arise from the absence of facts are far more numerous and more durable than those which result from unsound reasoning neglecting true data.[64]

The book was the product of travels and observations made over several years 'for the purpose of endeavouring to make myself acquainted with various resources of the mechanical art'.[65] In many ways it was a corollary of the investigations Babbage had necessarily undertaken into the materials, metals, and mechanisms required to build the Difference Engine. It was a broad and extensive synthesis of factory organization and workshop practice focused on the division of labour and the increasing use of machinery on the shop floor.[66] It bridged the gulf between the principles of academic political economy and 'the practical and local concerns of the provincial manufacturer'.[67] Babbage wanted to know how industry was actually organized in practice rather than in abstract, how mechanization was changing production, and how mental processes were themselves subject to the division of labour.[68] He wanted to present those key principles underlying mass production 'which struck me as the most important, either for understanding the actions of machines, or for enabling the memory to classify and arrange the facts connected with their employment.'[69] Researched and written while the Difference Engine was being constructed, the book was intimately related to his project of mechanizing calculations: 'The calculating engines were themselves products of

[62] Hyman, *Charles Babbage*, 111.   [63] Ibid., 104.
[64] Babbage, *On the Economy of Machinery and Manufactures* (1st edn. 1832), 156.
[65] Ibid. (4th edn., 1835), iii.
[66] Simon Schaffer, 'Babbage's Intelligence: Calculating Engines and the Factory System', *Critical Inquiry*, 21, Autumn 1994, 203–27; Richard M. Romano, 'The Economic Ideas of Charles Babbage', *History of Political Economy*, 14, 3, 1982, 385–405.
[67] Maxine Berg, *The Machinery Question and the Making of Political Economy 1815–1848* (Cambridge, 1980), 202.
[68] Schaffer, 'Babbage's Dancer and the Impresarios of Mechanism', 64.
[69] Babbage, *On the Economy of Machinery and Manufactures* (1832), iv.

the system of automatic manufacture which Babbage sought to model. They were some of that system's most famous and most visible accomplishments.'[70]

When the economist Thomas Tooke wrote to congratulate Babbage on the book's publication he focused on this empirical theme in Babbage's work: 'It presents a fund of curious & important facts exemplifying the force of disturbing causes in modifying the conclusions of absolute science.' Tooke accepted Babbage's argument that conventional political economy had emphasized theory at the expense of facts, but pointed out as well that 'practical men may with equal justice be charged with too small an employment of theory & a superabundance of facts'. Yet he accepted that Babbage had achieved a just balance: 'your work possesses the rare advantage of a correct theory adjusted by the use of a vast number of facts of the most interesting description to the actual course of human affairs; & leading to the most satisfactory practical conclusions'.[71]

Babbage was inspired by the prospect of using numbers as a way of explaining (and also for this particular radical Liberal) *changing* the world; 'his object was, by the systematic application of mathematical techniques and scientific method, to change and improve commercial and industrial practice'.[72] In this perspective the Difference Engine, which Babbage began constructing in the summer of 1823 and which was working as a prototype a decade later, and the much more sophisticated Analytical Engine, which was left in very detailed blueprints, were extensions of the statistical movement, instruments that might be used to process and analyse numerical data.[73] Other members of the movement thought about the collection of data; Babbage was far, far ahead of them in planning how to process and interpret that data. He sought to demonstrate 'the degree of assistance which mathematical science is capable of receiving from mechanism' as he wrote in 1837.[74] He looked forward to the many different outcomes from 'the complete control which mechanism now gives us over number'.[75] As he wrote later in his fragmentary autobiography, 'The circular arrangement of the axes of the Difference Engine round large central wheels led to the most extended prospects. The whole of arithmetic now appeared within the grasp of mechanism'.[76]

[70]   Schaffer, 'Babbage's Intelligence', 210.

[71]   Thomas Tooke to Charles Babbage, 25 June 1832, Babbage Papers, British Library, London, Add Mss. 37186, f. 497.

[72]   Hyman, *Charles Babbage*, 105.

[73]   The prototype of the Difference Engine, about a seventh of the size of the planned Difference Engine, constructed in 1832, is on display at the Science Museum in London and still functions. On the start of construction of the engine, see Buxton, *Memoir of . . . the late Charles Babbage*, 81.

[74]   [C. Babbage] 'On the Mathematical Powers of the Calculating Engine', 26 Dec. 1837, Mss draft. Buxton Collection, Oxford.

[75]   Ibid.

[76]   Charles Babbage, *Passages from the Life of a Philosopher*, 112–13. On the Difference Engine, see Laura J. Snyder, *The Philosophical Breakfast Club. Four Remarkable Friends who Transformed Science and Changed the World* (New York, 2011), 84–8.

Yet in the words of Doron Swade, the historian of Babbage's engines, 'the movement to build automatic calculating engines in the 19th century ultimately failed'.[77] Babbage's engines required technical skills and materials far beyond those commonly available in the 1820s and 1830s, which explained the very high costs and the years of labour involved: 'he lacked a whole set of supporting technologies'.[78] It was a failure Babbage acknowledged bitterly, though he placed the blame on the paucity of official funding and support: this 'irascible genius' was his own worst enemy who almost reflexively incited the opposition and hostility of officialdom.[79] Only a part of the Difference Engine was ever built and set in motion; work on it stopped in 1833 when Babbage fell out with his master engineer, Joseph Clement.[80] 'The first drawing of the Analytical Engine is dated [in] September, 1834' and plans for this much more powerful and sophisticated machine were drafted and refined between then and the mid-1840s but it was never constructed, though it is these plans, above all, that give rise to Babbage's claim to have been the first to imagine the modern computer.[81] Babbage had no patience to deal with politicians and bureaucrats. There were misunderstandings from the very first meeting he had with a minister, Goulburn, the Chancellor of the Exchequer, in 1828. The parsimony of governments, the lack of progress in building the engine, and Babbage's temper conspired against him. All hope of government support perished after a disastrous personal meeting with the prime minister, Sir Robert Peel, in November 1842.[82]

Charles Dickens knew Babbage well, and the 'HOW NOT TO DO IT OFFICE' in *Little Dorrit* is assumed to be inspired by his friend's dealings with government. The people chewed up by the office included 'Mechanicians, natural philosophers, soldiers, sailors, petitioners...'[83] The character of Daniel Doyce in the novel bears some resemblance to Babbage:

'This Doyce', said Mr. Meagles, 'is a smith and an engineer. He is not in a large way, but he is well known as a very ingenious man. A dozen years ago, he perfects an invention... of great importance to his country and his fellow-creatures.

[77] Doron Swade, 'Calculating Engines. Machines, Mathematics and Misconceptions', in R. Flood, A. Rice, and R. Wilson (eds.), *Mathematics in Victorian Britain*, (Oxford, 2011), 239–51.
[78] Francis Spufford. 'The Difference Engine and *The Difference Engine*', in Spufford and Uglow (eds.), *Cultural Babbage*, 267.
[79] For example: 'Throughout the whole of these labours connected with the Analytical Engine, neither the science, nor the institutions, nor the Government of his country have ever afforded him the slightest encouragement.' [Anon, Charles Babbage], 'Addition to the Memoir of M. Menabrea on the Analytical Engine', *Philosophical Magazine*, vol. 23 (1843), 235–9 in Charles Babbage, *Works*, vol. 2, 86.
[80] Doron Swade, *Charles Babbage and his Calculating Engines* (Science Museum, London, 1991), xi; Collier and MacLachlan, *Charles Babbage and the Engines of Perfection*, 60–1.
[81] Herschel P. Babbage, 'On the Mechanical Arrangements of the Analytical Engine' read at the BAAS Meeting in Bath, 1888, in Herschel P. Babbage, *Babbage's Calculating Engines* (London, 1889), 331–7. See Babbage, *Works*, vol. 2, 190.
[82] Buxton, *Memoir of the Life and Labours of the Late Charles Babbage*, 91–129.
[83] John Forster, *The Life of Charles Dickens* (London, 1928), 542.

I won't say how much money it cost him, or how many years of his life he had been about it, but he brought it to perfection a dozen years ago...He addresses himself to the Government. The moment he addresses himself to the Government he becomes a public offender! Sir...he ceases to be an innocent citizen, and becomes a culprit.[84]

In real life Babbage wrote thus to his friend, Alexander von Humboldt, in July 1841:

This Engine is unfortunately far too much in advance of my own country to meet with the least support. I have at an expense of many thousands of pounds caused the drawings to be executed, and I have carried on experiments for its perfection. Unless however some country more enlightened than my own should take up the subject, there is no chance of that machine ever being executed during my own life, and I am even doubtful to dispose of those drawings after its termination.[85]

Later, Babbage complained that 'if you speak to [an Englishman] of a machine for peeling a potato, he will pronounce it impossible; if you peel a potato with it before his eyes, he will declare it useless, because it will not slice a pineapple'.[86]

Babbage wrote relatively little about his calculating engines, and that which he did publish was essentially technical. By common—and by Babbage's own—consent, the most revealing and insightful account was published in French in October 1842. It was written by Luigi Menabrea, a young Piedmontese mathematician from the University of Turin and an officer in the Piedmontese military engineers, who would later become one of Garibaldi's generals and then prime minister of the new Italian state between 1867 and 1869.[87] As William Farr put it, 'Menabrea's paper is exceedingly clear and able: it indicates generally what the engine could do if constructed.'[88] It was translated, with extensive further notes added, by Ada, Countess of Lovelace, the poet Byron's only legitimate daughter, and it was published in England in the following year.[89] Byron had deserted Ada's mother when the baby was but a month old and never saw his daughter again. This strange combination of persons and subject requires explanation because, in Babbage's words, Menabrea's essay and Ada's notes 'taken together furnish, to

[84] Charles Dickens, *Little Dorrit* (London, 1857), Bk. 1, ch. 10, p. 13.
[85] Quoted in Swade, *Cogwheel Brain*, 131.
[86] Charles Babbage, 'Preface to the 3rd Edition', *Thoughts on the Principles of Taxation with Reference to Property Tax and Its Exceptions, Works of Charles Babbage*, 5, 41.
[87] L. F. Menabrea, 'Notions sur la machine analytique de M. Charles Babbage', *Bibliothèque universelle de Genève*, 82, Oct. 1842, 352–76.
[88] Farr, 'Inaugural Address', 414.
[89] Ada Lovelace, 'Sketch of the Analytical Engine Invented by Charles Babbage by L. F. Menebrea', passim.

those who are capable of understanding the reasoning, a complete demonstration—*That the whole of the developments and operations of analysis are now capable of being executed by machinery*.[90] This was later endorsed by Babbage's son who confirmed that his father 'considered the paper by Menabrea, translated with notes by Lady Lovelace...as quite disposing of the Mathematical aspect of the invention'.[91] Recent scholarship agrees: it was 'the definitive account of Babbage's programming ideas until modern times'.[92]

In August 1840 Babbage attended a scientific congress in Turin where he met privately with a group of Piedmontese mathematicians and savants. 'It was during these meetings that my highly valued friend, M. Menabrea, collected the materials for that lucid and admirable description which he subsequently published.'[93] Babbage was then acting as mathematics tutor to Ada; the two had met in 1833 and had become close friends. She had been taught mathematics from a young age and had achieved a high standard. It is said that she appreciated the Difference Engine at first sight:

> While the rest of the party gazed at this beautiful instrument with the same sort of expression and feeling that some savages are said to have shown on first seeing a looking glass or hearing a gun, Miss Byron, young as she was, understood its working and saw the great beauty of the invention.[94]

After the publication of Menabrea's article, Ada translated it. The idea had been suggested to her by the great engineer and inventor, Charles Wheatstone, a friend to Ada and Babbage both. As Babbage explained,

> I asked her why she had not herself written an original paper on a subject with which she was so intimately acquainted? To this Lady Lovelace replied that the thought had not occurred to her. I then suggested that she should add some notes to Menabrea's memoir; an idea which was immediately adopted. We discussed together the various illustrations that might be introduced: I suggested several, but the selection was entirely her own. So also was the algebraic working out of the different problems, except, indeed, that relating to the numbers of

---

[90] Babbage, *Passages from the Life of a Philosopher*, 136.

[91] Herschel P. Babbage, 'On the Mechanical Arrangement of the Analytical Engine', in Herschel P. Babbage, *Babbage's Calculating Engines* (London, 1889), 331–7. See Babbage, *Works*, vol. 2, 190.

[92] Alan G. Bromley, 'Table Making and Calculating Engines', in I. Bernard Cohen, 'Foreword', *Works of Charles Babbage*, vol. 1, 25.

[93] Ibid; Moseley, *Irascible Genius*, 140–4.

[94] Sophia De Morgan, *Memoir of Augustus De Morgan, by his wife Sophia Elizabeth De Morgan, with selections from his letters* (London, 1882), 89. Doris Langley Moore suspects that Lady Byron had heard this anecdote from another and had not been present herself. See Moore, *Ada, Countess of Lovelace*, 44.

Bernoulli, which I had offered to do to save Lady Lovelace the trouble. This she sent back to me for an amendment, having detected a grave mistake which I made in the process. The notes of the Countess Lovelace extend fully into almost all the very difficult and abstract questions connected with the subject.[95]

Because the essay is so important and because Ada's notes are so insightful, not to say original, the degree to which she acted alone, with Babbage's help, or at Babbage's dictation, has excited a lively controversy with feminist implications. Certainly there were 'countless letters back and forth' between the two as she worked on the translation and commentary.[96] Ada has her champions as a pioneer of computing, the 'first programmer' in fact, and her detractors.[97] Certainly Babbage referred to her as his 'enchantress of numbers' and his 'dear and much admired Interpretress.'[98] She was, he wrote to Michael Faraday, 'that Enchantress who has thrown her magical spell around the most abstract of Sciences and has grasped it with a force which few masculine intellects (in our own country at least) could have exerted over it.'[99] For our purposes, however, it is not necessary to enter the controversy but merely to note that Babbage was involved in the production of Ada's notes and he judged them very highly. It is their fidelity to his project and his approbation of her ideas which is important here, because Ada's notes are quite unlike the essays and assessments left by others, including by Babbage himself.

Ada certainly contributed important technical insights. In the last of her seven notes to Menabrea's text, each an essay in itself, she included what is sometimes claimed to be the first ever computer program to enable the Analytical Engine to compute the sequence of Bernoulli numbers (each of which is derived from the sum of all the previous numbers preceding it in the sequence). She also recognized that the Analytical Engine could be 'programmed' to work with symbols

---

[95] Babbage, *Passages from the Life of a Philosopher*, 136.

[96] Snyder, *The Philosophical Breakfast Club*, 272.

[97] See, for example, Bromley, 'Table Making and Calculating Engines', Babbage, *Works*, vol. 1, 25: 'A romantic myth has grown up surrounding Ada Lovelace to the effect that she originated the programming ideas discussed in her notes. This is discredited by modern biographical work, and the fact that all her examples are now known to have been prepared by Babbage.' See also Doron Swade, *The Cogwheel Brain*, 167: 'The conception and major work on the Analytical Engine were complete before Ada had any contact with the elementary principles of the Engines. The first algorithms or stepwise operations leading to a solution – what we would now recognize as a "program", though the word was not used by her or by Babbage – were certainly published under her name. But the work had been completed by Babbage much earlier.' According to Anthony Hyman, 'most writing about Ada belongs not to history but to the genre of faction - fiction presented as fact': H. W. Buxton, *Memoir of the Life and Labours of the Late Charles Babbage* (1872–80) (ed. A. Hyman, Cambridge, Mass., 1988), 'Introduction', xviii, note 7. For a different view of Ada's mathematical abilities see Christopher Hollings, Ursula Martin, and Adrian Rice, 'The Early Mathematical Education of Ada Lovelace', *BSHM Bulletin: Journal of the British Society for the History of Mathematics*, vol. 32, 3, 2017, 221–34.

[98] Charles Babbage to Ada Lovelace, 9 September 1843 quoted in Moseley, *Irascible Genius*, 191.

[99] Charles Babbage to Michael Faraday, 9 Sept. 1843, *The Correspondence of Michael Faraday* (ed. Frank A. James) (Stevenage, 1996), vol. 3, 164.

and other notation, as well as numbers. Ada's addition of mathematical commentaries and examples to Menabrea's text has been admired and celebrated, therefore. But the spiritual, metaphysical and mystical qualities to Ada's writing have received less attention, or have been dismissed for their grandiloquence.[100] These are what give the notes their resonance today, however, combined with her sense that Babbage's work would indeed have remarkable implications for, and impact upon, the future. What are described as 'notes' in fact compose another foundational text in the history of computing and its relationship to numbers.

Like Babbage and other members of the Cambridge Network, Ada intuited the role that numbers must play in the study of the natural world: 'it is by the laborious route of analysis that the man of genius must reach truth; but he cannot pursue this unless guided by numbers; for without numbers it is not given to us to raise the veil which envelopes the mysteries of nature.' To a nineteenth-century mind brought up on natural theology—the doctrine that nature was created for, and adapted to, mankind's requirements—it was mathematics above all that would allow an understanding of the creator's purpose: 'this science constitutes the language through which alone we can adequately express the great facts of the natural world…mathematical truth [is] the instrument through which the weak mind of man can most effectually read his Creator's works.'[101] An essentially similar observation would be made later by another woman crucial to the history of the statistical movement, Florence Nightingale, as we shall see. Meanwhile, Babbage's engines offered the only way in which to undertake the type of multivariate analyses of inter-related social phenomena that, as we have seen, the compilers of the *Statistical Illustrations* were attempting. As Ada explained,

> It must be evident how multifarious and how mutually complicated are the considerations which the working of such an engine involve. There are frequently several distinct sets of effects going on simultaneously; all in a manner independent of each other, and yet to a greater or less degree exercising a mutual influence. To adjust each to every other, and indeed even to perceive and trace them out with perfect correctness and success, entails difficulties whose nature partakes to a certain extent of those involved in every question where conditions are very numerous and inter-complicated; such as for instance the estimation of the mutual relations amongst statistical phaenomena, and of those involved in many other classes of facts.[102]

The phrase 'the estimation of the mutual relations amongst statistical phaenomena' suggests that Ada could readily appreciate the relationship between the

[100] Doris Langley Moore, *Ada, Countess of Lovelace*, 156.
[101] Ada Lovelace, 'Sketch of the Analytical Engine', note A.
[102] Lovelace, 'Sketch of the Analytical Engine', note D.

profusion of numbers in early Victorian society and the calculating engine which might make sense of them.

She could also see that calculating engines might be able to process what we now call 'Big Data', information too vast for human analysis and computation unaided by machinery:

> Herein may reside a latent value of such an engine almost incalculable in its possible ultimate results. We already know that there are functions whose numerical value it is of importance for the purposes both of abstract and of practical science to ascertain, but whose determination requires processes so lengthy and so complicated, that, although it is possible to arrive at them through great expenditure of time, labour and money, it is yet on these accounts practically almost unattainable; and we conceive there being some results which it may be absolutely impossible in practice to attain with any accuracy, and whose precise determination it may prove highly important for some of the future wants of science, in its manifold, complicated and rapidly-developing fields of inquiry, to aim at.[103]

Casting forward, Ada sensed the usefulness of a calculating engine to many different disciplines and subjects faced with very large data sets or vast numbers of calculations:

> Thus, the engine may be considered as a real manufactory of figures, which will lend its aid to those many useful sciences and arts that depend on numbers. Again, who can foresee the consequences of such an invention? In truth, how many precious observations remain practically barren for the progress of the sciences, because there are not powers sufficient for computing the results!

Historians of computing have often been accused and convicted of writing a version of whig history, meaning a present-minded and simplified account of the past, based on casting back in time to find the roots of present technologies and practices. But when assessing Babbage's foundational role in the development of the computer, and the interpolations of the Countess Lovelace, it is striking that both these people cast forward automatically, recognizing that whatever the technical and financial limitations of their own age, there would come a day when the calculating engine would be a requirement for the sciences. As early as 1822, in the first year of the calculating engine's projection, Babbage wrote presciently to David Brewster to

---

[103] Ibid, note F.

venture to predict, that a time will arrive, when the accumulating labour which arises from the arithmetical applications of mathematical formulae, acting as a constantly retarding force, shall ultimately impede the useful progress of science, unless this or some equivalent method is devised for relieving it from the overwhelming incumbrance of numerical detail.[104]

Later, in 1837, when in the thick of the designs for the Analytical Engine, Babbage cast forward to imagine a future in which natural scientists and researchers presented their data in numerical form precisely so that they could take advantage of the powers of a computer. Those who failed to do this would suffer the consequences: high costs, mistakes, and by implication, delay.

Whenever engines of this kind exist in the capitals and universities of the world, it is obvious that all those enquirers who wish to put their theories to the test of number, will apply their efforts so to shape the analytical results at which they have arrived, that they shall be susceptible of calculation by machinery in the shortest possible time, and the whole course of their analysis will be directed towards this object. Those who neglect the indication will find few who will avail themselves of formulae whose computation requires the expense and error attendant on human aid.[105]

He was echoed by William Farr in his later tribute to Babbage after his death. Farr used simple calculating machines in his work in the General Register Office and he envisaged computers set to work in social science:

I feel persuaded that ere many years an analytical machine will be at work, calculating accurately not only those elaborate numerical coefficients of the moon which puzzle the greatest adepts, but those still more complicated coefficients and variables which, it is easy to foresee, will be in requisition when future State problems are dealt with scientifically by a political Newton.[106]

Throughout Babbage's papers there are moments of foresight and intuition which strike a modern reader as remarkably accurate predictions. In Ada's notes there is a similar predictive power in association with a wider faith in the capacity of numbers to unlock nature's secrets. Menabrea himself sensed this years later, after Ada's death. In a short memoir in a major European scientific journal, he praised 'the notes which accompanied the translation of my little article' as 'remarkable';

---

[104] Charles Babbage, 'On the Theoretical Principles of the Machinery for Calculating Tables', 128.
[105] Charles Babbage, 'On the Mathematical Powers of the Calculating Engine', in Charles Babbage, *Works*, vol. 3, 60–1.
[106] Farr, 'Inaugural Address', 415.

they 'displayed, in their author, an extraordinary wisdom'. He gallantly gave praise to Ada, 'a lady as distinguished by the elevation of her mind as by her beauty, and who died, a few years ago, at the most brilliant period of her life'. He wished to draw attention, he explained, 'not to my modest work, but to the notes and commentaries in the translation and which are such as to make known the aim and power of the *analytical engine*'.[107]

As contemporaries as well as posterity recognized, Ada captured something beyond the technical and utilitarian in her commentaries, that sense of wonder that we feel every time we log on, surf the internet, or use a device for calculations, whether small or large. She felt this herself. Writing to Babbage in July 1843 as they were working together on her notes, and only half in jest, she described herself as a 'poor little fairy' and drew attention to the 'play and scope' of her imagination.

> That brain of mine is something more than merely mortal, as time will show…Before ten years are over, the Devil's in it if I haven't sucked out some of the life blood from the mysteries of *this* universe, & in a way that no purely mortal lips or brains could do…It is well for the world that my line & ambition is ever the *spiritual*, and that I have not taken it into my head to deal with the sword, poison and intrigue, in place of x, y & z.[108]

Ada's true strength 'lay in her extraordinary capacity to take on large, overarching ideas and interpret them not only boldly but imaginatively'.[109] According to William Farr, Ada's 'remarkable paper had a cheering effect on the Inventor's mind: he was now understood; and refreshed, he for the rest of his life never lost sight of the ideal analytical machine'. Farr went still further in his praise: 'if an analytical machine be ever at work—as it will be—in this age, it will be due in no small degree to a young and beautiful woman, the wife of a distinguished Fellow of this Society—Ada sole daughter of the house and heart of one of England's greatest poets.'[110]

The argument of this chapter is not that through Babbage the statistical movement gave rise directly to the modern electronic computer, for that would be crude and on the basis of the evidence, insupportable. But statistics and statistical organizations were important to Babbage, and he and his greatest interpreter envisaged the role that the calculating engines of the future might play in the

---

[107] L. F. Menabrea, 'Letter to the Editor of *Cosmos*', *Cosmos*, vol. 6, 1855, 421–2 in Babbage, *Works*, vol. 3, 171–4.

[108] Ada Lovelace to Charles Babbage, 5 July 1843, Babbage Papers, British Library, Add. Mss. 37192, ff. 350–3.

[109] Mark Bostridge, 'Was Ada Lovelace the True Founder of Silicon Valley?', *The Spectator*, 17 March 2018. (Review of Miranda Seymour, *In Byron's Wake: The Turbulent Lives of Lord Byron's Wife and Daughter: Annabella Milbanke & Ada Lovelace*, London, 2018.)

[110] Farr, 'Inaugural Address', 415.

analysis of numerical data. It would be false to claim that Babbage's statistical interests, dating back to the 1820s, played a direct role in leading him towards mechanical contrivance. He had imagined and begun designing a calculating engine some years before he displayed a serious interest in statistics, after all. Rather, for a savant fascinated by 'number', the statistical movement, which he had helped to create and institutionalize, was another element, though really rather an important one, in the rich intellectual context out of which 'Mr Babbage's engines' emerged. Supercomputers and the technology to process 'big data' allow us to do now what Babbage intuited and imagined then, but which could not be achieved within the existing technology. 'In the not-so-distant future, machines will be able to sift ever-greater data sets and offer conclusions that hitherto have been visible only to researchers who had spent years trawling through archives or decades making various kinds of experiments.'[111] As William Farr acknowledged more than 150 years before that was written, 'the field of statistics is immense. Its facts are innumerable, and the number of their possible combinations approaches the infinite. No one mind can marshall them all.'[112] But now we have 'analytical engines'—'minds'—of enormous power and sophistication to arrange and analyse the numbers for us.

John Burrow has argued that 'the intellectual world of the 1900s, though much of it now inevitably seems archaic, unacceptable, even ominous as well as portentous, is closer to ours in a number of ways, chiefly through the loss of various kinds of innocence, than is that of the 1850s'.[113] On the face of it such a judgment might seem uncontroversial, even self-evident. I would argue instead that the early Victorian statisticians and natural scientists are our contemporaries. The relationships they tried to create between numbers and welfare, the intellectual ambition they exhibited in these imagined technical innovations, and their ultimate scientific aspirations, were unrealizable in their own time. But all are now achievable by the technological means that Babbage and his illustrious pupil foresaw.

---

[111] Daniel Korski, 'Universities are about to Enter a New Era', *Daily Telegraph*, 24 Jan. 2017. (Korski was deputy head of the 10 Downing Street Policy Unit under Prime Minister Cameron).

[112] William Farr, *Report Submitted to the Organization Commission on the Programme of the Fourth Session of the International Statistical Congress* (London, 1860), 9.

[113] J. W. Burrow, *The Crisis of Reason. European Thought, 1848–1914* (New Haven and London, 2000), 67.

6

Richard Jones and William Whewell

Statistics, Induction, and Political Economy

Richard Jones and William Whewell were members of the Cambridge network of scientists and savants in the 1820s and 1830s. They had been undergraduates together, were both clergymen, served as each other's academic confidante, and were among the key founders of the Statistical Section of the British Association and of the Statistical Society of London. Whewell ranged across all scientific disciplines; Jones sought only to apply statistical approaches to the study of political economy. Together, they developed a critique of the orthodox methods of economics as it was then understood, and sketched a new approach. They cannot be said to have dislodged conventional views of the subject in the 1830s, but in retrospect, Jones has been held up as a pioneer of empirical and so-called 'historical' economics. The letters between the two friends, their publications of the early 1830s, as well as their roles founding new organizations, illustrate the creative intellectual potential of the statistical data newly available. By testing the doctrines of political economy against the statistics of actual economic behaviour they hoped to broaden the study of economic interactions into a genuine 'science of society'.

Orthodoxy was represented by the political economy of David Ricardo, the highly successful London stockbroker and member of parliament who published in 1817 *On the Principles of Political Economy and Taxation*, the summation of the subject for the next generation. As Whewell recollected that era in his old age,

> By the labours of various writers, culminating in the treatise of Mr Ricardo, Political Economy had become in great measure a deductive science: that is, certain definitions were adopted as of universal application to all classes of society, and from these definitions, and a few corresponding axioms, was deduced a whole system of propositions which were regarded as of demonstrated validity.[1]

'Ricardianism' denoted a supposedly value-free economics, based on a priori assumptions about human likes, desires, and needs, which were the basis for a series of deductions about economic behaviour that would hold everywhere, and

---

[1] William Whewell (ed.), 'Prefatory Notice', in *Literary Remains, Consisting of Lectures and Tracts on Political Economy of the late Rev. Richard Jones* (London, 1859), ix.

Victorians and Numbers: Statistics and Society in Nineteenth Century Britain. Lawrence Goldman, Oxford University Press.
© Lawrence Goldman 2022. DOI: 10.1093/oso/9780192847744.003.0006

at all times.² So to one Ricardian, Nassau Senior, in his inaugural lecture as the first holder of the Drummond chair in Political Economy in Oxford in 1826, the 'theoretic branch' of political economy rested 'on a very few general propositions, which are the result of observation, or consciousness; and which almost every man, as soon as he hears them, admits as familiar to his thoughts, or at least as included in his previous knowledge.'³ In opposition, members of the Cambridge network argued that this method and approach could not grasp economic reality, which varied everywhere according to context and conditions. As Jones put it,

> Mr Ricardo was a man of talent, and he produced a system very ingeniously combined of purely hypothetical truths; which, however, a single, comprehensive glance at the world, as it actually exists, is sufficient to show to be utterly inconsistent with the past and present condition of mankind.⁴

They believed that political economy had been prematurely formalized so as to prevent the consideration of economic diversity and difference. As Malthus had written in his textbook of economics, published three years after Ricardo's work, 'The principal cause of error, and of the differences which prevail at present among the scientific writers on political economy appears to me to be a precipitate attempt to simplify and generalize.'⁵ Whewell was to become an influential historian of science as well as a practitioner, the author of a vast *History of the Inductive Sciences* published in 1837.⁶ As he wrote in a published review of Jones's book in 1831, 'the reduction of any province of knowledge' to the form of 'a few general and simple laws...is the last grand act and denouement of its history'. That an academic discipline both so complex and so relatively young as political economy 'should have sprung at once to this ultimate condition, this goal and limit of its possible intellectual progress' was a truncation of natural academic development.⁷ Whewell's essay was an 'indictment of Ricardian doctrine through an assault on the political economists' terminology'.⁸ As he wrote of the Ricardians to Jones in the early 1830s,

² Mark Blaug, *Ricardian Economics: A Historical Study* (New Haven, 1958).
³ Nassau Senior, 'An introductory lecture on political economy, delivered before the University of Oxford, 6ᵗʰ December 1826', *The Pamphleteer*, xxix (London, 1828), 35.
⁴ Richard Jones, *An Essay on the Distribution of Wealth and on the Sources of Taxation. Part 1. Rent* (London, 1831), 7.
⁵ T. R. Malthus, *Principles of Political Economy Considered with a View to Their Practical Application* (London, 1820), 5–6.
⁶ William Whewell, *The History of the Inductive Sciences, from the Earliest to the Present Times* (3 vols, London, 1837). Richard Yeo, *Defining Science. William Whewell, Natural Knowledge and Public Debate in Early Victorian England* (Cambridge, 1993).
⁷ [W. Whewell], 'An Essay on the Distribution of Wealth and on the Sources of Taxation', *The British Critic and Quarterly Theological Review*, vol. x, no. xix, July 1831, 55.
⁸ Simon Schaffer, 'The History and Geography of the Intellectual World: Whewell's Politics of Language', in M. Fisch and S. Schaffer (eds), *William Whewell. A Composite Portrait* (Oxford, 1991), 222.

They have begun indeed with some inference of facts; but, instead of working their way cautiously and patiently from there to the narrow principles which immediately inclose a limited experience, and of advancing to wider generalities of more scientific simplicity only as they become masters of more such intermediate truths – instead of this, the appointed aim of true and permanent science – they have been endeavouring to spring at once from the most limited and broken observations to the most general axioms.[9]

Similar criticisms are made of economics today which has swapped the study of economic interactions as they actually occur for mathematical modelling and generalizations founded on the study of a narrow range of behaviours in market societies.

Richard Jones holds an uncertain position in the history of economics, largely because his publications represented only a fraction of his planned output. He intended to produce a work encompassing the four categories of distribution in classical political economy: wages, profits, taxation, and rent. Only his work on rent was ever completed and published. Until recently he was overlooked.[10] As a sceptical Joseph Schumpeter put it, 'he does not badly fit the role that historians of economics assigned to him, although it is not easy to be sure what either his programmatic pronouncements or his example amounted to'.[11] Some commentators have underplayed Jones's status as a critic of Ricardo.[12] Another, however, has compared him to the more famous 'German Historical School' of Roscher, Kneis, Hildebrand, and Schmoller later in the nineteenth century.[13] It is clear that some British 'historists', among them T. E. Cliffe Leslie, J. K. Ingram, W. J. Ashley, interested in the contexts of economic behaviour, took inspiration from Jones's work.[14] According to Alfred Marshall indeed, Jones's 'influence, though little heard of in the outside world, largely dominated the minds of those Englishmen who came to the serious study of economics after his work had been published by Dr. Whewell in 1859'.[15] But this, too, has been contested by a more recent historian of economic thought.[16]

[9] Whewell to Jones (n.d., 1830 or 1831?), Whewell Papers, Add MSS c. 51/92.
[10] He was not mentioned at all in the otherwise admirable essay by Joseph J. Spengler, 'On the Progress of Quantification in Economics', *Isis*, 52, 2, June 1961, 258–76.
[11] J. A. Schumpeter, *A History of Economic Analysis* (New York, 1954), 822.
[12] Blaug, *Ricardian Economics*, 153; W. L. Miller, 'Richard Jones: A Case Study in Methodology', *History of Political Economy*, iii, 1971, 198–207.
[13] Eric Roll, *A History of Economic Thought* (London, 1973 edn.), 311–18.
[14] A. W. Coats, 'The Historist Reaction in English Political Economy 1870–1890', *Economica* (1954), 145; G. M. Koot, 'T. E. Cliffe-Leslie, Irish Social Reform and the Origins of the English Historical School of Economics', *History of Political Economy*, vii (1975), 313.
[15] A. C. Pigou (ed.) *Memorials of Alfred Marshall* (New York, 1956 edn), 296.
[16] Rashid demonstrated the ignorance of Jones's work among late-Victorian economists and concluded 'that the "dominating influence" of Jones upon economists after 1859 is largely invisible. Marshall, it appears, has been misled by his love of continuity of thought'. Salim Rashid, 'Richard Jones and Baconian Historicism at Cambridge', *Journal of Economic Issues*, xiii, 1, March 1979, 159–73.

Karl Marx was undoubtedly impressed. He commended Jones's *Essay* in his manuscript of 1862–3 known as *Theories of Surplus Value* because Jones had demonstrated 'what has been lacking in all English economists since Sir James Steuart, namely a sense of the historical differences in modes of production'. For this reason, Jones's work 'distinguishes him from, and shows his superiority over, all his predecessors'.[17] Jones's approach was broad, holistic. Economic relations were intimately connected to political institutions, social behaviour and cultural life; indeed, they determined them. In Jones's words, he sought to investigate 'the mutual relations and influences of different orders of men as determined by different modes of producing and distributing public wealth'.[18] With language like this, it is little wonder that he influenced Marx: 'How the [economic] relations and consequently the social, moral and political state of nations change with the change in the national powers of production is very well explained', Marx commented.[19] He drew extensively on Jones's work on rural society in general and on Asian rents in particular for his own account of the Asiatic mode of production. His fourfold division of rents in *Capital* was based on Jones's categories.[20]

As early as 1825 Whewell was hoping to 'find the demolition of the Ricardites very forward if I see you, for it is a proper adventure for you to set out to kill such a dragon as that system'.[21] Jones envisaged an inductive economics:

If we wish to make ourselves acquainted with the economy and arrangements by which different nations of the earth produce and distribute their revenues, I really know of but one way to attain our object, and that is, to look and see. We must get comprehensive views of facts, that we may arrive at principles that are truly comprehensive.[22]

Those facts were to come from 'two sources of knowledge—history and statistics—the study of the past and detail of the present condition of the nations of the earth.'[23] Applying this method, and drawing on a variety of historical and geographical sources, Jones concluded that 'rent' was a non-unitary category of payment falling into five classes across the globe: serf, metayer, ryot, cottier, and farmer rents. The Ricardian system, based on the observation of only the final

---

[17] K. Marx, *Theories of Surplus Value*, Pt. III (Moscow edn, 1971, trans. J. Cohen and S. W. Ryazanskaya), 399.

[18] Jones, 'Introductory Lecture' in Whewell (ed.), *Literary Remains*, 575.

[19] Marx, *Theories of Surplus Value*, III, 430.

[20] Karl Marx, *Capital*, vol. III, ch. 47. https://www.marxists.org/archive/marx/works/1894-c3/ch47.htm See Lawrence Karder, *The Asiatic Mode of Production. Sources, Development and Critique in the Writing of Karl Marx* (Assen, 1975), 7, 54–5, 182, 307.

[21] Whewell to Jones, 24 May 1825, Whewell Papers, Add MSS c. 51/21.

[22] Richard Jones, 'An Introductory Lecture on Political Economy, Delivered at King's College, London, February 27th, 1833', in William Whewell (ed.), *Literary Remains*, 568–9.

[23] Ibid., 570.

class, was practically limited to the state of agriculture in England, the Netherlands, and part of the United States, but 'one hundredth part of the cultivated surface of the habitable globe'.[24] This narrow base was turned into a strict and universal definition of rent that Ricardian theory could then process and build on. But it followed from Jones's inductive approach, based on many different economic systems and positively seeking to incorporate behavioural diversity across the world, that precise definitions were impossible and inapplicable. As he explained in his *Essay*,

> It has been mentioned to me, that I have given no regular definition of the word Rent. The omission was not undesigned. On a subject like this, to attempt to draw conclusions from definitions is almost a sure step towards error...I have pointed out the origin of payments made to the owner of the soil. I have tracked their progress. If any reader, during this enquiry, is really puzzled to know what we are observing together, I shall be sorry: but I am quite sure that I should do him no real service, by presenting him in the outset with a definition to reason from.[25]

Jones's inductivism thus made possible a comparative analysis of land-tenure systems and their determining influence on social structures and forms of government under each system, without the restrictions which applied to conventional economic analysis. As he wrote to Whewell,

> What I value most in my book is the revelation (for it is one) of the intimate connections between the subsistence of the body of the people and the rents they pay, over almost the whole of the old world – By this fact, and its reaction on government and society, the past history, the present condition, the future progress of a vast majority of human communities, have been, are and will be for ages influenced.[26]

Among members of the Cambridge Network, Jones probably possessed the most imaginative grasp of the potential of inductivism to change political economy for the better. But his closest intellectual associates had publicly expressed disagreement with Ricardianism, as well. Whewell published two essays in 1829

---

[24] Jones, *Essay on the Distribution of Wealth*, 14. In accordance with Jones, and as more recently argued, 'the term "rent" is not an economic category with a unilateral application in agrarian economics, for the conditions of existence of rent as a relation linking land and labour and at the same time dividing them, are not eternally given.' Keith Tribe, *Land, Labour and Economic Discourse* (Cambridge, 1978), 33.

[25] Jones, *Essay on the Distribution of Wealth*, xlxi.

[26] Jones to Whewell, 4 March 1831, Whewell Papers, Add MSS c. 52/25.

and 1833[27]—'my mathematico-politico-economics' as he described it to Jones[28]—which had first been delivered to the Cambridge Philosophical Society, and which attempted to streamline Ricardian deductions by expressing them algebraically: 'to present in mathematical form some of the doctrines which have been delivered as part of the science of Political Economy'.[29] Indeed, these works amount to the first examples in the history of economic thought of the setting out of economic theory in mathematical form: the origins of econometrics.[30] But this didn't imply Whewell's assent to Ricardianism: as he wrote to James David Forbes, 'Pray recollect that I profess only to trace the consequences of Ricardo's principles – principles which I am sure are insufficient and believe to be entirely false'.[31] Rather, Whewell was attempting to consolidate a standard analytical procedure. The articles were saturated with criticism of the applicability of 'Mr Ricardo's system' to the actually-existing economic order. Moreover, Whewell, though from a humble background in the Lake District—he was the son of a master carpenter—was a confirmed Tory with paternalistic leanings, and had social and, as a clergyman, religio-ethical objections to the notorious 'whig science' and its 'strange supposition that the new economical instincts of individuals should but produce their effect of productiveness by acting unconnected and unconstrained'.[32]

These reservations were undoubtedly shared by Jones, another ordained Anglican attracted to notions of civic harmony who deplored the social implications of the Ricardian system that seemed to set the interests of the landed class against those of all other classes. Jones's stated aim was 'the establishment of sound political economic views in a subject which has hitherto only called itself a science to enforce a dogmatical philosophy of the most pernicious kind'.[33] His disagreements with Ricardianism were never purely methodological, therefore. He disputed the pessimism about natural checks to population growth built into Ricardian theory, for example, demonstrating that the growth of secondary wants led voluntarily to later marriage and fewer children. He also argued that 'improvements in the arts of production' contradicted the law of diminishing agricultural returns.[34] Neither Jones nor Whewell could have found a place easily

[27] William Whewell, *Mathematical Exposition of Some Doctrines of Political Economy* (Cambridge, 1829) reprinted in *Transactions of the Cambridge Philosophical Society*, vol. iii, 1831; 'Mathematical Exposition of some of the Leading Doctrines in Mr. Ricardo's *Principles of Political Economy and Taxation*', *Transactions of the Cambridge Philosophical Society* (Cambridge, 1833), vol. iv, 1833, 155–98.
[28] Whewell to Jones, 5 March 1829, Whewell Papers, Add MSS c. 51/62.
[29] Whewell, 'Mathematical Exposition', 191.
[30] Laura J. Snyder, *The Philosophical Breakfast Club. Four Remarkable Friends who Transformed Science and Changed the World* (New York, 2011), 103. Michael Ruse, 'William Whewell: Omniscientist', in Fisch and Schaffer (eds), *William Whewell. A Composite Portrait*, 102–6.
[31] William Whewell to J. D. Forbes, 14 July 1831 in Isaac Todhunter, *William Whewell D.D.: An account of his writings with selections from his library and scientific correspondence* (2 vols, London, 1876), II, 122.
[32] Whewell to Jones, 12 Jan. 1832, Whewell Papers, Add MSS c. 51/125. Yeo, *Defining Science*, 105.
[33] Jones to Whewell, 27 Sept. 1827, Whewell Papers, Add MSS c. 52/15.
[34] Jones, *Essay on the Distribution of Wealth*, 197–202.

in the Manchester Statistical Society whose members upheld a liberal political economy so different from these views: it is another example of the variety and divergence within the so-called 'statistical movement'. Babbage, as we have seen, reproached political economists for 'too small a use of facts and too large an employment of theory' in 1832 in *On the Economy of Machinery and Manufactures.*[35] As for Malthus himself, the experience of collecting data for subsequent editions of his *Essay on the Principle of Population* led to his growing awareness of variations in basic demography from place to place:

> The habits of most European nations are of course much alike, owing to the similarity of the circumstances in which they are placed; and it is to be expected therefore that their registers [of births, deaths and marriages] should sometimes give the same results. Relying however too much upon this occasional coincidence, political calculators have been led into the error of supposing that there is, generally speaking, an invariable order of mortality in all countries; but it appears, on the contrary, that this order is extremely variable; that it is very different in different places of the same country, and within certain limits depends upon circumstances, which it is in the power of man to alter.[36]

Malthus had begun his *Principles* of 1820 with a severe critique of the deductive method and an extended plea for induction: 'Before the shrine of truth as discovered by facts and experience, the fairest theories and the most beautiful classifications must fall.'[37] He was not as pronounced and public an inductivist as Jones and Whewell, but his very presence at the key meetings in the foundation of Section F and the Statistical Society in 1833–4 suggests his support for their approach, plans and projects.[38]

It is true that criticism of Ricardo's system was becoming widespread by the early 1830s. For example, the Political Economy Club in London twice discussed in early 1831 the question asked by one of its members, Robert Torrens: 'What improvements have been effected in the science of political economy since the publication of Mr Ricardo's great work; and are any of the principles first advanced in that work now acknowledged to be correct?'[39] No orthodox economist would

---

[35] Charles Babbage, *On the Economy of Machinery and Manufactures* (London, 1832), 119.

[36] T. R. Malthus, *An Essay on the Principle of Population* (1798) (6th edn, London, 1826), in *The Works of Thomas Robert Malthus* (eds E. A. Wrigley and David Soudan) (8 vols, London, 1986), 259–60.

[37] Malthus, *Principles of Political Economy*, 7.

[38] For this reason it would be incorrect to argue that Malthus 'was not inclined to align himself with Ricardo's inductivist critics'. See N. B. de Marchi and R. P. Sturges, 'Malthus and Ricardo's Inductivist Critics: Four Letters to William Whewell', *Economica*, 1973, 379–93. Cullen, meanwhile, mistakenly takes Malthus as a model of deductive orthodoxy. See M. J. Cullen, *The Statistical Movement in Early Victorian Britain: The Foundations of Empirical Social Research* (Hassocks, Sussex, 1975), 81.

[39] *Political Economy Club, Minutes of Proceedings, 1899–1920. Roll of Members and Questions Discussed, 1821–1920, with Documents bearing on the History of the Club*, vi (London, 1921), 223.

have agreed with Ricardo on all points, but disagreement concerned matters of detail, policy and deduction: only the 'Cambridge inductivists' attacked the very method of political economy, its reliance on deduction from 'self-evident truths' of human behaviour and its restriction to the existing economic state of Britain. And it was this methodological antagonism which encouraged Jones to lead in the formation of Section F which was to have 'wider views' than merely to serve as 'a subordinate to the inquiries of the political economist', as he told the gathering in his rooms in Cambridge in June 1833. Jones would exclude 'no classes of facts relating to communities of men' in the study of society, and he would use these empirical foundations as the only basis of 'general laws'.[40] In their correspondence and publications over the preceding decade the Cambridge network had discussed an alternative scientific and economic methodology. To them, Section F and the Statistical Society of London were means to the realization of a new critical intellectual project: to uncover and demonstrate social structures and processes by investigating, according to the *Third Annual Report* of the Society's council, 'the whole range of economical, political, medical, moral and intellectual statistics'.[41] Jones divined a disciplinary area 'beyond political economy, strictly so-called' which would unfold 'the shifting political and social influences which accompany the march of nations from rudeness and feebleness to power and civilisation'.[42] Whewell wrote to him to commend that very project: 'To bring the facts of the historical and economical conditions of nations under general laws, when once done, will never be forgotten, and the effects of such a view will be forthwith and forever operative.'[43]

In 1831 and 1832 Whewell referred to Jones in letters to him as a 'social economist' and to his style of analysis as 'social economy'.[44] We have already encountered 'social economy' as the term used some years earlier in the artisans' *Statistical Illustrations* to describe and define their critical economics. It was deliberately different from 'political economy' which the artisans opposed, just as Whewell and Jones did, because of its methodological limitations and anti-social doctrines. And we have seen how John Stuart Mill formalized the distinctions being made between these two approaches to economics in his 1836 essay 'On the Definition of Political Economy'. In the same article Mill made one of the first public references in English to 'social science', though he had used the term in his

[40] 'Adjourned Meeting, Friday 28 June 1833, Revd Richard Jones, Chairman', Transcript of the Note-book of Mr. J. E. Drinkwater. Minutes of the Committee of the Statistical Section of the British Association', Royal Statistical Society, London, Archives, f. 2.
[41] 'Third Annual Report of Council', 15 March 1837, 'Reports of Council and List of Fellows', Archives, Royal Statistical Society.
[42] Jones, 'Text book of Lectures', IV, 406.
[43] Whewell to Jones, 21 Dec. 1832, Whewell Papers, Add MSS c. 51/147.
[44] Whewell to Jones, 8 Nov. 1831 and 11 Oct. 1832, Whewell Papers, Add MSS c 51/118 and 51/142.

correspondence as early as 1829.[45] Bentham had used the term in 1812.[46] It also appears in the early 1830s in the writings of the socialists, Robert Owen and John Gray.[47] But 'social science' can be traced back further still to the French Revolution where and when it was used by the Parisian intellectual elite around the Marquis de Condorcet. And in the early 1830s there was another linguistic context for 'social science' among the French Saint-Simonians. They have secured a much more prominent and secure place among the founders of modern social thought than the members of the Cambridge network.

Henri de Saint-Simon and his disciples, including the first so-called sociologist, Auguste Comte, applied scientific methods drawn from biology and physics to the study of society. To the Saint-Simonians, science would inspire and direct social reform in favour of the working classes now emerging within the new industrial and technical society of the early nineteenth century. These groups included all workers, in fact, from entrepreneurs and technologists to manual labourers. They were pioneers of social change who would construct together an egalitarian technocracy, the utopian society that Saint-Simon and his followers envisaged and favoured.[48] It is assumed that Mill's contacts with them in the late 1820s provided for the introduction of the term 'social science' into English. As Mill recollected in his *Autobiography*,

> the writers by whom, more than any others, a new mode of political thinking was brought home to me, were those of the St Simonian School in France. In 1829 and 1830 I became acquainted with some of their writings...I was greatly struck with the connected view which they for the first time presented to me, of the natural order of human progress.[49]

Jones and Whewell shared his enthusiasm at almost the same time: in the first half of 1832 they exchanged letters extolling heterodox French philosophy. Far removed as they were from the political radicalism, technocracy, and utopian socialism of Saint-Simon, Jones was nevertheless enraptured: 'But oh! The Saint-Simonians!!! You must read and ponder and wonder and profit—*L'Organisateur* is, I am bold to say, one of the half-dozen most extraordinary and interesting books in the

---

[45] J. S. Mill to Gustave d'Eichtal, 3 Oct. 1829, cited in J. H. Burns, 'J. S. Mill and the Term "Social Science"', *Journal of the History of Ideas*, xx, 1959, 432.

[46] Bentham to Toribio Nunez, 21 April 1812, cited in K. M. Baker, 'The Early History of the Term "Social Science"', *Annals of Science*, XX, 1964, 225.

[47] G. G. Iggers, 'Further Remarks about the Early Uses of the Term "Social Science"', *Journal of the History of Ideas*, xx, 1959, 435.

[48] F. M. H. Markham (ed.), *Henri Comte de Saint-Simon 1760–1825: Selected Writings* (Oxford, 1952). G. G. Iggers (ed.), *The Doctrine of Saint-Simon: An Exposition: First Year 1828–9* (New York, 1958); Frank E. Manuel, *The Prophets of Paris: Turgot, Condorcet, Saint-Simon, Fourier and Comte* (Oxford, 1962).

[49] J. S. Mill, *Autobiography* (London, 1873), 163.

world—learned—logical—powerful' though also 'ignorant – unreasonable – feeble – mischievous and disgusting'.[50] Whewell caught the fever: 'I too have been reading the Saint-Simonian: who is the man that writes the *"Exposition"*? He must be a fine fellow: I am entirely charmed with the beauty and coherence of a great part of his theory'.[51] As Whewell recollected in the last months of his life in his review of Mill's essay, *Comte and Positivism*, 'Most readers at this time were deeply impressed by the largeness, subtlety and ingenuity of their views of society. Their doctrine of the alternation of critical periods and organic periods was really a startling theory, bringing together in a general view many historical facts'.[52]

Jones and Whewell would certainly have encountered the term 'social science' in their reading of both the *Exposition* of 1828–9, written by two of Saint-Simon's disciples, Amand Bazard and Barthélemy Prosper Enfantin, and Saint-Simon's own periodical *L'Organisateur*, published earlier in 1819–20, in which the very young Auguste Comte collaborated. They recognized a similarity of method, range and aim, as well: as Jones observed, 'The St. Simonites began with speculation on induction and . . . then applied their instrument of discovery to history and politics'.[53] Undoubtedly all the key features of Jones's 'social economy' had been developed fully by 1832; his work betrays no evidence of direct Saint-Simonian influence. But interest in, and acclamation of, the major intellectual tradition which generated the very concepts of a 'science of society' and of 'sociology', as Comte was to term it later, is surely indicative of the aspiration of some of the key founders of the statistical movement in Cambridge in the early 1830s to construct their own social science. Like Jones and Whewell, Comte was also a critic of the claims and scope of orthodox political economy.[54]

The Statistical Society of London, under the influence of its founders from Cambridge, was not far behind. In its fourth annual report in 1838 it noted the 'appearance of a general conviction that, in the business of social science, principles are valid for application only inasmuch as they are legitimate inductions from facts, accurately observed and methodically classified'.[55] In the following year it reported that

---

[50] Jones to Whewell, 17 June 1832, Whewell Papers, Add MSS c. 52/46.

[51] Whewell to Jones, 19 Feb. 1832, Whewell Papers, Add MSS c. 51/129. See also Whewell to Jones, 3 Feb. 1832 and Jones to Whewell 21 June 1832, Whewell Papers, Add MSS c. 51/128 and C. 52/47. During 1828 the Saint-Simonians held regular meetings in Paris to formulate their doctrine. At the end of the year Amand Bazard delivered a series of bi-weekly lectures entitled 'Exposition de la Doctrine St. Simonienne'. On religious and also economic themes, Barthélemy Prosper Enfantin was an influence over these. They were published as Amand Bazard, *Exposition de la doctrine de St Simon* (2 vols, Paris, 1828–1830). See Markham (ed.), *Henri Comte de Saint-Simon*, xxxv; Iggers (ed.) *The Doctrine of Saint Simon*, xxiii.

[52] W. Whewell, 'Comte and Positivism', *Macmillan's Magazine*, March 1866, 358. Richard Yeo, 'William Whewell's Philosophy of Knowledge and its Reception', in M. Fisch and S. Schaffer (eds), *William Whewell. A Composite Portrait* (Oxford, 1991), 183.

[53] Jones to Whewell, 17 May 1843, Whewell Papers, Add MSS c. 52/81.

[54] Raymond Aron, *Main Currents in Sociological Thought* (2 vols, Harmondsworth, 1968), I, 74.

[55] Fourth Annual Report of the Council of the Statistical Society of London, *JSSL*, May 1838, 8.

all the leading nations of Europe and the United States of America are striving to establish the true principles of social science, by prosecuting statistical labours with a continuously increasing conviction of the importance of collecting and publishing numerical facts affecting their social interests.[56]

Jones had seen the disciplinary space 'beyond political economy' and by the late 1830s it was being filled by a new numerical social science in these and other statements of intent, at least in theory.

Following Jones and Whewell, a minority of Victorian political economists in each successive generation continued to struggle against deductive orthodoxies, and usually their methodological differences with the rest of their academic colleagues were associated with wider political dissent, as well. This was certainly the case among the English Positivists of the 1860s, led by Frederic Harrison, whose commitment to an inductive economics, while taking some of its inspiration from the sociology of the founder of their movement, Auguste Comte, was the intellectual concomitant of their public support for organized labour and trades' unionism in that decade.[57] In the mid-1880s, Beatrice Webb, the Fabian socialist, was working as an assistant to Charles Booth, researching social and economic life in East London at the start of what became Booth's great study of *The Life and Labour of the People of London*, to be discussed below. She also criticized the 'self-contained science styled Political Economy, apart from the study of human behaviour in society' and argued that economic science should adopt a much wider perspective to include 'the birth, growth, disease and death of actual social relationships'.[58] In the 1890s, William Cunningham, who, like Ingram, Cliffe-Leslie, and Ashley, was positioned inside the English historical school of economics, took aim at deductive orthodoxy, personified by Alfred Marshall.[59] Both men taught at Cambridge, where Marshall was the senior, and both men participated in the Industrial Remuneration Conference, also to be discussed below.[60] But Cunningham, a clergyman and a Conservative, was a 'fair-trader', opposed to the continuation of free trade. His famous essay on 'The Perversion of Economic History', published in 1892, criticized Marshall for his 'neglect of serious study of facts' in the vague and generalized accounts of economic history he endorsed.

---

[56] Fifth Annual Report of the Council of the Statistical Society of London, *JSSL*, April 1839, 133.

[57] Frederic Harrison, *Autobiographic Memoirs* (2 vols, London, 1911), I, ch. xiv, 'Sociology and Economics', 245–99. Paul Adelman, 'Frederic Harrison and the "Positivist" Attack on Orthodox Political Economy', *History of Political Economy*, iii, 1971, 170–89.

[58] Beatrice Webb, 'On the Nature of Economic Science', Appendix D, *My Apprenticeship* (London, 1926), 373–5. See also ibid., 289–94.

[59] G. E. Koot, *English Historical Economics, 1870–1926. The Rise of Economic History and Neomercantilism* (Cambridge, 1988).

[60] John Maloney, 'Marshall, Cunningham, and the Emerging Economics Profession', *Economic History Review*, n.s., 29, 3, Aug. 1976, 440–51.

That neglect not only led to error but prevented 'the economist from finding out the narrow limits within which his generalizations are even approximately true'. Not only does this read as if written by Richard Jones; Cunningham even focused at one point on Marshall's treatment of the category of 'rent', the subject of Jones's best work, to make Jones's point again: that the Ricardian theory of rent, which Marshall had adopted, when applied to the past, was 'an anachronism, and shows a misunderstanding of the whole conditions of rural life.'[61]

Cunningham also criticized the work of the pioneer historian of living standards in British history, J. E. Thorold Rogers, who might have been thought an ally. If Marshall ignored the facts, Rogers piled them up in painstaking work assembling data on work, wages, and prices from past centuries.[62] But he then analysed these on the assumption that past economic practices and psychology were the same as in his own period. He took 'for granted that a motive which is often spoken as dominant in the present day was equally operative in the past'. Self-interest may not have been, always and everywhere, the mainspring of economic life, and even when it was, it may have varied in expression and type 'from age to age'. Rogers was 'misled from applying the habit of reasoning which is generally used for dealing with modern economic phenomena to distant periods'. He had investigated economic behaviour in isolation from social values, conventions, habits, and ways of thought—'from other sides of life'—that together provide the context for interpreting the data.[63]

> Society was different; people's wants were different; the things they prized were different; utility and disutility, if I must be technical, were different. We cannot isolate wealth from other social phenomena; we must view it in each age along with the social phenomena of the time.[64]

Cunningham's article, published with a riposte from Marshall himself,[65] 'precipitated a minor Methodenstreit in England' over the methodology of political economy.[66] It was an analogue to the larger, longer-lasting, and more profound conflict over method among German economists and sociologists of the late-nineteenth century.

---

[61] William Cunningham, 'The Perversion of Economic History', *The Economic Journal*, ii, Sept. 1892, 491–8.

[62] James E. Thorold Rogers, *Six Centuries of Work and Wages: The History of English Labour* (2 vols, London, 1884).

[63] Cunningham, 'The Perversion of Economic History', 498–506.   [64] Ibid., 505.

[65] Alfred Marshall, '"The Perversion of Economic History": A Reply', *The Economic Journal*, ii, Sept. 1892, 507–19.

[66] R. M. Hartwell, 'Cunningham, William', https://www.encyclopedia.com/social-sciences/applied-and-social-sciences-magazines/cunningham-william.

In each Victorian generation there were some practitioners of political economy in British public life who discovered for themselves the limitations of deductive political economy, just as Jones and Whewell had done in the 1820s and 1830s. Often, as in the case of Beatrice Webb, intellectual dissent was the product of actual social experience and investigation: the theory did not fit the facts she observed while tramping the streets of east London and interviewing the inhabitants in the late 1880s for her first essays in social science on 'dock life', the tailoring trade, the sweating system, and the Jewish community in the area.[67] In other cases, such as that of the Positivists, the theory—for example, the wage fund doctrine—taught that ultimately, nothing could be done to permanently improve working-class living standards. If that was the case, something was wrong with the theory, and also the methods used to arrive at such complacent and depressing conclusions. Through communication with leading trades' unionists, and through the study of contracts, wages, and the labour process, Frederic Harrison mapped out, in opposition, 'the limits of Political Economy', the title of a famous essay he published in 1865.[68] In Cunningham's case, his commitment to understanding economics historically, with reference to actual examples from the past and to changing social and institutional contexts, as well as his commitment to empire and Britain's imperial responsibilities, led him to take an unorthodox position on policy: that it was in the best interests of late-Victorian Britain to follow Germany and the United States by introducing tariffs. As these different examples suggest, Jones and Whewell did not form a long-lasting 'school'. There are few references to them in the work of these later inductivists, nor evidence that they had taken direct inspiration from the pair. Rather, like the proverbial wheel, economic induction was re-discovered and re-invented in each generation, often by economic dissidents, such as Positivists, Fabians, clergymen, and 'fair trade' imperialists. They were all, though for different reasons, opposed to the social consequences of laissez-faire orthodoxies. The Cambridge inductivists of the 1830s, pre-eminently Jones and Whewell, as well as the artisan networks that together compiled the *Statistical Illustrations* in the 1820s, deserve recognition as among the first to formulate a coherent intellectual critique of a purely deductive political economy, one that led them to collect statistics, create statistical societies, and envisage a holistic social science in which political economy would be a single component of a wider and more subtle synthesis. All those who have ever thrown data and numbers in the face of conventional economic theory have followed in their wake.

---

[67] Webb, *My Apprenticeship*, 311. The first three essays were published in the periodical the *Nineteenth Century* in October 1887, September 1888 and June 1890 respectively. Her essay on the Jewish community appeared in Booth's first volume, published in 1889.

[68] Frederic Harrison, 'The Limits of Political Economy', *Fortnightly Review*, vol. I, 1865, 356–76.

7

Adolphe Quetelet

Social Physics, Determinism, and 'The Average Man'

The statistician who exerted the most influence, personally and by example, on the intellectual development of the statistical movement came not from Britain but from the new Belgian state which had been established by the Treaty of London in 1830. Lambert Adolphe Jacques Quetelet, known as Adolphe Quetelet, was 'the grand old man' of the new science of statistics. He was born in Ghent in 1796 to a French father who had spent several years in Britain before he returned to the Low Countries. The son grew up in the midst of profound political instability. The southern Netherlands, hitherto controlled by Austria, were conquered by the first French Republic in 1794, became part of the United Kingdom of the Netherlands in 1815 at the end of the Napoleonic Wars, and were then transformed into Belgium after the 1830 Belgian Revolution established an independent, French-speaking, Catholic state. It was during this revolutionary era that official and governmental statistics were collected for the first time, and the practice continued after 1830.[1] It has been suggested that the political disruptions of Quetelet's youth, culminating in the military occupation in 1830 of the Royal Observatory, where he had recently become the resident astronomer, by volunteers defending Brussels from Dutch forces, disposed him to search for the rules of order and stability which mathematics might uncover in society as well as in the heavens.[2] As he wrote later,

> Dans un moment où les passions étaient vivement excites par les événements politiques, j'avais cherché, pour me distraire, à établir des analogies entre les principes de la mécanique et ce qui se passait sous mes yeux.[3]

---

[1] Nico Randeraand, 'The Dutch Path to Statistics (1815–1830)', in P. Klep and I. Stamhuis (eds), *The Statistical Mind in a Pre-Statistical Era: The Netherlands, 1750–1850* (Amsterdam, 2002), 99–123.

[2] Theodore M. Porter, *The Rise of Statistical Thinking 1820–1900* (Princeton, N. J., 1986), 46–7. Todd Rose, 'How the Idea of a 'Normal' Person Got Invented', *The Atlantic*, 18 Feb. 2016, https://www.theatlantic.com/business/archive/2016/02/the-invention-of-the-normal-person/463365/.

[3] 'At a time when strong feelings were stirred by political events, I sought, as a distraction, to establish analogies between the principles of mechanics and the things going on before my eyes.' L. A. J. Quetelet, *Du Système Social et des Lois qui le Régissent* (Paris, 1848), ch. vii, 'Analogies entre les lois physique et les lois morales', 104.

Victorians and Numbers: Statistics and Society in Nineteenth Century Britain. Lawrence Goldman, Oxford University Press.
© Lawrence Goldman 2022. DOI: 10.1093/oso/9780192847744.003.0007

Educated as a mathematician, by the early 1820s Quetelet was teaching at the Athenaeum in Brussels and in 1825 he founded the journal *Correspondance Mathematique et Physique*. But his prodigious output in physics and maths gradually shifted to the study of statistics and society. Asked to assist the work of the Belgian Royal Statistical Commission in preparation for a Belgian census, by 1829 he had become the census director. In 1841 he became president of the Belgian Commission Centrale de Statistique. From 1853 he presided over the International Statistical Congress which he had himself established.[4]

Quetelet was distinguished 'by his skill and acumen in applying the important science of numbers to every subject which he investigated'.[5] He was an associate of the Cambridge network. Babbage had met him at Laplace's house in Paris in 1826, an appropriate venue for it was Laplace who had issued a famous injunction to 'apply to the sciences of mankind the methods of observation and calculation which have served us so well in the natural sciences'.[6] Many of the ideas and concepts elaborated by Quetelet can 'be traced back to Laplace', moreover.[7] Whewell met Quetelet three years later in Heidelberg at the Gesellschaft Deutscher Naturforscher und Ärzte, also known as the Deutsche Naturforscher Versammlung, the annual meeting of German natural philosophers and the model for what became, in the early 1830s, the British Association for the Advancement of Science.[8] They invited Quetelet to the second and third meetings of the BAAS in Oxford in 1832 and Cambridge in 1833. As Whewell explained

You will probably have heard that we have begun holding in England annual meetings of the friends of science like that which brought us together at Heidelberg. The next of these meetings takes place at Cambridge and begins on the 24th of June. I should be extremely glad to believe that there was any probability of your attending this meeting, and I am persuaded that if you were to do so you would find there many persons and things which would interest you. If

---

[4] Kevin Donnelly, *Adolphe Quetelet. Social Physics and the Average Men of Science 1796–1874* (London, 2015); F. H. Hankins, 'Adolphe Quetelet as Statistician', *Columbia University Studies in History, Economics and Public Law*, xxxi, 1908; D. Landau and Paul F. Lazarsfeld, 'Quetelet', *International Encyclopedia of the Social Sciences*, xiii (New York, 1968), 247–56; Ian Hacking, 'How Should We Do the History of Statistics?', *I&C*, 8, Spring 1981, 16; Oscar Sheynin, 'Quetelet as Statistician', *Archive for the History of Exact Sciences*, 36, 4, Dec. 1986, 281–325.

[5] 'Publishers' Note' [William and Robert Chambers], L. A. J. Quetelet, *A Treatise on Man and the Development of His Faculties* (Tr. R. Knox) (Edinburgh, 1842), (1969 edn., Gainesville, Florida). This is a facsimile edition of the 1842 translation of Quetelet's *Sur L'Homme et le Développement de ses Facultés, ou Essai de Physique Sociale* (1835). All quotations from *Sur L'Homme* are taken from this edition.

[6] Babbage to Quetelet, 28 Oct. 1865, Quetelet Papers, Académie Royale de Belgique, file 267; Donnelly, *Adolphe Quetelet*, 28.

[7] Bernard-Pierre Lécuyer, 'Probability in Vital and Social Statistics: Quetelet, Farr, and the Bertillons', in Lorenz Krüger, Lorraine J. Daston, and Michael Heidelberger (eds), *The Probabilistic Revolution* (2 vols, Cambridge, MA), vol. 1, *Ideas in History* (Cambridge, Mass., 1987), 333.

[8] William Whewell to L. A. J. Quetelet, 15 May 1832, Quetelet Papers, file 2644.

you can visit us at that time pray do: we will give you apartments in the college
and do what we can to make your visit comfortable.[9]

Quetelet journeyed to Cambridge in 1833 as the official delegate of the Belgian
government. A recent study of his science and his role in co-ordinating international scientific activity, has emphasized his close connections in France and the
German states while neglecting his example and influence in Britain among the
leading scientists and reformers of the era.[10] When William Farr wrote him a new
year greeting at the end of 1863, he remarked that 'You have many friends in
many lands; but in no country more than in England.'[11] Two years later Farr
remarked that 'your name is still revered in England.'[12] According to Farr's biographer, Quetelet 'served as a catalyst' for the British statistical movement.[13]
Indeed, he was more than that, because Quetelet had a close and enduring relationship with no less than Albert, the Prince Consort himself, whom he had
tutored in mathematics in 1836 when Albert was a teenager, the lessons continuing by post for years afterwards. Quetelet's *Du Système Social et des Lois qui le
Régissent* (1848), was dedicated to Albert, and, as we have seen, it was Quetelet,
prompted by Farr, who persuaded the Prince to preside at the London Congress
of the ISC and give his famous opening address there.[14]

All accounts of the founding of Section F and the Statistical Society of London
start from Quetelet's presence in Cambridge at the end of June 1833 and the
desire to convene an audience to consider the 'statistical budget' of criminal statistics, a component of what the French then called 'statistique morale', that
Quetelet had brought with him. This 'budget' comprised 'the results of his inquiries into the proportion of crime at different ages and in different parts of France
and Belgium.'[15] As he explained a little later 'there is a *budget* which we pay with
frightful regularity—it is that of prisons, dungeons, and scaffolds. Now, it is this
budget which, above all, we ought to endeavour to reduce.'[16] In London in 1851
for the Great Exhibition, it was Quetelet who, with Babbage, suggested the formation of the International Statistical Congress which met for the first time in
Brussels in 1853. He was, indeed, an institutional innovator and networker,

---

[9] Ibid., 2 April 1833. There was a separate invitation in 1833 from the Scottish scientist, James
David Forbes. See J. Morrell and A. Thackray, *Gentlemen of Science. The Early Years of the British
Association for the Advancement of Science* (Oxford, 1981), 374.

[10] Donnelly, *Adolphe Quetelet*, passim, discusses at length Quetelet's sojourns in Paris in 1823 and
the German states in 1829, but neglects his visits to England in 1833, 1851, and 1860.

[11] Farr to Quetelet, 31 Dec. 1863, Quetelet Papers, file 990–1.

[12] Farr to Quetelet, 20 Jan. 1865, ibid.

[13] John M. Eyler, *Victorian Social Medicine. The Ideas and Methods of William Farr* (Baltimore, 1979).

[14] See Chapter 1, p. lii above.

[15] 'Transcript of the Notebook of Mr. J. E. Drinkwater', 'Minutes of the committee of the statistical
section of the British Association, June 27, 1833', f. 1, Archives of the Royal Statistical Society, London.

[16] Quetelet, *A Treatise on Man*, 'Introductory', 6.

skilled at bringing colleagues and associates together, sociable and naturally collaborative, and assiduous in founding new organizations to embody these
behaviours: 'it is hard to find an aspect of his career in which he was more successful, or one to which he was more dedicated, than in creating networks of scientific researchers'.[17] He was an entrepreneur of science who believed that to be
executed properly in the future, science would require large teams of researchers.
Like Babbage, he intuited a coming age of big data, collected and analysed on a
grand scale.[18]

After the Cambridge meeting in 1833 Quetelet travelled to London to attend
one of Babbage's famous Saturday evening soirees, which always attracted some of
the most notable intellectual, artistic, and political figures, and to meet with
Malthus who had sent Quetelet in advance some characteristic questions about
Belgian infant mortality, family size, and consumption of food.[19] To answer these,
Quetelet requested the Belgian Ministry of the Interior to authorize a special
inquiry by each of the Belgian provinces.[20] Quetelet then appeared as the 'prize
witness' before the 1833 Select Committee on Parochial Registration whose recommendations led to the 1836 Civil Registration Act and the foundations of the
General Register Office as the central British agency collecting and analysing
national demographic data.[21] Persuaded to give evidence by John Bowring, one of
Bentham's 'philosophic radicals', who also served as interpreter before the committee, Quetelet's interrogation, beginning with the discussion of the system of
civil registration in Belgium under the Code Napoleon, made it plain that Britain
was behind several European countries in the collection of vital statistics. The
point had been made to the founding statisticians of Section F by Richard Jones

---

[17] Donnelly, *Adolphe Quetelet*, 4.

[18] Lorraine J. Daston, 'Rational Individuals versus Laws of Society: From Probability to Statistics',
in Lorenz Krüger, Lorraine J. Daston, and Michael Heidelberger (eds), *The Probabilistic Revolution* (2
vols, Cambridge, MA), vol. 1, *Ideas in History* (Cambridge, Mass., 1987), 301.

[19] T. R. Malthus to L. A. J. Quetelet, 'Athenaeum July 183?(3)', Quetelet Papers, file 1697. As
Quetelet explained some days later 'For example, I am now desired, by a distinguished political economist here, to furnish certain population returns, to be made in a particular way. I shall obtain from
the Minister of the Interior, through the statistical department, an order; this is sent to the governors
of the different provinces, and in that way I shall be able to prepare a return, which is, in truth, a sort
of extra-official return.' *Report from the Select Committee on Parochial Registration*, P.P. 1833, vol. xiv,
119–122, 'M. Adolphe Quetelet, Director of the Brussels Observatory; Examined through the interpretation of Dr. Bowring', 10 July 1833, Q. 988.

[20] It would appear that Malthus's request 'underwent the fate of the many projects which sleep in
the ministerial portfolio'. See A. Quetelet, *Letters Addressed to H. R. H. the Grand Duke of Saxe Coburg
and Gotha on the Theory of Probabilities as applied to the Moral and Political Sciences* (London, 1849),
Letter xlvi, 245.

[21] D. V. Glass, *Numbering the People: The Eighteenth Century Population Controversy and the
Development of Census and Vital Statistics in Britain* (Farnborough, Hants, 1973), 127;L. A. J. Quetelet,
'Notes Extraites d'un Voyage en Angleterre, aux moins de Juin et Juillet, 1833', *Correspondance mathe-
matique et physique*, III, i (Brussels, 1835), 15.

in his 'Sketch of the objects of the Section',[22] and was confirmed by Quetelet in his evidence to the committee:

Lately, at the philosophical meeting in Cambridge, it was the subject of discussion; I heard from several distinguished persons that there was a general complaint of the imperfection of the elementary population documents of this country, and their imperfection led strangers, who wrote on England, into great mistakes. It is indeed a subject of wonder to every intelligent stranger, that in a country so intelligent as England, with so many illustrious persons occupied in statistical inquiries, and where the state of the population is the constant subject of interest, the very basis on which all good legislation must be grounded has never been prepared; foreigners can hardly believe that such a state of things could exist in a country so wealthy, wise and great.[23]

Quetelet's intellectual development exemplifies the early-nineteenth-century determination to construct a 'natural science of society'. Like the natural sciences, a true social science should be characterized by its use of mathematics, numbers, and calculation. His most important contribution to this new discipline was his famous study published in 1835, *Sur L'Homme et le Développement de ses Facultés: Physique Sociale*, a combination of anthropometry (the study of the dimensions of the human body), demography, and the statistical representation of social behaviour. A founding text in sociology, the historian of science George Sarton called it 'one of the greatest books of the nineteenth century'.[24] Reviewing it on publication, the *Athenaeum*, a London periodical, believed that it marked 'an epoch in the literary history of civilization'.[25] In this study Quetelet sought 'to collect in a uniform order, the phenomena affecting men, nearly as physical science brings together phenomena appertaining to the material world'.[26] He had initially named this subject 'social mechanics' in a memoir on crime he published in 1831. Indeed, in a letter to Babbage in June 1831 he had written of 'un projet de former une mécanique sociale'.[27] But he renamed it 'social physics' (*physique sociale*) to underscore that it was a search for order and regularity in the social realm: 'In giving to my work the title of Social Physics, I have had no other aim than to

[22] Rev. Richard Jones, 'Sketch of the objects of the Section', 28 June 1833, Transcript from the notebook of Mr. J. E. Drinkwater, f.3, Royal Statistical Society.

[23] Ibid., Q. 999. See: https://parlipapers-proquest-com.rp.nla.gov.au/parlipapers/result/pqpdocumentview?accountid=12694&groupid=100358&pgId=70ab6d40-0fa8-418b-b122-37ec228572e8&rsId=16D99141816#533

[24] 'Quetelet', in G. Sarton, *Sarton on the History of Science. Essays by George Sarton* (D. Stimson, ed.) (Cambridge, Mass., 1962), 229.

[25] 'On Man, and the Development of his Faculties', *Athenaeum*, 409, 29 August 1835, 661.

[26] Quetelet, *Sur L'Homme*, 'Preface' (1842 edn.), vii.

[27] L. A. J. Quetelet to Charles Babbage, 26 June 1831, Babbage Papers, British Library, London, Add Mss. 37185 f. 560.

collect, in a uniform order, the phenomena affecting man, nearly as physical science brings together the phenomena appertaining to the material world.'[28] Auguste Comte, the founding French sociologist of this period, had already used this term to describe his own oeuvre. Quetelet's statistical positivism has clear affinities also with the 'social mathematics' of Condorcet, the 'social physiology' of Saint-Simon, as well as the 'social calculus' of Bentham. In all these cases, it was assumed by philosophers that is was possible to understand social processes and structures by means of mathematics and/or natural science. As if to emphasize the point, *Sur L'Homme* was reissued in an amended form in 1869, at the end of Quetelet's career, with the subtitle, *Physique Sociale,* placed first.

According to a later president of the Statistical Society, Dr. William Guy, Quetelet's

> *Physique Sociale* was the first systematic attempt to apply the methods and formulae of the mathematician to the whole circle of observations which have the living man for their object, his physical growth and development, his intellect and morals, his vices and his crimes; in a word, all that he is, does, and suffers, from the cradle to the grave.[29]

Quetelet posited the construction of predictive laws of social behaviour based on the statistical regularities he demonstrated to exist in the 'moral' life of communities, specifically the roughly invariable annual rates of crime, suicide, and illegitimacy. 'We can say in advance how many individuals will sully their hands with the blood of their neighbours, how many will commit forgeries, and how many will turn poisoners with almost the same precision as we can predict the numbers of births and deaths.'[30] As one later admirer put it, 'The simple proposition that the moral nature of men...can best be determined by a statistical study of their actions was exalted by him into the foundation of an exact social science.'[31] To Quetelet the natural and social domains could both be studied by common procedures: 'All observations tend...to confirm the truth of the proposition which I long ago announced, that everything which pertains to the human species considered as a whole, belongs to the order of physical facts.'[32] *Sur L'Homme*'s aggressive and explicit positivism caught the mood of the age and established his immense reputation across Europe. In the words of the secretary of Statistical Society, Rawson W. Rawson, in 1839, 'surely there can be no reason for denying that moral, no less than physical phenomena may be found to be controlled and determined by particular laws [?]...Mankind is not exempt from these laws.'[33] As

---

[28] Quetelet, *Sur L'Homme*, vii.   [29] William Guy, 'Inaugural Address', *JSSL*, 37, 1874, 420.
[30] A. Quetelet, *Physique Sociale*, I (Brussels, 1869), 97.   [31] Hankins, 'Adolphe Quetelet', 33.
[32] Quetelet, *Sur L'Homme*, 96.
[33] R. W. Rawson, 'An Inquiry into the Statistics of Crime in England and Wales', *JSSL*, 2, 1839, 316.

historians of the social sciences have come to recognize the importance of mathe-matical and natural scientific models in generating the early sociological tradition in western Europe, so Quetelet's significance has grown. As Ian Hacking has put it, 'Today we see that Quetelet triumphed over Comte: an enormously influential body of modern sociological thought takes for granted that social laws will be cast in statistical form.'[34]

Quetelet was not the first theorist to focus on social-statistical regularities. As we have seen, they were interpreted by some of the early political arithmeticians as evidence of the divine order of creation.[35] In the 1820s, the French mathemati-cian Joseph Fourier noticed that the data in Paris on the number of births, deaths, marriages, suicides, and various crimes, showed stable frequencies from year to year. And before Fourier, in the 1780s, Immanuel Kant had also noticed that the vital statistics of life and death demonstrated what he called the 'constant laws of nature' in action. Kant sought to reconcile the random and irrational behaviour of individual humans, and their freedom of will, with his desire to construct 'a Universal History'.[36] Though each individual 'pursues its own aim in its own way, and one often contrary to another', they are nevertheless led 'as by a guiding thread, according to an aim of nature, which is unknown to them'. It was the phi-losopher's duty to attempt to discover this 'aim of nature' in the otherwise 'non-sensical course of things human'. The secret was in the scale of the historical analysis:

If it considers the play of the freedom of the human will *in the large*, it can dis-cover within it a regular course; and that in this way what meets the eye in indi-vidual subjects as confused and irregular yet in the whole species can be recognized as a steadily progressing, though slow development, of its original predispositions.

Kant therefore posited the possible composition of 'a history in accordance with a determinate plan of nature … even of creatures who do not behave in accordance with their own plan.' *Sur L'Homme* was in this intellectual tradition. It also attempted to demonstrate, using available social statistics, that a universal account of human society was possible, one that conjured deeper structures of social order, system, and predictability out of what appeared, at first sight, to be the inex-plicable acts of wilful, silly humans.

[34] Ian Hacking, 'How Should We Do the History of Statistics?', *I & C*, 8, Spring 1981, 16.
[35] See above, pp. 11–12.
[36] Immanuel Kant, 'Idee zu einer allgemeinen Geschichte in weltbürgerlicher Absicht' ('Idea for a Universal History with a Cosmopolitan Aim'), *Berlinische Monatsschrift*, iv (November 11, 1784). https://www.cambridge.org/core/books/kants-idea-for-a-universal-history-with-a-cosmopolitan-aim/idea-for-a-universal-history-with-a-cosmopolitan-aim/8B2BA346A82FA006AB982E3A941 E2A26

It was composed of four books building up from the first, 'The Development of the Physical Qualities of Man' (concerning 'the birth, life, reproduction and mortality of man'), to the second, the 'Development of Stature, Weight and Strength' (concerning anthropometry), to the third, the 'Development of the Moral and Intellectual Qualities of Man', concluding with analyses of 'The Social System' and 'the Law of Human Development'. Admittedly, the data was diverse, even random in nature, a bricolage of statistical information—'almost totally a scissors-and-paste assemblage' as one of his modern editors has described it.[37] Information was largely drawn from official as well as some unofficial sources from across Europe and the United States which Quetelet used as best he could to draw out trends, make comparisons, and generally show the potential of social statistics. But it was not a systematic treatment of any major issue; rather he used what information he could find in an era before any attempt had been made to give common form and unity to the collection of statistics across borders. To his credit, it was Quetelet himself who then set up the first international organization to do just that in 1853, the International Statistical Congress. He was conscious of trying to supply the architecture of new knowledge; the detail could then be filled-in by the labour of others. He considered *Sur L'Homme* 'but a sketch of a vast plan, to be completed only by infinite care and immense researches'.[38] He expected others 'to carry it farther and farther, and bring it more and more to the appearance of a science'.[39] Perhaps this explains William Whewell's slightly disarming praise for a work he received 'with great pleasure...Your researches are as curious and interesting as anything expressed by means of numbers can be'.[40]

Quetelet developed four basic ideas that had influence over his colleagues, his admirers, all Europe. First, he contended that the regularity of social phenomena was demonstrable by the use of statistical analysis. Second, that such regularities even in the 'moral' realm, were 'of the order of physical facts'.[41] As John Herschel summarized this in a review of his friend's work in 1850,

Men began to hear with surprise, not unmingled with some vague hope of ultimate benefit, that not only births, deaths, and marriages, but the decisions of tribunals, the results of popular elections, the influence of punishments in checking crime – the comparative value of medical remedies, and different modes of treatment of diseases...might come to be surveyed with that lynx-like scrutiny of a dispassionate analysis, which, if not at once leading to the discovery of positive truth, would at least secure the detection and proscription of many mischievous and besetting fallacies.

[37] Solomon Diamond, 'Introduction', *Sur L'Homme* (1842) (1969 edn.), v.
[38] Quetelet, *Sur L'Homme*, 9.   [39] Ibid., 102.
[40] Whewell to Quetelet, 2 Oct. 1835, Quetelet Papers, File 2644.
[41] Quetelet, *Sur L'Homme*, 96.

So far, so uncontroversial. But thirdly, Quetelet contended that these regularities were directly attributable to social conditions rather than the aggregation of individuals exercising choice. As he put it,

> Society includes within itself the germs of all crimes committed, and at the same time the necessary facilities for their development. It is the social state, in some measure, which prepared these crimes, and the criminal is merely the instrument to execute them. Every social state supposes, then, a certain order of crimes, these being merely the necessary consequences of its organisation.[42]

Contemporaries understood this to be determinism, or in the term of the age, 'fatalism', and objected to what appeared to be an apologia from Quetelet for criminality and suicide, setting off a major controversy about his work across Europe.[43] But Quetelet used the regularity of social behaviours, and their generation by social conditions, as rationales for a fourth principle, the commitment to social reform.

> This observation, so discouraging at first sight, becomes, on the contrary, consolatory, when examined more nearly, by showing the possibility of ameliorating the human race, by modifying their institutions, their habits, the amount of their information, and, generally, all which influences their mode of existence.[44]

In short, Quetelet was arguing that behaviours could be changed by changing the social conditions which gave rise to them. As he explained, 'In fact, this observation is merely the extension of a law already well known to all who have studied the physical condition of society in a philosophic manner: it is, that so long as the same *causes* exist, we must expect a repetition of the same *effects*.'[45] The very regularity and predictability of human activity offered possibilities for social improvement by the exercise of social reform. As Quetelet contended, the way to reduce suicides 'was to change social institutions instead of attending to the individual'.[46] *Sur L'Homme* is one of the clearest statements of the liberal environmentalism which characterized the statistical movement up to the 1870s, therefore. To its many adherents, however varied their backgrounds, the purpose of the collection and analysis of social data was to build a case for the reform of social conditions and institutions. Quetelet's conclusions seemed to supply a scientific basis for the changes that Victorian 'improvement' mandated and which were desired

---

[42] Ibid.
[43] Alain Desrosières, *The Politics of Large Numbers. A History of Statistical Reasoning* (1993) (Cambridge, MA, and London, 1998), 80–1. Ian Hacking, 'Nineteenth Century Cracks in the Concept of Determinism', *Journal of the History of Ideas*, 44, 3, 1983, 455–75.
[44] Ibid.   [45] Ibid.
[46] Donnelly, *Adolphe Quetelet*, 149. See L. A. J. Quetelet, *Du Système Social et des Lois qui le Régissent* (1848), 88.

by improvers like William Farr, Florence Nightingale and the Prince Consort, all of them devotees of 'blue books' and social statistics, and each of them an admirer of Quetelet himself, and of his 'synthesis of social concern and mathematical method'.[47] As Theodore Porter has put it, Quetelet 'aimed to create a science that would be indispensable to the legislator'.[48]

But not all would have followed Quetelet in his discussion of free will which continues thus:

> What has induced some to believe that moral phenomena did not obey this law, has been the too great influence ascribed at all times to man himself over his actions: it is a remarkable fact in the history of science, that the more extended human knowledge has become, the more limited human power, in that respect, has constantly appeared.[49]

This sounded like fatalism to many readers and to many more beyond who were otherwise sympathetic to the case Quetelet was building for reform, and they contested what seemed to be the abrogation of individual responsibility in his arguments. Quetelet denied that he was a determinist. In a subsequent preface to *Sur L'Homme*, composed for the 1842 English translation, he protested that 'the distinctions which I have already established with care in my work, ought to have proved, methinks, to some less prejudiced judges, how far I am from blind fatalism, which would regard man as unfit to exercise free-will, or meliorate the condition of his race.'[50] But, by his own admission, he excluded consideration of individual behaviour: 'It is the social body which forms the object of our researches, and not the peculiarities distinguishing the individuals composing it.'[51] There was no attempt made to analyse the motives and the psychology of the criminal, therefore, and not enough moral condemnation of criminal behaviour for some tastes. We can sense this in William Whewell's wary response to his erstwhile colleague somewhat later in the 1840s: 'Your statistical results are highly valuable to the legislator but they cannot guide the moralist. A crime is not less a crime because it is committed at the age of greatest criminality, or in the month of most frequent transgressions.'[52] Whewell was clear that the predictability of crime, though it might help the framing of public policy, in no way lessened or deflected the moral significance of the crime itself or exculpated the criminal. Quetelet's focus on collective rather than individual action, on mean values rather than the behaviour of sub-groups and individuals, led also to more technical criticisms of his approach. To work solely from average values risked overlooking

---

[47] Diamond, 'Introduction', *Sur L'Homme*, v.
[48] Theodore Porter, 'The Mathematics of Society: Variation and Error in Quetelet's Statistics', *British Journal for the History of Science*, 18, 1, March 1985, 54.
[49] Quetelet, *Sur L'Homme*, 6.   [50] Quetelet, 'Preface', x.   [51] Quetelet, *Sur L'Homme*, 7.
[52] William Whewell to L. A. J. Quetelet, 7 Oct. 1847, Quetelet Papers, file 2644.

and obscuring the significance of deviant or aberrant behaviour within any cohort or population. The variations from the average are often more revealing than the average itself.[53]

Left littered about *Sur L'Homme*, meanwhile, was quite enough material to turn unprejudiced readers into critics of an extreme environmentalism. For example, at the book's conclusion, Quetelet argued that

> experience proves as clearly as possible the truth of this opinion, which at first may appear paradoxical, viz., that society prepares the crime, and the guilty are only the instruments by which it is executed. Hence it happens that the unfortunate person who loses his head on the scaffold, or who ends his life in prison, is in some manner an expiatory victim for society. His crime is the result of the circumstances in which he is found placed.[54]

This was to misunderstand the 'law of large numbers' as discerned by Jakob Bernoulli at the beginning of the eighteenth century, and so named by Siméon-Denis Poisson in 1835: that in any random sample, the actual mean approaches closer to the theoretical mean as the size of the sample, or the number of throws of the dice etc., increases.[55] In any large and stable social sample the frequency of crime or suicide will be roughly equal from year to year, unless external conditions alter. But Quetelet jumped to a false conclusion on the basis of such regularities: 'the greater the number of individuals, the more does the influence of individual will disappear, leaving predominance to a series of general facts, dependent on causes by which society exists and is preserved'.[56] This fails to recognise that each individual criminal, and each individual suicide, will take their own decision, for their own specific and personal reasons, to commit a crime or take their life. No one is forced to do these things; all have a choice. But in exercising that choice, by the law of large numbers, in any group of sufficient size, comparable numbers will steal, murder and commit suicide in given periods of time, all other conditions remaining the same. To some contemporary readers Quetelet seemed confused, or put another way, his text was open to contradictory readings. As John Stuart Mill was to put it in 1862, in the 5th edition of his *Logic*, in his discussion of Thomas Buckle who, as we shall see, was influenced greatly by Quetelet,

> The very events which in their own nature appear most capricious and uncertain, and which in any individual case no attainable degree of knowledge would

---

[53] Theodore Porter, 'Lawless Society: Social Science and the Reinterpretation of Statistics in Germany, 1850–1880', in Lorenz Krüger, Lorraine J. Daston, and Michael Heidelberger (eds), *The Probabilistic Revolution* (2 vols), vol. 1, *Ideas in History* (Cambridge, MA, 1987), 362–4.

[54] Quetelet, *Sur L'Homme*, 108.   [55] Porter, *The Rise of Statistical Thinking*, 12, 52.

[56] Quetelet, *Sur L'Homme*, 96.

enable us to foresee, occur, when considerable numbers are taken into account, with a degree of regularity approaching to mathematical.[57]

As Mill saw, individual actions remained 'capricious', the product of the vagaries of the individual will, when taken singly. It is only *en masse* that they fall into a regular pattern and appear to be determined, therefore. Fascinated by the social regularities to which statistics gave witness, Quetelet found himself drawn to explanations of this regularity that depended on the action of social forces that could neither be evaded nor denied. This had the advantage, or so it seemed, of building a case for the possibility of social change, but only by treating people as ciphers acted on by forces they could not escape. And this left open the charge that he was denying personal responsibility, one he never shook off.

Quetelet is associated with another controversial and confusing idea, that of the 'average man' (*l'homme moyen*) a subject discussed in the fourth and final section of *Sur L'Homme*. In bringing together so much material on the physical, intellectual and moral attributes of men and women, it was possible, obviously, to calculate average values in any distribution, and from this build a composite portrait of 'l'homme moyen', the man who embodied all the qualities of a human at their average levels, and who would therefore, thought Quetelet, be representative of society as a whole.[58] The usually symmetrical distribution of moral and physiological qualities around the mean in a normal distribution or error curve encouraged this way of thinking. In 1831 Quetelet defined the average man as

the man whom I considered is the analogue in society of the centre of gravity in bodies; he is a fictional being for whom everything happens according to the average results obtained for society. If the *average* man were determined for a nation, he would represent the type of that nation; if he could be determined for all men, he would represent the type for the entire human race.[59]

But, over time, Quetelet's concept of the 'the average man', and his purpose in calculating and defining who that would be, began to change. Four years later in *Sur L'Homme* the definition was different:

I have said before that the average man of any one period represents the type of development of human nature for that period; I have also said that the average man was always such as was conformable to and necessitated by time and place;

[57] J. S. Mill, *A System of Logic Ratiocinative and Inductive. Being a Connected View of the Principles of Evidence and the Methods of Scientific Investigation* (London, 1843 *et seq.*) (Toronto, 1973, vols 7–8), 932.

[58] Desrosières, *The Politics of Large Numbers*, 73–81.

[59] L. A. J. Quetelet, *Recherches sur le penchant au crime aux differens ages*, (Brussels, 1831), 1.

that his qualities were developed in due proportion, in perfect harmony, alike removed from excess or defect of every kind, so that in the circumstances in which he is found, he should be considered as the type of all which is beautiful – of all which is good.[60]

This might be taken to mean that 'l'homme moyen' has ideal qualities and was not average in any manner.[61] But what Quetelet seemed to be encouraging was the redefinition of the beautiful and the good so that the qualities of the average person were better recognized and also admired. He idealized 'the mean as the standard of beauty and goodness'.[62] 'L'homme moyen' was 'the epitome of all qualities…the most moral, the most intelligent…the most artistic, the most beautiful'.[63] It was a profoundly democratic idea and meant as such. The average is not mediocre but of a type of goodness and quality we should collectively admire. As Quetelet continued, 'The natural consequence of the idea which I have just stated is, that an individual who should comprise in himself (in his own person), at a given period, all the qualities of an average man, would at the same time represent all that is grand, beautiful, and excellent'.[64] Virtues and qualities hitherto associated with aristocracy and elites were here redefined, reimagined, and recalibrated, so that they became the property, and embodied the experience, of the ordinary and the normal.

If this is correct, it helps us place Quetelet yet more firmly as a liberal and a democrat who employed the new intellectual discipline of social statistics to build a case for social change and moral revaluation. It also invites comparison with other movements and other works. The Saint-Simonians, Quetelet's contemporaries in France but very different from him, placed their faith in technocracy and 'emphasised the historical importance of the intellectual elite'.[65] It has been suggested that Quetelet took inspiration from the ideas of Victor Cousin, the French writer and educationist who, having suffered personally for his views in an era of political upheaval, developed 'eclecticism' – the synthesis of workable and pragmatic ideas and institutions—into a philosophy of social and intellectual moderation.[66] But arguably *Sur L'Homme,* published in 1835, is better compared to another notable political study published that year, the first volume of Alexis de

---

[60] Quetelet, *Sur L'Homme*, 100.

[61] 'For Quetelet, the Average Man was no figure of speech, but some kind of ideal that Nature aimed to produce with every individual', Timandra Harkness, *Big Data. Does Size Matter?* (London, 2016) (2017 edn.), 54.

[62] Porter, 'The Mathematics of Society: Variation and Error in Quetelet's Statistics', 65.

[63] Gustav Jahoda, 'Quetelet and the Emergence of the Behavioral Sciences', *SpringerPlus*, 4, 473 (2015), 2.

[64] Quetelet, *Sur L'Homme*, 100.   [65] Diamond, 'Introduction', vii.

[66] Maurice Halbwachs, *La Théorie de l'Homme Moyen: Essai sur Quetelet et la Statistique Morale* (Paris, 1913); Porter, *The Rise of Statistical Thinking*, 101–2.

Tocqueville's *Democracy in America*.[67] As a convinced social reformer and student of European and American institutions, Tocqueville was drawn to the study of social statistics, though never as the basis of a predictive social science. In the summer of 1835 during his sojourn in Britain he not only attended a meeting of the Manchester Statistical Society, as we have seen,[68] but made Babbage's acquaintance in London, and attended the deliberations of the Statistical Section at the Dublin meeting of the British Association.[69] The similarities with Quetelet, however, are historical and sociological rather than statistical: both were concerned to characterize the new type of mass society emerging in the 1830s, though they held different views on the desirability of the likely outcome.

In *Democracy in America* Tocqueville was writing about the new American republic, but also treating American society and government as representative of 'democracy' itself. He expected that in due course, the nations of Europe would follow the American example, and he supported such an outcome. Nevertheless, throughout *Democracy in America* Tocqueville also expressed reservations. They are caught in his comment that 'the nations of our time cannot prevent the conditions of men from becoming equal, but it depends upon themselves whether the principle of equality is to lead them to servitude or freedom, to knowledge or barbarism, to prosperity or wretchedness'.[70] Freedom and individualism might be threatened by the force of majority public opinion, by the political will of the majority deployed against a minority, or by the tendency towards cultural mediocrity intrinsic to the egalitarian mentality of democracies. Tocqueville feared the development of a powerful government providing material comfort and assistance to the many, and able thereby to dictate to the mass of equal individuals, unchallenged by any intermediate institutions or interest groups. This was what he called 'democratic despotism'. Unsurprisingly, his political legacy was ambiguous and Tocqueville was later claimed by both liberals and conservatives. He had foreseen this, writing in 1835 that 'Some will find that at bottom I do not like democracy and that I am severe toward it; others will think that I favour its development imprudently'.[71]

---

[67] The following discussion of Tocqueville is based on the discussion of him in Lawrence Goldman, 'Conservative Political Thought from the Revolutions of 1848 until the fin de siècle', in *The Cambridge History of Nineteenth Century Political Thought* (eds G. Stedman Jones and G. Claeys) (Cambridge, 2012), 700.

[68] See above, pp. 59–60.

[69] Michael Drolet, 'Tocqueville's Interest in the Social: Or How Statistics Informed His "New Science of Politics"', *History of European Ideas*, 31, 2005, 451–71. Tocqueville was present in Dublin for a controversial paper on popular education given by one of the founders of the Manchester Statistical Society, William Langton. See Laura J. Snyder, *The Philosophical Breakfast Club. Four Remarkable Friends who Transformed Science and Changed the World* (New York, 2011), 153.

[70] Alexis de Tocqueville, *Democracy in America* (2 vols, 1835, 1840), ed. Phillips Bradley (New York, 1945), 352.

[71] Alexis de Tocqueville to Louis de Kergolay, Jan. 1835, in Alexis de Tocqueville, *Selected Letters on Politics and Society* (ed. R. Boesche) (Berkeley and London, 1985), 95. It is worth noting that on 23 March 1835, the year of publication, De Tocqueville attended one of Babbage's famous Saturday

Quetelet certainly noticed trends towards social and cultural uniformity which were lamented not only by Tocqueville but also by J. S. Mill in *On Liberty*, published in 1859. But, unlike them in this matter, Quetelet wrote as a neutral observer. At the very end of *Sur L'Homme* he gave it as a conclusion

> that one of the principal facts of civilisation is, that it more and more contracts the limits within which the different elements relating to man oscillate...individual peculiarities tend to disappear more and more, and [that] nations assume a greater resemblance to each other...Even during the last half century, and within the limits of Europe alone, we see how great the tendency is for people to lose their national character and be amalgamated in one common type.[72]

Where Tocqueville and Mill worried about the inferior level of culture in democracies in comparison with aristocratic societies, and complained about social uniformity, Quetelet embraced wholeheartedly the virtues of the average man, even if the concept itself was left vague. For all its weaknesses, few ideas better summarise the emerging statistical movement than 'l'homme moyen'. Its later replacement, from the late 1860s, in the work of Francis Galton and others, by the statistical analysis of the hereditary intelligence and attributes of the elite, marks a crucial change in the social role of statistics, one that will be discussed at the end of this book. In 1846 Quetelet performed anthropometric measurements on a dozen Native Americans passing through Brussels.[73] Eight years later he repeated the exercise on a similar number of black Africans. Though the samples were far too small and random for genuine scientific study, he concluded that physically, the races were the same, and this underscored his faith in the unity of mankind: 'Les grandes linéaments de l'espèce humaine paraissent à peu près les mêmes pour les différents races.'[74] This was emphatically not the view of Francis Galton.

'Social physics' is open to a host of further objections. Quetelet kept hinting that it could be applied to all manner of moral characteristics, but he only ever applied it to crime, suicide and marriage. His conception of a social law was used promiscuously to describe a trend in a series of averages over time (for example, changes in average stature from century to century), to describe a regular pattern of correlations (for example, the propensity to commit crime related to age), indeed to describe any regular distribution in a predictable series. His work was subjected to searching mathematical and philosophical criticism in 1866 by the

evening soirees at his home in Dorset Square and there met Cavour. See Anthony Hyman, *Charles Babbage. Pioneer of the Computer* (Oxford, 1982), 175.

[72] Quetelet, *Sur L'Homme*, 108.

[73] L. A. J. Quetelet, 'Sur les indiens O-Jib-Be-Wa's et les proportions de leur corps', *Bulletin de L'Académie Royale des Sciences, des Lettres, et des Beaux-Arts de Belgique*, 15, 1, 1846, 70–6.

[74] 'The distinctive features of the human species appear to be very nearly the same among the different races'. L. A. J. Quetelet, 'Sur les proportions de la race noire', *Bulletin de L'Académie Royale des Sciences, des Lettres, et des Beaux-Arts de Belgique*, 21, 1 (1854), 96–100 (quotation at 100).

Cambridge logician, John Venn, in his book *The Logic of Chance*. Granting that Quetelet had collected the most varied and extensive datasets, Venn was 'convinced that there is much in what he has written upon the subject which is erroneous and confusing as regards the foundations of the science of Probability, and the philosophical questions which it involves.' Venn, who would later assist Francis Galton in his anthropometric research in the 1880s,[75] focused on Quetelet's frequent and often inapplicable use of the law of error to explain almost every set of statistical data that he collected, whether animate or inanimate, and whatever its distribution about the mean value computed. He criticized Quetelet's usual assumption 'that whenever we get a group of such magnitudes clustering about a mean, and growing less frequent as they depart from that mean, we shall find that this diminution of frequency takes place according to one invariable law, whatever may be the nature of these magnitudes, and whatever the process by which they may have been obtained.'[76] Quetelet's instinct was to construct mean values. But the values at the extremes of a distribution could be just as, if not more, revealing. As Stephen Stigler has put it, in discovering normal distributions and the same bell-shaped curves wherever he looked for them, Quetelet 'succeeded too well...The mere appearance of normality is not at all sufficient to conclude homogeneity.'[77]

Quetelet may be seen as a true, if neglected, founding father of sociology nevertheless: 'his conviction that a scientific study of social life must be based on the application of quantitative methods and mathematical techniques anticipated what has become the guiding principle of modern social research.'[78] Despite Quetelet's flaws and presumptions, his influence in Britain alone was conspicuous in its breadth: it is evident in his close personal and also intellectual relationships with Prince Albert, Babbage, Herschel, Whewell, and Farr. H. T. Buckle's 'statistical history', to be considered later, owes everything to Quetelet's arguments that human behaviour is not random. Herschel's 1850 review of Quetelet's work is said to have influenced the great physicist, James Clerk Maxwell, in his work on heat and 'the dynamical theory of gases', the kinetic gas theory.[79] This was 'the most

[75] Venn analysed physical data derived from Cambridge undergraduates who were differentiated into groups according to their class of degree. See Donald A. Mackenzie, *Statistics in Britain 1865–1930. The Social Construction of Scientific Knowledge* (Edinburgh, 1981), 236–7.

[76] John Venn, *The Logic of Chance. An Essay on the foundations and province of the Theory of Probability, with especial reference to its logical bearings on its applications to Moral and Social Science and Statistics* (London, 1866) (3rd edn., 1888), 27–31. For further criticism of this type, see Ian Hacking, 'Was There a Probabilistic Revolution 1800–1930?', in Lorenz Krüger, Lorraine J. Daston, and Michael Heidelberger (eds), *The Probabilistic Revolution* (2 vols), vol. 1, *Ideas in History* (Cambridge, MA, 1987), 47.

[77] Stephen Stigler, 'The Measurement of Uncertainty in Nineteenth-Century Social Science', in Krüger, Daston, and Heidelberger (eds), *The Probabilistic Revolution*, vol. 1, 290.

[78] Landau and Lazarsfeld, 'Quetelet', 250.

[79] [Sir John Herschel], 'Quetelet on Probabilities', L. A. J. Quetelet, *Letters addressed to HRH the grand-duke of Saxe-Coburg and Gotha on the theory of probabilities as applied to the moral and physical sciences* (London, 1849), *Edinburgh Review*, xcii (July–Oct. 1850), 1–56; P. M. Harman, 'James Clerk

successful application of statistical ideas to physics during the nineteenth century'
and it derived from Maxwell's application of the error law as employed by Quetelet
in social statistics and as explained by Herschel.[80] The anthropologist E. B. Tyler
used Quetelet's ideas in his early work on 'Primitive Society' in 1873 and then
later in his study of marriage in which he introduced the use of statistics to
anthropology.[81] Quetelet's most devoted and uncritical follower, meanwhile, was
Florence Nightingale who had met him at the London Congress in 1860. He sent
her a copy of *Physique Sociale* in 1872. In acknowledging the gift, Nightingale
wrote to him that social physics was 'l'étude la plus essentiellement necessaire aux
progrès de l'humanité.'[82] On receiving the news of Quetelet's death in February
1874 she wrote to William Farr, her lieutenant in many a statistical skirmish with
officialdom, that Quetelet had shown 'the path on which we must go if we are to
discover the laws of Divine Government of the Moral World'.[83] In an unpublished
essay on Quetelet that she wrote subsequently, she lamented that it was

> not understood that human actions are – not subordinate, but – reducible to
> general Laws...Of these, at present we hardly know any. Our object in life is to
> ascertain...*what they are*...if we work without the knowledge of these Laws,
> the best philanthropist of us all knows not but what he is doing [is] harm instead
> of good.[84]

Quetelet provided his many British admirers with a sketch—it was nothing
more—of a pure social science, based on numerical regularities, that he compared
to the laws of natural science, and that could be applied to the healing of social
pathologies. The dual aims of the statistical movement, intellectual and reformative, seemed reconcilable and achievable by following his example.

Maxwell 1831–1879', *ODNB*; Theodore Porter, 'A Statistical Survey of Gases: Maxwell's Social Physics',
Historical Studies in the Physical Sciences, 8, 1981, 77–116.
   [80] Porter, *The Rise of Statistical Thinking*, 118–9. See below, Chapter 11, p. 208.
   [81] E. B. Tylor, 'Primitive Society', *Contemporary Review*, Pt i, vol. 21, Dec. 1872, 701–18; Pt. ii, vol. 22,
June 1873, 53–72; 'On a Method of Investigating the Development of Institutions: Applied to Laws of
Marriage and Descent', *Journal of the Anthropological Institute*, 18, 1889, 245–72. Chris Holdsworth,
'Edward Burnett Tylor 1832–1917', *ODNB*.
   [82] F. Nightingale to L. A. J. Quetelet, 18 Nov. 1872, Quetelet Papers, File 1902. ['The subject most
essentially necessary to the progress of humanity.']
   [83] F. Nightingale to William Farr, 23 Feb. 1874, cited in Marion Diamond and Mervyn Stone,
'Nightingale on Quetelet', 3 parts, *JRSS* (series A), vol. 144, 1981, I, 73.
   [84] Diamond and Stone, 'Nightingale on Quetelet', III, 333.

8

Alexander von Humboldt

Humboldtian Science, Natural Theology, and the Unity of Nature

The Statistical Society of London elected Quetelet as its first foreign member. Its second, elected in January 1837, was the German naturalist, geographer, and celebrated scientific traveller, Alexander von Humboldt.[1] But unlike Quetelet, von Humboldt had nothing to do with the foundations of the Society; nor was he known as a social statistician. Why, then, was he honoured in this way? What did it signify?

Humboldt was the embodiment in the romantic era of the 'man of science' as hero. Born into a distinguished Pomeranian family—his brother, Wilhelm, was an influential minister in the early nineteenth-century Prussian governments of Frederick William III—and endowed with private wealth, he travelled extensively in the Americas between 1799 and 1804. This celebrated journey that took him to Cuba, Venezuela, the Andes, New Spain (Mexico), and the United States, where he met with President Thomas Jefferson in the White House on several occasions, was largely spent in the observation and measurement of geographical and natural phenomena. This he carried out with an array of scientific instruments that, following his example, would become *de rigeur* for the naturalist in the field thereafter. Humboldt, according to Byron was 'the first of travellers'. Much later, Charles Darwin judged him 'the greatest scientific traveller who ever lived...the parent of a grand progeny of scientific travellers'.[2] His trip to the Americas, which became world famous, was the basis of a long and influential career focusing on the interrelationships and unity in nature—the mutual associations of climate, landscape, geology, flora, and fauna—and the development of a holistic approach to the study of the natural world.[3] 'He saw the earth as one great living organism

---

[1] *Proceedings of the Statistical Society,* vol. 1 (1834–7), 16 Jan. 1837 (Archives of the Royal Statistical Society, London).

[2] Charles Darwin to J. D. Hooker, 6 Aug. 1881, *Life and Letters of Charles Darwin* (ed. Francis Darwin) (3 vols) (London, 1887), III, 227. https://www.darwinproject.ac.uk/alexander-von-humboldt.

[3] Byron, *Don Juan*, Canto IV, cxii. Byron's verses continued, however, at Humboldt's expense with chiding about 'an airy instrument, with which he sought / To ascertain the atmospheric state, / By measuring '*the intensity of blue*'. This was Humboldt's cynometer, used by him for measuring the blue of the sky.

Victorians and Numbers: Statistics and Society in Nineteenth Century Britain. Lawrence Goldman, Oxford University Press.

where everything was connected.'[4] The study of this connectedness was widely known by the term he himself coined, 'terrestrial physics', the complement, we might say, to Quetelet's 'physique sociale' in the human realm. Modern ecology and biogeography trace their origins back to Humboldt's example.

Humboldt was known personally to some of the founders of the statistical movement. He made six visits to Britain, three of them in the four years that followed Waterloo in 1815, and enjoyed celebrity and fame in London, though his attempts to win the permission of the East India Company to travel through India were always frustrated.[5] On a visit to England in 1817 he had stayed with the astronomer William Herschel and his family at their home in Slough.[6] Two years later, both John Herschel, William's son, and Charles Babbage, two friends recently graduated from Cambridge, visited von Humboldt in Paris.[7] They were in Paris with Richard Jones in 1821 and met him there again.[8] In 1827 they met him once more when he was in London.[9] In the following year Babbage visited him in Berlin and also attended the meeting of German natural philosophers at the Gesellschaft Deutscher Naturforscher und Ärzte.[10] They remained in contact by letter, with Babbage pouring out his contempt for the British government's lack of support at various points.[11] Babbage 'ever entertained the most profound respect and admiration for Humboldt'.[12] Quetelet had met him, as well, in the summer of 1829 in Berlin at the home of the composer, Felix Mendelssohn.[13] Later, William Farr would welcome the suggestion that the International Statistical Congress might meet in that city in 1863 just because of its proximity to von Humboldt: 'The idea of gathering round the great veteran of science— Humboldt—is admirable; the Association would be a homage to the man and an honour to statistics.'[14]

[4] Andrea Wulf, *The Invention of Nature. The Adventures of Alexander von Humboldt. The Lost Hero of Science* (London, 2015), 'Prologue', 2.

[5] Ibid., 162–7. W. H. Brock, 'Humboldt and the British: A Note on the Character of British Science', *Annals of Science*, 50, 4, 1993, 365–72.

[6] Wulf, *The Invention of Nature*, 165–6.   [7] Ibid., 176.

[8] Laura J. Snyder, *The Philosophical Breakfast Club. Four Remarkable Friends who Transformed Science and Changed the World* (New York, 2011), 71.

[9] Wulf, *The Invention of Nature*, 182.

[10] Babbage, *Passages from the Life of a Philosopher*, 198–202, 432.

[11] See Babbage to Alexander von Humboldt, July 1841: 'This Engine is unfortunately far too much in advance of my own country to meet with the least support. I have at an expense of many thousands of pounds caused the drawings to be executed, and I have carried on experiments for its perfection. Unless however some country more enlightened than my own should take up the subject, there is no chance of that machine ever being executed during my own life, and I am even doubtful to dispose of those drawings after its termination'. Quoted in Doron Swade, *The Cogwheel Brain: Charles Babbage and the Quest to Build the First Computer* (London, 2000), 131.

[12] H. W. Buxton, *Memoir of the Life and Labours of the Late Charles Babbage* (1872–80) (ed. Anthony Hyman) (Cambridge, Mass., 1988), 303.

[13] Kevin Donnelly, *Adolphe Quetelet, Social Physics and the Average Man of Science 1796–1874* (London, 2015), 105.

[14] William Farr to Adolphe Quetelet, 30 Sept. 1853, Quetelet Papers, Académie Royale to Belgique, Brussels, File 990.

The intellectual aims of the founders of the statistical movement in Britain can only be understood if we conceive in this period of a highly influential scientific methodology applicable to all forms of measurable phenomena which the historian of science, Susan Faye Cannon, memorably dubbed 'Humboldtian Science' after the example and procedures of Alexander von Humboldt. In making him an early member of the Society, the founders were signalling not only their associations with him, and admiration for him, but that they shared in this Humboldtian approach. It amounted to 'the accurate, measured study of widespread but interconnected real phenomena in order to find a definite law and a dynamical cause'.[15] It was employed across the full range of disciplines from astronomy to geology, and from geology to *statistique morale*. Indeed, Cannon explicitly associated Quetelet with Humboldtianism.[16] Most characteristically, it was best applied to *physique du globe*, the large-scale geographical sciences of this era. Hence 'the concept of "Humboldtian Science" contradicts conventional ways of looking at the history of science as developments in discrete special subjects, each with its own continuous comprehensive internal history'.[17] Indeed, it makes comprehensible the role that natural scientists played in the foundation of the social sciences because all academic disciplines were united by this method and approach. Humboldtianism was no flash in the pan, no sudden fashion soon exhausted, but 'the latest wave of international scientific activity' that was swiftly embraced by the scientific vanguard in the 1830s. This interpretation improves markedly on previous explanations of the rise of inductivism at this time which link it with a resurgent interest in Baconianism.[18] Indeed, William Whewell specifically set out to remove a simplistic concept of Baconian induction as the dominant 'ideology of proper science' at the British Association, and to replace it with an essentially Humboldtian method: 'Proper science was to be based on slowly cumulating inductive observations and hard won experimental results; only on this basis could true, mathematical generalisations be securely erected.'[19]

Humboldtian science denoted a focus on measurement and the mathematicization of science. As Quetelet had put it, 'we might even judge of the degree of perfection of which a science is capable of being carried out, by the greater or less facility with which it admits of calculation'.[20] Whewell consistently said the same and pursued a concerted 'Cambridge progamme' to reform mathematical analysis and place it at the head of scientific research and education.[21] 'When laws could

[15] Susan Faye Cannon, *Science in Culture: the Early Victorian period* (New York, 1978), 105.
[16] Ibid., 82.   [17] Ibid., 104.
[18] V. L. Hilts, '*Aliis Exterendum*, or the Origins of the Statistical Society of London', *Isis*, 1978, 69, 21–4; Salim Rashid, 'Richard Jones and Baconian Historicism at Cambridge', *Journal of Economic Issues*, xiii, 1, (March, 1979).
[19] J. Morrell and A. Thackray, *Gentlemen of Science. The Early Years of the British Association for the Advancement of Science* (Oxford, 1981), 291–6.
[20] L. A. J. Quetelet, *Instructions Populaire sur le Calcul des Probabilités* (Brussels, 1828), 230.
[21] Morrell and Thackray, *Gentlemen of Science*, 479–84.

be expressed in mathematical form Whewell was convinced that man had deciphered the language in which the Supreme Mind spoke to human minds.'[22] According to Herschel, 'Number, weight, and measure are the foundations of all exact science; neither can any branch of human knowledge be held advanced beyond its infancy which does not in some way or other, frame its theories or correct its practice by reference to these elements.'[23]

Humboldtianism was also characterized by international collaborations through which the study of broadscale natural phenomena like tides, weather systems, and strata, could be undertaken across extensive areas. As von Humboldt put it, 'observations from the most disparate regions of the planet must be compared to one another.'[24] The aim was to extract patterns of order from masses of seemingly random data, just as Quetelet sought evidence of the regularity of human affairs from the vital statistics of births, deaths, and the incidence of suicide. 'Humboldt plaited together the cultural, biological and physical world, and painted a picture of global patterns.'[25] Put into practice, this approach was associated with three processes: the recording of measurements; the search for relations within the data; and the generalization of those relations 'as the basis of new theory'.[26]

Babbage's suggestion that there be a collection of numerical 'Constants of Nature and Art'—a collection of datasets about features of the natural and human worlds expressed in numbers—was a classic example of Humboldtianism.[27] So was Whewell's research into tides, an international project of chronologically-coordinated global tidal observations and measurement that he masterminded. It was carried out across two decades, during which Whewell synthesized a mountain of local data.[28] It was a remarkable feat of organization as well as computation, resulting in tide tables and also 'cotidal maps' with 'contours' on them linking together places that experienced high and low tides simultaneously. Numbers were turned into maps, making navigation easier and safer.[29] John Herschel's career as an astronomer, in particular his mapping of the southern skies from his observatory at the Cape in South Africa, was derived, in part, from

[22] Richard Yeo, 'William Whewell, Natural Theology and the Philosophy of Science in Mid Nineteenth Century Britain', *Annals of Science*, 36, no. 5, 1979, 509.

[23] [Sir John Herschel], 'Quetelet on Probabilities', L. A. J. Quetelet, *Letters addressed to HRH the grand-duke of Saxe-Coburg and Gotha on the theory of probabilities as applied to the moral and physical sciences* (London, 1849), *Edinburgh Review*, xcii (July–Oct. 1850), 41.

[24] Wulf, *The Invention of Nature*, 91.

[25] Wulf, *The Invention of Nature*, 128.

[26] Morrell and Thackray, *Gentlemen of Science,* 13.

[27] Charles Babbage, 'On the advantage of a collection of numbers to be entitled the Constants of Nature and Art', *Edinburgh Journal of Science*, n.s. xii (1832), 334–40. See above pp. 107–9.

[28] Snyder, *The Philosophical Breakfast Club*, 169–179. Snyder attributes Whewell's study of the tides to his Baconianism, but von Humboldt was a more important, living influence and example on him.

[29] Ibid., 177–8.

the example of his father, William. But it was also the kind of vast empirical, observational and, ultimately, collaborative project that 'Humboldtianism' comprehended. 'Under Humboldtian influence, Herschel wanted to explore the world and measure, describe, and draw it. And the Cape provided him with the opportunity to do so.'[30]

When Herschel and Babbage went on an alpine adventure together in the summer of 1821 during which they climbed all over the Staubbach Falls in the Lauterbrunnen valley in Switzerland, and took temperature and barometric readings as they went, they were engaged in Humboldtian science, albeit on a small and personal scale.[31] They were in the sort of romantic landscape made famous in the paintings of Caspar David Friedrich, the German artist, with which Humboldtianism was intrinsically associated. Humboldt himself might have been the heroic model for Friedrich's most famous canvas, *The Wanderer Above the Sea of Fog*, set in a landscape of rock and mist.[32] The subsequent collaboration of Herschel and Babbage in organizing a system of world-wide meteorological observations on the two equinoxes and two solstices of 1835 was a much larger example of Humboldtianism in action. It was an origin of present-day systems of simultaneous weather reporting.[33] Meanwhile, Herschel's description of Quetelet's endeavours

in the collection and scientific combination of physical data in those departments which depend for their progress on the accumulation of such data in vast and voluminous masses, spreading out over many succeeding years, and gathered from extensive geographical districts – such as Terrestrial Magnetism, Meteorology, the influence of climate on the periodical phenomena of animal and vegetable life, and statistics in all branches of that multifarious science, political, moral and social[34]

is as precise a definition of Humboldtian science as exists. And Herschel recognized also that Humboldtianism required a specific type of personality to inspire and organize collaboration in science which Quetelet so evidently possessed:

[30] Steven Ruskin, *John Herschel's Cape Voyage. Private Science, Public Imagination and the Ambitions of Empire* (London, 2004), p. 29.
[31] John Herschel and Charles Babbage, 'Barometric Observations at the Fall of the Staubbach. In a Letter from Mr. Babbage to Dr. Brewster', *Edinburgh Philosophical Journal*, 6, 12, 1822, 224–7.
[32] Bruce Hevly, 'The Heroic Science of Glacier Motion', *Osiris*, 11, 1996, 66–86. *Der Wanderer über dem Nebelmeer*, painted in 1818, hangs in the Kunsthalle Hamburg.
[33] Adolphe Quetelet, 'Extracts from a Notice of Charles Babbage, by A. Quetelet of Brussels, translated from the *Annuaire de l'Observatoire Royale de Bruxelles* for 1873', *Annual Report of the Board of Regents of the Smithsonian Institution*, 1873, 184. The measurements took place on 21 March, June, September, and December 1835.
[34] [Herschel], 'Quetelet on Probabilities', 14.

The centre of an immense correspondence, he has moreover succeeded in inspiring numerous and able coadjutors, not only in Belgium but in other countries, with a similar zeal, and impress[ed] them with his views and secure[ed] their aid in carrying out a system of definite and simultaneous observation.[35]

The influence of Humboldtian science may also be traced in the belief of Richard Jones that

If…a spirit of statistical inquiry were fully spread across the globe, if the same phenomena were observed simultaneously in all the more civilised countries, with a common perception of their bearing on political questions, no very long period would elapse before such observations afforded grounds for safe and useful conclusions.[36]

Jones suggested that an international collaborative project be established to investigate the 'effects of different forms of wages on the movements of the population', explaining that 'men do not see that political knowledge ranged under general principles—that is political science—is attainable by the very same efforts that have spread so wide the dominion of physical science'.[37] In short, all of the members of the Cambridge network, the intellectual founders of the statistical movement in Britain—Babbage, Whewell, Jones, Herschel, and Quetelet—may be said to have been *Humboldtians*. Their methods were inductive; they were instinctive quantifiers; they attached numbers to natural and social phenomena; they believed that methods used in natural science could be applied successfully to social science; they looked for regularities, recurring patterns, and uniformity in both nature and society; they collaborated and built scientific institutions; they were influenced by scientists in other countries. The collaboration of Quetelet and Babbage in the creation of the International Statistical Congress in 1851 was Humboldtian as well. As William Farr wrote before its London Meeting,

Combined observation alone can solve the greatest problems of physical philosophy. In like manner statistical science, the basis of political philosophy, can only be successfully prosecuted to its utmost extent by combined systems of observation, in which all civilised nations of the earth join.[38]

---

[35] Ibid.

[36] R. Jones, 'Tract on the incidence of taxes and commodities that are consumed by the labourer. Pt. II. On the effect of fluctuations in the real wages of labour on the movement of population', in William Whewell D.D. (ed.), *Literary Remains consisting of lectures and tracts of the late Rev. Richard Jones* (London, 1859), 181–2.

[37] Ibid.

[38] William Farr, 'Report on the Programme of the Fourth Session of the Statistical Congress', *Programme of the Fourth Session of the International Statistical Congress to be held in London on July 16th and Five Following Days* (London, Her Majesty's Stationery Office), 14.

Beyond all of these similarities in method and approach, there was also a similarity in fundamental assumptions about the natural world. Humboldt's *Personal Narrative of Travels to the Equatorial Regions of the New Continent During the Years 1799–1804*, a compendious account running to 34 volumes of all that he saw and examined, was required reading among the savants of Europe for many years after his return. Darwin took some of the early volumes with him on *his* great voyage aboard the *Beagle* in the 1830s. He had his brother send him yet more volumes of Humboldt that he picked up when the *Beagle* docked in South American ports, and Darwin referenced Humboldt frequently in his subsequent published account, the *Voyage of the Beagle*.[39] As he wrote in 1874, 'My admiration of his famous personal narrative (part of which I almost know by heart) determined me to travel in distant countries, and led me to volunteer as naturalist in her Majesty's ship *Beagle*.'[40] Later, in Darwin's *Autobiography*, he recalled his reading of Humboldt as an undergraduate:

> During my last year at Cambridge I read with care and profound interest Humboldt's *Personal Narrative*. This work and Sir J. Herschel's *Introduction to the Study of Natural Philosophy* stirred up in me a burning zeal to add even the most humble contribution to the noble structure of Natural Science. No one or a dozen other books influenced me nearly so much as these two. I copied out from Humboldt long passages about Teneriffe, and read them aloud...[41]

The two greatest naturalists of the nineteenth century met just once in January 1842 in London, and Darwin, tongue-tied, struggled to get a word in as von Humboldt talked at speed and at length.[42]

Humboldt's last major work, *Cosmos: A Sketch of the Physical Description of the Universe*, published between 1845 and 1862, emphasized the order, long-run regularity, equilibrium, pattern, harmony, and beauty of nature. The universe exhibited a dynamic equilibrium of natural forces, held in opportune balance.[43] Thus Humboldt 'effortlessly combined a commitment to empiricism and the experimental elucidation of the laws of nature with an equally strong commitment to holism and to a view of nature which was intended to be aesthetically and spiritually satisfactory'.[44] Humboldt was criticized, however, for failing to mention, let alone find a specific place for 'God the creator' in this and in his other works, and he was known to disparage aspects of organized religion, which led to speculation

[39] Wulf, *The Invention of Nature*, 133, 229–30.
[40] Charles Darwin to D. T. Gardner (August 1874), *New York Times*, 15 Sept. 1874.
[41] *The Autobiography of Charles Darwin 1809–1882* (ed. Nora Barlow) (London, 1958), 67–8.
[42] Wulf, *The Invention of Nature*, 241–3.
[43] Michael Dettelbach, 'Humboldtian Science', in N. Jardine, J. A. Secord, and E. C. Spary (eds), *Cultures of Natural History* (Cambridge, 1996), 287–304.
[44] Malcolm Nicolson, 'Alexander von Humboldt, Humboldtian science and the origins of the study of vegetation', *History of Science*, 25, 2, June 1987, 180.

that he was an atheist. His heterodoxy might be expected to have adversely affected the reception of his work in early-nineteenth-century England. But his emphasis on the unity of nature, and on natural order, made his outlook compatible with 'natural theology', the domestic religious and intellectual tradition in which the Cambridge savants, and all English scholars of their generation, were educated.[45]

Natural theology, as opposed to revelation in scripture, found evidence for the deity and for His beneficence to mankind in the perfect order and harmony of creation. Its most influential exponent in this period was William Paley, author of *A View of the Evidences of Christianity*, published in 1794, and *Natural Theology; or, Evidences of the Existence and Attributes of the Deity*, published in 1802. Both used the good order and beneficent design of the natural and social realms to build a case for the existence of a God whose providence towards mankind had given us a world so well-adapted to our needs. *Natural Theology* is famed for Paley's metaphor of the watch, found by chance, which must have had a watch-maker to have been constructed so excellently for its function to tell the time. Whewell was a lifelong opponent, in fact, of Paley's ethical ideas as expressed in his earlier work, *Principles of Moral and Political Philosophy*, published in 1785, which came far too close to endorsing utilitarianism for Whewell's tastes.[46] But for clergymen-scientists like Whewell and Jones, and for the wider community of the 'Gentlemen of Science' in the 1830s, including figures like Herschel and David Brewster, very few of whom had broken with the Christianity in which they were schooled, Paley's work on natural theology provided a framework through which to appreciate the natural world. As Cannon expressed it, 'The God of Whewell's *Astronomy* was the God of Paley's *Natural Theology*.'[47] The concepts of design and order could be found in Humboldt's works as well, though shorn of the formal theology and any attribution to the deity. Though it emanated from a different national philosophical tradition, and found no place for God as a 'first cause', Humboldt's central idea that nature was harmonious and unified—that 'all forces of nature are interlaced and interwoven' as he wrote to a colleague in 1799[48]—was nevertheless familiar and reassuring, and hence assimilable by scholars for whom natural theology was the very foundation of their view of nature. As William

[45] Douglas Botting, *Humboldt and the Cosmos* (New York, 1973), 258–62.
[46] Richard Yeo, *Defining Science. William Whewell, Natural Knowledge and Public Debate in Early Victorian England* (Cambridge, 1993), 180–2. Harvey W. Becher, 'William Whewell's Odyssey. From Mathematics to Moral Philosophy' and Perry Williams, 'Passing on the Torch. Whewell's Philosophy and the Principles of English University Education', in Menachem Fisch and Simon Schaffer (eds), *William Whewell. A Composite Portrait* (Oxford, 1991), 17–20; 140–2.
[47] Walter Cannon, 'The Problem of Miracles in the 1830s', *Victorian Studies*, 4, 1, Sept. 1960, 15. The reference is to Whewell's third Bridgewater Treatise, *Astronomy and General Physics considered with reference to Natural Theology* (London, 1834).
[48] Alexander von Humboldt to David Friedlander, 11 April 1799, quoted in Wulf, *The Invention of Nature*, 45.

Brock has argued, Humboldt's potent combination of romanticism and generalised religiosity, which was embodied in a vision of nature as a unity, thrilled and consoled his British audience.[49]

In Whewell's version of natural theology, 'the laws which man detected in the universe were the laws by which God had ordered the creation'.[50] God and nature were not identical but were nevertheless tied together by common principles and could be appreciated in the human mind using common intellectual procedures and approaches. As Whewell put it, 'man . . . is capable, so far, of understanding some of the conditions of the Creator's workmanship. In this way, the mind of man has some community with the mind of God'.[51] Human reason was a reflection of a universal and Divine Reason, but the emphasis was on the word 'some'.[52] The excesses of German idealism in the early nineteenth century, in which mind, nature, and God were held to be identical and interchangeable, the distinctions between them dissolved, were rejected by Whewell, notably in his later works on *The Plurality of Worlds* (1853) and *The Philosophy of Discovery* (1860). Humboldt was acceptable to natural theologians in the British tradition, however, precisely because he was silent about God and did not make these equivalences. They could assimilate his concept of the unity of nature—a concept both material and also spiritual, which was equally important to them—without being tied to any further, heterodox beliefs about the intrinsic identity of the natural and the divine, which was the error of pantheism.

Babbage, we are told, 'was no idolater'. Rather, in the tradition of natural theology which emphasized the links between God and the natural world, but not their identity, he believed 'he saw in the unbounded magnificence of the aspects of Nature, the true reflection of the image of their Almighty Creator'.[53] Not only could Humboldt be used to confirm a Christian conception of 'man's place in nature', therefore: because he was also an enthusiastic observer and experimentalist rather than an idealist believing in the *a priori* origins of knowledge, his methods were heartily acceptable to the 'gentlemen of science' of the 1830s who believed so strongly in the empirical method of scientific procedure.[54] If their aim was to understand 'the language in which the book of nature is written', as Whewell had expressed it in his notebook in 1825, Humboldt was a very good scientific guide who could be followed without doing damage to the fundamental

[49] Brock, 'Humboldt and the British', 372.
[50] Yeo, 'William Whewell, Natural Theology and the Philosophy of Science', 509.
[51] William Whewell, *On the Plurality of Worlds* (London, 1853), 109.
[52] Yeo, 'William Whewell, Natural Theology and the Philosophy of Science', 510.
[53] H. W. Buxton, *Memoir of the Life and Labour of the Late Charles Babbage* (1872–80) (ed. Anthony Hyman) (Cambridge, Mass., 1988), 359.
[54] R. M. Young, 'The Historiographical and Ideological Contexts of the Nineteenth-Century Debate on Man's Place in Nature', in M. Teich and R. M. Young (eds), *Changing Perspectives in the History of Science* (London, 1973), 344–438.

principles of faith and natural theology.[55] Indeed, on the face of it, even the Statistical Society of London seemed to subscribe to natural theology. In 1838, the introduction to the first number of its *Journal*, in asserting the disciplinary claims of the new science of statistics, set the case being made (though in defiance of logic) within a classic definition of natural theology:

> As all things on earth were given to man for his use, and all things in creation were so ordained as to contribute to his advantage and comfort, and as whatever affects man individually affects also man in a state of society, it follows that Statistics enter more or less into every branch of Science, and form that part of each which immediately connects it with human interests.[56]

Darwin, who was educated at Paley's Cambridge college, Christ's, was among many admirers of the great natural theologian. Having finished writing *The Origin of Species* he wrote to his friend Sir John Lubbock that 'I do not think I hardly ever admired a book more than Paley's *Natural Theology*: I could almost formerly have said it by heart.'[57] In his autobiography he wrote of the *Evidences of Christianity* and *Natural Theology* that 'the careful study of these works [at Cambridge] . . . was the only part of the Academical Course which . . . was of the least use to me in the education of my mind'.[58] Darwin's admiration for both Humboldt and Paley fixes the association we are making between two thinkers, one a voyaging German aristocrat and naturalist, and the other an English clergyman and philosopher with a parish in the Lake District, whose views of nature were conformable and complementary in the minds of the intellectual founders of the statistical movement. For these philosophical as well as methodological reasons it was entirely natural, and to be expected, that Alexander von Humboldt be granted honorary membership of the Statistical Society of London and of the statistical movement more generally.

---

[55] Isaac Todhunter, *William Whewell D. D., An Account of His Writings* (2 vols, 1876, London), I, 363; Yeo, 'William Whewell, Natural Theology and the Philosophy of Science', 511.
[56] 'Introduction', *JSSL*, 1, 1, May 1838, 3.
[57] Charles Darwin to Sir John Lubbock, 22 Nov. 1859, Letter 2532, Darwin Project, Cambridge. https://www.darwinproject.ac.uk/letter/DCP-LETT-2532.xml.
[58] Darwin, *Autobiography*, 59.

9

The Opposition to Statistics

Disraeli, Dickens, Ruskin, and Carlyle

The artisans in the London Statistical Society made their case by deploying the latest information and the most up-to-date technology. Statistics were too precious and told too good a story to be left under the control of the governing class. Although it is usually argued that inspectors and investigators were 'met with suspicion',[1] it is a surprising aspect of the history of social investigation in Victorian Britain that the subjects of many inquiries, the working classes, often welcomed the investigators, eager to expose their lives and their case to scrutiny, and keen to co-operate with the statisticians, officials, journalists, and other explorers of the slums so that the conditions in which they lived should be more widely known. During a strike in 1845 on the Durham coalfields, the miners' committee of a dozen men 'were greatly pleased' to be questioned carefully by the mines' inspector H. S. Tremenheere, the architect of the 1850 Mines Act. They asked '"Why don't gentlemen such as you come among us and listen to what we have got to say? There wouldn't be half so much trouble if they did."'[2] When the Christian Socialist, John Malcolm Ludlow, investigated labour relations on the West Yorkshire coalfields in the late 1850s, he noted the contrast between the miners' willingness to share information with him and the contrasting unwillingness of the coal owners. He concluded that the miners knew more than the owners: 'The master practically knows little more than the working of his own pit or pits, to which he is mostly tied for life; the men on the contrary, shifting from pit to pit, sometimes even from district to district, have a much wider field of experience.'[3] When Charles Babbage was in Bradford he was shown all the courtesies by the workingmen he met there: led round a factory by a worker who had even read *On the Economy of Machinery and Manufactures*, taken to the

[1] Tom Crook and Glen O'Hara, 'The "Torrent of Numbers": Statistics and the Public Sphere in Britain, c. 1800–2000' in Tom Crook and Glen O'Hara (eds), *Statistics and the Public Sphere. Numbers and the People in Modern Britain, c. 1800–2000* (New York and Abingdon, 2011), 4.

[2] Brian Harrison, 'Finding Out How the Other Half Live: Social Research and British Government Since 1780', in Harrison, *Peaceable Kingdom. Stability and Change in Modern Britain* (Oxford, 1982), 286.

[3] J. M. Ludlow, 'Account of the West Yorkshire Coal-Strike and Lock-Out of 1858', in *Trades' Societies and Strikes. Report of the Committee on Trades' Societies appointed by the National Association for the Promotion of Social Science* (London, 1860), 17.

Victorians and Numbers: Statistics and Society in Nineteenth Century Britain. Lawrence Goldman, Oxford University Press.
© Lawrence Goldman 2022. DOI: 10.1093/oso/9780192847744.003.0009

local co-operative society, and given 'a most cordial welcome' at all the local workingmen's clubs. The secretary of these 'expressed great anxiety to give [him] the fullest information.'[4]

Distrust of statistics and their collection, on the other hand, has a long history and is woven into the traditions of liberty, localism, personal independence, and hostility to central government that have run through British history since the early seventeenth century, if not before. The first attempt to hold a national census dates from 1753, but the bill then brought before Parliament was opposed in the House of Commons and eventually voted down in the Lords by propertied gentlemen who resented state scrutiny and saw a census as subversive of liberty.[5] William Thornton, MP for York, who had raised his own regiment to see off the Jacobites in the rebellion of 1745, spoke for them all when he inveighed against the loss of freedom and privacy, and the 'French' method of maintaining social order.

> Can it be pretended, that by the knowledge of our number, or our wealth, either can be increased?...And what purpose will it answer to know where the kingdom is crowded, and where it is thin, except we are to be driven from place to place as graziers do their cattle? If this be intended, let them brand us at once: but while they treat us like oxen and sheep, let them not insult us with the name of men.[6]

Thornton would refuse to give information to any official: 'I would order my servants to give him the discipline of the horse-pond.'[7] There was more precision in his condemnation of the bill on its second reading:

> We are to entrust petty tyrants with the power of oppression, in confidence that this power shall not be executed; to subject every house to a search; to register every name, age, sex, and state, upon oath; record the pox as a national distemper, and spend annually £50,000 of the public money – for what? To decide a wager at White's![8]

---

[4] Charles Babbage, *Passages from the Life of a Philosopher* (London, 1864), 228–30.

[5] D. V. Glass, *Numbering the People: The Eighteenth-Century Population Controversy and the Development of Census and Vital Statistics in Britain* (Farnborough, 1973), 17–21; Peter Buck, 'People Who Counted: Political Arithmetic in the Eighteenth Century', *Isis*, 73, 1, March 1982, 32–3.

[6] William Cobbett (ed.), *Cobbett's Parliamentary History of England. From the Earliest Period to the Year 1803*, vol. 14, 1747–1753 (London, 1813), 1320.

[7] Ibid., 1320.

[8] Ibid., 1326. See also S. D. Bailey, 'Parliament and the Prying Proclivities of the Registrar-General', *History Today*, 31, 4, April 1981; Robert Harris, *Politics and the Nation: Britain in the Mid-Eighteenth Century* (Oxford, 2002), 250–1; David Boyle, *The Tyranny of Numbers. Why Counting Can't Make Us Happy* (London, 2000), 62–3.

Later, there was a similar response to the new procedures under the 1836 Registration Act, on which *The Times* reported in January 1837. The paper threw its weight behind 'a poor woman, the wife of a day labourer' resident in the village of Thurlaston in Leicestershire, who had refused to answer the questions of the local registrar about her new-born infant. Her name was Mary Shaw and she referred the official to the local church and its parish register: when asked 'for the required information... she replied that she would not, for she had had her child baptized at the church, and would tell him nothing'. According to *The Times*, the government saw this as 'a pattern case, in order to establish a principle and exhibit a warning to the public' that they must answer and comply when confronted by any 'coarse and greasy district registrar'. The authorities brought the woman before the Quarter Sessions to answer for her crime of 'refusing to register the birth of a child'. The jury found her guilty, though the prosecution 'did not wish to press for punishment; their object was to make the law known' and they were therefore 'satisfied by the chairman giving the defendant a reprimand'. According to *The Times* this was 'a tyranny leading to the violation of the decencies of domestic life' and an illustration of 'the tendency of Whig legislation'.

> Surely our countrymen will not be found prepared to submit, whatever may be the statistical advantages of an universal register, without remonstrance, to the provisions of an act of Parliament, if they shall be found capable of being ratified to such an extent as this under the authority of a court of justice? It has been hitherto the boast of Englishmen that their constitutional history exhibits a progressive improvement of those securities by which their liberties are guarded from invasion and their rights and immunities ascertained and protected.

For good measure *The Times* added a tale from the reign of Richard II concerning the intrusive questioning of 'a tyler at Deptford' by an official who wanted to know the date of birth of his daughter, and whether she was therefore liable for the poll tax. The official did not take the father's word, whereupon 'the father with his hammer knocked out his brains'. This was no ordinary tyler: this was Wat Tyler, leader of the Peasant's Revolt. The implication was clear.[9] The vicar of St Paul's, Herne Hill, was not quite so irate when asked in the 1890s to provide information in person for Charles Booth's survey of the religious life of London. But he was not over-enthusiastic, either: 'As you have the consent of the Bishop of Rochester and Southwark I suppose I must consent to see your representative ... But I am weary and sick of the incessant "numbering of the people" for one cause

---

[9] 'The Registration Act', *The Times*, 7 Jan. 1839, 3.

or another.'[10] Another clergyman flatly refused to assist Booth: 'I am not much in sympathy with the tabulating and pigeonholing of our people.'[11]

Distrust of statistics also sprang from cynicism, boredom, sometimes ignorance. It was of the type made famous in the remark of Mark Twain, 'There are three types of lies: Lies, Damn Lies, and Statistics.'[12] Twain believed he was quoting an outburst by Disraeli, though there is no evidence that it was ever written or said by him. Yet it seems an altogether apt witticism from such a figure.[13] There was— and still is—a personality type in public life disinclined to worry over details, dismissive of those who do, interested more in high principle or partisan advantage, and attracted to performative politics rather than long hours of wearying analysis. When, in December 1853, Gladstone savaged Disraeli's budget for its mistakes and innumeracy in a three-hour speech that confirmed their mutual loathing, two contrasting personality types, and two contrasting attitudes to numbers and quantities were evident in the theatre of the House of Commons.[14] It was almost inevitable that Disraeli would join the long list of politicians whom Charles Babbage held in contempt for their failure to recognise the value of his work and support it financially. It was another example of Babbage's enduring ill-fortune in dealing with officialdom that when a new proposal to revive his work on the Difference Engine was put to the Liberal government of the day in 1852, that government should fall and the matter come to rest on the desk of the new Conservative Chancellor of the Exchequer, one Benjamin Disraeli. Knowing that Babbage had never finished any of his projects, he turned him down, flat: 'That Mr. Babbage's projects appear to be so indefinitely expensive, the ultimate success so problematical, and the expenditure certainly so large and so utterly incapable of being calculated, that the Government would not be justified in taking upon itself any further liability.'[15] Babbage had some later pleasure in his autobiography of 1864 in questioning Disraeli's powers of comprehension, skill with figures, use of public funds, and his all-round political abilities. But Disraeli proved him, and so many others, wrong: he was prime minister twice, in 1868 and 1874–80.[16]

[10] Boyle, *The Tyranny of Numbers,* 121–2.
[11] Rev. H. P. Denison of St. Michael's, North Kensington, in 1899. Booth Collection, Notebooks, Religious Influences series, British Library of Political and Economic Science, London School of Economics, B261, f. 173.
[12] Mark Twain, 'Chapters from my Autobiography' (Ch. XX of XXV), *North American Review,* 185, 618, 5 July 1907, 465–74, quotation at 471.
[13] The phrase has been associated with a host of equally likely candidates: the Liberal politician Charles Dilke, the Conservative A. J. Balfour, the naturalist T. H. Huxley, and even the Duke of Wellington. See Michael Wheeler, *Lies, Damned Lies, and Statistics: The Manipulation of Public Opinion in America* (New York, 1976). 'Lies, Damned Lies and Statistics', Department of Mathematics, University of York: https://www.york.ac.uk/depts/maths/histstat/lies.htm.
[14] Robert Blake, *Disraeli* (London, 1966), 345–8. See *Hansard's Parliamentary Debates,* 3rd series, cxxiii, 1666–93, 16 Dec. 1853.
[15] Charles Babbage, *Passages from the Life of a Philosopher,* 107.   [16] Ibid., 108–11, 231–2.

Gladstone's brother-in-law, Lord Lyttelton, the Liberal politician and educationist with whom the prime minister corresponded in classical Greek, divided mankind into 'those who are fond of Statistics, and those who detest them'. He 'lament[ed] to say that I belong to the latter class'.[17] Meanwhile, Anthony Trollope, wise to the ways of the British governing elite, depicted in *Phineas Redux* the emptying of the House of Commons as the club bore rose to make an earnest speech in favour of reform and found that he swiftly lost his audience: 'That gentleman whose statistics had been procured with so much care, and who had been at work for the last twelve months on his effort to prolong the lives of his fellow-countrymen, was almost broken-hearted.'[18] Trollope's Duchess of Omnium had only limited patience for her husband's attention to numbers, though it made him prime minister in the end, a lesson in the attributes and talents required for public life that Trollope wanted his readers to understand and applaud.

> The Duke ought to be here to welcome you, of course,' said the Duchess; 'but you know official matters too well to expect a President of the Board of Trade to do his domestic duties. We dine at eight; five minutes before that time he will begin adding up his last row of figures for the day. You never added up rows of figures, I think. You only managed colonies.[19]

There is a kind of ritualized badinage and humour about figures and numbers that runs through all places, cultures and periods. As William Farr remarked in 1860, 'statistics has been put to the question by hostile critics. The Census itself has been derided: and the sarcasm – "you can prove anything by figures" – suffices in the opinion of some people to refute an argument.'[20]

Decades earlier in the foundational text of modern Conservatism, Edmund Burke had lamented that 'the age of chivalry is gone. That of sophisters, economists, and calculators, has succeeded; and the glory of Europe is extinguished forever.'[21] Subsequently, a profound suspicion of the statistical vocation and the statistical mind characterized some of the most influential and prominent critics of Victorian modernity. To Charles Dickens, John Ruskin and Thomas Carlyle, statistics, numbers, and measurement were contrary to the spirit of social harmony, kindness, co-operation and fellowship which, almost by definition, keep no tally and know no limits. To all three writers, the replacement of sympathy by exactitude was a defining evil of the new industrial society.

---

[17] Lord Lyttelton, 'Address on Education', *Transactions of the National Association for the Promotion of Social Science*, 1868 (London, 1869), 40.

[18] Anthony Trollope, *Phineas Redux* (London, 1874), vol. ii, 298.   [19] Ibid., 302.

[20] William Farr, 'Report on the Programme of the Fourth Session of the Statistical Congress', *Programme of the Fourth Session of the International Statistical Congress to be held in London on July 16th and Five Following Days* (London, Her Majesty's Stationery Office, 1860), 11–12.

[21] Edmund Burke, *Reflections on the Revolution in France* (London, 1790, 5th edn.), 113.

Dickens's lighter-hearted satire on statistics was a product of his early work, an article published in *Bentley's Miscellany* in 1837 entitled 'Full Report of the First Meeting of the Mudfog Association for the Advancement of Everything', a send-up of the British Association.[22] The puffed-up professors and self-satisfied savants in their different sections were fair game for banter and fun, none more so than the devotees of the statistical section where Mr Slug, 'so celebrated for his statistical researches',[23] delivered a paper on the education of infants within three miles of the Elephant and Castle in London. These benighted children knew nothing of mathematics, their ignorance was 'lamentable', and they had the temerity to read imaginative literature. They liked the rhyme of 'Jack and Jill' but this had 'one great fault, *it was not true*'. Worse, one young man believed in dragons and giants, and collectively the children 'considered Sinbad the Sailor the most enterprising voyager that the world had ever produced'. In response, several members of the Mudfog Association 'dwelt on the immense and urgent necessity of storing the minds of children with nothing but facts and figures'. Dickens ended with more stupidity. Mr Slug regaled the meeting with his calculations regarding the daily feeding of dogs and cats in London, and then Mr Ledbrain counted legs. He counted the number of legs belonging to the manufacturing population in a town in Yorkshire, and the number of chair and stool legs they collectively possessed. He calculated that 'ten thousand individuals (one half of the whole population) were either destitute of any rest for their legs at all, or passed the whole of their leisure time in sitting upon boxes'.[24]

The same ideas, when developed later in *Hard Times*, which was published in 1854 and was Dickens's only novel set in the industrial north of England, created an altogether darker and more menacing satire.[25] The Mudfog professoriate was simply ludicrous, but Mr. Gradgrind was the caricature of the statistical mind and the purely utilitarian sensibility who lacked imagination, understanding, fellow-feeling, generosity of spirit, an iota of compassion.

> Thomas Gradgrind, sir. A man of realities. A man of fact and calculations... With a rule and a pair of scales, and the multiplication table always in his pocket, sir, ready to weigh and measure any parcel of human nature, and tell you exactly what it comes to. It is a mere question of figures, a case of simple arithmetic.[26]

---

[22] Charles Dickens, 'Full Report of the First Meeting of the Mudfog Association for the Advancement of Everything', *Bentley's Miscellany*, vol. 2, 1837, 397–413, reprinted in Charles Dickens, *The Mudfog Papers* (London, 1880).

[23] Ibid., 53.   [24] Ibid., 85–90.

[25] For the argument that *Oliver Twist*, serialized between 1837 and 1839, represents another example of Dickens's hostility to statistics, see Maeve E. Adams, 'Numbers and Narratives. Epistemologies of Aggregation in British Statistics and Social Realism, c. 1790–1880', in Crook and O'Hara (eds.), *Statistics and the Public Sphere*, 103–20.

[26] Charles Dickens, *Hard Times* (London, 1854) (Vintage Books edn, London, 2009), 2.

As before in 1837, Dickens counterposed the statistical mentality with the innocence, sympathy and imagination of the child. Gradgrind's opposite, his anathema, is little Sissy Jupe, the circus girl (known as 'Girl number twenty' in school) who had 'a very dense head for figures' and who couldn't pronounce 'statistics' calling them 's-s-s-stuterers' instead. When told by her teacher, Mr M'Choakumchild, that Britain was a thriving nation worth 'fifty millions of money', Sissy could not agree 'unless I knew who had got the money'. To the proposition that only 25 out of a million inhabitants starve in a year, Sissy thought 'it must be just as hard upon those who were starved, whether the others were a million, or a million million'. And when told that out of a hundred thousand passengers who went to sea only 'five hundred of them were drowned or burnt to death', Sissy's computation of the percentage of fatalities was 'Nothing…to the relations and friends of the people who were killed. I shall never learn.'[27]

The axiom 'to know the price of everything and the value of nothing' might have been invented for Grandgrind, M'Choakumchild and their ilk. But Gradgrind's grasp of facts, and thereby of the certainties of life, was of little help to his son, Tom Junior, whose distinctive wail of distress when discovered to be a thief suggests that Dickens had at least a passing familiarity with the ideas and language of Quetelet. As Tom says to his father, 'So many people are employed in situations of trust; so many people, out of so many, will be dishonest. I have heard you talk, a hundred times, of its being a law. How can *I* help laws?'[28] Dickens gloried in individuality and variety, in the diversity of people, their characteristics and their experiences. His soul rebelled against uniformity, authority, bureaucracy, and also fatalism—in novelistic form, the idea that his characters could have no control over their lives.

In Ruskin's writing, the critique of the statistical way of thinking was focused more on the false goals of precision, exactitude and regularity in art and life. Ruskin was contemptuous of the materialism and cupidity that drove many of his contemporaries and thereby distorted morality and taste in Victorian Britain. In his most controversial and also most influential work, *Unto this Last,* published in 1862, he tried to engineer a moral redefinition of key economic concepts including 'wealth' itself.[29] But his admiration for medieval society, art and culture was driven not only by the attraction of pre-industrial and pre-capitalist values, but by the absence of 'engine-turned precision' in the architecture and visual arts of that age. In the famous chapter on 'The Nature of Gothic' in the second of the three volumes of *The Stones of Venice*, published in 1853, he set himself to explain the essence of medieval craftsmanship. This depended on the free and independent

---

[27] Ibid., 52–4.   [28] Ibid., 269.
[29] Lawrence Goldman, 'From Art to Politics: William Morris and John Ruskin', in John Blewitt (ed.), *William Morris and John Ruskin. A New Road on which the World Should Travel* (Exeter, 2019), 130.

creative spirit of the workman and designer. That spirit could not be constrained, limited, or measured in any way. Nor was it precise; indeed, its beauty and distinctiveness depended on its imprecision and irregularity for 'in all things that live there are certain irregularities and deficiencies which are not only signs of life, but sources of beauty . . . All admit irregularity as they imply change; and to banish imperfection is to destroy expression, to check exertion, to paralyse vitality.'[30] In striving for exactitude and perfection Ruskin contended that we de-humanize ourselves and the things that we make:

> You must make either a tool of the creature or a man of him. You cannot make both.

> Men were not intended to work with the accuracy of tools, to be precise and perfect in all their actions. If you will have that precision out of them, and make their fingers measure degrees like cog-wheels, and their arms strike curves like compasses, you must unhumanize them.[31]

Would Babbage have agreed? The Difference Engine was a delicate assemblage of myriad cogs, wheels, and pulleys, that depended in their manufacture on an exactitude and precision hitherto unknown in British engineering. The specifications and tolerances required were more exacting than for anything manufactured thus far in the industrial revolution. The task, and Babbage's demands, led to frequent arguments and the downing of tools. Without this initial human precision in the making of this 'engine' and many other 'engines' since the 1820s, there could be no machine calculation. But once achieved and built it was the machine itself which was supposed to work with exactitude, time after time, and not the human maker or human calculator. Human ingenuity would construct the tool that would calculate repetitively for us: we would thus escape the fate Ruskin foretold of men becoming tools to industry and capitalism. Precision would be attained, but not, it could be argued, at the expense of humanity. The cogs and compasses would enhance our creativity and set us free.

It was Thomas Carlyle who captured epigrammatically the dilemma of this era—of the impact of industrialism on psychology, behaviour and creativity—in his famous observation in one of his early essays, *Signs of the Times*, published in 1829, that 'Men are grown mechanical in head and in heart, as well as in hand.'[32] It was echoed three years later in James Phillips Kay's observation in his study of the Manchester working classes that 'the persevering labour of the operative must rival the mathematical precision, the incessant motion, and the exhaustless power

[30] John Ruskin, *The Stones of Venice* (3 vols, 1853) (New York, 1867 edn), 189.
[31] Ibid., 161.
[32] Thomas Carlyle, 'Signs of the Times', in *Thomas Carlyle: Selected Writings* (ed. Alan Shelston) (Penguin edn, London, 1971), 67.

of the machine'.[33] In what Carlyle called 'the Age of Machinery, in every outward and inward sense of that word',[34] it was Carlyle who also mounted the most vehement assault on statistical thinking, devoting the second chapter of his extended essay on *Chartism*, the great working-class movement of the late 1830s and 1840s, to 'Statistics'.[35] He was concerned with the living standards of the people, the issue that he himself had already termed 'the Condition-of-England question'. But that led him on to a critique of statistics in themselves and of their manipulation, so that the 'question' itself was never answered.

> We have looked into various statistic works, Statistic Society Reports, Poor Law Reports, Reports and Pamphlets not a few, with a sedulous eye to this question of the Working Classes and their general condition in England: we grieve to say, with as good as no result whatever... The condition of the working men in this country, what it is and has been, whether it is improving or retrograding, – is a question to which, from statistics hitherto, no solution can be got.[36]

Carlyle threw up many more issues to which no authoritative answers seemed to be forthcoming:

> And then, given the average of wages, what is the constancy of employment; what is the difficulty of finding employment; the fluctuation from season to season, from year to year? Is it constant, calculable wages; or fluctuating, incalculable, more or less of the nature of gambling? This secondary circumstance, of quality in wages, is perhaps even more important than the primary one of quantity.[37]

The same point—that for many workers employment and wages came irregularly and were not to be relied upon—was to be made at the end of the century by workers themselves at the Industrial Remuneration Conference, the subject of a later discussion in this book.

The lack of answers was in some manner owing to the limitation of statistics in themselves: 'Statistic Inquiry, with its limited means, with its short vision and headlong extensive dogmatism, as yet too often throws not light, but error worse than darkness.'[38] Numbers could not capture the realities of life: 'Tables are like cobwebs, like the sieve of the Danaides; beautifully reticulated, orderly to look upon, but which hold no conclusion. Tables are abstractions . . .'[39] Nor could

---

[33] James Phillips Kay, *The Moral and Physical Condition of the Working Classes Employed in the Cotton Manufacture in Manchester* (London, 1832), 10.

[34] Carlyle, 'Signs of the Times', 64.

[35] Thomas Carlyle, 'Chartism' (1839) in *Carlyle: Selected Writings*, 157–61.

[36] Ibid., 157–8.   [37] Ibid., 159.   [38] Ibid., 159.

[39] Ibid., 157. According to Greek legend, the daughters of Danaus killed their husbands, the sons of Aegyptus, and for this they were condemned to the perpetual filling of bottomless vessels with water.

figures adequately express the sensibility and the experience of the things they enumerated:

> The labourer's feelings, his notion of being justly dealt with or unjustly; his wholesome composure, frugality, prosperity in the one case, his acrid unrest, recklessness, gin-drinking, and gradual ruin in the other, – how shall figures of arithmetic represent all this?[40]

Yet despite these failings, numbers had attained a spurious social authority which encouraged their abuse:

> With what serene conclusiveness a member of some Useful-Knowledge Society stops your mouth with a figure of arithmetic! To him it seems he has there extracted the elixir of the matter, on which nothing more can be said. It is needful that you look into his said extracted elixir; and ascertain, alas, too probably, not without a sigh, that it is wash and vapidity, good only for the gutters.[41]

If social statistics were therefore unreliable, though endlessly manipulable, Carlyle's final point concerned the uses that might be made of them by government:

> A Legislature making laws for the Working Classes, in total uncertainty as to these things, is legislating in the dark; not wisely, nor to good issues. The simple fundamental question, Can the labouring man in this England of ours, who is willing to labour, find work, and subsistence by his work? is matter of mere conjecture and assertion hitherto; not ascertainable by authentic evidence...[42]

Carlyle's urgent questions for the statisticians were never answered. He ridiculed their disembodied and untrustworthy responses. He was contemptuous of the grand claims made for numbers in the analysis of social issues. Beyond this, his critique manifested a profound intellectual pessimism. He had no faith in this fresh instrument for collecting and presenting social knowledge or for enhancing social administration. Held up in the 1830s as new, objective and definitive, Carlyle's natural scepticism of all such claims on behalf of statistics led to a profound suspicion of numbers in themselves and of the statisticians who deployed them. Dickens showed his readers the moral and psychological shortcomings of a world made of facts but without feelings. ('In this life, we want nothing but Facts, sir; nothing but Facts!').[43] Ruskin recoiled when confronted by the new but inhumane civilization growing up around him, based on the attainment of exactitude, regularity and perfection, which turned men into tools. Carlyle directly questioned

---

[40] Carlyle, 'Chartism', 159–60.   [41] Ibid., 157–8.
[42] Ibid., 161.   [43] Dickens, *Hard Times*, 9.

the claims made on behalf of numbers to perfectly represent society and provide answers to the most pressing of questions.

When William Cooke Taylor, historian and commentator, wrote approvingly of statistics in 1835, he counterposed the 'incontrovertible facts' deployed by manufacturers when answering their critics over factory conditions to 'pathetic tales, more than sufficient to supply a whole generation of novelists'.[44] There were 'still people in the world, who prefer the figures of speech to the figures of arithmetic'. No social history of statistics would be complete without reference to this literary counter-culture, not least because of its profound influence over generations of readers, right up to the present. The social and political consciousness of the late-Victorian labour movement was largely composed by the reading of exactly these authors, Carlyle, Ruskin, and Dickens. They imparted a strong spiritual and ethical anti-capitalism to the leadership and institutions of the working class in the period 1880–1914. When the initial Labour MPs who had been elected to the 1906 parliament were asked by the campaigning journalist, W. T. Stead, about their reading and the influences on them, it was Ruskin whom they mentioned first before any other author, with Dickens and Carlyle not far behind.[45]

Yet caution is required: as the next section will show, numbers were used by many social and medical reformers to try to understand and eradicate the greatest afflictions in working-class life: disease, high infant mortality, industrial conditions, early death. The mortality statistics for the successive outbreaks of cholera in the mid-Victorian period were no mere 'cobwebs' or 'tables of abstraction' but the way in which the deaths of thousands could be tabulated and represented. The misuse of statistics requires censure, just as Carlyle saw and made clear. But the 'Condition-of-England Question' could only be answered by careful analysis of wage rates, prices, rents, living standards, calorific intake, and the like. Thomas Carlyle is a corrective and caution to the story of Victorian statistics, for sure— but only that. As this book argues, numbers carry no political or moral freight in themselves; all depends on how they are interpreted and used. They have the potential to liberate as well as constrain.

Nor were all 'men of letters' hostile to numbers. It has been argued that Gerard Manley Hopkins 'encrypt[ed] mathematical relationships into the fabric of his own literary works', notably in the structure and meaning of his great poem, *The*

[44] W. Cooke Taylor, 'Objects and Advantages of Statistical Science', *Foreign Quarterly Review*, 16, 31, Oct. 1835, 103–16. (This was a review of Quetelet's *Sur L'Homme*.)

[45] [W. T. Stead], 'The Labour Party and the Books that Helped to Make It', *Review of Reviews*, 33, June 1906, 568–82. Of 45 Labour and Lib-Lab MPs who replied to Stead's inquiry in early 1906, seventeen mentioned Ruskin while Dickens and Carlyle were mentioned by 13 in each case. These were the first three authors mentioned, ahead of Shakespeare, Mill, Scott, Bunyan, Burns, Morris and Marx et al. See Lawrence Goldman, 'Ruskin, Oxford and the British Labour Movement 1880–1914', in Dinah Birch (ed.), *Ruskin and the Dawn of the Modern* (Oxford, 1999), 58–9.

Wreck of the Deutschland.[46] As Samuel Taylor Coleridge's social outlook changed from radical to conservative, so his attitude to numbers changed, as well. At first, in 1812, he objected to the alarmism and exaggeration in Burke's 'curious assertion that there were 80,000 incorrigible jacobins in England'; to Patrick Colquhoun's 'equally precise' calculation of 'the number of beggars, prostitutes, and thieves in the city of London', and to the calculation that there had been '50,000 incorrigible ATHEISTS in the city of Paris' who had helped cause the French Revolution.[47] All were canards against the cause of the people. Five years later, at a moment of high social tension and distress in Britain, Coleridge cautioned that men 'ought to be weighed, not counted. Their worth ought to be the final estimate of their value.'[48] But ten years after that, in 1827, writing in his notebook, Coleridge divined a universal identity that linked together material reality, words, symbols, and also numbers. All of creation, and all systems, notations, figures and languages within it, were interchangeable; all embodied 'the Word', the Christian *logos*, the spirit of the universe.

> What is Nature? Multeity coerced into Number and Rhythm…How?…by the WORD, and by Every Word that proceedeth in and thro' the same – Call the Words Numbers numerant, Living Numbers; or Ideas; or Laws; or Spirits; or ministrant Angels; - if you please & which you please. The terms are all equivalent.[49]

Some nineteenth-century thinkers were instinctively hostile to the reductionism intrinsic to the numbering of people and things. Others, however, deciphered a higher message and the key to God's purposes in the numbered order and regularity of nature and human behaviour. The greatest of these, as we shall see later, was Florence Nightingale.

---

[46] Imogen Forbes-Macphail, 'The Enchantress of Numbers and the Magic Noose of Poetry: Literature, Mathematics, and Mysticism in the Nineteenth Century', *Journal of Language, Literature and Culture*, 2013, 60, 3, 141.

[47] Samuel Taylor Coleridge, '128. Statistics', in Robert Southey, *Omniana or Horae Otiosiores*, vol. 1 (London, 1812).

[48] Samuel Taylor Coleridge, 'A lay sermon addressed to the higher and middle classes on the existing distresses and discontents', in *Biographia Literaria, Or Biographical Sketches of My Literary Life and Opinions* (London, 1817), 429.

[49] Samuel Taylor Coleridge, *The Notebooks of Samuel Taylor Coleridge* (ed. K. Coburn and A. J. Harding) (Abingdon, 2002), vol. v., 1827–1834, 5522, June 1827.

PART IV

STATISTICS AT MID-CENTURY

10

Mapping and Defining British Statistics

10.1 Statistics at Mid-Century: William Farr's Survey in 1853

In a fantasy published in 1988 by William Gibson and Bruce Sterling entitled *The Difference Engine*, which is set in Britain in 1855, Charles Babbage's computers have been built successfully and widely deployed.[1] They are to be found in the 'Central Statistics Bureau' in particular, whirring and clinking away, as the lives of the people are monitored, their vital data stored, and the health and welfare of the nation are calculated. It is a fantasy of steam, mechanism, and bureaucracy that, of course, bears little relation to the confused realities of the mid-Victorian age. There were no calculating engines; the collection and analysis of national information were in their infancy; no central office existed to track the lives of the people. Indeed, the statistical community was spread across a wide range of organizations, some inside the state, some in communication with the state, and some others keen to uphold their academic and political independence from the state.

Twenty years after the foundation of the statistical movement in Cambridge in 1833, William Farr, the Superintendent of the Statistical Department at the General Register Office was delegated by the British government to attend the first meeting of the International Statistical Congress in Brussels and report there on the organization of statistics in Britain. From the 1840s to the 1870s Farr was the most active member of the British statistical community, a government officer who used his position to push for the needed reforms in public health, and also a key figure in statistical research and discussion outside government in organizations like the Statistical Society of London and the Social Science Association. Farr's speech in Brussels offers a 'tour d'horizon' of the development of statistics inside government with some discussion, also, of its place in society more generally.

With Quetelet, his friend, in the chair, Farr presented to the first Congress 'un résumé…de l'organisation actuelle de la statistique en Angleterre'—a summary of the then current organisation of English statistics.[2] He explained that he spoke with the official sanction of government and also on behalf of English statisticians. His remarks, he explained, would describe the agencies that existed to collect and analyse statistical information for both national administrative, and scientific

---

[1] William Gibson and Bruce Sterling, *The Difference Engine* (London, 1990).
[2] For Farr's speech, see 'Congrès Général de Statistique, Séance du 19 Septembre 1853', *Compte Rendu des Travaux du Congrès Général de Statistique, réuni à Bruxelles, 1853* (Brussels, 1853), 35–40.

Victorians and Numbers: Statistics and Society in Nineteenth Century Britain. Lawrence Goldman, Oxford University Press.
© Lawrence Goldman 2022. DOI: 10.1093/oso/9780192847744.003.0010

purposes. He began by making clear that statistics were collected and processed in government by separate departments. This meant a dispersal of function at all levels: central government, local government at both county and municipal levels, the armed services, the customs and excise administrations, officials collecting taxes and running the Post Office, the Poor Law bureaucracy. Farr mentioned in this context the registration system established in 1836 for births, marriages, and deaths, which was divided into districts and sub-districts. As he explained, all of these different agencies produced reports 'more or less statistical', among them the annual financial budget presented by the Exchequer to the House of Commons.

Parliament had the right to request information from all departments and to establish commissions—Royal Commissions of Inquiry—composed not only of public officials but of 'des savants and des hommes spéciaux'—learned specialists. These published their results as 'blue books', so named after the soft blue covers of official reports at this time. This was, in fact, the peak age for Royal Commissions: in the 1830s, 42 royal commissions were appointed by government; in the 1840s the number was 54, and in the 1850s it rose to 74.[3] Thereafter the number began to fall as government developed a permanent civil service capable of enquiring for itself and therefore easier for government to control.[4] Farr mentioned the role of the press in publishing key statistics of public interest. He spoke of the annual commercial reports published by the Board of Trade. Criminal statistics came to the Home Office from the courts and the police. There were named officials who were responsible for the statistical returns of the army and navy. The Poor Law Board published information on paupers and pauperism.

Unsurprisingly, Farr dwelled at some length on the work of the General Register Office. By statute, the GRO had to provide parliament with annual statements of scheduled information, but Farr mentioned as well the annual reports that it published. Farr was the author of these 'Registrar General's Reports' and he made each one into a *tour de force*, an account not only of the state of the nation and its health, but a manifesto for change. It is on these, above all, that Farr's great reputation is based. He explained the system of registration in England and Wales under the 1836 Civil Registration Act, including the penalties for non-compliance. Scotland and Ireland were not covered by this legislation and registration there was so defective, contended Farr, that it could not be said that either had a system at all: 'si défectueux qu'on ne saurait dire, à proprement parler, qu'il y existe.'

Farr spent considerable time describing the organisation of the decennial census of 1851, providing 'un court résumé des faits relatifs au dernier recensement

---

[3] H. M. Clokie and J. William Robinson, *Royal Commissions of Inquiry* (Stanford, Calif., 1937), 72–80.

[4] On the political implications of social investigations as undertaken by government, see Brian Harrison 'Finding Out How the Other Half Lives: Social Research and British Government since 1780', in Harrison, *Peaceable Kingdom. Stability and Change in Modern Britain* (Oxford, 1982), 260–308, esp. 263–4, 271, 289, 299–300.

dans le Grande-Bretagne, et de la manière don't il sont été recueillis au bureau central'—a short summary of its organisation and of the way it was brought together by central government. The British census in 1851 had three commissioners: the Registrar General, Major George Graham, who, when appointed in 1842, was brother-in-law of the then Home Secretary, Sir James Graham; Horace Mann, the son of the chief clerk at the General Register Office; and Farr himself. The Irish census of the same year had two commissioners: the Registrar General for Ireland and 'M. Wylde' (sic), whom we have met before at the London Congress of 1860, Oscar Wilde's father. Farr described the logistics of census-taking on 31 March 1851 on the British mainland and on 175 inhabited islands from the Scillies to the Shetlands, which was 'heureusement un beau jour'. In England there were 2190 registrars for the task and another 30,610 enumerators ('compteurs') employed for local enumeration. In Scotland the system was organized by the 32 sheriffs, with 8130 enumerators. Significantly, in Ireland, in the period just after the Famine, the census had to be conducted by the police: 'la population de l'Ireland a été recensée par la gendarmerie, sous la direction du registraire générale pour l'Irelande et de M. Wylde'. Farr listed the questions asked of the population from name, relationship to the head of the family, marital status, age and sex, to profession, place of birth, and the presence in the household of the deaf, mute or blind. The returns on individuals and their households were finalized locally, revised centrally, and sent on to the Home Office. A subsequent published volume provided an overview of vital statistics across the nation with details of the numbers married, single and widowed; the age of marriage in each district and town; and the numbers and ages of people in each major occupational group.[5] Farr was able to tell the Congress that the population of Great Britain was 21,121,967, while that of the United Kingdom (including Ireland, therefore) was 27,724,849. This was, said Farr, at least a million fewer than in 1845 as a consequence of death, falling birth rates, and emigration caused by the Irish Famine. The population of London in 1851 was 2,362,226, about a quarter of what it is today. Seventy British cities had more than 20,000 inhabitants. Britain's armed services amounted to 210,474 personnel (as compared with approximately 150,000 regulars today, and another 40,000 voluntary reserves). Her merchant marine totalled 124,744 people. Out of these two categories, 162,490, or nearly half, were living abroad in India, other British colonies, and foreign ports. In the thirty years since 1821 the number of people emigrating from Great Britain, for whom there was an official return, was 2,685,746.

Farr also discussed the simultaneous religious census, devised by Horace Mann and overseen and conducted by the same local officials. It is the only such census ever taken in Britain. This counted the number of churches, chapels and other

---

[5] *Census of Great Britain, 1851.Population Tables. 1. Numbers of Inhabitants. Report and Summary Tables* (London, HMSO, 1852).

places of worship in Britain and the number of attendees at all religious services on Sunday 30 March 1851, 'Census Sunday'. At the same time, and because education was still largely conducted by the religious denominations, the number of schools, the number of pupils taking lessons, and the systems of instruction in use, were also enumerated between 29 and 31 March. Both reports were published in 1854.[6]

At the end of his survey Farr discussed those organizations and societies beyond government that were also involved in statistical research. He referred at this point to the British traditions of voluntarism and of independence from the state which we have met before in this story: 'dans les questions qui présentent un intérêt d'une grande importance, il n'y est pas d'usage de laisser complètement les choses aux mains du Gouvernement'. Leaving highly significant matters in the hands of the government was not the British way. First in the list was the Statistical Society of London which 'undertook much independent research'. He also mentioned in this context the notably clear work of G. R. Porter, 'remarquable par la clarté de ses écrits', and J. R. McCulloch who, following his work in political economy had turned his great abilities 'à la diffusion, la discussion, l'analyse et la généralisation des faits observés, c'est-à-dire de la statistique'—to the study and promotion of statistics.

10.2 Edward Jarvis Among the Statisticians in 1860

A second way of enumerating the statistical apparatus of mid-Victorian Britain— the societies, organizations, projects, and key figures that together constituted the statistical movement—is to follow the account left by Dr. Edward Jarvis in 1860. We have already met Jarvis at the London meeting of the International Statistical Congress in 1860. As president of the American Statistical Association he was long familiar with British authors on social medicine and statistics, publishing a review of the British public health movement as early as 1848.[7] He organized an exchange of books and periodicals with the Statistical Society of London from 1854;[8] and he began corresponding with two of the most important investigator-bureaucrats of the era, William Farr and Edwin Chadwick, in 1850 and 1853

---

[6] Horace Mann, *Census of Great Britain 1851. Religious Worship in England and Wales* (London, HMSO, 1854); *Census of Great Britain 1851. Education.England and Wales. Report and Tables* (London, HMSO, 1854).

[7] Edward Jarvis, 'Sanitary Reform', *American Journal of Medical Sciences*, 15, April 1848, 419–50. See Gerald N. Grob, *Edward Jarvis and the Medical World of Nineteenth Century America* (Knoxville TN, 1978), esp. 157–62.

[8] Edward Jarvis to Edward Cheshire, assistant secretary of the Statistical Society of London, 5 May 1854, Jarvis Collection, Letterbooks, vol. 3, B. Ms. B. 56. 4.3, ff. 224–6, Francis A Countway Library of Medicine, Harvard Medical School, Boston. Mass.

respectively.[9] His five-month sojourn in Britain from March to August 1860, ostensibly accompanying an American patient taking the water cure at Malvern, was a professional tour of the country in which he delighted in the exchange of views with fellow savants. He visited government departments, learned societies, asylums, hospitals, prisons, and schools, exchanging views and information while making new contacts. He was at home in the overlapping worlds of political economists, sanitarians, educationists, and statisticians, the new experts of mid-Victorian society. Feted by his peers, Jarvis discovered the extent of his eminence on the other side of the Atlantic. He wrote to his wife that

> I have been the especial subject of interest and kindness from the men who were engaged in the three great matters that have occupied me most – insanity, mortality and statistics. I found that I was known to them in advance and they received me as an old friend, as one with whom they had before communed freely, and whose principles and life, whose wishes and sympathies, they understood and knew how to meet and were desirous to gratify.[10]

If Farr described the centres and outposts of this community in his address to the ISC in Brussels in 1853 we might say that in 1860 Edward Jarvis 'walked the course', visiting in person many of the key institutions that employed the statisticians and produced the statistics of the era. His letters home and his manuscript autobiography provide a different type of evidence of the extent of British statistics at mid-century.

In London Jarvis frequented the offices of the Statistical Society of London in St James's Square where 'he was received as a friend and made welcome to visit as much as he could while in the city'.[11] He dined with the Society's council—'scholars, mathematicians, political economists, philanthropists'[12]—before listening to a paper on the British electoral system by Thomas Hare, the leading advocate of its reform, whose scheme for amended voting, now known as the 'single transferable vote', naturally interested statisticians and ultimately converted John Stuart Mill into a committed supporter.[13] As Jarvis reported, 'A gentleman (Mr Hare) read an exceedingly ingenious dissertation on a new method of voting after which a

[9] Jarvis to Farr, 10 June 1850, Jarvis Collection, Countway Library, B. Ms. b. 56. 4.1, ff. 133–4; Jarvis to Chadwick, 10 June 1853, ibid., B. Ms. b. 56. 4.3, ff. 15–20.

[10] Edward Jarvis to Almira Jarvis, 11 August 1860 in Edward Jarvis papers, 'European Letters 1860', Concord Free Public Library, Concord, Mass., (3 vols.), III, f. 289. All the letters in these three letter books were written to his wife, Almira.

[11] Edward Jarvis, 'Autobiography', unpublished mss, Houghton Library, Harvard University, Ms. Am. 541, f. 275.

[12] Ibid., f. 277.

[13] Jennifer Hart, *Proportional Representation. Critics of the British Electoral System 1820–1945* (Oxford, 1992), 24–85; Lawrence Goldman, *Science, Reform and Politics in Victorian Britain. The Social Science Association 1857–1886* (Cambridge, 2002), 283–6. Jarvis wrote that 'At the meeting Mr Thomas Hare first read his celebrated Treatise on the more equal distribution of influence and power of voting.' Jarvis, 'Autobiography', f. 278.

general discussion arose.'[14] Jarvis contributed to this with remarks on the various franchises in American states. He also met there for the first time, in person, Chadwick, Farr, and William Newmarch. He was 'introduced to many of these statisticians who were men of mark in England, and some of whom are well known, by reputation, in America, as well as at home'.[15] He visited Chadwick subsequently and they talked of 'the effect of good and bad houses, ventilation, drainage, intemperance etc. etc. on health'.[16] Newmarch encouraged him to compose a paper 'On the System of Taxation prevailing in the United States and especially in Massachusetts' which was delivered after Jarvis had returned to the United States to Section F, the Statistical Section, at the 1860 meeting of the British Association for the Advancement of Science in Oxford.

At various times Jarvis visited the General Register Office in Somerset House, the Poor Law Commissioners, and the Office of the Commission in Lunacy.[17] As he wrote home, 'These Lunacy Commissioners and the Statistical men and the Sanitary men and others seem to take upon themselves the responsibility of making me comfortable and having me obtain what I want.'[18] The Privy Council Medical Office was the centre of the rudimentary public health bureaucracy from where it commissioned and sponsored innovative research into the relations between environment and health. 'Dr. Simon, the head, received me as others had, seemed glad to talk with me' and 'offered me all the reports of the office'.[19] On three occasions Jarvis breakfasted at the home of Florence Nightingale where he sat with 'the elite of the philanthropists, the men of higher culture of England and of Europe'.[20] Invited to speak at a dinner for the members of the Law Amendment Society, many of them leading figures in the English legal profession and judiciary, he discussed the nature of American democracy and its basis in 'the universal intelligence there cultivated, the developed power of all the individuals'.[21]

Jarvis met Dr. Forbes Winslow, the 'celebrated physician in lunacy' who led in the diagnosis and treatment of the insane in this period and in the founding of private asylums.[22] Jarvis visited several asylums himself, including the Bethlem Hospital in London. He attended a meeting of the Royal College of Physicians. He discussed the Health of Towns Commission of the 1840s, a turning point in the history of public health in Britain, with Robert Slaney, MP, who had written on the sanitary state of Birmingham as one of its investigating commissioners.[23]

---

[14] Jarvis, 22 June 1860, 'European Letters', III, f. 57.
[15] Jarvis, 'Autobiography', f. 279.   [16] Jarvis, 8 July 1860, 'European Letters', III, f. 156.
[17] Jarvis, 'Autobiography', ff. 275, 300.   [18] Jarvis, 18 July 1860, 'European Letters', III, f. 178.
[19] Jarvis, 22 June 1860, 'European Letters', III, f. 76.
[20] Ibid., 28 July 1860, f. 209.
[21] Jarvis, 'Autobiography', f. 309; 8 July 1860, 'European Letters', III, f. 142.
[22] Jarvis, 31 May 1860, 'European Letters' II, f. 231.
[23] Jarvis, 8 July 1860, 'European Letters', III, f. 144. Jarvis called him 'McSlaney' in his letter home. Ernest Clarke, 'Robert Aglionby Slaney 1792–1862', ODNB.

He saw the founder and president of the British Medical Association, Sir Charles Hastings.[24] He spent much time with Sir Charles's son, George Woodyatt Hastings, the organizer and secretary of the Social Science Association, and later a Liberal MP.[25] The SSA had been founded in 1857 to bring together all the new specialists on social questions into a single organization for the discussion of 'social science' and for the implementation of its findings through its extensive contacts in parliament and the political establishment. All the notable statisticians of this era were active members of the SSA. Farr, Simon, and Newmarch were among the 51 founders of the SSA at Lord Brougham's London home on 29 July 1857.[26] Florence Nightingale contributed seven papers to the Association. Chadwick was the single most prolific paper-giver at its meetings over the coming three decades. Jarvis twice visited Lord Brougham, whom we have also met before at the London meeting of the ISC, and discussed with him Anglo-American cooperation, and slavery and its abolition.[27] Brougham, now of advanced age, was president of the Social Science Association. Jarvis returned home to Massachusetts impressed by what he had seen and heard of the SSA as a centre of social research and debate in Britain. The Civil War intervened, but as chairman of its Committee of Arrangements, it was Jarvis who presented the motion in the State House in Boston on 4 October 1865 which brought into being an equivalent American Social Science Association. This was to serve as centre of social research and discussion in the north eastern states for nearly half a century, though it was never as influential as its British counterpart.[28]

In the summer of 1860 Jarvis was the guest of the British social-scientific, medical, and statistical communities. His letters and memoirs make clear that these groups intersected and shared memberships at many points and in many organizations. Jarvis was welcomed into professional and expert groups as a fellow worker, found that he was known already as a pioneer of these disciplines in the United States, and not surprisingly, took back with him models and ideas which he tried to replicate. His perambulations around London, meeting professionals and experts in many emerging medical and social disciplines in their learned societies, ministries, and commissions, which reached their climax at the International Statistical Congress, delineate very effectively the borders, contours and strong points of mid-Victorian statistics. In the offices he visited and the organizations he attended, statistics were being collected and harnessed in the service of improved public administration.

---

[24] Jarvis, 'Autobiography', f. 263; Jarvis, 11 April 1860, 'European Letters', I, ff. 260–1.
[25] Jarvis, 'Autobiography', f. 312; Jarvis, 5 July 1860, 'European Letters', III, f. 133.
[26] Goldman, *Science, Reform and Politics*, 29–30.
[27] Jarvis, 'Autobiography', 291; Jarvis, 5, 8 July 1860, 'European Letters', III, ff. 132, 145.
[28] On the founding of the ASSA and its connections with the SSA in Britain see Lawrence Goldman, 'A Peculiarity of the English? The Social Science Association and the Absence of Sociology in Nineteenth Century Britain', *Past & Present*, 114 (Feb. 1987), 133–71, esp. 154.

10.3 Statistical Compilation: G. R. Porter and J. R. McCulloch

A third way of mapping the extent of statistics in mid-Victorian Britain is to consider the two most influential and famous compendia of national data that were mentioned by Farr in his 1853 survey in Brussels, G. R. Porter's *The Progress of the Nation*, first published in 1836 and J. R. McCulloch's *Statistical Account of the British Empire*, published a year later.[29] So fast was the production of new social data, so significant the changes in British society during this period, and so great the interest in, and demand for this kind of information, that *The Progress of the Nation* was updated three times by 1853 and the fourth edition of the *Statistical Account* was published a year later. The two works differ very considerably. Porter was the sole author of *The Progress of the Nation* which is the less scholarly and more popular of the two. McCulloch, on the other hand, drew together some notable experts to compose a more dependable and objective compendium, among them Robert Bakewell on Britain's geology, Charles Neate on business, Herman Merivale on education, and William Farr himself on 'Vital Statistics'. This composition was Farr's first break, and it won him his job in the General Register Office. In a work of greater intellectual and social range, McCulloch even included a chapter on the 'Origin and Progress of the English Language', a cultural marker of the development of the nation as shown by the growth of an English-speaking world. McCulloch's compendium was not a work of theory or higher speculation. Nor did it expressly seek to be a calculation and summation of the 'state of the nation' which might be used in political argument or social discussion. It was a reliable and professional work written by specialists.

The Progress of the Nation was divided into eight sections on population, production, interchange (i.e. communication), public revenues, consumption, accumulation (i.e. property and wealth), information on the colonies, and 'moral progress' which was demonstrated by statistics on crime, manners, education and even the use of the postal services. The format remained the same, though the information became more complete and sophisticated with each edition. Porter was always a partisan for his country and for free trade. Britain, he explained in 1836, had, within the present generation, 'made the greatest advances in civilization that can be found recorded in the annals of mankind'. Thus 'to inquire into the progress of circumstances which have given pre-eminence to one's own country would seem to be a duty'. His aim was to illustrate, using 'well-authenticated facts' the 'progress of the whole social system in all its various departments, and as

---

[29] G. R. Porter, *The Progress of the Nation in its Various Social and Economical Relations, From the Beginning of the Nineteenth Century* (London, 1836, 1847, 1851 *et seq.*). J. R. McCulloch (ed.) *A Statistical Account of the British Empire: Exhibiting its Extent, Physical Capacities, Population, Industry, and Civil and Religious Institutions* (2 vols, London, 1837, 1839, 1847, 1854 *et seq.*).

affecting all its various interests'. He limited himself to the period since 1801 for a political reason: that this would allow the inclusion of Ireland since the Act of Union in a study of the United Kingdom as a whole. But there was also a question of sources to be considered: only in recent decades were the data plentiful and reliable: 'the materials which can be brought in aid of a labour of this kind, and which relate to the occurrences of the present century, are vastly superior in amount and value to those that are to be collected from any existing records of earlier date'.[30] Twelve years later, Porter's introduction to the second edition was even more direct: 'all the elements of improvement are working with incessant and increasing energy' and the aim of the work was to 'faithfully record the onward progress of England'. Improvement, for Porter, included not just material advances but 'all that relates to the moral condition of society'. He castigated 'the reign of ignorance in this country' and strongly endorsed the commitment of the new Whig prime minister, Lord John Russell, to popular education.[31] He was even more outspoken in his support for the very recent repeal of the corn laws and the transition to free trade:

> The corn duties and the sugar duties being thus disposed of in a manner fatal to the continuance of monopolies, it may be looked upon as certain, that the principle contended for in the following pages – that of not imposing any Customs' duties except for the purpose of obtaining revenue – must, ere long, be universally acted upon by Parliament.[32]

Five years later he could contend that his third edition contained such evidence of 'general prosperity' as to be 'clear and conclusive in favour of a free-trade policy'.[33] So far from being a neutral compendium of national statistics, Porter had composed, as he saw it, a hymn to progress in British policy and legislation as represented in the statistics of consumption and national income.

McCulloch echoed Porter in justifying his *Statistical Account* by the growth of data and the changes to the nation:

> A vast mass of materials has consequently been collected that may be employed to illustrate the statistics of the empire; and the time seemed to be at length arrived when it might be attempted to compile a work that should give a pretty fair representation of the present condition of the United Kingdom.[34]

---

[30] Porter, 'Introduction', *The Progress of the Nation*, 1st edn, 1836, 1–2.
[31] Porter, 'Preface to the Second Edition, 30 Nov. 1846', *The Progress of the Nation*, 2nd edn, 1846, xvii–xxii.
[32] Ibid., xx.
[33] Porter, 'Preface to the Third Edition, 1 Jan. 1851', *The Progress of the Nation,*, 3rd. edn, 1851, xxv.
[34] McCulloch (ed.), 'Preface', *Statistical Account of the British Empire*, 1st edn, 1837, vi.

Like Porter as well, McCulloch and his authors had

> not refrained from freely stating our opinions...We can, however, assure the reader that we have not advanced any statements for which we did not suppose we had good grounds; and that the opinions we have expressed, whether right or wrong, have been honestly formed.[35]

McCulloch, like Porter, saw no reason to refrain from praise for free trade, in this case by offering a dedication in the fourth edition to Sir Robert Peel, the architect of 'repeal', whose 'devotion to the NATIONAL INTEREST has ensured [him] the highest place in the public estimation'.[36] The overall tone, however, was more balanced, modest, professional, and scholarly. It was also less chauvinistic: McCulloch admitted using a French model for his work, for example.[37] The aim of the *Statistical Account* was to furnish thereby 'a tolerably correct view of the present state and resources of the empire'. Unlike Porter, McCulloch admitted that many aspects of national life lacked 'authentic and trustworthy information on various important points'.[38] And unlike Porter, McCulloch, as a scholar, was much more aware of the relationship of these compendia of national information to the development of the infant discipline of statistics itself, to which *The Statistical Account* was a contribution:

> Few, indeed, would imagine, *a priori*, how ill-supplied British writers are with the means necessary to throw light on some of the most interesting departments of statistical inquiry. Latterly, indeed, the public attention has been, in some degree, awakened to the state of this long-neglected department of science, and a few steps, though of no great importance, have been taken to supply the deficiencies in question. But much remains to be done. And we may be permitted to hope that the circulation of this work, by bringing those deficiencies under the public view, and drawing attention to them, will, in this respect, if none else, contribute to the advancement of the science.[39]

We should note that Porter would have disagreed with the note of national self-criticism here. In his view, the British had records 'such as no other country or government in the world has ever brought together'.[40]

---

[35] McCulloch (ed.), 'Preface to the Third Edition', *Statistical Account of the British Empire*, 1847, ix.
[36] McCulloch (ed.), 'Dedication', *Statistical Account of the British Empire*, 4th edn, 1854.
[37] McCulloch (ed.), 'Preface', *Statistical Account of the British Empire*, 1st edn, 1837, vi. The work concerned was Jacques Peuchet, *Statistique Élémentaire de la France. Contenant les Principes de cette Science et leur application à l'analyse de la Richesse, des Forces et de la Puissance de l'Empire Français* (Paris, 1805).
[38] McCulloch (ed.), 'Preface', *Statistical Account of the British Empire*, 1st edn, 1837, viii.
[39] Ibid.   [40] Porter, 'Introduction', *The Progress of the Nation*, 1st edn, 1836, 2.

Engels and Marx, who were immersed in empirical social research on Britain from all sources, would have agreed with Porter. In the preface to the first edition of the German edition of *The Condition of the Working Class in England in 1844*, Engels' great study of Manchester, he wrote that 'only in England is adequate material available for an exhaustive enquiry into the condition of the proletariat'. His book, he added, was based not only on 'personal observation' but also on 'authentic sources'. And the combination of 'official documents [and] *Liberal* sources' could be employed, he wrote, 'to defeat the liberal bourgeoisie by casting their own words in their teeth'.[41] The first and most advanced industrial society was also the most well-documented nation, giving ammunition to its critics as well as its supporters. Making similar comparisons, Marx judged in 1867 that 'the social statistics of Germany and the rest of Continental Western Europe are, in comparison with those of England, wretchedly compiled'. He went on to praise 'men as competent, as free from partisanship and respect of persons as are England's factory inspectors, her medical reporters on public health, her commissioners of inquiry into the exploitation of women and children, into conditions of housing and nourishment, and so on'.[42] It is difficult to compare the quality of national statistics in different countries at this formative stage in the mid-nineteenth century when states and their agencies were haphazardly engaged in establishing bureaucracies capable of collecting and processing social data. To set international standards and assist national governments was the raison d'être of the International Statistical Congress, after all. But the testimony of two such authors is significant: the case for socialism could only be made on the basis of the reliable, accurate, impartial, and copious data that mid-Victorian Britain was generating about the economy and society of industrial capitalism.

Nothing did more to advance statistical science than the choice of 'William Farr, Esq., Surgeon' as 'the author of the celebrated and original article on Vital Statistics' in McCulloch's compendium.[43] But in the absence of the sort of reliable and comprehensive data that only came with the establishment of civil registration under the 1836 Act and the formation of the General Register Office, Farr's essay in 1837 depended on haphazard sources. There was material on special groups of men such as soldiers and sailors, as well as employees of the East India Company, which Farr used for his section on sickness. He did his best with some local life tables and surveys of vital statistics for an assortment of English towns. He could use the records of insurance companies and also of benefit societies, though the latter had not reached the state of professional organization of the

---

[41] Friedrich Engels, 'Preface' to the 1st German edn, *The Condition of the Working Class in England in 1844* (Leipzig, 1845), translated and edited by W. O. Henderson and W. H. Chaloner (Oxford, 1958), 3.

[42] Karl Marx, 'Preface to the First Edition', *Capital* (1867 edn) (Harmondsworth, 1976, Pelican Marx Library, ed. Ernest Mandel), vol. I, 91.

[43] McCulloch (ed.), 'Preface', *Statistical Account of the British Empire*, 1st edn, 1837, vii.

large friendly societies which provided services for millions of workers in the late-Victorian era.[44] There was also evidence collected in investigations and enquiries by government. For his section on the causes of death he still relied on patchy and unreliable records such as the London Bills of Mortality, the same source used by John Graunt nearly two centuries before to inaugurate 'political arithmetic'. Though Farr managed remarkable feats of ingenuity and synthesis in squeezing sense out of this diverse and inaccurate material, nothing could better illustrate the need for systematic registration of the population's key life events and the analysis of this data by skilled statisticians like Farr himself. In commissioning the essay, McCulloch gave Farr the opportunity to write his own job description. He joined the General Register Office two years later in 1839.[45]

10.4 The Order of Things

There remains another way of defining statistics, related not to its institutional and bureaucratic geography—the places and spaces occupied by statisticians in government and civic life—as described by Farr and Jarvis, but to a succession of attempts to define the subject categories of the new discipline, which were a feature of its first fifty years. These efforts, usually setting out the different heads of inquiry to be covered by the statistical movement, involved constructing taxonomies for the subject. As such, they bear close resemblance to the taxonomic endeavours that Michel Foucault uncovered in intellectual life at the end of the eighteenth and beginning of the nineteenth centuries in his famous study of *The Order of Things*. The 'science of statistics' as it was often then called, epitomized the new study of men and women which, according to Foucault's scheme, became part of human knowledge at this time.[46] 'The mode of being of things, and of the order that divided them up before presenting them to the understanding was profoundly altered', creating thereby the modern demarcation of subjects and the map of knowledge that we have inherited.[47] The refining and codifying of the scope of a subject, its sub-divisions, and its definition in relation to other subjects, was, and is, the preliminary stage in the genesis of a new science.[48] Early and

---

[44] Arthur Downing, 'The Friendly Planet. Friendly Societies and Fraternal Associations around the English-speaking World, 1840–1925', D. Phil Thesis, 2015, University of Oxford.

[45] John M. Eyler, *Victorian Social Medicine. The Ideas and Methods of William Farr* (Baltimore and London, 1979), 8–9.

[46] See, for example, 'Introduction', *JSSL*, Vol. 1, no. 1, 1838, 1: 'It is within the last years only that the Science of Statistics has been at all actively pursued in this country'.

[47] Michel Foucault, *The Order of Things: An Archaeology of the Human Sciences* (*Les Mots et Les Choses*, Paris, 1966) (1970 edn, London), 'Preface', xxiv.

[48] On the classificatory endeavours of Enlightenment and post-Enlightenment science, see John V. Pickstone, *Ways of Knowing. A New History of Science, Technology and Medicine* (Manchester, 2000), 60–82.

mid-Victorian statisticians were trying to find order in nature and society. Reflexively, they sought order in their own studies, as well. The need to establish statistics as intellectually separate from, and critical of, the dominant social discourse of political economy, gave added impetus and importance to this exercise: 'statistics was inextricably linked to classifactory reasoning and the ceaseless generation of new taxonomies and descriptive categories'.[49]

At the very first opportunity, therefore, during those irregular meetings in Cambridge in June 1833 of what became Section F of the British Association, Richard Jones told his select audience that statistics could be divided into four heads: economic, political, medical, and moral and intellectual statistics.[50] Writing to Whewell in February 1834 to tell him about the events leading to the creation of the Statistical Society of London in meetings at Babbage's house, Jones added flesh to those bare bones:

The divisions I propose are these and I want their first work to be drawing up plain interrogatories to distribute:

Economical Statistics 1.agriculture. 2. manufactures. 3. commerce and currency. 4. distribution of wealth, i.e. rent, wages and profits.

Political Statistics 1.statistics of elements of institutions—electors etc. 2. legal statistics—number of national and local tribunals, nature of cases tried etc. etc. 3. Finance—taxes, expenditure, public establishments etc. etc.

Medical Statistics 1.general medical statistics. 2. population (the doctors say they shall want subdivisions).

Moral and Intellectual Statistics 1.crime. 2. education and literature. 3. ecclesiastical statistics.[51]

This was 'back-of-the-envelope' musing without much preparation, for sure. But it formed the basis for the formal Prospectus of the Society, which was published on 23 April 1834, and which employed the same four-part taxonomy and many of the same subdivisions to define the Society's remit.[52] Committees were then

---

[49] Tom Crook and Glen O'Hara, 'The "Torrent of Numbers": Statistics and the Public Sphere in Britain, c.1800–2000', in Tom Crook and Glen O'Hara (eds), *Statistics and the Public Sphere. Numbers and the People in Modern Britain, c. 1800–2000* (New York and Abingdon, 2011), 18.

[50] 'Minutes of the Committee of the Statistical Section of the British Association, June 27, 1833', in Transcript from the note-book of Mr. J. E. Drinkwater, Archives of the Royal Statistical Society, London, f. 3.

[51] Richard Jones to William Whewell, 18 February 1834, Whewell Papers, Trinity College, Cambridge, Add Mss. c. 52/60.

[52] 'Prospectus of the Objects and Plan of Operation of the Statistical Society of London', in volume 'Reports of Council and List of Fellows', Archives of the RSS. The Prospectus was reprinted in William Newmarch, 'The President's Inaugural Address on the Progress and Present Condition of Statistical Inquiry', *JSSL*, 32, Dec. 1869, 359–84. See also James Bonar and Henry W. Macrosty, *Annals of the Royal Statistical Society 1834–1934* (London, 1934), 24.

established for each of the four themes to describe them more precisely. As we have seen, Jones and Whewell were assigned to the committee on Moral and Intellectual Statistics, and it was Whewell's outline of its responsibilities and coverage that was adopted by the Society some weeks later.[53] As the Society developed and as its membership and their interests pushed it more towards the concerns of government, so the taxonomy evolved. By 1840, in the sixth annual report from the Society's council, 'the whole field of our labours' was divided into five quite different sections. First, there were the 'statistics of physical geography', of the natural environment. Next, the 'statistics of production' focused on manufacturing and commercial life. This was distinguished from the 'statistics of consumption and enjoyment', a broad category comprising demographic, social, and also medical data. The 'statistics of instruction' included education as well as religious, scientific, and literary life. The 'statistics of protection' was the name given to information on the constitution, politics and law.[54]

Inevitably, these plans and projections had little influence over the structure of the work of the Society, which was much more varied and eclectic. Nor did they guide the wider development of statistics in general. Twenty-five years later, in 1865, Dr William Guy, one of the most active and celebrated of its Victorian members, analysed the Society's *Journal* and divided all the published contributions into eight classes. There were papers on the state of a particular nation; or on a province or region of the nation; or on a key social theme within that nation such as 'education, crime, industry, health, wealth'. There were reports of social investigations; 'polemical papers' contesting a point; papers on 'the numerical method'; discursive rather than statistical contributions on history and political economy; and papers relating to other sciences which were 'rich in facts and figures'. A taxonomy of the subjects that the Society *should* study had been replaced by an account of what *had been* studied, and they were very different.[55] Slightly later, in 1869, William Newmarch, speaking from the President's chair, defined eight different subjects which the Society had pursued actively. These comprised vital statistics; census statistics; 'statistics of pauperism, crime and police', which, in its very title, treated poverty and indigence as forms of social deviance; fiscal and financial statistics; data on 'conveyance', by which was meant transportation; and statistics on trade, education, and government, respectively.[56] But he enumerated a further 18 more specific areas of social investigation where knowledge was still meagre. These included several sub-divisions on the statistics

[53] See above, Chapter 3, 'Cambridge and London', pp. 49–50.
[54] 'Sixth Annual Report of the Council of the Statistical Society of London, Session 1839–40', *JSSL*, vol. 3, 1, 1840, 4–5.
[55] William Guy, 'On the Original and Acquired Meaning of the term "Statistics", and on the Proper Functions of a Statistical Society; also on the Question Whether there be a Science of Statistics; and if so, what are its Nature and Objects, and what is its relation to Political Economy and "Social Science"', *JSSL*, 28, 1865, 484–5.
[56] Newmarch, 'President's Inaugural Address', 361.

of consumption, incomes, and standards of living; data concerning public institutions like charities and hospitals; on the different religious denominations; on trade and the effect of tariffs; on the colonies, especially India; and finally on the subject of statistics itself: 'investigations of the mathematics and logic of statistical evidence'.[57]

By this stage, the taxonomy of statistical endeavour looked very different from that devised by Jones in 1834. Indeed, taxonomy was itself becoming redundant. This was because of a fundamental, though slow, recognition across the course of the period, that statistics did not form a discipline in itself—akin to biology, or physics, or economics—the sub-divisions of which might be enumerated as by some Linnean system of classification applied to the knowledge of society. This had, in fact, been the initial conception of statistics as the 'Introduction' to the *Journal of the Statistical Society* made clear in 1838:

> It is unnecessary to shew how every subject relating to mankind itself, forms a part of Statistics; such as, population; physiology; religion; instruction; literature; wealth in all its forms; raw material; production; agriculture; commerce; finance; government; and, to sum up all, whatever relates to the physical, economic and moral or intellectual condition of mankind.[58]

Statistics encompassed them all at the outset. Yet, by the late-Victorian period, experience and practice had shown that the subject of Statistics was both wider and narrower than this. It was wider in the sense that the analysis and use of numerical data was a growing requirement of all the sciences and subjects, to each of which statistics could make an obvious, continuous, and fundamental contribution. It was narrower because of the recognition that statistics was not, therefore, *the* social science, or even *a* social science, but a set of techniques and procedures that could be applied in all spheres, academic, civil, and political. As one fellow of the Statistical Society of London put it in a paper to Section F of the British Association in Oxford in 1860,

> Statistics has no facts of its own; in so far as it is a science, it belongs to the domain of Mathematics. Its great and inestimable value is, that it is a "method" for the prosecution of other sciences. It is a "method of investigation" founded upon the laws of abstract science; founded on the mathematical theory of probabilities; founded upon that which has been happily termed the "logic of large numbers".[59]

The author, Mr J. J. Fox, accepted that his audience at this relatively early stage was 'not prepared to admit [his] views as to Statistics not being a science in

---

[57] Ibid., 366.   [58] 'Introduction', *JSSL*, vol. 1, 1, May 1838, p. 2.
[59] J. J. Fox, 'On the Province of the Statistician', *JSSL*, 23, 3, 1860, 331.

itself'.[60] But from the 1880s, with the advent of mathematical statistics which allowed the application of much more sophisticated techniques to the analysis of data, such a view became commonplace. One might now be trained in the use and analysis of statistics, but statistical taxonomies were no longer compiled because no longer needed: there was no branch of knowledge, as opposed to technique, to be called *Statistics*.

In the 1830s, in an age struggling to find new ways to understand and demonstrate social change, and at a time when much social knowledge consisted of *a priori* assumptions taken on trust, it must have seemed as if the collection and comprehension of numerical data was a new and necessary division of intellectual life *in itself*, one that could better represent personal and social reality than the then existing social disciplines. Half a century later, the subdivisions of social science—sociology, psychology, anthropology, neo-classical economics among them—which were in the process of emerging and subdividing in turn, had made statistical procedures intrinsic to all sciences, but had removed the rationale for statistics as a science in itself. Foucault was insistent that 'the order of things' he uncovered and described was synchronic rather than diachronic: that to appreciate it, one should look across the piece, at the different sectors and divisions of knowledge in diverse fields, at the same moment in time rather than through time. In his view, the nature of late-eighteenth century natural history, therefore, was better appreciated by comparing it to contemporary ideas of a general grammar, or to contemporary thinking on money and wealth, than to the later biological theories of Cuvier or Darwin.[61] His approach and insights in explaining the crystallization of the human sciences only seem to gain in their persuasive force, however, when we are enabled to see a new order emerging over a period of time, during which disciplines tested themselves against rival explanations, procedures, taxonomies, and subjects. In the case of statistics, the process by which it found its true disciplinary role and space took several decades. It was in this way that the 'archaeology of the human sciences', another Foucauldian expression, was reordered.[62]

---

[60] Ibid., 336.   [61] Foucault, *The Order of Things*, xxiv.
[62] For further discussion of Foucault's theoretical influence, see the discussion in the Conclusion at pp. 315–18 below.

11

Buckle's Fatal *History*

Making Statistics Popular

At roughly the same time that the personal and published surveys of Jarvis and Farr of the extent of British statistical culture were made, the publication in June 1857, at his own expense, of the first volume of the *History of Civilization in England* by Henry Thomas Buckle, brought the findings of the statistical movement to a new and broad readership. Yet Buckle was not a statistician: he was a self-taught and independent scholar, somewhat shy and weak in health, almost unknown up to that point, whose views on the past and historical method were intrinsically controversial, and whose use of statistics, picked over by the critics, demonstrated the grand claims but also the intellectual limitations of the statistical movement.

Buckle had a close friend—he may even have proposed marriage to her—in Emily Shirreff, the women's educationist and writer on women's issues.[1] She was very briefly the Mistress of what became Girton College, Cambridge, the first institution for the higher education of women in Britain. She also established the Girls' Public Day School Company in 1872: it founded many first-rate girls' schools. In the insightful and touching memoir of Buckle that she composed after his early death in 1862 while he was travelling in the Middle East, Shirreff recalled the stir on publication of his first volume:

> He sprang at once into celebrity; and singularly enough, considering the nature of the book, he attained not merely to literary fame, but to fashionable notoriety. To his own great amusement, he became the lion of the season; his society was courted, his library besieged with visitors, and invitations poured in upon him, even from houses where philosophical speculation has surely never been a passport before.[2]

The book met with 'an almost instantaneous success' and 'a world-wide curiosity'.[3] It has been described as 'one of the great books of the nineteenth century'.[4] The

---

[1] Philippa Levine, 'Emily Shirreff 1814–1897', *ODNB*.

[2] Emily Shirreff, 'Biographical Notice', in Helen Taylor (ed.), *Miscellaneous and Posthumous Works of Henry Thomas Buckle* (3 vols, London, 1872), vol. I, xl.

[3] Alfred Henry Huth, *The Life and Writings of Henry Thomas Buckle* (2 vols, London, 1880), I, 140, 159.

[4] Henry Thomas Buckle, *On Scotland and the Scottish Intellect* (H. J. Hanham, ed.) (Chicago and London, 1970), xiii.

Victorians and Numbers: Statistics and Society in Nineteenth Century Britain. Lawrence Goldman, Oxford University Press.
© Lawrence Goldman 2022. DOI: 10.1093/oso/9780192847744.003.0011

famous Oxford don and college head, Mark Pattison, welcomed it as 'certainly the most important work of the season; and it is perhaps the most comprehensive contribution to philosophical history that has ever been attempted in the English language'.[5] According to Lord Acton, the great historian of the next generation, 'it must have powerfully appealed to something or other in the public mind...in order to have won so rapid a popularity'.[6] That 'something' was Buckle's attempt to give systematic order to History, based on the examples of natural science and social statistics. If social statistics now demonstrated the predictability of human behaviour, then, in Buckle's view as expressed in the opening pages of the book, the writing of History could assume the posture of a science, one that would demonstrate the regularity and predictability of social and intellectual development in the building of 'civilization'. This appealed to the mid-Victorian reading public at a time when 'science' was in vogue and debates over determinism, be they in theology, philosophy, or history, were common. Buckle's version of determinism, or fatalism as it was then known, which tried to make sense of human history in broad generalizations and sought historical laws of universal applicability, appealed to the spirit of the age as Acton intuited. Buckle wrote that he hoped 'to accomplish for the history of man something equivalent, or at all events analogous, to what has been effected by other inquirers for the different branches of natural science'.[7] One notable physician giving a paper on statistics at the Social Science Association in 1875 recalled the effect of the book's publication:

> Few of those who were grown up at the time can forget the sensation which that book excited when first published or the enthusiasm which it aroused, and certainly one of the most remarkable portions of it is the passage in the first chapter which deals with statistical investigations and their results.

Nevertheless, Dr G. M. Child, the Medical Officer of Health for Oxford, also considered it 'the most ridiculously over-rated book which has ever been published in my life-time'.[8]

Buckle only managed to publish two volumes of his *History*, the second coming out in 1861 just before his death. In twenty extant chapters he wrote much more on the histories of France and Spain than on England, in fact. But contemporaries and subsequent scholars have been interested above all in the six chapters on the philosophy of history that start the work. Buckle was clear that he

---

[5] [Mark Pattison], 'History of Civilization in England', *Westminster Review*, vol. XII n.s., Oct. 1857, 375.

[6] John Emerich Edward Dalberg-Acton (Lord Acton), 'Mr Buckle's Thesis and Method', *Historical Essays and Studies* (eds J. N. Figgis and R. V. Laurence) (London, 1908), 305.

[7] Henry Thomas Buckle, *History of Civilization in England* (2 vols, London, 1857, 1861), I, 6.

[8] G. M. Child, 'On the Necessary Limits of the Applicability of the Method of Statistics, with Especial Reference to Sanitary Investigation', *Sessional Proceedings of the National Association for the Promotion of Social Science 1875–6* (London, 1876), 50.

was writing 'philosophical history', and was consistently critical of what he saw as the trivialization of history by other historians, among whom 'a strange idea prevails that their business is merely to relate events, which they may occasionally enliven by such moral and political reflections as seem likely to be useful'.[9] This reads like a criticism of Macaulay, then at the height of his fame as the historian of England in the seventeenth century and of liberal progress in general. Buckle's ambition was at once grander, more integrated, more synthetic. As Shirreff admirably summarized,

> His seemingly desultory studies were co-ordinated to one definite purpose, and that purpose, the gigantic project of setting forth in one connected view the various paths through which the human intellect has worked its way, and won for our practical life that fulness and freedom which we call civilization; seeking through the records of history to make manifest in the march of human progress that same empire of law which physical science discloses in the material universe.[10]

In Buckle there was a confluence of science, history, and statistics in a synthesis designed to bring moral and social questions within intellectual structures hitherto reserved for the explanation of the material world: 'The unfortunate peculiarity of the history of man is, that although its separate parts have been examined with considerable ability, hardly anyone has attempted to combine them into a whole, and ascertain the way in which they are connected with each other.'[11] In this synthetic history, statistics seemed to offer a way of imposing uniformity and order, based on the best evidence of human behaviour in the aggregate, on descriptions of the otherwise messy, chaotic, and arbitrary past as it was then presented to a Victorian audience.

Buckle's predominant theme, the key to understanding the development of civilization itself, was intellectual progress, the march of ideas and knowledge. He refuted the claims of religion, government, or literature as 'the prime movers of human affairs'. Instead, 'the growth of European civilization is solely due to the progress of knowledge, and [that] the progress of knowledge depends on the number of truths which the human intellect discovers, and on the extent to which they are diffused'.[12] He pledged himself 'to show that the progress Europe has

---

[9] Buckle, *History of Civilization*, I, 3–4. See T. W. Heyck, *The Transformation of Intellectual Life in Victorian England* (London, 1982), 135.

[10] Shirreff, 'Biographical Notice', xxviii.

[11] Buckle, *History of Civilization*, I, 35. According to Shirreff, 'Literature, science, philosophy, however engrossing singly, occupied him as a part of a great whole; and the mode of coordinating all those various branches of knowledge was his chief concern.' Shirreff, 'Biographical Notice', xxx.

[12] Buckle, *History of Civilization*, I, 232, 265.

made from barbarism to civilization is entirely due to its intellectual activity'.[13] As he wrote to Shirreff,

> My habits of mind accustom me to consider actions with regard to their conse-quences; you are more inclined to consider them with regard to their motives. You, therefore, are more tender to individuals than I am, particularly if you think them sincere; and you hold that moral principles *do* hasten the improvement of nations. I hold that they do *not*...[14]

Buckle relegated moral progress, and hence Christianity, to, at best, a secondary role in the civilizational process, and this was controversial in itself.[15] As Acton commented, Buckle wrote 'the history of modern civilisation without taking into account the two elements of which it is chiefly composed—the civilisation of antiquity and the Christian religion'.[16] These were sins of omission. Buckle's con-sistent hostility in his *History* to the malign influence of authoritarian churches and their hierarchies was a sin of commission, however. His chapters on Spain demonstrated the suppression of reason, evidence and science by the Roman Catholic church, admittedly a common enough theme in nineteenth-century British historiography, but in Buckle's case it was part of an assault on all churches and, in particular, on ecclesiasticism in every form.

It did not help Buckle's cause in the eyes of orthodox opinion that he took inspiration from Auguste Comte, the French thinker who first coined the very term 'sociology'.[17] Comte emphasized the unity of all knowledge, whether of nature or society. He popularized a conception of human progress through three stages of history, from the theological to the metaphysical, and from the meta-physical to the scientific or 'positive', in which the human intellect broke free from the shackles of superstition and religion to attain the positive age of science and reason. In that crowning era, human society itself would become the object of investigation, and sociology the highest intellectual endeavour. The organization of Comte's followers into a secular 'religion of humanity' cast not only his ideas but his supporters in both Britain and France as heterodox.[18] Buckle was not an uncritical admirer of Comte's *Cours de Philosophie Positive*, published in 1830, but

[13] Ibid., 204.
[14] Buckle to Shirreff, 13 May 1859 quoted in Huth, *Life and Writings of Buckle*, 313.
[15] See, for example, the critical remarks of the editor of *The Spectator*, Richard Holt Hutton: that Buckle 'made the evolution of the purely intellectual principle the sole governing element in the his-tory of civilisation' while seeking to 'demonstrate that faith and feeling have no important influence in determining the course of the world's progress', *The Spectator*, 16 May 1874, 623.
[16] Lord Acton, 'Mr Buckle's Philosophy of History', 334.
[17] For an influential introduction to Comte's thought see Raymond Aron, *Main Currents in Sociological Thought*, vol. 1, (1965) (2019 edn, Abingdon, Oxfordshire), 60–118.
[18] T. W. Wright, *The Religion of Humanity. The Impact of Comtean Positivism on Victorian Britain* (Cambridge, 1986).

he contended, nevertheless, that Comte had 'done more than any other to raise the standard of history'.[19]

Like Comte, and under the influence of natural science, Buckle saw regularity and order when other historians and churchmen emphasized chance, personal agency, and the miraculous:

> In regard to nature, events apparently the most irregular and capricious have been explained, and have been shown to be in accordance with certain fixed and universal laws. This has been done because men of ability, and, above all, men of patient, untiring thought, have studied natural events with the view of discovering their regularity; and if human events were subject to a similar treatment, we have every right to expect similar results.[20]

This rather suggests that those who seek regularities will find them, which is one of the weaknesses of Buckle's work in general. Like another intellectual synthesizer of this era with whom he shares many features, Herbert Spencer, of whom it was said by his friend Thomas Henry Huxley that he 'merely picks up what will help him to illustrate his theories', Buckle looked for what he wanted to find and ignored evidence or interpretations to the contrary.[21] Buckle concluded that 'the actions of men, being determined solely by their antecedents, must have a character of uniformity, that is to say, must under precisely the same circumstances always issue in precisely the same results'.[22] This led him to seek evidence for uniformity in statistics, specifically the statistics compiled on annual rates of suicide, murder, illegitimacy, and other social behaviours that showed stable frequencies year after year. Murder had been shown to be as regular 'as the movements of the tides and the rotations of the seasons', and suicide, seemingly 'so eccentric, so solitary, so impossible to control by legislation' was 'merely the product of the general condition of society...the individual felon carries into effect what is the necessary consequence of preceding circumstances. In a given state of society a certain number of persons must put an end to their own life'.[23] That the annual number of unaddressed letters at the main post office in Paris was always roughly similar was yet another instance of the regularity of social life. Buckle quoted and cited the work of Quetelet at this point, drawing extensively on the evidence for

---

[19] Buckle, *History of Civilization*, I, 5n: Buckle added that 'there is much in the method and in the conclusions of this great work with which I cannot agree'.

[20] Ibid., I, 6.

[21] Beatrice Webb, *My Apprenticeship* (London, 1926), 30. Beatrice Webb herself pictured Spencer 'sitting alone in the centre of his theoretical web, catching facts and weaving them again into theory', ibid., 28. See Lawrence Goldman, 'Victorian Social Science: From Singular to Plural', in Martin Daunton (ed.), *The Organisation of Knowledge in Victorian Britain* (Oxford, 2005), 87–114.

[22] Buckle, *History of Civilization*, I, 18.

[23] Ibid., I, 23, 25.

behavioural regularities in *Sur L'Homme* to make his case.[24] G. R. Porter's discovery in *The Progress of the Nation* that marriages 'bear a fixed and definite relation to the price of corn' was also worthy of comment by Buckle.[25]

In the opening pages of the *History of Civilization* Buckle drew attention to the recent growth of interest in the collection of social statistics, to the sophisticated analysis of reliable vital statistics, and to the amassing of facts on human behaviour as well as human physical characteristics. He gloried in 'the immense value of that vast body of facts which we now possess', describing statistics as 'a branch of knowledge which, though still in its infancy, has already thrown more light on the study of human nature than all the sciences put together'.[26] In a letter to a clergyman soon after the publication of his first volume, Buckle described 'the fundamental principle of my method—viz., that Political Economy and Statistics form the only means of bridging over the chasm that separates the study of nature from the study of mind'.[27] Statistics provided the best evidence available, he thought: 'The most comprehensive inferences respecting the actions of men which are admitted by all parties as incontestable truths, are derived from this or analogous sources: they rest on statistical evidence, and are expressed in mathematical language'.[28] Buckle's historical method required the comparison of different cultures to determine the degree to which they promoted intellectual freedom and the progress of the mind. He needed 'to ascertain whether or not there exists a regularity in the entire moral conduct of a given society'.[29] In this task, 'statistics supply us with materials of immense value': statisticians had been 'the first to bring forward' proofs 'of the existence of a uniformity in human affairs'.[30]

Like many in the tradition we are studying, Buckle was an environmentalist: if statistics could unlock the uniformity and regularity of human functions and behaviours they could also, through evidence of perturbation and variance, demonstrate when and under what circumstances those same behaviours changed.

> If it can be demonstrated that the bad actions of men vary in obedience to the changes in the surrounding society, we shall be obliged to infer that their good actions, which are, as it were, the residue of their bad ones, vary in the same manner; and we shall be forced to the further conclusion, that such variations are the result of large and general causes, which, working upon the aggregate of

---

[24] Hanham argues that the greatest influence on Buckle was Quetelet. See Hanham, 'Introduction', in Buckle, *On Scotland and the Scottish Intellect*, xx.

[25] Buckle, *History of Civilization*, I, 29. See G. R. Porter, *The Progress of the Nation in Its Various Social and Economical Relations, From the Beginning of the Nineteenth Century* (2 vols, London, 1836), II, 244–5.

[26] Buckle, *History of Civilization*, I, 3, 31.

[27] Buckle to Mr Capel, 10 Oct. 1857, in Huth, *Life and Writings of Henry Thomas Buckle*, 151.

[28] Buckle, *History of Civilization*, I, 20.   [29] Ibid., 21.   [30] Ibid., 20.

society, must produce certain consequences without regard to the volition of those particular men of whom the society is composed.[31]

Behaviour changes as conditions change, in short. Though Buckle took no part in politics and espoused no social cause, the implications of his views of history and society for practical and reformist purposes were clear:

> To solve the great problem of affairs; to detect those hidden circumstances which determine the march and destiny of nations; and to find, in the events of the past, a key to the proceedings of the future, is nothing less than to unite into a single science all the laws of the moral and physical world.[32]

Like so many thinkers on social science in the nineteenth century, to Buckle the science of society denoted both understanding the world and setting it to rights.

Buckle's book was received enthusiastically by the critics when first published. Though James Fitzjames Stephen writing in the *Edinburgh Review* did not think he had achieved his aims, credit was given for 'the first attempt which has been made in this country to treat History as a Science'.[33] *The Westminster Review* questioned the easy elision Buckle was making between history and social science but also recognized 'the most comprehensive contribution to philosophical history that has ever been attempted in the English language'.[34] *The Times* believed Buckle 'to be frequently mistaken' on matters of historical detail and interpretation, but recognized the work's strongest claims were in its methodology: 'The gathering together and the connecting of such vast and various materials upon any principles of arrangement, or with a view to any definite conclusions, is the achievement to which rather we ought to direct attention.'[35] *Blackwood's Magazine* was unrestrained: 'We have met with no book for a long time which we have read with so much interest, from which so much information and so many novel views are to be derived, or which is altogether so worthy of studious perusal.'[36] The *Saturday Review* spoke for them all in its appreciation of Buckle's synthetic ambitions: 'It is the scientific conception of history, the conviction of a universal order of the internal and external world, the co-ordination of the different branches of human knowledge, that constitute the chief merit of the book.'[37] We have already encountered Lord Stanley's praise for Buckle's *History*.[38] Charles Darwin thrilled to Buckle's approach and argument, perhaps because he, too, had reached a

---

[31] Ibid, 21.   [32] Buckle, *History of Civilization*, II, 327.

[33] [James Fitzjames Stephen] 'Buckle's History of Civilization in England', *Edinburgh Review*, April 1858, vol. cvii, 465.

[34] [Mark Pattison] *The Westminster Review*, 1 Oct. 1857, n.s., vol. xii, 375.

[35] *The Times*, 13 Oct. 1857, 5.

[36] *Blackwood's Edinburgh Magazine*, Nov. 1858, vol. lxxxiv, 535.

[37] [T. C. Sanders], *The Saturday Review*, 11 July 1857, 40.   [38] See above, pp. xlvii–xlviii.

universal theory through years of painstaking observation and comparison. If he admired the first volume 'which with much sophistry as it seems to me, is *wonderfully* clever & original & with astounding knowledge',[39] he admired the second volume even more:

> Have you read Buckle's 2nd vol: it has interested me greatly; I do not care whether his views are right or wrong, but I shd. think they contained much truth. There is a noble love of advancement & truth throughout, & to my taste he is the very best writer of the English language that ever lived, let the other be who he may.[40]

The most interesting and significant appreciation came from John Stuart Mill. In 1843 Mill had published his *System of Logic* on the philosophy and methodology of the sciences.[41] The sixth and last section of the work was on 'The Logic of the Moral Sciences' and was designed to vindicate a social science—to show that it was possible—and to set out a plan for such a subject, a model that might be followed that would be faithful to the canons of scientific reasoning. Mill intended to demonstrate 'that the collective series of social phenomena, in other words the course of history, is subject to general laws, which philosophy may possibly detect'.[42] Such an idea, if current in Europe for a generation, wrote Mill, 'was almost a novelty' in Britain. But Buckle 'with characteristic energy, flung down this great principle, together with many striking exemplifications of it, into the arena of popular discussion, to be fought over by a sort of combatants, in the presence of a sort of spectators'. This had led to a considerable amount of controversy, 'tending not only to make the principle rapidly familiar to the majority of cultivated minds' but also to clear from it 'confusions and misunderstandings'.[43] Mill was so encouraged by the publication of Buckle's *History* that he added a new chapter to the 5th edition of the *System of Logic*, published in 1862, to praise Buckle and prove their intellectual kinship.[44] Mill accepted the grounds for Buckle's claim that social regularities, as demonstrated in statistics when considered *en masse,* provided a foundation for a scientific history. He rehearsed, as had so many writers since Quetelet, the regularities observed in annual rates of murder, suicide and illegitimacy, and commented that 'The facts of statistics, since

[39] Darwin to J. D. Hooker, 23 February [1858]. https://www.darwinproject.ac.uk/letter/?docId=letters/DCP-LETT-2222.xml.

[40] Darwin to J. D. Hooker, 7 March 1862, https://www.darwinproject.ac.uk/letter/DCP-LETT-3468.xml.

[41] J. S. Mill, *A System of Logic Ratiocinative and Inductive. Being a Connected View of the Principles of Evidence and the Methods of Scientific Investigation* (London, 1843).

[42] Mill, *A System of Logic*, Ch. xi, 'Additional Elucidations of the Science of History', in *The Works of John Stuart Mill* (8th edn, 1872) (ed. J. M. Robson, Toronto, 1973 edn), 931.

[43] Ibid., 931.

[44] Ch. xi, 'Additional Elucidations of the Science of History'.

they have been made a subject of careful recordation and study, have yielded conclusions, some of which have been very startling to persons not accustomed to regard moral actions as subject to uniform laws.'[45] But he was careful to allow for free will and collective action: 'The subjection of historical facts to uniform laws also does not imply the inefficacy of the characters of individuals and of the acts of government.'[46] Buckle and Mill were unknown to each other when the *History of Civilization* was published but Mill came to the support of an author with the courage to lay out a *schema* for History which, at least in spirit, approached Mill's aim of remedying 'the backward state of the Moral Sciences...by applying to them the methods of Physical Science, duly extended and generalised'.[47]

Whether Buckle succeeded in that is debatable, however. Not every reader was impressed, even during the initial wave of admiration. The leading judge and jurist, Lord Hatherley, formerly Sir William Page Wood, wrote to a friend to dismiss Buckle in the book's first season:

It will be severely received, and justly, for it is in the conceited intellectual style, and goes out of the way to favour infidelity...He is of independent fortune, but, most unhappily, did not go to a university, where he would have found men superior to himself.[48]

William Newmarch, the economist and statistician, questioned the identity Buckle had tried to establish between laws of physical science and supposed laws in social science:

We have heard a great deal lately of these so-called 'statistical laws'. We have heard a great deal of the Necessarian conclusions which are said to flow inevitably from the evidence of a certain class of statistical results, or 'laws'. It appears to me, with all deference, that the term 'law' as applied to any statistical result whatever, is a misapplication of the term. The utmost that Statistics can do is to express numerically the average result of any given series of observations or occurrences taking place under particular conditions among human beings.[49]

Newmarch makes evident here the very limited technical conception of the scope and potential of statistics before Galton. But he also punctures the wrong-headed claims of those who projected statistics as a predictive science of human

---

[45] Ibid., 932.

[46] Ibid, 936. On Mill's defence of free will, see above, Ch. 7, pp. 149–50.

[47] Ibid., 833.

[48] William Page Wood to the Rev. W. F. Hook, 8 Oct. 1857 in W. R. W. Stephens, *A Memoir of the Rt. Hon. William Page Wood, Baron Hatherley* (London, 1883), 149.

[49] William Newmarch, 'The Progress of Economic Science during the Last Thirty Year', *JSSL*, 24, 1861, 452–3. (Opening Address at the 31st Meeting of the BAAS, Manchester, September 1861.)

behaviour. A year later, as Mill was adding his chapter praising Buckle, John Herschel, a founder of the intellectual statistical project in Cambridge, was also scornful. Writing to Quetelet to thank him for the gift of his *Physique Sociale*, an updated and reworked version of *Sur L'Homme*, just published in 1862, Herschel was concerned about the reputation of statistical social science:

> A certain discredit has been thrown on the subject in the opinion of superficial readers & thinkers by the way in which Mr Buckle has envisaged the results of the statistics of life, accident, crime etc. as if indicative of an *absence of free agency* in human beings and the presence of some sort of impelling reality.[50]

As the years went by, later readers, many of them highly sophisticated, pondering Buckle's arguments and use of evidence, came to similar conclusions. Seduction by the 'law of large numbers', which we have encountered before in the statistical movement, was a frequent criticism. Buckle, like many others, had mistaken something interesting, suggestive, but essentially trivial for the basis of social science. Buckle protested that he had no intention of writing about individuals:

> My inquiry has nothing to do with the individual, but is solely concerned with the dynamics of masses. Thus, for instance, when I say that the marriages annually contracted by a nation are uninfluenced by personal considerations, I am surely justified in a scientific point of view in making this statement; because, although each individual is moved by such considerations, we find that they are invisible in the mass…[51]

But this was no answer to his critics. That the frequency of certain events and actions was predictable could provide no explanation of individual motivation and volition, nor be accounted a scientific 'law' of human life: it is just what occurs in groups of a large size, a feature of large datasets. Recurrent uniformity is not a law of nature or society, merely a statistical probability.

Even Buckle's greatest admirer, the philosopher J. M. Robertson, who published a book in Buckle's defence, recognized that he had used exaggerated language in commending 'the prodigious energy of those vast social laws' to his readers and had failed to define what he meant by a 'law' based on statistical regularity: 'he never fully realised the importance of a scientific, that is an exact, terminology'.[52] Other critics worried about Buckle's fatalism not just as it affected his understanding of individual free will but in relation to social administration, as

[50] John Herschel to L. A. J. Quetelet, 18 Aug. 1862, Papers of L. A. J. Quetelet, Académie Royale de Belgique, Brussels, File 1289.
[51] Buckle to William Page Wood, 31 Oct. 1857 in Huth, *Life and Writings of Henry Thomas Buckle*, 144.
[52] John Mackinnon Robertson, *Buckle and His Critics. A Study in Sociology* (London, 1895), 18–19.

well. Though Buckle's environmentalism seemed to favour social reform, it was in tension with the statistical evidence on which so much of his structure relied, for this evidence seemed to suggest that social evils were bound to be repeated again and again, at the same frequency, year after year. Both statisticians and historians objected to the combination of their disciplines as laid out by Buckle at the start of his work. What relevance to the study of the past were contemporary social statistics from the early and mid-nineteenth century? Why should evidence for some few recurrent social behaviours encourage the remarkable claim that history was a science, as predictable as physics or mathematics? Even if a social science could be demonstrated to exist and be established on a firm foundation, what would be its relation to the study of the past and of cultures where attitudes and behaviours were different? The list of such questions could be much extended. As Mark Pattison intuited in the *Westminster Review* in 1857,

> On the whole it appears as if Mr Buckle was not quite free from a confusion which prevails over minds far inferior from his, between the Science of Society, and History, as it is, and must be written. That fixed laws of social change exist, we believe. That we possess a collection of observations sufficient to establish those laws, is very doubtful. That those laws have not, as yet, been established, is certain. But the history of any particular state, or system of states, such as that of Western Europe, is not that Social Science.[53]

Emily Shirreff had tried to defend Buckle's methods, notably his aversion to primary sources:

> It was not his province to examine into the accuracy of this or that particular document, or to search for proofs for or against the received version of individual conduct or national transactions. All he wanted was the great outline of history which furnished him with the data for some of his speculations and the proof of others. It was the broad history of nations that he sought to illustrate, and erudite researches would have afforded him no assistance.[54]

But among those scholars of the 1860s and 1870s who began the professionalisation of History in Britain and sought to establish the credentials of a new discipline grounded, above all, in historical sources, Buckle was no model at all, but an irritant who would lead readers astray. As Sir John Seeley, Regius Professor of History in Cambridge and one of the founders of the academic study of History in these decades, put it later, 'That book had indeed somewhat more success with the public than with students. It was much talked of, and opened a new view to

---

[53] [Pattison], 'History of Civilization in England', *Westminster Review*, 393.
[54] Shirreff, 'Biographical Memoir', xxxv.

the public, but it had no great effect on the course of speculation. It is not now very often referred to.'[55] In Germany, where Buckle's *History* served as an introduction to the ideas of Quetelet, much academic opinion was dismissive of a work that showed so much untutored and unsystematic thinking on the nature of free will and social 'laws': this was 'dilettantismus'.[56]

Buckle himself had mused that 'before another century is elapsed . . . it will be as rare to find an historian who denies the undeviating regularity of the moral world, as it now is to find a philosopher who denies the regularity of the material world'.[57] Yet he was wrong on both counts. No historian today would venture to predict the future based on the past.[58] Modern physics, meanwhile, has thrown up many cases of irregularity, indeterminacy, and inconstancy in nature. The influence of Victorian social statistics led James Clerk Maxwell to the novel use of statistics in physics and to the conclusion that some physical phenomena could only be studied by 'the statistical method', which involved thereby 'an abandonment of strict dynamical principles, and an adoption of the mathematical methods belonging to the theory of probability'.[59] Newton's laws were certain; but the behaviour of gases was intrinsically uncertain and to explain it required a different methodology. Maxwell was a pioneer of statistical analysis within physical science, therefore, and he demonstrated that, in his own words, 'the 2nd law of Thermodynamics has only a statistical certainty'.[60] He had been stimulated by reading Buckle's *History*: it was, he wrote, 'a bumptious book...but a great deal of actually original material, the true result of fertile study, and not mere brainspinning'.[61] Maxwell could not agree with Buckle's statistical determinism, however. Buckle thought that statistical analysis demonstrated the invariance of individual and social behaviour. On the contrary, by employing statistics, Maxwell made it possible to understand probabilistic events and systems in nature where variance was intrinsic.[62]

Buckle wrote the history of masses rather than individuals. The march of civilisation in his mind was not an unbroken and contingent story but a series of demarcated stages. Those stages replaced religion as the guiding force of history

---

[55] J. R. Seeley quoted in Deborah Wormell, *Sir John Seeley and the Uses of History* (Cambridge, 1980), 123.

[56] Theodore Porter, 'Lawless Society: Social Science and the Reinterpretation of Statistics in Germany, 1850–1880', in Lorenz Krüger, Lorraine J. Daston, and Michael Heidelberger (eds), *The Probabilistic Revolution* (2 vols, Cambridge, MA), vol. 1, *Ideas in History* (Cambridge, Mass., 1987), 356–7.

[57] Buckle, *History of Civilization*, I, 31.

[58] For a wider discussion of these themes, see Ian Hesketh, *The Science of History in Victorian Britain. Making the Past Speak* (London, 2011).

[59] James Clerk Maxwell, 'Introductory Lecture on Experimental Physics', in W. D. Niven (ed.) *Scientific Papers of James Clerk Maxwell* (2 vols, Cambridge, 1890), 253. Theodore Porter, *The Rise of Statistical Thinking, 1820–1900* (Princeton, NJ, 1986), 150, 200–2.

[60] Maxwell to Peter Guthrie Tait quoted in P. M. Harman, 'James Clerk Maxwell', *ODNB*.

[61] James Clerk Maxwell to Lewis Campbell quoted in Porter, *The Rise of Statistical Thinking*, 294–5.

[62] Ibid., 113, 194–5.

with science and reason; indeed, religion was not the key to moral improvement but the very obstacle to enlightenment and progress. There was a structure and telos to Buckle's history and it wasn't Christian. Some of the most notable contemporary English historians, though holding quite different political and religious positions among themselves, could unite in opposition to this historical positivism, of which Buckle was the prime mid-Victorian exemplar.[63] His work was largely ignored in the universities, therefore, where Christian and national narratives dominated. But academic recognition was never Buckle's objective. As he wrote to Emily Shirreff in 1858, 'I want my book to get among the Mechanics' Institutes and the people: and to tell you the honest truth, I would rather be praised in popular and, as you rightly call them, *vulgar* papers, than in *scholarly* publications.'[64] That was his fate: his two volumes were admired and read widely by working-class autodidacts and in workers' educational organizations in Britain and far beyond.

When Daniel Deronda was taken by Mordecai to one such fictional workers' circle, called 'The Philosophers', Buckle's influence seems to echo through George Eliot's lines:

> But tonight our friend Pash, there, brought up the law of progress; and we got on statistics; then Lily, there, saying we knew well enough before counting that in the same state of society the same sort of things would happen...Lily saying this, we went off on the causes of social change, and when you came in I was going upon the power of ideas, which I hold to be the main transforming cause.[65]

Though in his *Notes from the Underground*, published in 1864, Dostoyevsky held up to ridicule Buckle's view that militarism and warfare were on the wane as civilization advanced,[66] on the other side of Europe Buckle was very popular among Russian radicals in the late nineteenth century.[67] Marxism was another type of scientific history, another type of determinism. Though it was driven by materialism rather than mind and intellect, it also gave a structure and purpose to History. Buckle seemed to be cut from the same cloth and found eager readers among earnest socialists in Europe and the United States who were primed to see history unfolding in Buckle's way, sweeping away superstition, religion, and obscurantism, marching onward to the present and a better future. It was entirely appropriate,

---

[63] Christopher Parker, 'English Historians and the Opposition to Positivism', *History and Theory*, vol. 22, 2, May 1983, 120–145.

[64] Huth, *Life and Writings of Henry Thomas Buckle*, 142.

[65] George Eliot, *Daniel Deronda* (1876) (New York, Penguin edn, 1967), 582–3.

[66] In Dostoyevsky's words, according to Buckle, 'through civilisation mankind becomes softer, and consequently less bloodthirsty and less fitted for warfare. Logically it does seem to follow from his arguments...Only look about you: blood is being spilt in streams, and in the merriest way, as though it were champagne.' F. Dostoyevsky, *Notes from the Underground* (1864), Ch. 7.

[67] Hanham, 'Introduction' in Buckle, *On Scotland and the Scottish Intellect*, xxiv–v.

therefore, that Arnold Ruge, the former Young Hegelian and associate of Marx, who was by then in exile in Britain, should have translated Buckle into German.[68] As George Bernard Shaw told a Buckle devotee in 1894, 'Out of the millions of books in the world, there are very few that make any permanent mark on the minds of those who read them. If I were asked to name some nineteenth-century examples, I should certainly mention Marx and Buckle among the first.'[69] This was a somewhat unlikely destination, pairing, and fame, for a liberal author who, at the end of the 1850s, seemed to be making the study of statistics popular among the English middle classes, and who believed so fervently in the possibility of a scientific history founded upon numerical regularities. Indeed, in the various errors of inference and method that his use of statistics demonstrated, and in the hostility he incited among traditional, less ambitious scholars, Buckle may have set back the cause of Victorians and Numbers.

---

[68]   H. T. Buckle, *Geschichte der Civilisation in England* (tr. Arnold Ruge) (2 vols, Leipzig, 1864–5).
[69]   Dan H. Lawrence (ed.), *Bernard Shaw: Collected Letters, 1876–1897* (London, 1965), 456–8.

12

Medicine and Statistics at Mid-Century

12.1 William Farr: Statistician and Statist

On the first page of William Farr's 1837 essay on 'Vital Statistics' as published in McCulloch's compendium of national information, he wrote that

> No one, contemplating a solitary individual of the human species...can foretell the period when some mortal derangement will occur in his organisation...The same uncertainty is extended in the popular thought to families, nations, and mankind, considered in collective masses; but observation proves that generations succeed each other, develop their energies, are afflicted with sickness, and waste in the procession of their life according to fixed laws; that the mortality and sickness of a people are constant in the same circumstances, or only revolve through a prescribed cycle, varying as the causes favourable or unfavourable to health preponderate.[1]

The contrast Farr was making between the chance events that might afflict an individual and the entirely predictable experiences of a generation as a whole, which unfold according to 'fixed laws', was a well-rehearsed trope of the statistical movement and, more than that, an idea that fascinated mid-Victorians in general, on which were built many claims and schemes for a science of society. This would be founded on the predictability of vital statistics but might incorporate a successively wider spectrum of regular and predictable social behaviours. Farr's statement is interesting for another reason, one that characterizes his own contributions to Victorian statistics and to the movement as a whole: the contrast between a conception of statistics as the basis for social science, and of statistics as tools in the administration and regulation of society, weapons in the fight against disease, and the raw material from which better policies and effective reforms could be made. The contrast was caught in the debate in London in 1860 at the International Statistical Congress over the use of the terms 'statistician' and 'statist'. The former preserved the pure and academic identity of Victorian statistics; the latter denoted an applied, utilitarian and political function.[2] These two purposes and definitions

---

[1] William Farr, 'Vital Statistics; or, the Statistics of Health, Sickness, Diseases and Death', in J. R. McCulloch (ed.), *A Statistical Account of the British Empire* (London, 1837), 567.
[2] See above, Prologue, pp. lvii–lviii.

Victorians and Numbers: Statistics and Society in Nineteenth Century Britain. Lawrence Goldman, Oxford University Press.
© Lawrence Goldman 2022. DOI: 10.1093/oso/9780192847744.003.0012

need not have been at odds, of course: in the person of William Farr they were, in fact, conjoined. No one did more with statistics for the health and welfare of the people than Farr. But throughout his career he gave glimpses of a higher intellectual and spiritual calling as a statistician.[3]

Farr's status as the most prominent and active of mid-Victorian statisticians brought him professional reputation, political influence within government, and international celebrity. As a leading member of the British delegations to the various International Statistical Congresses he met many of the crowned heads of Europe, and he advised Prince Albert before the London Congress. But the most remarkable aspect of Farr's life was surely his birth and early years. Born in Kenley, a village in Shropshire, in 1807, William Farr was the first child of an agricultural labourer who entered the service of an elderly bachelor, John Pryce, in the village of Dorrington. Pryce became William's sponsor and patron. When the family moved on, William, still a young child, remained with him to be educated and was then given a rudimentary training in medicine in the local town, Shrewsbury, where he was apprenticed to a physician. On Pryce's death, Farr inherited a legacy and at the end of the 1820s attended medical lectures in Paris, including those by the French physician, social researcher, and exponent of *hygiene publique*, Louis René Villermé. He was also in the audience for lectures at the newly-founded University College, London.[4] In 1832 he passed the examination of the Society of Apothecaries, Farr's only formal academic qualification, which placed him on the lowest rung of the British medical profession. He tried and failed to practise medicine in London before turning to medical journalism. He published in *The Lancet*, already an established medical journal, and of a radical outlook, and was an editor for the *British Medical Almanack,* before editing his own short-lived publication, the *British Annals of Medicine, Pharmacy, Vital Statistics and General Science,* for some months in 1837.[5] The journal was to survey 'all the laws of vitality capable of being observed in masses of men, expressible in numbers'.[6] His early articles give a clear sense of the views and outlook he would carry through the rest of his career, those of a political radical and a medical reformer opposed to the metropolitan elite of his profession. He favoured the restructuring and democratization of medicine to unlock its potential as an agent of social improvement. His essay on 'Vital Statistics' for McCulloch brought him to the attention of the new General Register Office which he joined in July 1839. As 'Compiler of Abstracts' and then, from 1842, as 'Superintendent of the

---

[3] Everyone who writes on William Farr and on Victorian social statistics in general is indebted the single best book on the subject as a whole. This section draw on the brilliant work of John M. Eyler, *Victorian Social Medicine. The Ideas and Methods of William Farr* (Baltimore and London, 1979).

[4] John Eyler, 'William Farr', *ODNB*; Simon Szreter, *Fertility, Class and Gender in Britain 1860–1940* (Cambridge, 1996), 87.

[5] Eyler, *Victorian Social Medicine*, 2.

[6] 'The State of Medical Science', *British Annals of Medicine*, 1, 1837, 25, quoted in Eyler, *Victorian Social Medicine*, 8.

Statistical Department', he served there for forty years.[7] From the outset, his tenure saw the GRO add to its core function of data collection an equivalent role as part of the Victorian public health movement with 'primarily medical aims and objectives.'[8]

Farr took a major part in the design and execution of three censuses in particular in 1851, 1861, and 1871 out of the five conducted in 1841–81. From census and registration returns he compiled three life tables for England and Wales across his career.[9] He also devised, and twice revised, a statistical nosology, or classification of the diseases causing death, which was the standard national classification and which long influenced international practice as well.[10] He was an active member and President of the Statistical Society of London and one of the founders of the Social Science Association in July 1857. A member of the SSA's famous investigation into trades' unionism at the end of the 1850s, published as *Trades' Societies and Strikes* in 1860, he presided over the Association's Public Health Section at its Manchester Congress in 1866 and delivered three papers to the SSA in all. He was made president of the British Medical Association's committee on State Medicine and in that role, in alliance with the SSA, helped to secure the 1870 Royal Sanitary Commission which led to the re-writing of British public health law and administration in legislation passed in 1872 and 1875.[11] He wrote two special reports on the cholera epidemics of 1848–9 and 1866. His most famous work was reserved each year for his 'Letter of the Registrar General', included in each *Annual Report of the Registrar General* to parliament, which gave him scope to discuss all aspects of public health and preventive medicine.[12]

The 'Letters' provided much of the ammunition for the campaign for central intervention to improve the public's health and Farr is justly famed for his pioneering endeavours in social medicine, especially in relation to the occupational and geographical factors affecting mortality and morbidity. In his 'Letter' in the first *Annual Report*, for example, Farr set out to demonstrate the relationship between mortality and population density, always one of his most insistent themes. He ranked annual female mortality across the 32 registration districts in London and cross-tabulated it with measures of crowding. Female mortality in Whitechapel was very nearly 4% per annum; in St. George's, Hanover Square, it was only 1.785%.[13] For the same year, 1841, he compared male life expectancy at birth: it was 24.2 in Manchester, 25 in Liverpool, 31 in Whitechapel, 37.4 in

[7] Ibid., 9.
[8] Edward Higgs, 'Disease, Febrile Poisons, and Statistics: The Census as a Medical Survey, 1841–1911', *Social History of Medicine*, 4, 3, Dec. 1991, 465–78. Quotation at 478.
[9] Eyler, *Victorian Social Medicine*, 77–80.
[10] Ibid., 53–60. See below, Chapter 13, p. 248.
[11] Lawrence Goldman, *Science, Reform and Politics in Nineteenth Century Britain. The Social Science Association 1857–1886* (Cambridge, 2002), esp. 174–200.
[12] Eyler, *Victorian Social Medicine*, 50–1.   [13] Ibid., 131.

St. George's, Hanover Square, and 44 in the county of Surrey.[14] After the 1851 Census he used the data to analyse group-specific mortality and in the *Fourteenth Annual Report of the Registrar General*, published in 1853, Farr broke down mortality by age and occupation, comparing a farmer with a shoemaker, a weaver, a grocer, a blacksmith, a carpenter, a tailor *et al*.[15]

As this suggests, Farr was not in any sense a modern civil servant. No conventional bureaucrat would have written to *The Lancet*, as Farr did in 1840, soon after joining the GRO, to compare the victims of smallpox to infants being thrown off London Bridge.[16] After the mid-Victorian reforms, from the 1870s the British civil service developed its recognizable features: recruitment of the cleverest products of the universities, strict impartiality and objectivity, and limited, controlled contact with expert opinion and civil society outside government. Before then Farr moved easily between the GRO and the leading medical and social-scientific societies of the capital, trying out ideas as he went and developing a dual identity as a civil servant and an academic expert. He was also from the mid-1850s to the late-1860s, a key ally of Florence Nightingale, and one of the conduits for her influence over hospital and army sanitary reform in government.[17] She referred to him as her 'sheet anchor'.[18] Where she was passionate and wilful, Farr was measured and patient, more adept at the ways in which British government then worked. Men like Farr, and figures analogous to him such as Edwin Chadwick and John Simon, all three engaged with public health, had the knowledge and contacts required for better public administration. But the loss of political support could spell the end of a career; and if experts were required to make aristocratic government work, they never supplanted the aristocracy in this period. At the end of Farr's career, despite the desire of the whole medical profession that he be made Registrar General in succession to George Graham, Farr was overlooked in favour of Sir Brydges Henniker, a cavalry officer and the brother-in-law of a government minister, who possessed neither medical nor statistical qualifications.[19] The medical profession protested loudly and in public against the injustice and Farr resigned in disappointment in January 1880. It was not, in fact, a personal snub: government remained the preserve of a social elite, even in the 1880s, though it was an elite progressively more aware of their reliance on men like Farr and on expertise in general.

---

[14]   Taken from Table 9.1, 'Life Expectancies in Different Parts of England and Wales, 1838–1983', in Richard Stone, *Some British Empiricists in the Social Sciences 1650–1900* (Cambridge, 1997), 264.

[15]   William Farr, 'Mortality of males engaged in different occupations', *Fourteenth Annual Report of the Registrar General* (HMSO, London, 1853).

[16]   William Farr, 'Note on the Present Epidemic of Small-Pox, and on the Necessity of Arresting Its Ravages', *Lancet*, 1840–41, i, 353 cited in Eyler, *Victorian Social Medicine*, 199.

[17]   Mark Bostridge, *Florence Nightingale. The Woman and Her Legend* (London, 2008), 313–14.

[18]   Florence Nightingale to William Farr, 28 Sept. 1861, Farr Papers, Wellcome Library, London, 5474/45/2.

[19]   Eyler, *Victorian Social Medicine*, 190.

Farr displayed an early commitment to statistics and the role of mathematics in medicine. As a journalist in the 1830s he called for the reform of medical education to include 'chemical analysis, weighing, measuring phenomena, determining their relations, and by applying that mighty instrument of natural science—arithmetic, mathematics'.[20] But as John Eyler makes clear, though he knew some calculus and higher mathematics, Farr did little himself beyond very basic analysis of the data.[21] In this he was entirely representative of the statistical movement as a whole at this time. In his defence it might be argued that for the tasks at hand, notably persuading parliamentarians and ministers to effect simple sanitary measures, sophisticated mathematics were neither required nor likely to succeed. Farr never let up in arguing the utilitarian case for statistics. 'This I know', he wrote to Nightingale in 1857, 'Numbers teach us whether the world is well or ill-governed'.[22] Later, in 1864, he asked, in a letter to her, 'And what are figures worth if they do no good to men's bodies or souls?'[23] By 'investigating the things of the greatest interest to nations' the purpose of statistics was 'to promote their prosperity and to benefit mankind'.[24] As he explained in his inaugural address as president of the Statistical Society in 1872, statistics

was never in such demand as it is in the present day; and the supply promises to be equal to the demand. Politics is no longer the art of letting things alone, nor the game of audacious Revolution for the sake of change; so politics, like war, has to submit to the spirit of the age, and to call in the aid of science: for the art of government can only be practised with success when it is grounded on a knowledge of the people governed, derived from exact observation.[25]

There could be no doubt about Farr's environmentalism, his firm and declared belief that by altering conditions, lives could be lengthened, improved and enhanced. He told the Social Science Association in 1866 that 'no variation in the health of the states of Europe is the result of chance: it is the direct result of the physical and political conditions in which they live.'[26] As he explained in a speech to the organising committee of the 1860 London International Statistical Congress,

The conditions remaining the same, the lives of successive generations of men, as they succeed each other, describe the same average number of years; just as

[20] *British Medical Almanack*, 2, 1837, 61, quoted in Eyler, ibid., 6.   [21] Ibid., 66–7.
[22] William Farr to Florence Nightingale, 17 July 1857, Nightingale Papers, British Library, Add Mss 43398, f. 14.
[23] William Farr to Florence Nightingale, 2 September 1864, B.L. Add Mss 43399, f. 154.
[24] William Farr, 'Report on the Programme of the Fourth Session of the Statistical Congress', *Programme of the Fourth Session of the International Statistical Congress to be held in London on July 16th and Five Following Days* (London, Her Majesty's Stationery Office), 10.
[25] William Farr, 'Presidential Address', *JSSL*, 35, 1872, 417.
[26] William Farr, 'Address on Public Health', *Transactions of the National Association for the Promotion of Social Science*, 1866 (Manchester congress) (London, 1867), 70.

under the same winds the waves would break in equal numbers on the shores of the ocean. That is the law of life. The lifetimes also vary. The chances of life change. So also do the conditions; and the determination of those conditions on which the variations depend, is the great problem which statistical inquiry seeks to solve...As men then have the power to change the conditions of life, and even to modify their race, they have the power to change the current of human actions, within definite limits, which statistics can determine.[27]

Few would have disagreed at this time that the purpose of statistics was to help make the case for social change. But Farr went further than this in ways that made him, in the arresting phrase of one historian, a 'statistical poet'.[28] The phrase echoes the appreciative comment of Francis Galton that Farr's reports revealed 'what might be called the poetical side of statistics'.[29] He became so skilled in manipulating this new language, in playing with it, condensing it, embellishing it, that he could use it not merely in the analysis of society, but to move hearts and minds in its reform as well. Farr glimpsed a deeper meaning and other possibilities in the numbers that he used daily. Statistics were, in themselves, 'the science of States – the science of men living in political communities'.[30] From numbers would come enlightenment and the construction of a social science that would explain social behaviour and provide clues to its efficient regulation. As Farr told his distinguished audience in 1860,

> The sublimest considerations arise out of the numerical laws expressing either the relations of the universe, or the relations of the successive generations of men...the field of statistics is immense. Its facts are innumerable, and the number of possible combinations approaches the infinite. No mind can comprehend them all.[31]

The term 'numerical law' is especially relevant to Farr's efforts to create a genuine science of social statistics which would also heal 'men's bodies or souls'. His medical training, computational skills, epidemiological knowledge, and interest in vital statistics and demography were all focused on an intellectual enterprise, the construction of a predictive social science based on relationships between variables which could be expressed numerically. He wrote often of 'laws' in statistics, by which he meant the patterns and regularities that statistics would yield on close

[27] Farr, 'Report on the Programme', 11.
[28] Victor Hilts, 'William Farr (1807–1883) and the "Human Unit"', *Victorian Studies*, 14, 2, 1970, 145–6. On the language of statistics, see also Lawrence Goldman, 'Statistics and the Science of Society in Early Victorian Britain: An Intellectual Context for the General Register Office', *Social History of Medicine*, 4, 3, Dec. 1991, 433–4.
[29] Francis Galton, *Memories of My Life* (London, 3rd edn, 1909), 292.
[30] Farr, 'Presidential Address', *JSSL*, 35, 4, 1872, 417.
[31] Farr, 'Report on the Programme', 10.

examination and which would, once discerned, explain the course of a disease or some key aspect of social life. Farr appreciated, as a statistician rather than a doctor, that what he called 'fixed laws' could be found regulating many aspects of human behaviour and social life.

> Despite the accidents of conflagrations, the unstableness of the winds, the uncertainties of life, and the variations of men's minds and circumstances, on which fires, wrecks and deaths depend, *they are subject to laws as invariable as gravitation*, and fluctuate within certain limits, which calculus of probabilities can determine beforehand. Upon this basis the great system of insurance rests securely. And all statistical facts have the same character; the phenomena observe a definite order, and the same acts of bodies of men are found by experience to recur, with the same intensity, under the same circumstances. This holds of crimes, and other acts of will; so that volition itself is subject to law.[32]

Farr added, however, that he was not a fatalist and there were no grounds for such an attitude because wise policy, effective regulation, and skilful action—be it good seamanship, good hygiene, or the well-regulated construction of buildings—can change outcomes and limit the effects of fate. It was in the hands of society, through responsible personal conduct and good policy, to moderate chance and alter what might otherwise occur. If this was so, why not look for those fixed laws that determined human health and disease and intervene as necessary?

In accordance with this, throughout his career, Farr posited 'laws giving an individual's chances of recovery or death during an illness, laws describing the course of an epidemic over time, and laws illustrating the effect of some environmental circumstances on the course of disease'.[33] He hoped to demonstrate that disease was not random but followed a discernible pattern which, once clear, would assist in its eradication. Quite simply, in an age before bacteriology, virology, and the understanding of disease causation, let alone prevention, Farr and others hoped that statistical analysis might provide insight into the origins of disease, its spatial and social diffusion, and its treatment. By pinning numbers on different elements and factors, be they age or sex or income or occupation or density or location, and relating those numbers to the statistics of disease mortality, Farr hoped to unlock a pattern that would allow for successful intervention in one country and in *every* country. For at his most expansive and ambitious, Farr imagined a universalism to the statistical enterprise from which all would benefit: as he stated in 1860 'a deduction from a given series of observations admits of general application. The past enables us to forecast the future; and the statistical discoveries of one nation are the lights of all nations.'[34]

---

[32] Ibid.   [33] Eyler, *Victorian Social Medicine*, 108.
[34] Farr, 'Report on the Programme', 10.

In 1837–8 Farr published three articles on 'laws of recovery and death in smallpox' showing the regular incidence of the disease at different ages. It was at a peak of virulence in infancy; was least dangerous in adolescence; and then small-pox mortality grew in a regular frequency with each succeeding decade of life. In this case, Farr's 'law' was a confirmation of what was known and observed empir-ically, that disease mortality followed an established pattern over the human life cycle.[35] His 'elevation law of cholera mortality' in London, based on the analysis of the cholera outbreak of 1848–9, posited an inverse relationship between deaths from the disease and the height of habitations above the level of the River Thames.[36] Using data from London's 38 registration districts, Farr calculated that the mortality rate, at over 10 per 1000, was highest within 20 feet of the river level. It fell to 6 per 1000 approximately, at 30–40 feet. Of those who lived more than 340 feet above the Thames, the death rate from cholera was less than one in a thousand. The 'law' merely depicted the geographical dispersion of cholera across the city, in other words, and it could not explain it. Once again, Farr confirmed what had long been known: in this case, that disease mortality would decrease the further and higher from the centre of a city one looked. This was why salubrious suburbs like Hampstead, on a hill to the north of the city, had long been fashionable. But Farr, working with a 'miasmatic' theory of disease, believed that cholera was air-borne and caused by foul atmosphere, whereas it would shortly be shown by Budd and Snow that it was a waterborne disease distributed through the city by some of the London water companies drawing infected water. There may have been some sort of relationship between elevation and disease, but it was not a causal one, for sure.[37] Had Farr accepted the waterborne origins of cholera and thoroughly investigated the impact of water supply on London's population he would have discovered stark differences between the relatively low mortality among those taking their water from above Hammersmith to the west of the city, and the high mortality among those further east, drinking water from below Battersea Bridge.[38]

Farr continued to study disease statistically in his later work. By the time he wrote his report on the 1866 cholera outbreak he accepted that the disease was transmitted through water but he continued to count and measure in pursuit of a numerical distribution, or a sequence, or a relationship that might provide the

---

[35] William Farr, 'On a Method of Determining the Danger and Duration of Diseases at Every Period of Their Progress, Article I', *British Annals of Medicine*, 1, 1837, 72–9; 'On the Law of Recovery and Dying in Small-Pox, Article II', ibid., 134–43; 'On Prognosis', *British Medical Almanack*, 1838, 199–216. See Eyler, *Victorian Social Medicine*, 108–9.

[36] On Farr's elevation law of cholera, see John M. Eyler, 'William Farr on the Cholera. The Sanitarian's Disease Theory and the Statistician's Method', *Journal of the History of Medicine and Allied Sciences*, 28, April 1973, 79–100. See also Fred Lewes, 'William Farr and the Cholera', *Population Trends*, Spring 1983, 8–12.

[37] William Farr, *Report on the Mortality of Cholera in England, 1848–9* (HMSO, London, 1852) and 'Influence of Elevation on the Fatality of Cholera', *JSSL*, 15, 1852, 155–83.

[38] Timandra Harkness, *Big Data. Does Size Matter?* (London, 2016) (2017 edn), 63. Eyler, *Victorian Social Medicine*, 117–22.

key to understanding the cause and duration of the disease.[39] Yet the mystery of cholera would be solved in the laboratory where, in 1883, Robert Koch isolated the waterborne comma bacillus as the cause of the disease. The study of pathogens by microbiological techniques rather than the discovery of a numerical pattern among the disease fatalities, held the key to the future. In the late 1870s and early 1880s not only the comma bacillus but the gonococcus, tubercle, and tetanus bacilli were also isolated and shown to be the cause of their respective diseases.[40] Farr, like so many other doctors and sanitarians of this age, was searching for environmental clues to disease, which were amenable to statistical analysis, when the answers lay elsewhere and in other sciences.[41] We can sense the general frustration at the inability to understand cholera in Nightingale's comment to Farr about the 1866 epidemic, 'I am thankful to see that the cholera is declining (mainly through your exertions). But ought it not to decline faster, now the tide is turned?'[42] Later, in 1878, Farr announced his 'density law' to the Social Science Association, which demonstrated yet another long-observed feature of spatial demography: that life expectancy varied according to the mean density of the population.[43] No one could have doubted, least of all Farr himself, that lives were more likely to be cut short in the most densely populated parts of the cities.

These were not 'laws' at all. They did not act constantly in all circumstances and all places, and they could not be used to predict the course of an epidemic, but simply expressed in mathematical form the relation of certain environmental variables in unique and discrete situations. The 'laws' were essentially descriptive of a distribution, nothing more. The 'elevation law' could neither locate nor explain the cause of the London epidemic, nor be used as an analytical tool in regard to other affected centres of population. Farr's work could calculate the average duration of an infection, for instance, and it could show how resistance to disease declined at different ages, but much of this was of secondary importance or observed already. It was useful information but it was not the understanding of the cause, transmission and cure of disease that Farr and others were seeking. This is not to dismiss such heroic early efforts in statistical epidemiology: they are true ancestors of the modelling of diseases today. Farr's grasp of the environmental factors that affected health, shared with many others in his generation, and

---

[39] [William Farr] 'Report on the Cholera Epidemic of 1866 in England'. *Supplement to the 29th Annual Report of the Registrar General, Parliamentary Papers, 1867–8, xxxvii.*

[40] J. K. Crellin, 'The Dawn of the Germ Theory: Particles, Infection and Biology' in F. N. L. Poynter (ed.), *Medicine and Society in the 1860s* (London, 1968), 57–76.

[41] Simon Szreter, 'The GRO and the Public Health Movement in Britain, 1837–1914', *Social History of Medicine*, 4, 3, Dec. 1991, 456.

[42] Florence Nightingale to William Farr, 29 Sept. 1866, William Farr papers, Wellcome Library, London, 5474/1042.

[43] William Farr, 'Density or Proximity of Population: Its Advantages and Disadvantages', *Transactions of the National Association for the Promotion of Social Science, 1878* (London, 1879), 530–35.

demonstrated in figures and graphs, deserves the very highest respect. But his hope in the 1830s that medical statistics would in themselves unlock the secrets of disease was unfounded; they had a role to play for sure, but in association with other and more fundamental biological sciences. Farr's carefully crafted statistical laws of disease, correlating the incidence of illness with specific environmental factors, are emblems of the statists' project at mid-century. But they were outmoded within a generation, and are today relics of a bygone system of socio-medical belief and practice, though beautiful relics for all that, the carefully coloured maps of London's districts and topography a tribute to Farr's methodology.

Farr was drawn to devise and compose these laws because he subscribed to both the utilitarian and academic conceptions of mid-Victorian statistics. In the first of these, numbers would improve the lives of men and women by improving their conditions. It was a perfectly straightforward position for a political liberal like Farr to adopt. This aspect of his work is summarized in the letter of Leone Levi to *The Times* on his death: 'Dr. Farr was a recorder of the common facts of births, deaths, and marriages, but by a wide induction he made those facts impart lessons which almost created the science of sanitation, while they enlarged and established the principles of medical science.'[44] Farr contributed to an academic project, as well, however: through careful analysis of numerical data about society taken *en masse* (rather than about people as individuals), a predictive social science would be constructed. What better way to join these two conceptions of the purpose of statistics than to construct, from observation and analysis, statistical laws that both contributed to the immediate improvement of health and to the deeper understanding of disease? This helps explain why Farr moved between many different types of statistical project 'with perfect ease': he was involved in 'the construction of a numerical science of health and human progress' which united all the various different uses of statistics at this time.[45] Numbers would both safeguard society and unlock its deeper structures.

The fellowship between statisticians came to Farr's assistance in his personal life in the 1860s. Farr had eight children with two wives. At his retirement a testimonial fund was begun to assist three of his unmarried daughters.[46] His last child, Florence Farr, born in 1860 and named after Florence Nightingale, was a remarkable and famous woman: an actress and actor-manager, feminist, novelist, headmistress, theosophist, lover of George Bernard Shaw, muse to Yeats and Pound, and the subject of lively historical interest today.[47] However, one of Farr's sons, Frederick, born in 1844, is forgotten, perhaps because on stowing away across the Atlantic in February 1863 and arriving in Portland, Maine, he assumed the name of Frederick Clark. It was under that alias that he joined the 7th Maine

[44] *The Times*, 23 April 1883.   [45] Eyler, *Victorian Social Medicine*, 200.
[46] Eyler, 'William Farr'.
[47] Virginia Crosswhite Hyde, 'Florence Beatrice Farr 1860-1917', *ODNB*.

Volunteers in the Union Army to fight in the American Civil War. William Farr used his professional and statistical connections in the United States to trace his son and try to bring him home. He wrote to colleagues in the United States Sanitary Commission to ask for their help. Farr knew Joseph Kennedy, twice Director of the US Census in 1850–3 and 1860–65. Kennedy had represented the United States at the Great Exhibition in 1851 and then at the first International Statistical Congress in Brussels in 1853. He was enlisted to write to the Secretary of State, William H. Seward, and to the British minister, Lord Lyons, to secure Frederick's release from the army.[48] Farr also wrote to his friend Edward Jarvis in April 1863 to ask for his help, and in the correspondence between the two leading statisticians on each side of the Atlantic the story unfolded.[49] As Farr explained in June,

> He was enrolled in April at Portland, state of Maine, under the name of Frederick Clark. He is 19 years of age. He was one of our volunteers in England; & was studying for the Civil Service; when, without consulting me, he started off for Portland on a steamer & joined the above regiment. He has written home expressing his deep regret at having acted without consent & stating that he will do whatever I wish; but he can only get his discharge through the kind offices of American friends. As Lord Lyons says, this is more likely to lead to a successful issue than diplomatic interference.[50]

By that stage the regiment had left Portland 'for the seat of war. The regiment was on the Potomac.' William Farr learnt that at the beginning of July 1863 his son had fought at Gettysburg in Pennsylvania, and 'had escaped all harm'. Much later, Frederick Clark of 'London, England' was listed in a roster of those who fought in his regiment in this, the largest battle of the Civil War.[51] But the last the family heard from Frederick was in November 1863.[52] At the end of the war they received official notification, as Farr wrote to Jarvis, that he had

> died in the Hospital at Richmond of fever March 23 1864. He fought at Gettysburgh – of which he gave me a full description – and afterwards under Meade – along the Potomac – but in that excessively cold night of November 30th 1863 after a fatiguing march – fell asleep by the camp fire & was taken prisoner by the enemy's cavalry. He was young – 19 - & at that age boys

[48] https://www.census.gov/history/www/census_then_now/director_biographies/directors_1840_-_1865.html.
[49] William Farr to Edward Jarvis, 5 letters, 1863–65, Edward Jarvis Papers, Francis A. Countway Library for the History of Medicine, Harvard University, B MS c 11.2.
[50] Farr to Jarvis, 9 June 1863.
[51] Rosters from 'Maine at Gettysburg: Report of Maine Commissioners' (Portland, Maine, The Lakeside Press, 1898) at www.segtours.com/files/maineatgettysburg.pdf.
[52] Farr to Jarvis, 15 June 1864.

cannot do the work of men at their full vigor. He rests in the ground where his comrades now triumph & he would have rejoiced had he lived to see these days.[53]

It seems likely that Frederick was taken prisoner in the minor battle known as Mine Run (or Payne's Farm) fought around the Rapidan River in Virginia.[54] A final letter from Farr to Jarvis, written as the news of Lincoln's assassination reached England, expressed his 'deep sympathy with you & the citizens of the Union'. Farr sent English newspapers to Jarvis so he could see 'how the loss of a good man is felt here, and how the hearts of our race beat together'.[55] When Florence Nightingale learnt of Farr's loss she also sent her sympathies: 'I am sorry to hear of your griefs. I do not find that mine close my heart to those of others— and I should be more than anxious to hear of yours—you who have been my faithful friend for many years.'[56]

Who can tell why a nineteen-year-old would run away to fight for the Union? By February 1863 Lincoln's two emancipation proclamations had changed the war from one fighting against Confederate secession to one fighting also for the liberty of the slaves. Perhaps that moved Frederick. Admittedly, when interviewed by a lawyer sent to try to engineer his discharge, Frederick Farr/Clark told him that he had always wanted to be a soldier rather than a civil servant and was dis-inclined to leave the army.[57] But support for the North was an article of faith in many British liberal households in the 1860s, and Frederick's apparent impetuos-ity would not have been out-of-keeping with his father's known radicalism. Stowing away and joining up may tell us something more about the politics of the statistical movement and of its key members, therefore. That William Farr turned reflexively for help from statistical colleagues in the United States is certainly fur-ther evidence of the movement's internationalism. But despite William's best efforts, his son became a casualty of war and a statistic himself. 1505 men served in the 7th Maine regiment during the Civil War, of whom 152 were killed in battle or died of their wounds, 403 were wounded, 212 died of disease, and 19 perished as prisoners of the Confederacy. Frederick Farr was in this last category.[58]

---

[53] Farr to Jarvis, 24 April 1865.

[54] Farr's letter to Jarvis on 15 June 1864 refers to his son being 'taken prisoner on the Rappahannock', but this was further to the north, away from the epicentre of the battle.

[55] Farr to Jarvis, 29 April 1865. For more on British reactions to Lincoln's death, see Lawrence Goldman, "'A Total Misconception". Lincoln, the Civil War, and the British, 1860–1865', in Richard Carwardine and Jay Sexton (eds), The Global Lincoln (Oxford, 2011), 107–22.

[56] Florence Nightingale to William Farr, 5 Aug. 1864, William Farr papers, Wellcome Library, London, 5474/73.

[57] Amanda Foreman, A World on Fire. An Epic History of Two Nations Divided (London, 2010), 469–70.

[58] https://en.wikipedia.org/wiki/7th_Maine_Volunteer_Infantry_Regiment.

12.2 William Augustus Guy: Statistics as Social Science

The statistician who contributed more papers than any other to the *Journal of the Statistical Society of London*, which he himself edited for four years between 1852 and 1856, was Dr. William Guy, Professor of Forensic Medicine at King's College, London from 1838 until 1869. To the medical profession in Victorian Britain he was well known as the author of the standard work on this subject, *Principles of Forensic Medicine*, first published in 1844. But as one of his obituaries explained, 'the general public probably know him better through the attention which he devoted for many years to questions of sanitary reform and social science'.[59]

Guy was born in Chichester in 1810, educated in Cambridge, London, Heidelberg, and Paris, and he took his MB degree in 1837.[60] Like William Farr he combined liberal politics, social reform and medical expertise: 'his views in politics, in religion, and in social science [were] far in advance of his age'. Like Farr, he possessed 'a mind of great breadth', reminding us that there were many, physicians in particular, who shared Farr's expansive grasp of the province of statistics and the potential of statistical analysis.[61]

Guy was a founder of the Health of Towns Association in the 1840s and an activist in its support thereafter. He wrote several studies to further this work including, later on, *Public Health: A Popular Introduction to Sanitary Science,* published in the 1870s.[62] In 1844 he gave evidence to the Health of Towns Commission and its first report included his inquiry into the health of printers 'and others following indoor occupations'. Later, he undertook a commission from John Simon when Simon was directing the Privy Council Medical Office to produce a report on the problem of arsenic poisoning in manufacturing industry, a subject on which Guy was already expert. In 1860 he delivered the Croonian Lectures of the Royal Society, of which he was a fellow, on 'The Numerical Method, and its application to the Science and Art of Medicine'. Late in his life he served on the Royal Commissions on Penal Servitude and Criminal Lunatics in 1878 and 1879 respectively. He was President of the Statistical Society of London in 1873–5. The Guy Medal, the Society's highest annual award for research, was founded in 1891 'in honour of the greatest benefactor the Society has ever had, who was at the same time eminent as a statistician'.[63] Guy 'blazed the trail of medical statistics with quite inadequate recognition through the middle of the nineteenth century', integrating statistical analysis in all his varied socio-medical interests—sanitary and prison reform, occupational disease, insanity, and forensic medicine.[64]

---

[59] *British Medical Journal*, 19 Sept. 1885, 573.
[60] G. T. Bettany, 'William Augustus Guy 1810–1885', *ODNB*.
[61] *The Lancet*, 19 Sept. 1885, 554.
[62] W. A. Guy, *Public Health: A Popular Introduction to Sanitary Science* (2 vols, London, 1870, 1874).
[63] 'Report of the Council. Session 1890–91', *JSSL*, vol. 54, 3, Sept. 1891, 418.
[64] The best account of Guy's life is in C. Fraser Brockington, *Public Health in the Nineteenth Century* (Edinburgh and London, 1965), Appendix 1, 238–40, on which this account is based.

In 1865, in mid-career, reflecting on his various papers to the Statistical Society of London, Guy explained his belief that he had been contributing

to the gradual, slow growth, not of heaps of facts without reference to their use or application, but of a veritable science, social and political – a science with a definite aim and orderly classification of subjects, a numerical method with its strict rules of synthesis and analysis.[65]

Four years later in his presidential address to the Statistical Society, the economist William Newmarch paid tribute to Guy, who had led, he said, in 'the most fundamental enquiry' into the theoretical aspects of statistics. These Newmarch defined as 'investigations of the mathematics and logic of Statistical Evidence; that is to say, to the true construction and uses of averages, the deduction of probabilities... and the discovery of the laws of such phenomena as can only be exhibited by a numerical notation.'[66] In his early work for the Statistical Society Guy was the most eloquent contributor to its journal on the nature and purposes of statistics. The 'first and great object' of the society, he wrote, was 'collecting facts bearing upon man's social condition, and expressing the results of those facts in the simple and concise language of figures.'[67] His 1839 article on 'The value of the numerical method as applied to Science' might be read as the movement's manifesto. In it he set out the foundational principle that 'the certainty of a science is exactly proportioned to the extent to which it admits of the application of numbers.'[68] It is taken directly from Quetelet who had written in 1828 that 'the perfection reached by a science can be judged by the approximate ease with which it may be pursued through calculation. More and more we are confirming the ancient wisdom that numbers rule the world.'[69]

For that very reason, one of the characteristics of the age was the 'growing disposition to apply calculation to the phenomena of life.'[70] And from this came the 'science of statistics' which Guy defined as 'the application of the numerical method to the varying conditions and social relations of mankind.'[71] As he continued,

---

[65] William A Guy, 'On the Original and Acquired Meaning of the term "Statistics", and on the Proper Functions of a Statistical Society; also on the Question Whether there be a Science of Statistics; and if so, what are its Nature and Objects, and what is its relation to Political Economy and "Social Science", *JSSL*, 28, 1865, 493.

[66] William Newmarch, 'Inaugural Address on the Progress and Present Condition of Statistical Inquiry', *JSSL*, 32, Dec. 1869, 359–84, quotations at 366, 373.

[67] William Guy, *On the best method of collecting and arranging facts: with a proposed new plan of a common-place book* (London, 1840), 2.

[68] William A. Guy, 'On the Value of the Numerical Method as Applied to Science, but especially to Physiology and Medicine', *JSSL*, 2, 1839, 34.

[69] L. A. J. Quetelet, *Instructions Populaire sur le Calcul des Probabilités* (Brussels, 1828), 233. 'On pourrait même, comme je l'ai déjà fait observer ailleurs, juger de degré de perfection auquel une science est parvenue par la facilité plus ou moins grande avec laquelle elle se laisse aborder par le calcul, se qui s'accorde avec se mot ancient qui se confirme de jour en jour: *mundum numeri regunt*.'

[70] Guy, 'On the Value of the Numerical Method', 35. [71] Ibid.

Man, considered as a social being, is its object; the mean duration of his life, and the probable period of his death; the circumstances which preserve or destroy the health of his body, or affect the culture of his mind; the wealth which he amasses, the crimes which he commits, and the punishments he incurs – all these are weighed, compared and calculated; and nothing which can affect the welfare of the society of which he is a member, or the glory and prosperity of the country to which he belongs, is excluded from its grand and comprehensive survey.[72]

Guy wrote under the clear influence of Quetelet, impressed by the evidence of the constancy of human experiences like marriage rates, murders and suicides, the regularity of crimes committed and the proportion of the accused found guilty that *Sur L'Homme*, so recently published, had presented.[73] According to Quetelet, 'It is the social body which forms the object of our researches and not the peculiarities distinguishing the individuals that compose it.'[74] According to Guy, 'It is from groups of persons that we obtain our knowledge: it is to like groups that we apply it.'[75] To Guy and to many others, Quetelet was 'the founder of that science of man himself, which finds its only fitting expression in figures of arithmetic.'[76] In Guy's view, the regularity and predictability demonstrated by Quetelet made social science possible: 'However numerous and however various may be the causes to which an event owes its existence, these causes will be accurately reproduced in equal intervals of time, so long as the same circumstances exist, and provided that the number of facts observed is sufficiently great.'[77] Hence, through numerical comparison, by looking for long-run similarities or for significant differences in a run of figures, we can determine 'the permanence or variation of the causes which contribute to the production of any given event'. In this way, thought Guy, it would be possible to attain 'the highest aim and best achievement of statistics... the discovery of general laws'. Though his own mathematical abilities were limited,[78] Guy was another of those Victorian statisticians persuaded that statistical regularities would allow for the construction of a social science comparable with the certainties of natural science and based on similar procedures:

---

[72] Ibid.   [73] Ibid., 36–8.
[74] L. A. J. Quetelet, *Sur L'Homme et le Développement de ses Facultés. Physique Sociale* (Brussels, 1835). English edn, translated by Dr J. Knox as *A Treatise on Man and the Development of His Faculties* (Edinburgh, 1842), 7.
[75] Guy, 'On the Original and Acquired Meaning of the term "Statistics", and on the Proper Functions of a Statistical Society', 487.
[76] William Guy, 'Inaugural Address' (as President of the Statistical Society), *JSSL*, 37, 1874, 420.
[77] Guy, 'On the Value of the Numerical Method', 36.
[78] Bernard-Pierre Lécuyer, 'Probability in Vital and Social Statistics: Quetelet, Farr, and the Bertillons', in Lorenz Krüger, Lorraine J. Daston, and Michael Heidelberger (eds), *The Probabilistic Revolution* (2 vols, Cambridge, MA), vol. 1, *Ideas in History* (Cambridge, Mass., 1987), 325.

The mean results which [the science of statistics] obtains are almost as constant as the numerical values determined by the experiments of the natural philosopher, and the former are reproduced with almost as much certainty by a recurrence of the same circumstances as the latter are by a repetition of the same experiments.[79]

Guy contributed studies to statistical journals on such subjects as the effect of occupation on health, the life expectancy of different social classes, and mortality in London's hospitals.[80] But he may have been more important as one of the chief propagandists for statistics, a writer of manifestos and interpreter of the work of others. As he himself explained this role, 'I trust that I may lay claim to some success in my attempt to give increased dignity and importance to this society, and a new interest to the labours of its members.'[81] Over time Guy devoted more space in his articles to the history of statistics, the meaning of the term itself, its deployment in literature, and to the different genres of statistics as represented in different types of article published in the *Journal of the Statistical Society of London*. As the term 'social science' became more common in mid-Victorian Britain and entered everyday discourse, so Guy claimed that it originated in the statistical movement which alone possessed the evidence and methodology for its development as an academic discipline: 'The original prospectus of this Society... did really establish a Social Science when it stated as its object the procuring, arranging, and publishing of "facts calculated to illustrate the condition and prospects of society".' Only statisticians offered 'the services of a social and political science, slowly and painfully constructed on the basis of facts laboriously brought together'. Only they, in 'the collection, arrangement, tabulation and analysis' of this material 'bring constantly to bear the pure light of scientific method.'[82] To Guy, statistics was 'an all-encompassing social science aimed at identifying the laws of society.'[83]

However, by the end of Guy's life, his approach had become more defensive as a number of intellectual challenges to Victorian statistics, as he understood the discipline, emerged to contest its scientific credentials. It was no longer just a question of claiming primacy in social science: indeed, in the use of the phrase 'statistics, sociology or social science' that Guy employed himself in his final

---

[79] Guy, 'On the Value of the Numerical Method', 39.

[80] For example: 'On the Health of Nightmen, Scavengers and Dustmen'; 'On the Duration of Life as affected by the Pursuits of Literature, Science and Art: with a summary view of the Duration of Life among the Upper and Middle Classes of Society'; 'On the Mortality of London Hospitals: and incidentally on the deaths in the Prisons and Public Institutions of the Metropolis', *JSSL*, 1848, 11, 72–81; 1859, 22, 337–61; 1867, 30, 293–322.

[81] Guy, 'On the Original and Acquired Meaning of the term "Statistics"', 492–3.

[82] Ibid., 492.

[83] Libby Schweber, *Disciplining Statistics. Demography and Vital Statistics in France and England, 1830–1885* (Durham, N.C., 2006), 202.

article in 1885, he was accepting that there were other competitors by the 1880s to the intellectual terrain he had once claimed for the Statistical Society of London alone.[84] Now the very claim to be a science at all was subject to scrutiny. So in this final paper he was keen to distance statistics from descriptive 'sciences of identification' like natural history, and to establish its more respectable kinship with harder sciences like physiology and medicine.[85] Quetelet was admired, as ever, but now also criticized for seeking evidence that corroborated his findings rather than considering 'facts which ran counter to those first observed'. Had he done this, Quetelet would have discovered the limitation of the aggregate figures with which he largely worked, which often contained unrecognized or unremarked fluctuations with 'a wide range between a high maximum and low minimum of intensity'. Averages are easy to work with, and society 'is most interested' in them because they are easy to understand, but they obscure significant internal variations: these Quetelet ignored.[86] Guy started out making the strongest case for the new science of statistics, but over time the nature of that case changed as competitors emerged to claim the mantle of 'social science' and as early-Victorian methods and outlooks were shown to be inadequate.

12.3 John Simon at the Privy Council Medical Office, 1858–71

In 1862 William Guy was commissioned by his friend, Dr. John Simon, Medical Officer at the Privy Council, to investigate the illnesses associated with the use in several trades of a vibrant green pigment, largely made from arsenic, known as 'emerald green' or 'Scheele's green'. Reports had reached doctors in London especially, of illness among workers who used the compound to colour artificial flowers, which were very popular at this time, as well as cakes and confectionery, fancy papers and wrappings, wallpapers for the home and other ornaments. The issue had received national attention after a letter from the Ladies Sanitary Association entitled 'The Dance of Death' was published in The Times on 1 February 1862 drawing attention to the abuse. One of its two signatories, a secretary of the Association, was Mrs William Cowper, the wife of the minister who had helped to arrange the International Statistical Congress in 1860 in London and who had spoken with such passion about the internationalism and brotherhood of statistics.

Guy set to work, visiting many workshops and factories and interviewing workers and their employers, while collecting evidence from medical colleagues.[87]

[84] William A. Guy, 'Statistical Development, with Special Reference to Statistics as a Science', JSSL, Jubilee Volume (1885), 72.
[85] Ibid., 76.   [86] Ibid., 82.
[87] 'Dr. Guy's Report on Alleged Fatal Poisoning by Emerald Green; and on the Poisonous Effects of that Substance as used in the Arts', Appendix 3, Fifth Report of the Medical Officer of the Privy Council, PP 1863, xxv, 126–62.

He concluded that there was a genuine problem, though not one with overwhelming mortal consequences: he could only find unambiguous evidence that a single young woman had died from handling emerald green. But dozens of others who worked with the green dye, especially in its powdered form, developed rashes, sores, headaches, indigestion, sickness, and vomiting. In some workers, such as the young woman, the arsenic had penetrated the lungs and organs, leading to severe illness. Guy suggested the taking of simple precautions as a first step: ventilating workshops, keeping them clean, not overcrowding the workspace, following procedures that minimized the spread of the compound, wearing protective clothing and gloves, washing frequently. Premises making and using the pigment should be registered and inspected, he advised. But Guy was against banning the manufacture and use of emerald green both on principle and because the trade would just move overseas to less scrupulous countries. He hoped that publicity and full disclosure would alert workers to the risks they ran in working with the compound and make employers act responsibly. Awaiting a much more comprehensive inquiry into the health of the workplace, Guy therefore recommended what he called 'a policy of expectation' in which all were encouraged to step up to their responsibilities. It was a classic mid-Victorian liberal solution which placed onus on informed individuals behaving wisely, fairly, and considerately.[88]

Guy's investigation was one of a large number of inquiries undertaken by the Privy Council Medical Office (PCMO) in a 13-year period between the 1858 Public Health Act, which established the Medical Officer at the heart of government and the Local Government Board Act in 1871 which merged the PCMO with the Poor Law administration in a single department. Thereby, the legislation subordinated medical specialists to a lay secretariat largely drawn from the Poor Law bureaucracy, and brought to an end what is widely seen as the most creative phase in the administration of public health in the Victorian period.[89] The PCMO had risen to prominence after the fall of the General Board of Health in 1854. This had lasted only six years under its head, Edwin Chadwick, perhaps the most famous and ubiquitous of all the 'statesmen in disguise'. It was the first attempt to centralize public health administration, but centralization and, indeed, the improvement of public health, were too much to bear for a variety of vested interests and for those legislators of a localist persuasion. On losing support in parliament, the General Board fell.[90] But Simon was different: he was an eminent physician in his own right, widely connected and especially admired within his

---

[88] Ibid., 161–2. Mid-Victorian social thought and institutional reform is treated most fully and sensitively in W. L. Burn, *The Age of Equipoise. A Study of the Mid-Victorian Generation* (London, 1964).

[89] On this period in the history of central medical administration in Britain see Lawrence Goldman, *Science, Reform and Politics in Victorian Britain. The Social Science Association 1857–1886* (Cambridge, 2002), 174–200.

[90] S. E. Finer, *The Life and Times of Edwin Chadwick* (London and New York, 1952).

profession. He had powerful allies in government and in the Social Science Association of which he was, like William Farr, a founder.[91] Farr at the General Register Office provided a flow of information on which Simon and his associates could depend in their investigations.[92] And Simon brought with him into government the claims of science, using the PCMO as an initiator and sponsor of a host of research projects to better assess and understand the nation's health, of which Guy's inquiry into arsenic was just one. Those projects were characterized by their use of statistics and their search for precision. As Simon would later describe their *modus operandi*,

> Confident that, if the knowledge were got, its utilization would speedily follow, we had to endeavour that all considerable phenomena of disease prevalence in the country should be seen and measured and understood with precision, - should be seen as exact quantities, be measured without fallacious admixture, be understood in respect of their causes and modes of origin.[93]

Simon had an eye for talent which he usually scouted at meetings of the Epidemiological Society in London. William Guy was already long-established as a leading toxicologist but Simon brought into government or used, as well, a group of young men who, over the next thirty years or so, 'gathered to themselves so many distinctions that their names stand out like a galaxy of stars in the sky'.[94] As Simon's biographer commented, 'the team began...to explore the nation's health with a scientific thoroughness of method and a comprehensiveness of approach never before attempted'.[95] By organizing the different inquiries 'into small, manageable pieces that he then farmed out to the various researchers', Simon's patronage created a first community of research pathologists in Britain, for example.[96] Among the key salaried staff of the Office who were personally close to Simon and used by him repeatedly on a wide variety of investigations were three physicians: E. H. Greenhow, for twenty years a consultant at the Middlesex Hospital, 1861–81; E. C. Seaton, Simon's deputy from 1871 and successor from 1876; and Edward Smith, later a medical officer at the Poor Law Board.

[91] Goldman, *Science, Reform and Politics*, 31.

[92] Simon Szreter, 'The GRO and the Public Health Movement in Britain, 1837–1914', *Social History of Medicine*, 4, 3, Dec. 1991, 441.

[93] John Simon, *English Sanitary Institutions, Reviewed in their Course of Development, and in some of their Political and Social Relations* (London, 1890), 291.

[94] C. Fraser Brockington, 'Public Health at the Privy Council 1858–71', *Medical Officer*, xci, 1959, 176–7; 211–14; 243–6; 259–60; 278–80; 287–90. Quotation at 176. See also C. Fraser Brockington, *Public Health in the Nineteenth Century* (Edinburgh and London, 1965), 192–232.

[95] Royston Lambert, *Sir John Simon 1816–1904 and English Social Administration* (London, 1963), 317–18. Like all historians of this era I am indebted to the work of Royston Lambert who wrote probably the best biography of any nineteenth-century medical figure and one of the finest of any modern bureaucrat.

[96] Terrie M. Romano, *Making Medicine Scientific. John Burdon Sanderson and the Culture of Victorian Science* (Baltimore and London, 2002), 55, 74.

The group included at least five further leading hospital practitioners from London. Charles Murchison, an expert on diseases of the liver, and a young William Miller Ord, who led the investigation into the health effects of the infamous 'great stink' in London in 1858 'when the offensive condition of the Thames was exciting considerable alarm among the public', were both from St. Thomas's.[97] Henry Stevens was physician to St. Luke's Hospital for the Insane. George Whitley was Medical Registrar at Guy's Hospital. Timothy Holmes was surgeon at St. George's Hospital. At a time when the local Medical Officers of Health in London included doctors of the highest ability and ambition, Simon found work for Thomas Hillier of St. Pancras; Conway Evans of the Strand; Andrew Barclay of Chelsea; George Buchanan from St. Giles; John Bristowe from Camberwell; and John Burdon Sanderson from Paddington. Buchanan would go on to hold Simon's position as Medical Officer at the Local Government Board from 1879 to 1892. Bristowe collaborated with Holmes in the Office's national survey of hospital provision and practice in 1863.[98]

The physician and chemist J. L. Thudichum (formerly Ludwig Johann Wilhelm) was chemist to the Local Government Board from 1864 to 1883. Occasional expertise was co-opted from Sir William Gull, consulting physician at Guy's Hospital, Alfred Taylor the medical jurist and toxicologist, and John Netten Radcliffe who worked sometimes for the PCMO on cholera. Richard Thorne Thorne followed Buchanan as Principal Medical Officer at the Local Government Board: when a young doctor, he was also employed occasionally by Simon. Henry Julian Hunter had been one of Simon's personal pupils at King's College, London and took on key investigations. John Gamgee was a veterinary surgeon with 'a passionate interest in matters of public health and contagious disease' who was the first to recognize the 1865 'cattle plague' as rinderpest. It is a mark of their eminence that the large majority of these figures has an entry in the *Dictionary of National Biography*. As the *Social Science Review* remarked in 1864, 'Mr Simon is possessed of the rare genius requisite to give the unsystematic medical responsibilities of the Privy Council a systematic development.'[99]

Their work can be broken down into different types of investigation, most of which were published as detailed parliamentary papers attached as appendices to the *Report of the Medical Officer of the Privy Council* which Simon submitted each year. There were inquiries into epidemic diseases such as diptheria and typhoid. Greenhow and Burdon Sanderson investigated 17 counties where diptheria was prevalent in 1858–9, and Sir William Gull reported on diptheria in London in 1860.[100] The general relationship between health and occupation was a

---

[97] For a summary of Ord's famous report on the state of the Thames in 1858–9 see *Second Report of the Medical Officer of the Privy Council*, PP 1860, vol. xxix, 54–6 (quotation at 54).

[98] PP 1864, xxviii, 467–753.

[99] *Social Science Review*, n.s., 1, 1864, 55–6.

[100] For Gull's inquiry, see PP 1860, xxiv, 363–529.

major theme: Greenhow studied lung diseases in different industries in 33 towns and Edward Smith undertook investigations into sweated trades like tailoring and printing in 1862.[101] There were specific inquiries into health and the environment, including George Buchanan's study of the relationship of phthisis and damp in 1867, Richard Thorne Thorne's examination of the relation of foot and mouth disease to milk for human consumption in 1869, and Radcliffe's analysis of the quality of London's drinking water in the same year.[102] As we have seen, dangerous toxins were investigated: beyond arsenic, the PCMO investigated the effects of phosphorous, lead, and mercury on human health. Infant mortality in factory towns was examined by Greenhow in 1861 and in five rural districts of East England by Henry Julian Hunter in 1863. They came to similar conclusions. Greenhow found a strong relationship between high infant mortality and the early return of mothers to factory jobs. Hunter found that land reclamation had led to extensive employment of mothers in the fields. In both cases the results were poor care, inadequate nutrition, and the dosing of babies with opiates.[103]

Edward Smith was a pioneer in the study of food and nutrition, one of the first investigators to estimate the intake of food required for healthy life. His inquiry of 1864, published as a blue book and summarized in *The Present State of the Dietary Question*, was based on an examination of the nutrition of more than 600 families.[104] In the same year he published a *Practical Dietary for Families*.[105] The housing of the poor, involving the survey of more than 5000 habitations in England, was undertaken in 1864 by Hunter, who concluded that much of the worst housing was in country districts rather than in the towns and cities.[106] There were surveys in the early 1860s into the extent of hospital and pharmaceutical provision.[107] An attempt was also made by George Buchanan in the mid-1860s to gauge the progress made in safeguarding the public's health in 25 towns which had adopted and enforced the Public Health Acts.[108] As *The Times* commented in 1868, these studies were 'chiefly characterised by a systematic endeavour to lay the foundation of a true scientific knowledge of the sanitary state of the requirements of the English population'.[109] When disease broke out in specific areas there were local reports as well, as with typhus in Lancashire in 1861 and yellow fever in Swansea in 1865, written by George Buchanan in both cases.[110] When bubonic

[101] For Greenhow's inquiry see PP 1861, 444–536 and 1862, 602–51.
[102] For Buchanan's study, see PP 1867–8, xxxv, 427–30; for Thorne's see PP 1870, 920–5; for Radcliffe's report see ibid., 767–93.
[103] See PP 1862, xxii, 496–8 and 651–9 (Greenhow); PP 1864, xxviii, 458–66 (Hunter).
[104] See 'Report on the Food of the Poorer Labouring Classes in England', PP 1864, xxviii 216–329 and Edward Smith, *The Present State of the Dietary Question* (London, 1864).
[105] Edward Smith, *Practical Dietary for Families, Schools and the Working Classes* (London, 1864).
[106] PP 1865 xxvi, 126–302.
[107] For Dr. Alfred Taylor's investigation into the keeping and distribution of poisons, see PP 1864, xxviii, 753–69.
[108] PP 1867, xxxvii, 40–208.   [109] *The Times*, 27 Aug. 1868.
[110] PP 1863, xxv, 299–319 (Lancashire); PP 1866, xxxiii, 430–65, 862–98 (Yellow Fever).

plague and cholera emerged in Russia and the Mediterranean respectively in 1865, Simon dispatched investigators to both regions to assess the outbreaks and their likely impact on Britain.[111]

Undertaken by trained physicians, chemists and pharmacologists, such studies were recognized for their merits at the time. According to *The Lancet* in 1863, 'These investigations are so planned and effected as to give a permanent value to what might have proved of transitory interest.'[112] In no other period of the nineteenth century was government so active and so creative in the study of the nation's health and medical needs. As this list of the types of project undertaken suggests, the medical men Simon employed linked medicine to social conditions automatically: occupation, hours of labour, conditions, location, habitation, nutrition—all these aspects of health were taken fully into account in empirical inquiries that took it for granted that health and disease were intimately linked to the environment, and that investigation, publicity, legislation, and inspection could lead to changes to the poor states in which people lived and worked. In unmasking unhygienic practices, unsafe substances, and sub-standard conditions, Simon and his co-workers used statistics at every turn, counting every variable. As Simon wrote later, 'We had to aim at stamping on public hygiene a character of greater exactitude than it had hitherto had.'[113] There were few more interesting and important statistical ventures in the Victorian period than these studies, though they were undertaken by people who did not self-identify as statisticians. Many of them attended the Public Health section of the Social Science Association, which gave Simon vital political support in the late 1850s and tried to save his Office from merger and downgrade in the early 1870s, it is true. But most of these practitioners moved in medical and scientific circles. They used statistics as a matter of course to count the incidence of chronic illness, occupational exploitation, nutritional inadequacy, and excess mortality, providing evidence thereby of the degree to which the statistical approach to disease had been normalized as one of the basic medical methodologies at mid-century.

Of them all, John Scott Burdon Sanderson was both representative and also the most eminent: from 1895 to 1904 he was Regius Professor of Medicine in Oxford. Born in 1828 and raised in an intensely evangelical household, he escaped to study medicine at Edinburgh University and in Paris. He took a position at St Mary's Hospital, Paddington in 1853 and then became medical officer for the district. As Burdon Sanderson's biographer explains, his local reports on west London 'were full of statistics—births, deaths, causes of death, repairs done, among others'.[114] Able, hard-working and focused on scientific research, Burdon

---

[111] PP 1866, xxxiii, 445–66 and 675–727 (plague in Russia); 728–813 (Cholera in the Mediterranean).
[112] *The Lancet*, 1863, ii, 257 quoted in Lambert, *Sir John Simon*, 330.
[113] John Simon, *English Sanitary Institutions reviewed in their course of development and in some of their political and social relations* (London, 1890), 267, 286.
[114] Romano, *Making Medicine Scientific*, 40.

Sanderson came to the attention of Simon who made him an inspector for the PCMO, and he was used in many inquiries during these years. 'Burdon Sanderson's hopes to develop a research career matched Simon's aspiration to expand his department's mandate to include medical research.'[115] In the early 1860s he investigated vaccination in Britain. In 1865 he was sent to Danzig to investigate an outbreak of meningitis, and to the Lower Vistula to investigate the plague. In that year also he was a researcher for the Royal Commission on the cattle plague. Further work for the PCMO on infectious diseases followed, Burdon Sanderson undertaking studies of the transmission of tuberculosis, cholera, cow-pox, and smallpox. During the 1860s he had held posts at the Brompton and Middlesex Hospitals, but freed from general medical work by an inheritance, and with time for research, he built a reputation as an experimental pathologist who 'advanced the acceptance of germ theory in Britain'. He subsequently developed interests in mainstream physiology as well, and held chairs in physiology at University College, London before embarking for Oxford in 1882.[116]

As his biographer concluded, without Simon's patronage at a formative stage in his career, 'Burdon Sanderson most likely would have become yet another inter-ested clinician who had stopped doing serious research'[117]—yet another frus-trated Dr. Lydgate as depicted in George Eliot's *Middlemarch* which was published at exactly this time, 1871–2. Through this long and distinguished career, and based on the methodology and approach to science he had imbibed as a student and young practitioner in the 1850s, Burdon Sanderson retained a commitment to exactitude, figures and statistics. As he said in 1876, 'The study of the life of plants and animals is in very large measure an affair of measurement.'[118] The 'sci-entific study of nature' was defined by the production of quantitative data, by measurements taken and compared to 'known standards', those pre-existing results that had hitherto held the field. For these reasons Burdon Sanderson 'avoided discussions that moved beyond his data'.[119] It is said that Burdon Sanderson inspired the painting by Henry Stacy Marks entitled 'Science is Measurement', first exhibited in 1879 and held today by the Royal Academy. Somewhat incongruously, this depicts an eighteenth-century savant, in his wig, measuring the skeleton of a very large heron.

Simon worked alongside William Farr, often employing his data and ideas to make the case for improved public health. When, in the mid-1850s, Farr devised the model of a 'healthy district', experiencing a mortality rate of 17 per thousand or less, which could be used as a standard baseline from which to judge the health

---

[115] Ibid., 47.
[116] Steve Sturdy 'Sir John Scott Burdon Sanderson 1828–1905', *ODNB*.
[117] Romano, *Making Medicine Scientific*, 74.
[118] John Burdon Sanderson, 'Opening Address to Biology Section of Instrument Exhibition', *Nature*, 1 June 1876, 117–19, quoted in Romano, *Making Medicine Scientific*, 126.
[119] Romano, *Making Medicine Scientific*, 127–8.

of all other districts in real life, Simon adopted the measure and technique. The *Papers Relating to the Sanitary State of the People of England* were issued by Simon in June 1858. They marked the start of Simon's stewardship, a 'significant turning point in the development of public health administration' which restored to it 'the spirit of inquiry'.[120] They began with an introductory essay by Simon which adopted Farr's measure of health, pointing out that on this basis 'nearly nine-tenths of the registration districts of England show death rates which are in excess of 1,700 [per hundred thousand]', with some exhibiting rates as high as 3,600. As Simon continued 'No one pretends that people live too long' in those districts with rates below 1,700. 'That life is artificially shortened in the other 564 districts, seems the necessary alternative.'[121] The power of numbers was deployed throughout Simon's 13 annual reports, the first of which, in 1858, was entitled 'On the State of Public Health in England'. He wished to emphasize, he wrote, two issues 'first, *the present very large total of deaths*; and secondly, *the inequality with which deaths are distributed in different districts of the country*'.[122] Starting from the Registrar General's last quarterly report, the work of William Farr again, deaths in 1858 amounted to 450,018. Simon and 'the best authorities' considered this 'at least 300 deaths a day above...the natural and attainable death-rate of England'. Thus a 'saving of more than 100,000 lives' a year would be possible if the 'chief sanitary evils' could be addressed. To do so would require cleaning up the country's urban areas: 'Seven-tenths of the excess of deaths were suffered by the inhabitants of large towns, who do not constitute half the population of England. The remaining three-tenths were suffered by the inhabitants of small towns and country parishes.' The excess of deaths in the cities was caused by 'the very great development of preventable diseases'. In some districts where conditions were notably poor, mortality rates were 'from 50 to 100 per cent higher than the mortality of other districts' and the combined effect of the deaths in these neighbourhoods was 'to raise the death-rate for the whole country 33 per cent above the death rate of its healthiest parts'. But even here there was room for improvement: 'the healthiness of those districts, as compared with perfection, is but of moderate excellence'. In this way Simon, by presenting what he called 'the broad facts', well-chosen and simple to grasp, of the 'present average state of public health', made his point.

Thirteen years later, in his final report, written in the knowledge that the PCMO was going to lose its independence and be yoked to a quite different part of central government with a quite different, and parsimonious ethos, Simon

---

[120] Lambert, *Sir John Simon*, 261.

[121] John Simon, 'Introductory Report of the Medical Officer of the Board', *Papers Relating to the Sanitary State of the People of England*, PP. 1857–8, xxiii, iii–iv. Eyler, *Victorian Social Medicine*, 71–2.

[122] 'On the State of Public Health in England', *First Report of the Medical Officer of the Privy Council*, PP 1859 (i), xii, 26–8.

could only end as he had begun.[123] Once more he reminded the Council, and through them, parliament 'that the deaths which occur in this country are fully a third more numerous than they would be if our existing knowledge of the chief causes of disease were reasonably well applied throughout the country'. The number of preventable deaths a year he now computed at 120,000, higher than in 1858 because, most likely, of population growth. He reminded his audience that for each death in the registers there was 'a larger or smaller group of other cases in which preventable disease, not ending in death, though often of far-reaching ill-effects on life, has been suffered'. He added that death and disease fell disproportionately 'upon the poor, the ignorant, the subordinate, the immature' whose ill-health encouraged moral weakness also, and who therefore had the strongest claims on parliamentary protection.[124]

Simon and his teams set an example of state-sponsored investigation, led by the latest science, which was applied by those with genuine expertise. Yet as he knew only too well, for all the research and investigation of his office and his protégés, the state of public health had not generally improved across these years, nor would it for some years to come. Death rates fell at last during the final two decades of the nineteenth century as a consequence of improvements in working-class housing and nutrition as living standards rose, from the better regulation of the workplace, and through general social advance. Simon's kind of contextual medical investigation, trying to link disease with specific environmental factors, had its limitations. The initial inquiry into the diptheria epidemic at the end of the 1850s, which was undertaken by Greenhow, Burdon Sanderson, and Gull, was 'disappointing: though the inquiry increased knowledge of the symptoms and predilections of the disease, it made no fundamental discoveries as to its etiology'.[125] The same could be said of Greenhow's subsequent study in 1859-60 on diarrhoeal diseases which, in its ignorance of how the disease was spread, was 'cardinally defective'.[126] Overall, the PCMO's inquiries, at least up until the mid-1860s, had 'negligible impact'.[127] They should not be dismissed, however, because of the model they established of activism in central government through the employment of genuine ability. They are also luminous examples of studies of the relationship of disease to environment that are undertaken all the time today, but now with the benefit of much greater medical knowledge as well as powerful statistical techniques and digital computers. The kind of research that depended on 'counting' as it was undertaken by Farr, Simon, Guy, and others, had its limitations, and this was becoming clear by the 1860s and 1870s. Establishing the numbers, discerning the patterns, and linking them to one external condition or another

[123] On the background to the undermining of medical control of medical research in central government in this period, see Roy M. Macleod, 'The Frustration of State Medicine 1880–1899', *Medical History*, 11, 1967, 15–40.
[124] *Thirteenth Report of the Medical Officer of the Privy Council*, PP 1871, xxxi, 771–2.
[125] Lambert, *Sir John Simon*, 319.   [126] Ibid., 320.   [127] Ibid., 365.

(but without the benefit of statistical correlation at this stage for that was a technique that came later) was not without benefit and purpose, but it did not amount to the understanding of disease required for its eradication. The hope of the statistical movement that the numbers would reveal all, had not come to pass. It led to a decline in the prestige of statistics in this non-technical form, and to an assault on its pretensions as set out by the founders in the 1830s. We will turn to the decline of the statistical movement in the next section.

Conclusion: Statistics and Medicine

William Guy had articulated the most ambitious of programmes for the new statistical movement in the 1830s, but as his career developed he became less confident and less assertive about the capacity of numbers to explain the laws of health and disease. William Farr tried to construct those laws in relation to epidemic diseases like smallpox and cholera, but analysing the distribution of a disease in a specific place in the absence of a scientific understanding of its origin and the means of its transmission, though of interest, was without general applicability. Taking into account the meagre funds then available, the lack of political support, and the ignorance of disease across society at all levels, John Simon oversaw perhaps the most impressive programme of environmental medicine ever undertaken in Britain, but it did not save him and his department. He resigned in despair in 1876, believing that the influence of the Poor Law bureaucracy over central medical services, such as they were, would put back the improvement of the public's health for at least a generation.[128] Although Simon was knighted in 1889, the honour came only a year before he died and long after his services to the state had been forgotten. Like William Farr and before them both, Edwin Chadwick, Simon left public service in disappointment and with the fear that his work would not be preserved, let alone built on. Something of this fate also awaited one of the most important figures in the early provincial statistical movement, Dr James Phillips Kay, a founder of the Manchester Statistical Society and its most famous investigator.[129] As we have seen, between 1839 and 1849 he was secretary to the Committee of Council for Education and had an excellent working relationship with the Whig leader and prime minister, Lord John Russell, who usually followed his advice. On his retirement Russell nominated him for a knighthood. But he bitterly resented his official subordination to a 'gossiping dilettante' like Charles Greville, grandson of the Duke of Portland, who was Clerk to

[128] Ibid, 544; John Simon, *Personal Recollections of Sir John Simon K. C. B.* (London, 1894, printed privately), 23; Simon, *English Sanitary Institutions*, 392.
[129] Frank Smith, *The Life and Work of Sir James Kay-Shuttleworth* (London, 1923), 20–8.

the Privy Council. Although James Phillips Kay married into an old, established family, and so gained an estate worth £10,000 a year—and accordingly changed his name to Sir James Kay-Shuttleworth—he could not escape the discrepancy between his high income and elevated private status, and his lowly official position as a functionary of state.[130] Experts in government usually had short-lived careers and many grievances over their anomalous social status.

If we return to Edward Jarvis, in 1860 he believed he was mixing in Britain with a confident, esteemed and secure community of experts. In comparison with the United States where such a community would not emerge for at least another generation, it will have seemed so. As he wrote to William Farr after his return to the United States,

> I look upon my visit to London and my intercourse with the statisticians there, the Statistical Society, the Registrars' Office, and its learned & courteous officers, with unalloyed satisfaction. The Statistical Congress was the crowning pleasure of my London visit & indeed of my life; for there I met with those with whom, in all the world, I had the most sympathy and whose minds were bent upon and represented the interests that had occupied my intensest thoughts ever since I entered upon the world's busy stage.[131]

Before the expansion, reform, and professionalisation of the civil service in the last generation of the Victorian era brought expertise within government and reduced the role of experts outside it, the economists, social scientists, social investigators, and statisticians were at the height of their independent influence. However, had Jarvis returned to Britain in the year of his death, 1884, he would have had greater difficulty in identifying, let alone securing access to, the dominant policy-making experts. There were more salaried bureaucrats now, but far fewer 'statesmen in disguise'. Writing about the fledgling educational bureaucracy, Richard Johnson has pointed to the anomalous position of the expert in a society in which only birth and the possession of land conferred authority and status. The expert generally had no property, depending instead on 'particular knowledge, particular skills, a particular sense of commitment' as his title to employment.[132] But these talents, because so alien to aristocratic politics, were suspect in British public life, and the expert was vulnerable to changing political whims and fortunes. Expertise was required but distrusted and usually undervalued. In other words,

---

[130] R. J. W. Selleck, 'Sir James Phillips Kay-Shuttleworth 1804–1877', *ODNB*; Richard Johnson, 'Administrators in Education before 1870: Patronage, Social Position and Role', in Gillian Sutherland (ed.), *Studies in the Growth of Nineteenth Century Government* (London, 1972), 123–6.

[131] Edward Jarvis to William Farr, 26 Nov. 1860, Jarvis Letter Books, Jarvis Papers, Francis A Countway Library for the History of Medicine, Harvard University, B MS b. 56.4, v.6, ff. 138–40.

[132] Johnson, 'Administrators in Education', 122.

although the decline of 'the statist' in the later-Victorian period reflects, to some degree, the analytical limitations of statistics before the intellectual revolution the subject underwent in the late nineteenth century, there is also a sociological explanation for the decline of the old-style statistician whose position in government was generally precarious. And beyond sociology, there were also international, political, and ideological changes that undermined the statistical movement from the 1870s. These will be discussed in the next section.

PART V
LIBERAL DECLINE AND REINVENTION

13

The International Statistical Congress, 1851–1878

Conservative Nationalism versus Liberal Internationalism

The Victorian statistical movement denoted outlooks and practices that were liberal, environmentalist, reformist, and descriptive rather than technical. It reached its height in the 1860s, but was in decline within a decade after that. Its ideological assumptions were undermined by wider social and intellectual changes, and its simple methodologies were surpassed by the invention of mathematical statistics, which made available powerful tools for the study of social and natural phenomena like correlation and regression. The process took two decades at least, during which the three founders and friends—Babbage, Quetelet, and Farr—all died, in 1871, 1874, and 1883, respectively. The chapters to follow examine different elements of this decline and transition, focusing in particular on the collapse of the International Statistical Congress in the 1870s, and the remarkable technical innovations coupled to highly controversial ideas developed by Francis Galton. Galton's genius established mathematical statistics, and served as an inspiration to a generation of young professional statisticians then graduating from the universities. His commitment to Darwinian ideas linked the new discipline to the emerging eugenics movement which Galton himself had named and founded. He also led public opposition to the 'old statistics' of the statistical movement. He took aim at Section F, the statistical section of the British Association for the Advancement of Science; he offered technical and mathematical papers to the Statistical Society of London in a break with past practice and convention; in his correspondence with Florence Nightingale over the endowment of a first chair of statistics in Britain he demonstrated the difference between the descriptive statistics of the past and the new discipline he had created.

Each of these developments is significant; each might be taken to mark the end of the statistical movement as it had been established in the 1830s. But taken together, which is more appropriate for a transition that took years, they mark the completion of a cycle of intellectual history which was co-extensive with Victorian liberalism in its hey-day. This, too, was in decline by the 1880s, now in competition with a reinvigorated conservatism on the right, which had found new supporters in the suburbs and new causes such as imperialism, and the rise of a labour

Victorians and Numbers: Statistics and Society in Nineteenth Century Britain. Lawrence Goldman, Oxford University Press.
© Lawrence Goldman 2022. DOI: 10.1093/oso/9780192847744.003.0013

movement to the left, which was better able to represent the interests of a mass working class. Nevertheless, there were genres and styles of collection and analysis that endured. The type of critical social statistics pioneered by the artisans who compiled the *Statistical Illustrations,* with the purpose of defending working-class living standards, was a constant component of Victorian social investigation and debate. In the Industrial Remuneration Conference in London in 1885, which is the subject of the final chapter in this section, they were deployed again by workers and their advocates, in ways that remind us of the 1820s but which point forward, also, to all the intervening social history and contestation up to our own age. We argue, in similar ways and with similar statistics, about the same issues.

The International Statistical Congress was founded in 1851 to compare social institutions and standardize social administration across borders. It is said of the ISC that it was the first international organization to link together practitioners of a single academic discipline. It has not been overlooked in the history of social policy and statistics, but its story has never been told in much detail and may have been misunderstood in biographies of its founder and patron, Adolphe Quetelet, and in earlier, pioneering histories of statistics.[1] It has been presented at face value according to its own professions and propaganda.[2] It has also been suggested that its eventual failure 'can be explained by the difficulties in realizing effective knowledge transfer, in other words effective communication, in an age that was not fully prepared for truly international activities'.[3] But that is to underestimate both the ISC's commitment to internationalism and the interrelated set of intellectual, institutional, and, above all, political tensions throughout its history which eventually wrecked the organization. There were indeed great difficulties of transnational collaboration in social affairs in the mid and late nineteenth century at a time when the internationalism of the 1850s and 1860s was giving way to the nationalism of the 1870s and 1880s. But these were ideological rather than technical in nature. The history of the ISC is a study in the frustration of the political liberalism which had once inspired the statistical movement. The technical difficulties of international communication are but a side issue.

The ISC was founded in 1851 at the Great Exhibition in London in discussions between savants from France, Germany, Britain, and the United Sates. Among them were Adolphe Quetelet, the leading figure in nineteenth-century social

---

[1] Frank H. Hankins, *Adolphe Quetelet as Statistician* (Columbia Studies in History, Economics and Public Law, New York, 1908), 30–1. Harald Westergaard, *Contributions to the History of Statistics* (London, 1932)

[2] Libby Schweber, *Disciplining Statistics. Demography and Vital Statistics in France and England, 1830–1885* (Durham N.C. and London, 2006), pp. 31–2.

[3] Nico Randeraad, 'The International Statistical Congress (1853–1876): Knowledge Transfers and their Limits', *European History Quarterly,* 41, 1, Jan. 2011, 50–65 (quotation at 51). Randeraad's slightly earlier study of the failure of the ISC can be read as evidence that the problems which undermined the organization were more profound than merely questions of communication. See Nico Randeraad, *States and Statistics in the Nineteenth Century. Europe by Numbers* (Manchester, 2010).

statistics, and Charles Babbage.[4] The reprise of their friendship and alliance in London in 1851, assisted by Auguste Visschers, another member, like Quetelet, of the Belgian Statistical Commission, led to the first congress of the ISC, held in Brussels in September 1853.[5] As William Farr had written encouragingly to Quetelet the year before, 'I hope that time will ripen that project of a statistical congress which occurred to you and to [M. Visschers] & which I think might be advantageously sanctioned by the statistical (including all the civilized) countries of Europe and America.'[6] Farr wrote again to praise the 'admirable arrangements' and 'the hospitable and friendly reception' in Brussels, a city then in its heyday as the capital of a new, vigorous, and independent liberal state.[7] Eight more congresses followed: in Paris (1855), Vienna (1857), London (1860), Berlin (1863), Florence (briefly the capital of the new, but still incomplete Italy, in 1867), The Hague (1869), St Petersburg (1872), and Budapest (1876). There was even a joint resolution of both houses of the United States Congress in 1873 requesting President Ulysses S. Grant 'to invite the International Statistical Congress to hold its next session in the United States'.[8]

The congresses were generally attended by several hundred participants. There were 541 present at Vienna in 1857, 595 in London in 1860, 477 in Berlin in 1863, and 717 in Florence in 1867. Approximately a hundred of these in each case were official delegates from foreign countries. Two-thirds of those regularly attending were members of the general public, and this was to become a problem. At its core were leading figures in the collection and analysis of official statistics from the most notable European nations, men like Baron Czoernig, Director of the Habsburg Statistical Department in Vienna, an expert in ethnographic statistics as befitted a statistician from a multi-ethnic state;[9] Professor Luigi Bodio, Director General of the Statistical Department in Rome; Dr C. N. David, Director of the Statistical Department in Copenhagen; Dr F. B. W. Hermann, Director for the Bavarian Statistical Department; and the ubiquitous Dr. Ernst Engel, '*echt* Prussian bureaucrat'. He had begun his career in the collection of state statistics in Saxony in the 1850s and was afterwards director of the Prussian Bureau of

---

[4] *Bulletin de la Commission Centrale de Statistique*, Tome vi, p. 3 (Brussels, 1855). *Report of the Proceedings of the Fourth Session of the International Statistical Congress, Held in London July 16th, 1860 and the Five Following Days* (London, 1861), 207; 'Letter from Charles Babbage, esq. F.R.S' in ibid., pp. 505–7; Fred Lewes, 'The Letters Between Adolphe Quetelet and William Farr 1852–1874', *Bulletin de la Classe des Lettres et des Sciences Morales et Politique*, 5 series, Tome 69, 1983, 420.
[5] *Comte Rendu des Travaux du Congrès Générale de Statistique, réuni à Bruxelles les 19, 20, 21 et 22 Septembre 1853* (Brussels, 1853).
[6] William Farr to L. A. J. Quetelet, 17 Aug. 1852, Quetelet Papers, Académie Royale de Belgique, Brussels, File 990–1.
[7] Ibid., 30 Sept. 1853.
[8] *Congressional Globe*, 8 March 1873, 2005–6. There is no evidence that Grant did extend an invitation.
[9] Karl von Czoernig, *Ethnographie der österreichischen Monarchie* (2 vols, Vienna, 1855, 1857).

Statistics for 22 years.[10] The congress was therefore dominated by leading officials from state bureaucracies, generally located in central statistical bureaux in the capital cities of Europe. Indeed, its quasi-official status drew state patronage and remarkable royal favour: in 1853 the Congress dined with the King of the Belgians; in 1855 it was received by Emperor Louis Napoleon in the Tuileries, Paris; in 1857 by the Austrian Emperor, Franz Joseph, at the Burg Palace in Vienna. The members were entertained by William I, King of Prussia, in Berlin in 1863; by Victor Emmanuel II in the Pitti Palace in Florence in 1867; and by Alexander II of Russia at the palace of Tsarkoe-Selo at St Petersburg in 1872. But none of the crowned heads of Europe did more than formally welcome the delegates: only Prince Albert addressed them as an equal.

It is not difficult to imagine the aim of such a body: to help collect complete, uniform, systematic, and comparable sets of national statistics from all the major nations. In the process the Congress would spread best practice, disseminating the latest techniques of data collection, tabulation, and analysis as a prelude to policy formation. According to one early participant, the Italian-born British lawyer, political economist, and statistician, Leone Levi, 'we must settle on the nomenclature of things; we must, so to say, adopt a universal language for the purpose, and simplify the tables which are to be the basis of comparison.'[11] The object of the congresses was 'the establishment of a complete body of national statistics in all countries, which shall not only be complete in the items of information, and scientifically classified, but so prepared as to be comparable among themselves.'[12] When setting out the rationale of the London Congress of 1860, William Farr went further: 'The statistical discoveries of one nation are the lights of all nations... The heads of the statistical departments of Europe, and private inquirers....can point out sources of information to each other, and aid each other in carrying out great combined inquiries.'[13] Led by Ernst Engel, the ISC projected in 1869 the *Statistique Internationale de L'Europe*, a 'grand idea of an authentic comparison of the actual social and economic condition of all the nations of Europe', subdivided into 24 heads, to be investigated by teams of researchers across the continent.[14]

Albert, the Prince Consort, had drawn out the political implications of this comparative project in his speech to the 1860 Congress: the work of the ISC, he

[10] Lorraine J. Daston, 'Introduction', in Lorenz Krüger, Lorraine J. Daston, and Michael Heidelberger (eds), *The Probabilistic Revolution* (2 vols, Cambridge, MA), vol. 1, *Ideas in History* (Cambridge, Mass., 1987), 1.

[11] Leone Levi, 'Resume of the Statistical Congress, held at Brussels, September 11th 1853', *JSSL*, 17, March 1854, 4.

[12] Leone Levi, 'Resume of the Second Session of the International Statistical Congress held at Paris, September, 1855', *JSSL*, 19, March 1856, 2.

[13] William Farr, 'Report on the Programme of the Fourth Session of the Statistical Congress', in *Programme of the Fourth Session of the International Statistical Congress to be held in London on July 16th 1860 and Five following days* (London, 1860), pp. 4, 8.

[14] Samuel Brown, 'Report of the Seventh International Statistical Congress, held at the Hague, 6-11 September 1869', *JSSL*, 32, 1869, 402.

explained, 'will afford to the statesman and legislator a sure guide in his endeavours to promote social development and happiness'.[15] The Congress was a manifestation of the mid-Victorian doctrine of improvement, its liberal credentials secure in its support for rational enquiry, international interchange, and social advance. It was evidence in itself of the recent 'progress and advancement of civil society' as Levi put it,[16] and according to the leading British actuary of this era, Samuel Brown, its deliberations would 'make prosperity, goodwill, and peace everywhere take the place of misery, distrust, and war'.[17] Speaking on behalf of the British delegates at the Hague in 1869, the writer and savant Richard Monckton Milnes, Lord Houghton, who had once courted Florence Nightingale, 'alluded to the rapid growth and cultivation of such studies in countries where freedom and independence led to open discussion of subjects so important to the progress and prosperity of the people'.[18] One British observer referred to the Congresses as 'great international peace gatherings,' though in the event they were frequently interrupted by wars.[19]

To read its transactions is to enter the world of mid-nineteenth century internationalism and to encounter the advocacy of pan-European projects. The ISC favoured a uniform metric system of weights, measures, and currency; it considered a system of international postage; it wanted to extend international commercial law. In 1869 Edwin Chadwick even persuaded it to investigate the scale, expense and burden of Europe's military establishments, though nothing came of the plan. The subject had been broached in London in 1860 as well, and the outcome was similarly inconclusive.[20] Leone Levi summarized all these sentiments in his report to the Statistical Society of London on the first Congress in Brussels:

> The readiness evinced of late by all Governments to co-operate in the promotion of science and of subjects of general utility, is one of the most remarkable features of the age in which we live. The affinity of interests which binds all nations of the earth is better understood and appreciated; the study of natural laws, in their relation to society, is more expanded and intelligent; the institutions of all countries are closely scrutinised, and rather than be wedded to antiquated systems, each is eager to profit by the experience of the other.[21]

[15] 'Address of the Prince Consort on opening as President the Fourth Session of the International Statistical Congress', *JSSL*, 23, 1860, 281.

[16] Levi, 'Resume of the Second Session', 1.

[17] Brown, 'Report of the Seventh...Congress', 410.   [18] Ibid., 393–4.

[19] Frederic J. Mouat, 'Preliminary Report of the Ninth International Statistical Congress, held at Buda-Pesth, from 1st to 7th September 1876', *JSSL*, vol. 39, Dec. 1876, 635.

[20] Brown, 'Report of the Seventh...Congress', 409; Brown, 'Report on the Eighth International Statistical Congress held at St. Petersburg, August 1872', *JSSL*, 35, Dec. 1872, 435.

[21] Levi, 'Resume of the Statistical Congress, held at Brussels, September 11th 1853', 1.

Alas, fine sentiments like these were not enough to prevent the ISC from fracturing along national lines in the 1870s.

The ISC's transnational project was not written across a blank sheet: it had to contend with contrasting national traditions of statistical investigation. Theodore Porter and Ian Hacking have compared the ambitions of nineteenth-century French and British statisticians to discern regularities in social behaviour and social structure with the narrower Prussian focus on the collection of data as an official function of the state for purposes of better administration, and the Prussian denial of any more analytic and academic potential in social statistics.[22] Meanwhile Libby Schweber has probed the Anglo-French tradition more closely still and demonstrated the differences between British vital statistics and the development of demography in France in this period.[23]

The ISC was faced with a severe external constraint, as well: European warfare. It was an international organization, one of the very first, seeking to unite savants from across the continent in an age of national upheaval and violent state formation. The first congress was delayed by Louis Napoleon's coup d'état against the French Second Republic of December 1851. Subsequent problems can be traced through William Farr's correspondence with Quetelet. This starts on that characteristic note of early-1850s liberal optimism: 'One advantage of this Congress will be the tendency it must have to promote a friendly feeling among the statistical brotherhood all over the world.'[24] But within a year—the year, in fact, that the Crimean War erupted, 1854—Farr could not disguise his contempt for both the Russians and Prussians, and his opposition to meeting in St. Petersburg: 'Russia isolated herself by sending an insulting message to the statistical conference & neither science nor civilization nor freedom can advance in the world so long as the power of those barbarians preponderates as it does in Germany that has lain paralyzed at her feet.'[25] Farr had always been keen to bring the meeting to London: 'I am anxious that the Commission should be aware that in England we should be very glad to receive the Congress. And I have no doubt that the Government & the Statistical Society would give it a warm reception.' In the event, the 1860 London Congress should have been held the year before, and planning was already well-advanced before the Franco-Austrian War of 1859 (sometimes called the Second Italian War of Independence) ended all hopes of a pan-European gathering that year: 'Our government felt that they could not well invite our peaceful Congress to meet during the war which unhappily raged this year.'[26] It is interesting to reflect that in 1859, Luigi Menabrea, the mathematics professor

[22] Theodore Porter, 'Lawless Society: Social Science and the Reinterpretation of Statistics in Germany, 1850–1880' and Ian Hacking, 'Prussian Numbers, 1860–1882' in L. Krüger, L. Daston, and M. Heidelberger (eds), *The Probabilistic Revolution* (Cambridge, Mass., 1987), vol. 1., 351–75, 377–94.
[23] Schweber, *Disciplining Statistics, passim.*
[24] William Farr to L. A. J. Quetelet, Quetelet Papers, 9 Aug. 1853.
[25] Ibid., 31 Aug. 1854.   [26] Ibid., 27 Sept. 1859.

from the university of Turin who became Babbage's expositor in 1842, now a Major-General, was in command of the Italian engineers in Lombardy and used his skills to superintend siege works, repair fortifications, and flood the north Italian plains during the conflict, making them impassable to Austrian armies.

Disagreement between France and Prussia over a free trade treaty ostensibly led to the postponement of the scheduled 1862 congress for a year, though it is more likely that the Prussian constitutional crisis over the military estimates that brought Bismarck to power was to blame.[27] Farr then complained at the apparent unwillingness of the Prussian government to properly support the 1863 Congress in Berlin.

> The opinion here is – and I concur in it – that the responsibility of sustaining the Congress & paying all expenses on attending it, direct and indirect, should be borne by the government of the country in which the Congress meets. This has hitherto been done in Brussels, in France, in Austria & England.[28]

So meagre was the official hospitality that Bismarck's personal banker, Gerson Bleichröder, 'organised a banquet for the conferees at his own expense, thus saving the organisers from embarrassment'.[29] In the event, however, Farr was pleased to report that 'our Congress at Berlin went off much better than might have been expected; and must, I think, have done good there'.[30] Three years later there was the Austro-Prussian War to contend with:

> I am afraid that there is no prospect of our peaceful gathering at Florence. With the cries of battle about our ears, issuing from the mouths of our dear friends on both sides, we could not sit down calmly to deliberate. When Fox was asked to speak on a great question, the evening before the death of his great rival Pitt, he replied to express his inability: *mentem mortalia tangunt*. We should be overpowered by the same or similar feelings. Let us hope that the war will be soon over.[31]

And still it went on, however, for next came the Franco-Prussian War in the autumn of 1870, with the subsequent encirclement of Paris and the rising of the Paris Commune: 'Poor Paris!' wrote Farr to Quetelet: 'We all deeply sympathise

---

[27] Randeraad, *States and Statistics*, 104.
[28] William Farr to L. A. J. Quetelet, 18 Aug. 1863.
[29] Randeraad, *States and Statistics*, 124. On Bleichröder, see Fritz Stern, *Bismarck, Bleichröder and the Building of the German Empire* (New York, 1977).
[30] William Farr to L. A. J. Quetelet, 31 Dec. 1863. The 'good' done probably refers to the liberalizing influence of the ISC while in Berlin.
[31] Ibid., 3 July 1866. The Latin text is from Virgil, Book 1, line 426 of *The Aeneid*. It can be translated as 'mortal things touch the mind'.

with the nation in its sufferings and hope for peace on reasonable terms. In this hope we look forward to a time when our Congress may meet again.'[32]

War, pitting national delegations against each other, and the disruption of continental travel and communication that war brought, were two reasons why the ISC met infrequently rather than every three years. But problems of national difference also affected internal co-operation within the Congress itself. This is evident in one small example of collaboration' sponsored by the ISC, the attempt to devise a standard nosology, or 'a uniform nomenclature of the causes of death, applicable to all countries'. The cause of death of those who die in the United Kingdom today is determined according to a nosology originally devised by William Farr, though changed much over the decades as medical knowledge has advanced. But it isn't the nosology he was asked to construct by the Brussels Congress of 1853 in collaboration with Dr Marc D'Espine of Geneva. They laboured in vain to 'come to an understanding from a distance' and hence submitted separate reports. D'Espine categorized deaths into eight classes and classified diseases by their duration, whether chronic or acute. Farr paid more attention to the specific causes of death, arranged anatomically according to the parts of the body affected, and divided diseases into 'epidemic' and 'sporadic' (i.e. of common occurrence). As one observer noted, 'D'Espine's classification would be useful in medical jurisprudence; Dr. Farr's has a more direct reference to sanitary science.'[33] The difference between the bureaucratic-legal and medical-scientific approaches to the regulation of public health which demarcated continental and British practice in the mid-nineteenth century is here made evident.[34] Farr's nosology has been revised continually since the 1860s but the general structure he provided, based on the principle of identifying diseases by their anatomical location, has survived as the basis of what is now the internationally accepted classification of disease and cause of death.[35]

The bureaucratic and official ethos of the Congress—its 'governmentality'—was a problem for the British delegates and eventually compromised the ISC itself. But it was a problem for the British in two distinct and contradictory ways: initially, because it pointed to the failure of British statisticians to institutionalize their discipline in alliance with the state, and secondly because it made the ISC into the plaything of politics rather than an independent scientific organization. These discordant messages tell us something about tensions within Victorian intellectual life in Britain; they also give reasons for the demise of the Congress itself.

Early British commentators and participants at the Congress lamented the national failure to centralize statistical collection and analysis in a single

---

[32] Ibid., 1 Jan. 1871.   [33] Levi, 'Resume of the Second Session', *JSSL*, 19, 1856, p. 8.
[34] F. Lewes, 'Dr Marc d'Espine's Statistical Nosology', *Medical History*, 32, 1988, 301–13.
[35] World Health Organisation, *International Classification of Diseases* (11th revision, 18 June 2018) https://www.who.int/classifications/icd/en/. On Farr's nosology see Simon Szreter, 'The GRO and the Public Health Movement in Britain, 1837–1914', *Social History of Medicine*, 4, 3, Dec. 1991, 444.

government department (which was one of the declared aims of the Congress) and were evidently envious of the close associations between statistical collection and the state overseas. Farr himself worried that in comparison with other European nations 'we have no statistical board, and there is want of co-ordination in our publications'.[36] Samuel Brown complained in 1867 that 'the great questions... are on too large a scale to be entrusted to individual enquirers'.[37] In 1860 in London, Prince Albert's speech also noted 'certain defects in our returns which must be traced to the want of such a central authority or commission as was recommended to the congress at Brussels and Paris'.[38] Conversely and simultaneously, however, the British also defended the traditions of voluntarism and independence. For six mornings in succession at the London Congress in July 1860 the delegates sat through interminable reports on the workings of the central statistical bureaucracies of each country represented. But when it was Britain's turn, there being no such body to offer a report, the floor was given over to representatives of different organisations engaged with statistics including the Institute of Actuaries and the Statistical Society of London. As we have seen, William Newmarch then declared that it was 'the merit of the Statistical Society not to be official at all'.[39] Later, when the ISC's 'governmentality' had so compromised its actions as to destroy the organization, Dr Frederic Mouat, the first historian of the Statistical Society of London, delivered himself of a homily on the distinctions to be maintained between science and authority: 'They had much better be independent of each other, as they are in England, and act with the unfettered freedom which can alone secure satisfactory results.'[40]

There was certainly evidence to support this view. In Berlin in 1863, for example, the Prussian government intervened directly to remove liberals from the organising committee of the Congress. They included Rudolph Virchow, the great pathologist and a leading liberal who was reputedly challenged to a duel by Bismarck two years later; Salomon Neumann, Virchow's friend, a physician, medical reformer and statistician; and the economist and leading member of the Progressive Party, Hermann Schulze-Delitzsch.[41] Looking on from London, Florence Nightingale was moved to ask William Farr 'why Germans cannot succeed at liberal institutions as we do [?]...We in England should just lay our heads together, as we did in James II's time, & say, this shall not be. And the objective

[36] William Farr, 'Reports of the Official Delegates from England at the Meeting of the International Statistical Congress in Berlin', *JSSL*, 26, 1863, 412.
[37] Brown, 'Report on the Sixth International Statistical Congress', *JSSL*, 31, 1868, 11.
[38] 'Address of the Prince Consort', 283.
[39] *Report of the Proceedings of the Fourth Session of the International Statistical Congress*, 115–16. See also Farr's similar remarks at 114.
[40] Frederic J. Mouat, 'History of the Statistical Society', in *Jubilee Volume of the Statistical Society* (London, 1885), 43.
[41] Randeraad, *States and Statistics*, 113, 117. Rudolf von Gneist, the anglophile jurist and leading National Liberal, also came under pressure from the authorities.

actions would follow directly.'[42] Ian Hacking has therefore contrasted 'the British way with numbers' that 'reflected a resistance to centralized management' and the establishment by the Prussians of 'an office of numbers-in-general', their Bureau of Statistics, used by all departments of government and thus under government control.[43]

Differences between British and continental practice were evident also in related institutional issues. The Congress enforced a distinction between 'official delegates' of the various European governments and 'the subordinate position assigned to all other members of the Congress'.[44] It was not a distinction respected by British attendees who were excluded from office in the organization because they went in their own capacity or as delegates of learned societies and professional bodies, rather than as representatives of the state. British commentators complained that the ISC's excessive ambitions, procedural errors, and impractical resolutions conspired to limit its effectiveness, and that it accomplished rather little. Newmarch called the congresses 'international picnics'.[45] One of the reasons for this was the popularity and openness of the early Congresses: just about anybody and everybody could and did attend. It was reported that the Viennese Congress in 1857 had been swamped by the public for the simple reason that the emperor had offered a free Sunday excursion to all the participants.[46]

William Farr wanted the ISC to be an expert forum composed of genuine statisticians: 'There is a strong tendency in all our Congresses to divest themselves of their scientific character in which their strength essentially consists. Unfortunately, few scientific men—except yourself—have entered the field of statistics and the statists have not, in some instances, the indispensable mathematical aptitude required for the satisfactory execution of their work.'[47] He agreed with Quetelet 'as to the necessity of making it exclusively an official congress – of the government delegates – in order to give the decisions due weight & efficacy'.[48] In the weeks before the London Congress he wrote to Quetelet that it was 'more than ever important to apply the strictest methods to statistical inquiries. We must, I think, ask you to speak to this subject at the opening of the Congress.'[49] Frederic Mouat, writing in 1876, recognized that the broad and popular congress covering a multitude of subjects and issues, which was a distinctive institutional feature of British

[42] Florence Nightingale to William Farr, 7 Aug. 1863, Wellcome Library, 5474/61.
[43] Ian Hacking, *The Taming of Chance* (Cambridge, 1990), 33.
[44] Frederic J. Mouat, 'Second and Concluding Report of the Ninth International Statistical Congress', *JSSL*, 40, 1877, 548.
[45] Newmarch is quoted in Mouat, ibid., 554.
[46] The anecdote is taken from C. J. Willcox, 'The Relation of the United States to International Statistics', in the *Journal of the American Statistical Association*, 1924, and is presented in J. W. Nixon, *A History of the International Statistical Institute 1885–1960* (The Hague, 1960), 10. See also Walter F. Willcox, 'Development of International Statistics', *Milbank Memorial Fund Quarterly*, 27, 2, 146.
[47] Farr to Quetelet, 7 Jan. 1859, Quetelet papers.   [48] Ibid., 22 Dec. 1858.
[49] Ibid., 13 April 1860.

and European culture in the mid-nineteenth century, when science sought public support, was no longer relevant in a later age of disciplinary specialization:

> Many persons doubt the usefulness of these international congresses, consider that academies and scientific societies are the only really successful cultivators of science, that the time for nomadic gatherings for purposes of research and inquiry has passed away; that special journals are the soundest propagators of scientific truth…[50]

The Congress experimented with different solutions to this problem, trying to concentrate influence among the official delegates and holding special pre-meetings for them alone (the *avant congrès*) in advance of the full congress at Berlin in 1863 and subsequently. Ultimately the solution to the problem of the 'mixed multitude'[51] as first proposed by Ernst Engel at the Berlin Congress in 1863, was to establish, at the St. Petersburg Congress, a Permanent Commission, an executive body bringing together the leading official statisticians from the constituent nations that would give direction, impetus and unity to the organization. The two British members of the Commission were Robert Giffen of the Board of Trade and Frederic Mouat, Inspector of the Local Government Board.[52]

Yet the very attempt to galvanize the organization brought it down. There were immediate disputes between members over the authority of the new Commission. Engel, speaking for Germany, argued that the Commission derived its mandate from the several governments rather than the Congress; others contended that the Commission was simply the servant of the Congress which was ultimately sovereign in the organization's affairs.[53] In 1876 the Congress agreed that the Commission should be based in Paris. Meeting there in 1878 it proposed that it would put ISC resolutions to member governments and be bound by the decisions of the majority of them. While most official representatives approved of this idea, 'some, and particularly the representatives of the Statistical Bureaux of the German Empire reserved to themselves absolute freedom of action as to acceptance or otherwise of the decisions arrived at, at the Paris Commission'.[54] Germany would not be bound by the decisions of an international body in short, especially one based in the capital of France. German representatives from the several German states, later joined by the Swiss, refused the invitation to the next session of the Commission in 1879 in Rome, and to the projected tenth Congress of 1880.

---

[50] Mouat, 'Preliminary Report of the Ninth International Statistical Congress', *JSSL*, 39, 1876, 645.
[51] 'The Permanent Commission of the International Statistical Congress', *JSSL*, 41, 1878, 549–50.
[52] Ibid.
[53] 'The Permanent Commission of the International Statistical Congress',*JSSL*, 37, 1874, 116.
[54] Professor F. X. von Neumann-Spallart, 'Résumé of the Results of the International Statistical Congresses and Sketch of Proposed Plan of an International Statistical Association', *Jubilee Volume of the Statistical Society* (London, 1885), 301.

The president of the ISC, Károly Keleti, who was director of the Royal Hungarian Statistical Bureau in the new Hungarian state established under the Dual Monarchy of 1867, resigned, and the organization collapsed.[55]

There is evidence that behind German opposition lay the will of Bismarck himself. According to a later report in the Viennese *Allgemeine Zeitung*,

> When in 1876 this Congress passed a resolution that a permanent International Statistical Commission should be appointed which should have its permanent seat in Paris, Bismarck protested against this decision and forbade the Prussian statisticians from participating in the Congress from henceforth. In consequence the other German states also did not take part in the Congress which, since that point, albeit still formally in existence, did not engage in further activity.[56]

As one historian concluded, 'German opposition to tendencies in the Statistical Congress and especially in the Permanent Commission was mainly responsible for their death.'[57] As Mouat put it, 'The cardinal defect of the congress was its attempting too much, and to make the resolutions passed at the general assemblies binding on the several Governments.'[58] The semi-official character of the ISC hamstrung it in two related ways: in one direction it inhibited the Congress's freedom of action and independence because its key members were representatives of their government, and in the other it raised political concerns that 'the Governments should be bound by the decisions of the Congress or Permanent commission', which Germany, at least, would not accept.[59]

The opposition of the German state to the ISC also extended to the Prussian Statistical Bureau itself, which from the late 1870s lost the confidence and support of the Bismarckian authorities. Hitherto a centre for social investigation associated with liberal reform, the Statistical Bureau was reduced to purely academic functions in the same way that the Verein für Sozialpolitik, the German analogue to the British Social Science Association which had been founded in 1873, lost any influence over government policy after 1879. In that year, Bismarck turned decisively against a liberal future and consigned German academics and their organizations to the margins.[60] Ernst Engel himself, a National Liberal representative

---

[55] See the letter of September 1879 from Keleti to members of the Permanent Commission reproduced in Nixon, *History of the International Statistical Institute*, 9–10.

[56] *Allgemeine Zeitung* (Vienna), 14 June 1885 quoted in Nixon, *History of the International Statistical Institute*, 9–10. For other brief secondary accounts of these events see Friedrich Zahn, *50 Année de L'Institut International de Statistique* (1934) and R. Latter, *The International Statistical Institute 1885–1985* (Voorburg, Netherlands, 1985), 2.

[57] Willcox, 'Development of International Statistics', 147.

[58] Mouat, 'History of the Statistical Society', 43.

[59] von Neumann-Spallart, 'Résumé of the Results of the International Statistical Congresses', 304.

[60] Lawrence Goldman, 'The Social Science Association and the Absence of Sociology in Nineteenth Century Britain', *Past & Present*, 114, Feb. 1987, 161–66.

in the Prussian Landtag in the late 1860s and a free-trader opposed to the state's protectionist policy from 1879, was forced out of his position in 1882 and 'the bond of trust between the government and the Prussian bureau of statistics was broken'.[61] It has been argued that the history of economic statistics in modern Germany 'reveals the limits of the supposedly 'strong' German state'.[62] This may be so after 1900. Nevertheless, over a period of two decades, the various attempts of Prussian and German governments, before and after national unification, to control liberalism internally and promote German interests and autonomy externally, served to undermine the collaboration of statisticians across Europe.

At this point the British came back into the picture: efforts were made from London in the early 1880s to revive international statistical collaboration, though on the British model. The Jubilee meeting of the Statistical Society of London, postponed from 1884 until 1885, drew statisticians and savants to London and provided the opportunity. The Professor of Statistics and Political Economy at the University of Vienna, F. X. von Neumann-Spallart, was invited to take the lead. He summarized the history of the ISC before the assembly, and set out a plan for a new International Statistical Association, which was promptly formed in London on 24 June 1885, the last day of the Jubilee meeting.[63] That it immediately took the name of the International Statistical *Institute* is not without significance, for this was to be a professional organization for specialists and not a general association for the crowd. It was also to be 'a free association, divested of any official character, but which would endeavour to establish a basis for the uniformity of official statistics', and it was to exert influence not through its bureaucratic connections and status but 'owing to the great personal influence of the members of which it would be composed, and to their valuable labours'.[64] It would operate not from within bureaucracies but 'by *inviting* [my italics] the attention of Governments to the various problems capable of solution by statistical observation'[65] According to then president of the Statistical Society of London, Rawson W. Rawson, the Institute was

purely a private and scientific body. It is altogether different from the International Statistical Congresses or their successor, the Permanent International Commission.

[61] Alain Desrosières, *The Politics of Large Numbers. A History of Statistical Reasoning* (1993) (Cambridge, MA, and London, 1998), 183-5. On Engel's biography, see Ian Hacking, 'Prussian Numbers 1860-1882', in Lorenz Krüger, Lorraine J. Daston, and Michael Heidelberger (eds), *The Probabilistic Revolution* (2 vols), vol. 1, *Ideas in History* (Cambridge, Mass., 1987), 379-81.
[62] J. Adam Tooze, *Statistics and the German State, 1900-1945. The Making of Modern Economic Knowledge* (Cambridge, 2001), 33.
[63] *Jubilee Volume of the Statistical Society*, 320-31.
[64] F. X. von Neumann-Spallart, 'Résumé of the Results of the International Statistical Congresses', 305.
[65] Article I, 'Rules and Regulations of the International Statistical Institute', *Jubilee Volume of the Statistical Society*, 322.

It has no official character. It seeks to exercise no official authority or influence, though it is not without hope that its labours will be useful in furnishing information and suggestions to the Governments, as well as to the statisticians of the world.[66]

Limited to 150 members only, these would contribute in their own right and not as representatives or functionaries. The International Statistical Institute duly held its first meeting in 1886 with von Neumann-Spallart as its president.[67] Until 1914 the ISI retained the original focus of the ISC on official statistics and their international comparability, eschewing the study of the new mathematical statistics, but from a position independent of governments, as a learned society of freely associating professionals and specialists on the British model.[68]

This institutional history has several strands to it. We could pick out those strictly organizational reasons for the Congress's demise, most notably the transition between the third and fourth quarters of the nineteenth century in the characteristic form of cultural and intellectual association in Europe. In the 1870s and 1880s the large congresses that at mid-century had engaged the public alongside practitioners, collapsed, as in the case of the British Social Science Association;[69] or went into temporary decline, like the British Association for the Advancement of Science;[70] or were rededicated as specialist, professional forums as occurred to the Verein für Sozialpolitik, after its members divided in their response to Bismarck's introduction of tariffs on trade in 1879.[71] By the 1880s the map of knowledge was more complex and more specialized and was being filled-in by learned, single-discipline societies. Equally, there is a tension in this story between the effects on intellectual life of voluntarism on the one hand and bureaucracy on the other. The close connections between continental statisticians and their governments made British statisticians spin with envy at first: Leone Levi's communique to the Statistical Society of London about the initial meeting of the ISC in 1853 was a lament for the absence of such connections. But, in time, they came to appreciate the limitations placed on free enquiry and also on international collaboration itself, caused by dependence on the state. In this reading, the history

---

[66] *Bulletin de L'Institut International de Statistique*, Tome 1, 1886, 33, quoted in H. Campion, 'International Statistics', *JRSS*, series A, 1949, no. 2, 105–43, quotation at 109.

[67] F. X. von Neumann-Spallart, 'La fondation de l'institut international de statistique', *Bulletin de L'Institut International de Statistique*, 1, 1886, 1–32.

[68] For a similar view, see Schweber, *Disciplining Statistics*, 209–10. The ISI became a forum for mathematical statistics in the inter-war period.

[69] Lawrence Goldman, *Science, Reform and Politics. The Social Science Association 1857–1886* (Cambridge, 2002), 349–67.

[70] Roy MacLeod, 'Introduction. On the Advancement of Science', and Philip Lowe, 'The British Association and the Provincial Public' in Roy MacLeod and Peter Collins (eds), *The Parliament of Science. The British Association for the Advancement of Science 1831–1981* (London, 1981), 31–2 and 131–35.

[71] Goldman, 'The Social Science Association and the Absence of Sociology', 161–6.

of the ISC proved William Newmarch right, and offers an example of the advantages of scholarly independence.

The importance of long-run national bureaucratic and academic traditions in the history of statistics has been noted by Ian Hacking and is evident in the story of the International Statistical Congress. The development of government statistics via the creation of ad hoc boards and committees by the British state 'reflected a resistance to centralized management'. In the British tradition a central bureau would have been an 'anomaly'. But the Prussian tradition was different. In Berlin was created 'an office of numbers-in-general' to 'guide policy and inform opinion', though there was no Prussian equivalent of the attempt by Quetelet, Babbage, and Farr to construct social laws and a social science based on statistical regularities. Thus 'every country was statistical in its own way' and there were many 'parallel developments', but not intersecting ones, across Europe. 'National conceptions of statistical data varied.'[72] For that very reason the ISC was compromised: the very thing it set out to abolish ultimately brought it down.

There were irreducible tensions between nationalism and transnationalism, as well. We may note how the internationalism of the ISC in the 1850s and 1860s, which was so characteristic of this era, and which promoted the idea of swapping 'best practice' among the nations of Europe, was followed by increasing national tensions between member delegations in the 1870s as the European mood and international politics changed for the worse. The tensions across Europe caused by the creation of first Italy and then Germany, and the enduring national rivalries thereafter, could not be kept out of this pioneer international organization. We may also note how, in this story of nations collaborating and then dividing, each played its expected part. The Swiss stood upon longstanding national independence and autonomy. German nationalism refused to be bound by any agreements that might fetter national action, and German interests placed loyalty to government above professional and intellectual solidarity. The British, always ambivalent about the role played by European states in the Congress, championed political and intellectual independence and offered a different model for the organization of knowledge when the opportunity arose in the mid-1880s. Transnationalism had flourished briefly in propitious international circumstances in the Congress's early years in the 1850s and 1860s when it was enough merely to set out many diverse plans of action for the future. But this spirit was fragile and broke easily when, in the understandable desire to actually achieve something, the Congress trenched upon national privileges and independence in the 1870s. The International Statistical Congress has been lauded as an admirable example of international social collaboration; in truth, it may be more interesting as an example of why such collaborations frequently failed, and still fail.

---

[72] Hacking, *The Taming of Chance*, 28–35.

In its demise was written the demise of liberalism across Europe, brought low by institutional tensions certainly, but also undermined by the rise of conservative nationalism across the continent in general. As a political outlook and ideology, late-nineteenth century conservatism was intrinsically suspicious of internationalism per se. It was also resistant to the social ideas of liberal professionals among whom were numbered the statisticians of this era, right across Europe, from London to St Petersburg. The decline of the statistical movement was, in this perspective, a component of a much wider political and intellectual defeat for liberalism, environmentalism, and internationalism from the 1870s onwards.[73] It was also caused by internal intellectual changes within the discipline of statistics itself, and to these we must now turn.

---

[73] Lawrence Goldman, 'Conservative Political Thought from the Revolutions of 1848 until the Fin de Siècle', in Gareth Stedman Jones and Gregory Claeys (eds), *Cambridge History of Political Thought vol. iv: The Nineteenth Century* (Cambridge University Press, 2011), 691–719.

14

The End of the Statistical Movement

Francis Galton, Variation, and Eugenics

From the 1830s to the 1870s the Statistical Movement was broadly defined by its political liberalism, its commitment to social reform, and its environmental approach to social problems and their solution. It was focused on communities—on the patterns that might be discerned from the analysis of statistics at a societal level—rather than on the behaviour of individuals. It was inductive in its procedures, collecting data from the world as it was, and then generalizing from actual experience as discerned in the numbers, rather than a deductive discipline based on a priori assumptions. For this reason so many of its advocates were critical of the dismal science of political economy. From the outset it was believed that statistics would uncover 'facts relating to communities of men which promise when sufficiently multiplied to indicate general laws'.[1] According to Quetelet, numbers would liberate, 'showing the possibility of ameliorating the human race, by modifying their institutions, their habits, the extent of their knowledge, and generally, all which influences their mode of existence'.[2]

This resume is required because the movement's aims and assumptions were to be contested in the later decades of the nineteenth century in the work of the most innovative and significant statistician of the nineteenth century, Francis Galton. From the 1870s he revolutionized the role that statistics could play in social research—indeed, in every academic discipline and practical application. While extending the range and capability of statistical analysis, which was transformed into an immensely powerful tool for understanding the natural and social worlds, Galton associated the discipline with quite contrary values. From the 1880s, statistics were frequently used to divide communities and groups within them; to emphasize difference and variation between individuals, rather than to uncover features held in common; and to build a case that favoured 'nature over nurture', a term coined and popularized by Galton himself,[3] in stressing the importance of inheritance rather than the impact of the environment in the

---

[1] Richard Jones quoted in 'Transcript from the notebook of Mr J. E. Drinkwater, 28 June 1833, Cambridge', f. 4, Archives of the Royal Statistical Society, London.

[2] L. A. J. Quetelet, *Sur L'homme et le Développement de ses Facultés: Physique Sociale* (Brussels, 1835), *A Treatise on Man and the Development of his Faculties* (Tr. R. Knox), (Edinburgh, 1842) (1969 edn, Gainesville, Florida), 6.

[3] Francis Galton, *English Men of Science. Their Nature and Nurture* (London, 1869), 9.

Victorians and Numbers: Statistics and Society in Nineteenth Century Britain. Lawrence Goldman, Oxford University Press.
© Lawrence Goldman 2022. DOI: 10.1093/oso/9780192847744.003.0014

understanding of human behaviour. Making a case for political intervention *against* the poor and underprivileged, whose afflictions were blamed on their own inadequacies rather than seen as the consequence of the conditions in which they lived, became the norm for Galton and many of his followers. There is no more profound and historically-significant change of intellectual approach and social outlook in the whole of the nineteenth century, in Britain and elsewhere, and statistics were at its heart. To what extent the transformation in statistics was the cause or effect of this wider change of social philosophy is a complex intellectual problem in itself.

14.1 Galton, Darwinism, and the Revolution in Statistics

Francis Galton was a true 'gentleman of science', the friend and associate of other leading savants like Darwin, Wallace, Hooker, Tyndall, Huxley, and Herbert Spencer.[4] He had enormous impact on a later generation of mathematicians and natural scientists. His protégé, scientific follower, co-worker, and biographer, Karl Pearson, paid tribute to 'what Galton's work meant for some of us, in the 'eighties, when fresh from Cambridge we encountered his papers'.[5] Galton was born in Birmingham in a house on the site of the former home of Joseph Priestley, the political radical and discoverer of oxygen. The Galtons were originally Quakers, but by the time of Francis's birth had crossed over to the established church. Their wealth came from banking and also from the manufacture of firearms. Both his grandfathers, Samuel John Galton and Erasmus Darwin, were founder members of the famous Lunar Society of Birmingham, the monthly gathering of provincial businessmen, technologists, and savants to discuss the science and technology of the Enlightenment, and the embodiment of the progressive, learned, and enquiring entrepreneurial culture which made the Industrial Revolution.[6] This lineage made Francis a cousin to Charles Darwin. Like Charles, Francis was sent away to study medicine, and like his cousin he threw it up and went instead to Cambridge where he studied mathematics at Trinity College.[7] During his years there William Whewell, one of the founders of the Cambridge network, was appointed Master. Galton's closest undergraduate friend, Harry Hallam, was the son of the historian

[4] Ruth Schwartz Cowan, 'Francis Galton 1822–1911', *ODNB*; Nicholas Wright Gillham, *A Life of Sir Francis Galton. From African Exploration to the Birth of Eugenics* (Oxford, 2001).

[5] Karl Pearson, *The Life, Letters and Labours of Francis Galton* (4 vols, Cambridge, 1914–30); vol. II, *Researches of Middle Life*, 358.

[6] Robert E. Schofield, *The Lunar Society of Birmingham. A Social History of Provincial Science and Industry in Eighteenth Century England* (Oxford, 1963). Jenny Uglow, *The Lunar Men: Five Friends whose Curiosity Changed the World* (New York, 2002).

[7] It is not the case that 'Galton had no mathematical training' as stated by Desrosières. Nor did he study medicine at Cambridge. See Alain Desrosières, *The Politics of Large Numbers. A History of Statistical Reasoning* (1993) (Cambridge, MA, and London, 1998), 116, 128.

Henry Hallam, one of the founders of the Statistical Society of London. Brilliant though he was, a nervous collapse led Galton to take a pass degree only, without honours. With a private fortune behind him he turned to exploration and geography. At various times he travelled through eastern Europe to Constantinople; to Egypt and the Sudan; to Beirut, Damascus and the Jordan valley; and on behalf of the Royal Geographical Society he led a celebrated expedition to south-west Africa, now Namibia, between 1850–2, mapping it and establishing a national reputation as an explorer. His account of the last of these expeditions, published in 1853 as *The Narrative of an Explorer in Tropical South Africa*, was a great success. It displayed the clear, casual, and accepted racism of the age in a tale of exploits related in anecdotal and laconic style. Galton became a well-known figure in London clubs and societies in the 1850s.

Galton's widely-admired passion in his youth for collecting data on longitude, latitude, altitude, temperature, and so forth—for 'quantitative geography' executed in the field—bears close resemblance to the Humboldtian science that inspired the founders of the statistical movement. His contribution to an 1854 article entitled 'Hints to Travellers' in the *Journal of the Royal Geographic Society*, which included a description of the equipment that the serious traveller required in forest, bush, or swamp, was Humboldtian in style.[8] His first scientific papers, published in the 1860s, were on that classically Humboldtian subject of meteorology, and Galton has the distinction of naming the 'anticyclone' and designing the first public weather map, published in *The Times* on 1 April 1875.[9] He could also be moved to the kind of rapture over numbers that inspired figures like William Farr, Florence Nightingale, and Ada Lovelace:

> Some people hate the very name of statistics, but I find them full of beauty and interest. Whenever they are not brutalised, but delicately handled by the higher methods, and are warily interpreted, their power of dealing with complicated phenomena is extraordinary. They are the only tools by which an opening can be cut through the formidable thicket of difficulties that bars the path of those who pursue the Science of Man.[10]

Galton 'frequently consulted' Farr, whose advice he valued highly and who, he noted 'keenly appreciated what might be called the poetical side of statistics, as shown by his Annual Reports and other publications'.[11] Galton evinced constant

---

[8] Francis Galton, 'Letter addressed by Francis Galton, Esq., to the Secretary', in Henry Raper and Robert Fitzroy, 'Hints to Travellers', *Journal of the Royal Geographic Society*, vol. 24, 1854, 328–58. (Galton's Letter was published at pages 345–58.)

[9] Martin Brookes, *Extreme Measures. The Dark Visions and Bright Ideas of Francis Galton* (London, 2004), 133.

[10] Francis Galton, *Natural Inheritance* (1889), 62–3.

[11] Francis Galton, *Memories of My Life* (London, 1908) (3rd. edn, 1909), 292.

enthusiasm for quantification: 'Until the phenomena of any branch of knowledge have been submitted to measurement and number it cannot assume the status and dignity of science.'[12] He was a man obsessed by measurement, in fact, well beyond the traditions of anthropometry that had launched the statistical movement. According to his memorial tablet, 'the dominant idea of his life's work was to measure the influence of heredity on the mental and physical attributes of mankind.'[13] He would count the number of pretty women he passed in the streets of the towns and cities he visited. He even counted the number of people fidgeting while listening to him lecture.[14]

Galton's harmless eccentricity could become serious methodological and procedural error when applied indiscriminately, however. In a lecture at the Royal Institution in 1874 on the 'Nature and Nurture of Men of Science' which *The Spectator* found generally unremarkable and simplistic, Galton suggested that biographers should be encouraged to rank 'a man among his contemporaries, in respect to every quality that is discussed, and to give ample data in justification of the rank assigned to him'. In this way, thought Galton, 'social and political science would be greatly raised in precision'. *The Spectator* demurred, accusing him, in trying to count the qualities of writers and statesmen, and in placing them in rank order of ability, of using 'the physico-scientific method' inappropriately. There could be no such 'statistical scale of mental and moral qualities' it countered; it was impossible to assign places in order of greatness to figures like John Stuart Mill, Matthew Arnold and H. T. Buckle, or to Gladstone, Trollope, or Browning. Galton's lecture had 'plunge[d] literary men into wonder and awe at the clumsiness, uselessness, inefficacy, and complete inapplicability to the subject of the tests and gauges suggested'. It did not help Galton that in his lecture he had compared methods for judging the qualities of great men with sorting out the properties of nuts. Hence the title of the critical article: 'Mr Galton on Nuts and Men'. *The Spectator* concluded that his attempt to add science to the study of biography 'would spoil literature, without enriching science'.[15] By implication it found him naïve, unworldly, hopelessly out of his depth, the Mr. Gradgrind of literature and the arts. Galton replied, arguing that judgements as between people and between things are required in any number of everyday situations. But the magazine would not back down and its editor, the notable journalist, critic, and anti-materialist, Richard Holt Hutton, who criticized Darwin, Huxley, Spencer, and Mill in similar terms, wanted the last word:

We can imagine no more profitless or idle task than the attempt to draw out a Statistical Scale (say) of Candour or of Power of Repartee, and to arrange the

---

[12] Ibid., 334.
[13] The memorial is in the church of St. Michael and All Angels, Claverdon, Warwickshire. Galton, along with his parents, is buried in the churchyard.
[14] Peter L Bernstein, *Against the Gods. The Remarkable Story of Risk* (New York, 1996), 152.
[15] 'Mr Galton on Nuts and Men', *The Spectator*, 16 May 1874, 623–4.

public men of this generation in it, except indeed doing the same thing for a considerable number of qualities, and giving the reasons for the places assigned in the biographies, which would be rendered unreadable in the process.[16]

The exchanges are evidence that Galton's ideas met, initially, with scepticism; that the mid-Victorian sensibility was nothing like as receptive to hereditarian arguments as the next age was to be; that many contemporaries considered Galton's methods and arguments unconvincing; that Galton himself lacked judgement, a quality he hoped he could replace with rankings and numbers; and that he was 'intellectually isolated until quite late'.[17] Galton's lecture and his defence also betray an enduring and highly controversial aspect of the whole eugenics movement: that non-physical human characteristics, including intelligence and overall moral worth, could be measured and quantified.

Galton's intellectual awakening was the consequence of the publication of the *Origin of Species* in 1859. This

made a marked epoch in my own mental development, as it did in that of human thought generally. Its effect was to demolish a multitude of dogmatic barriers by a single stroke, and to arouse a spirit of rebellion against all ancient authorities whose positive and unauthenticated statements were contradicted by modern science.[18]

He 'devoured its contents and assimilated them as fast as they were devoured'. In 1886 in a speech at the Royal Society, Galton was more precise and also more effusive about the impact of his cousin on him:

Few can have been more profoundly influenced than I was by his publications. They enlarged the horizon of my ideas. I drew from them the breath of a fuller scientific life, and I owe more of my later scientific impulses to the influence of Charles Darwin than I can easily express. I rarely approached his genial presence without an almost overwhelming sense of devotion and reverence, and I valued his encouragement and approbation more, perhaps, than that of the whole world besides.[19]

At the very end of his life, he paid public tribute to Darwin once more, and also to Alfred Russel Wallace, the other discoverer of the theory of natural selection, this time for the freedom of mind that natural selection encouraged, because it liberated Galton and others from inherited paradigms like the argument from design.[20]

---

[16] 'Nuts and Men' in ibid., 30 May 1874, 689. Harold Orel, 'Richard Holt Hutton 1826–1897', *ODNB*.
[17] Donald A. Mackenzie, *Statistics in Britain 1865–1930: The Social Construction of Scientific Knowledge* (Edinburgh, 1981), 9.
[18] Galton, *Memories of My Life*, 287.
[19] *The Times*, 1 Dec. 1886, quoted in Pearson, *Life… of Francis Galton*, II, 201.
[20] *The Darwin-Wallace Celebration… by the Linnean Society of London* (1908), 25–6, quoted in ibid., II, 201.

The *Origin of Species* encouraged Galton to investigate subjects which 'clustered around the central topics of Heredity and the possible improvement of the Human Race'.[21] Darwin had written:

We cannot suppose that all breeds were suddenly produced as perfect and as useful as we now see them; indeed in many cases, we know that this has not been their history. The key is man's power of accumulative selection: nature gives successive variations; man adds them up in certain directions useful to him.[22]

Galton was to argue that human intelligence and abilities were no different from traits in animals and plants that had been bred selectively: in consequence these human characteristics should be selected themselves. According to Pearson his 'creed' thus became 'the study of the heredity of the mental and moral characters as a basis for Race improvement'.[23] In this view, improvements to the environment were less effective as means to improve 'the race' than improvements engendered by selective breeding to accentuate the best characteristics while preventing the inheritance of the worst. Indeed, environmental reforms, by easing the lives of those who were physically or mentally unfit, might do more harm than good by encouraging procreation by the so-called sub-normal and defective. The 'creed' was set out first in an article published in *Macmillan's Magazine* in the summer of 1865, 'Hereditary Talent and Character'.[24] It was taken much further in four major works: *Hereditary Genius* (1869), *English Men of Science, their Nature and Nurture* (1874), *Inquiries into Human Faculty and its Development* (1883) and *Natural Inheritance* (1889). Galton pursued the theme of the inheritance of mental and moral qualities quantitatively in all these works, and this led him to make startling innovations in statistical analysis.

In his quest to improve the race, Galton was the originator of 'eugenics', and coined the term himself in the 1883 *Inquiries into Human Faculty*:

We greatly want a brief word to express the science of improving stock, which is by no means confined to questions of judicious mating, but which, especially in the case of man, takes cognisance of all influences that tend in however remote a degree to give the more suitable races or strains of blood a better chance of prevailing speedily over the less suitable than they would otherwise have had.

[21] Galton, *Memories of My Life*, 288.
[22] Charles Darwin, *The Origin of Species By Means of Natural Selection, Or the Preservation of Favoured Races in the Struggle for Life* (London, 1859), 30.
[23] Pearson, *Life…of Francis Galton*, II, 87.
[24] F. Galton, 'Hereditary Talent and Character', *Macmillan's Magazine*, June and August 1865, vol. xii, 157–66, 318–27.

Taken from *eugenes*, the Greek for 'good in stock, hereditarily endowed with noble qualities' and 'equally applicable to men, brutes, and plants', *eugenics* as named by Galton, covered 'both the science and the practice of improving human stock'.[25] It was espoused in different forms, and with different degrees of commitment, by some natural and social scientists, statisticians, sundry social movements and their publications, political organizations of the left as well as of the right, and by some of their leaders, for the next half century.[26] The use of eugenic expressions and rhetoric was widespread by the early twentieth century, and it is contended that the professional and intellectual classes in Britain, in both the mid and late-Victorian periods, very largely subscribed to these ideas, and that Galton was a representative figure, therefore.[27] These contentions are questionable, however. Interest in eugenics was unenthusiastic and slow to develop in the more liberal mid-Victorian decades of the 1860s and 1870s when Galton first developed his core concepts; it was only later in his life that such ideas became a part—and by no means the whole—of the cultural mainstream. It was only in Galton's last decade—the Edwardian era—that he found himself feted by societies and organizations committed to 'the improvement of the race', for these years marked the peak of the eugenics movement.[28] The Eugenics Education Society, the most notable propagandist, was only founded in 1907. As Jose Harris has argued moreover, though the language of social and racial 'degeneration' was common in this period and inscribed in many novels and tracts, few of those who worked with the poor, or made policy for them, believed wholeheartedly in the concept.[29] To argue, furthermore, that Galton represented the views of his class is to ignore the very many thinkers and institutions from the 1860s onwards who were relatively untouched by eugenics. It imposes an ideological consensus and intellectual homogeneity on the British elite which it never possessed. The traditional influences of Christianity and humanitarianism, the prudential calculation of the benefits of social solidarity, the power exercised through the ballot box by an expanding electorate after the Reform Acts of 1867 and 1884, and new thinking about the nature of society derived from philosophical idealism—the tradition of T. H. Green and his followers emanating from Oxford from the 1870s—were more potent forces among the liberal elite than eugenics.[30]

---

[25] F. Galton, *Inquiries Into Human Faculty and Its Development* (London, 1883), 24–5.

[26] Greta Jones, *Social Darwinism and English Thought. The Interaction between Biological and Social Theory* (Brighton, Sussex, 1980).

[27] Simon Szreter, *Fertility, Class and Gender in Britain 1860–1940* (Cambridge, 1996), 148–81. See especially 159, 165, and 180. Mackenzie, *Statistics in Britain*, 32–3, 221.

[28] Ibid., 235.

[29] Daniel Pick, *Faces of Degeneration. A European Disorder c. 1848—c. 1918* (Cambridge, 1989); Jose Harris, *Private Lives, Public Spirit: Britain 1870–1914* (Harmondsworth, 1993), 241–5.

[30] On the idealist tradition, see Melvin Richter, *The Politics of Conscience: T. H. Green and His Age* (London, 1964).

The statistical movement had sought the improvement of the environment as a precondition for human betterment; Galton and his followers, many of them more brilliant statisticians than any who had gone before, sought more directly the improvement of humans. In that quest, the role played by statistics in proving the importance of inheritance was crucial. At several points and places Galton explained that his statistical innovations followed shifts in his thinking and were required as ways of proving these new paradigms and beliefs. Found among Galton's papers after his death was this 'manuscript note':

> About the time of the appearance of Darwin's *Origin of Species* I had begun to interest myself in the Human side of Geography, and was in a way prepared to appreciate his view. I am sure I assimilated it with far more readiness than most people, absorbing it almost at once, and my afterthoughts were permanently tinged by it. Some ideas I had about Human Heredity were set fermenting and I wrote *Hereditary Genius*. In working this out I forced myself to become familiar with the higher branches of Statistics, and, conscious of the power they gave in dealing with populations as a whole, I availed myself of them largely.[31]

Darwinian biology was focused on the crucial role played by variation within populations. This necessitated a new form of mathematical statistics because, in the preceding era of Quetelet, statistics were characterized by the study of whole populations and were focused on the construction of means. Galton gently mocked his predecessors, therefore:

> It is difficult to understand why statisticians commonly limit their inquiries to averages and do not revel in more comprehensive views. Their souls seem as dull to the charm of variety as that of the native of one of our flat English counties, whose retrospect of Switzerland was that, if its mountains could be thrown into its lakes, two nuisances would be got rid of at once. An average is but a solitary fact, whereas if a single other fact be added to it, an entire Normal Scheme, which nearly corresponds to the observed one, starts potentially into existence.[32]

In the former age of natural theology and Humboldtianism, natural philosophers expected to find unity, order and perfection in nature, as we have seen. Now, in the age of Darwin, if they were to find evidence of natural selection at work and of species' formation, they had to be able to demonstrate, explain and interrogate

---

[31] 'Manuscript Note of Francis Galton in the handwriting of Mrs. Galton found among his papers', Pearson, *Life... of Francis Galton*, II, 70.
[32] Galton, *Natural Inheritance*, 62–3.

'difference', and 'it was difference, not homogeneity, that Galton was pursuing'.[33] Variations were not aberrations; to eugenicists, 'human variability was the potential source of racial progress'.[34] Statisticians who preceded Galton were looking for the average and expected to use Gaussian error theory to fit biological phenomena into normal distributions. But as Galton explained in his autobiography,

> The primary objects of the Gaussian Law of Error were exactly opposed, in one sense, to those to which I applied them. They were to get rid of, or to provide a just allowance for errors. But those errors or deviations were the very things I wanted to preserve and to know about.[35]

As Eileen Magnello has explained,

> Biological Darwinism precipitated a paradigmatic transition in the measurement and analysis of statistical data when individual *variation* became the principal unit of statistical investigations. The shift in the locus of statistical units of measurement from averages to variation brought about a new way of analysing and managing statistical data on a larger scale than previously possible.[36]

'The measurement and analysis of statistical variation' became 'the lynch-pin of modern mathematical statistics'.[37] According to Pearson in his first editorial for the new journal that carried forward this intellectual revolution in statistics, *Biometrika*, 'every idea of Darwin – variation, natural selection, sexual selection, inheritance, prepotency, reversion – seems at once to fit itself to mathematical definition and to demand statistical analysis'.[38] Darwinism 'shifted the direction of statistics', a shift begun by Galton and taken much further by Pearson and the Darwinian biologist, W. F. R. Weldon, who together founded *Biometrika* with a generous endowment provided by Galton himself.[39] The journal focused on the application of mathematics to problems of biological inheritance. Pearson believed, indeed, that the life sciences could be made mathematically exact, and his life's work was devoted to developing statistical techniques and ideas to try to effect this.[40] In the development of statistical biology Pearson did more than merely extend understanding or re-order knowledge: 'he took Galton's insights

---

[33] Bernstein, *Against the Gods*, 162.   [34] Mackenzie, *Statistics in Britain*, 58.
[35] Galton, *Memories of My Life*, 305.
[36] M. Eileen Magnello, 'Darwinian Variation and the Creation of Mathematical Statistics', in *Mathematics in Victorian Britain* (eds Raymond Flood, Adrian Rice, and Robin Wilson) (Oxford, 2011), 299.
[37] Ibid., 284.
[38] [Karl Pearson], 'Editorial (II.) The Spirit of *Biometrika*', *Biometrika*, 1, 1, 1 Oct. 1901, 4.
[39] Magnello, 'Darwinian Variation and the Creation of Mathematical Statistics', 280.
[40] Theodore M. Porter, *Karl Pearson: The Scientific Life in a Statistical Age* (Princeton, N.J., 2004).

and made out of them a new science'.[41] The biometric school founded by Pearson quantified biology, and did so in the ideological context of eugenics. Eileen Magnello is at pains to explain that however much Pearson and his project owed to the example of Galton, Pearson was much too original a thinker and practitioner to be accounted merely a follower of his idol and mentor.[42] Indeed, unlike Galton, Pearson believed in eugenics and planned reproduction as key aspects of a future technocratic *socialist* state where the intellectuals and scientists would supposedly govern rationally in the interests of all. But like Galton he was a consistent critic of what had gone before and he 'disparaged old-school statisticians... for employing a crude style of vital statistics that consisted simply of comparing percentages in order to make inferences about the causes of diseases and social problems'.[43] This was the *modus operandi* of William Farr, William Guy, and John Simon.

This transition, indeed revolution, in the nature of statistics was prompted by the genuine intellectual challenge to confirm Darwin's hypotheses. It had a profound, internal academic motive behind it, therefore, and cannot be explained merely as a response to changing modes of thought and social values, external to the changes within the discipline itself.[44] The academic focus on the study of 'difference' influenced social thought in turn, and contributed to that change in outlook and philosophy in the late-nineteenth century that emphasized the supposed differences between social groups and races. This change of outlook was already underway for other, independent reasons. It was a less generous age of rising nationalism and international competition on the one hand, and on the other, within states, of increasing struggles between social interests and classes for power, resources and cultural authority. In these struggles, Darwin's ideas were applied promiscuously and without authority to justify competition between groups, the subjugation of the 'different' and the inferior, and to uphold the prevailing order. For in the 'survival of the fittest', that crude term coined by Herbert Spencer, those at the top of the social pile were, by definition, those most fit for survival.

This classically circular argument was 'Social Darwinism', so-called, and it was often promoted in association with eugenics. Its devotees were rarely faithful to Darwin's ideas, however, because, as Darwin explained, natural selection was a random and erratic process in which physical prowess or moral worth were irrelevant. To most Social Darwinists, however, the supposed superior abilities of socially-dominant groups were usually presented as the reason and reward for their dominance, an idea quite contrary to the way that natural selection works. In the view of Social Darwinists, in the struggle between different races and

---

[41] Mackenzie, *Statistics in Britain*, 88.   [42] Ibid., 296–8.
[43] Joanne Woiack, 'Karl Pearson 1857–1936', *ODNB*.
[44] For an argument driven by externalism, see Mackenzie, *Statistics in Britain*, *passim*.

classes, at whatever level, whether global or local, the characteristics of the 'master race' should be conserved and passed on, while those from lower orders controlled and inhibited. Statisticians may, or may not, have approved of eugenics and Social Darwinism, but their discipline, called into being to investigate and explain natural variation, became intertwined with movements and ideologies that exploited difference and variation for political and geopolitical ends. Indeed, statistics may have provided spurious scientific justification for unjustifiable, purely ideological views and frameworks. All this was very different from Babbage's innate curiosity about numbers, the liberal optimism of Quetelet, and the environmentalism of the doctors Guy and Farr.

14.2 Francis Galton and Charles Booth: Social Differentiation and Eugenics

The intertwining of social statistics and eugenics, and the abuse of social research by those who would use it for their own ideological ends, is made evident in Galton's attempted assimilation of the findings of the social investigator, Charles Booth, in one of the clearest formulations of eugenic ideas, the second Huxley Lecture of the Anthropological Institute which Galton delivered in October 1901. Entitled 'The possible improvement of the human breed under the existing conditions of law and sentiment', Booth's empirical division of the population of London into eight socio-economic sub-classes from A at the bottom to H at the top, was co-opted and abused by Galton.[45]

Charles Booth, the Liverpool-born businessman, was by 1901 a celebrated public figure because of his research into the extent of poverty in London published as *The Life and Labour of the People of London*.[46] Though included in the 'aristocratic liberal meritocracy' by one historian and the 'intellectual aristocracy' by others, Booth actually demonstrates the heterogeneity, complexity and openness of the various elites in late-Victorian Britain.[47] There was nothing intellectual, aristocratic or even very liberal in his biography, in fact. The son of a Liverpool corn merchant who was brought up in the close world of the city's Unitarian community, Booth was educated locally and never went to a university. He certainly did not share the 'common educational formation' of other members

[45] The following discussion of Booth and Galton draws on Mackenzie, *Statistics in Britain 1865–1930*, 16–18, and Simon Szreter, 'The Genesis of the Registrar-General's Social Classification of Occupations', *British Journal of Sociology*, 35, 4, Dec. 1984, 525–7.

[46] Charles Booth, *The Life and Labour of the People of London* (17 vols in 3 series: Poverty, Industry, Religion) (London, 1889–1903). T. S. Simey and M. B. Simey, *Charles Booth. Social Scientist* (Oxford, 1960).

[47] Szreter, *Fertility, Class and Gender in Britain*, 114n, 178; Rosemary O'Day and David Englander, *Mr. Charles Booth's Inquiry. Life and Labour of the People of London Reconsidered* (London, 1993), 15–16.

of the supposed liberal elite, for the years when he might have been at one of the ancient universities, the early 1860s, were spent in New York trying to start a business against the background of the American Civil War.[48] He was known later as the owner of an Atlantic steamship line, which had a certain romance and social cachet, and he did marry the niece of the great Whig historian and statesman, Thomas Babington Macaulay. The foundation of his business, however, and the source of its recurrent profits, was actually something decidedly ungentlemanly, unmeritocratic, and plain dirty, the animal skins' trade: Booth's company transported pelts and hides across the Atlantic and manufactured leather from them in New York and New Jersey.[49] A confirmed Conservative in politics from his early middle age, and a distinct opponent of organized labour and trade unions, he nevertheless mixed in high professional circles and became President of the Royal Statistical Society in 1892.

Booth's great social investigation had begun in 1886 and would not finish until 1903, but by 1901 his key findings were already known and Booth himself, while finishing the survey, was already campaigning for the introduction of a state-funded old age pension to support the aged poor when they could no longer work.[50] Booth was an empiricist who collected a vast range of data, some statistical and some impressionistic, to build up a picture of wealth and income across London. The level of detail was such that it allowed for judgments to be made on the social class of the inhabitants of individual streets, and thus for the composition of so-called 'poverty maps' of the city.[51] Booth's initial surveys and results had been communicated to the Royal Statistical Society in essays in 1887 and 1888.[52] It was in the first of these, on the sociology of Tower Hamlets in London, that he set out the 8-class division of the city's population. His presidential address to the RSS was on 'Dock and Wharf Labour', and the London project as a whole was always closely associated with the Society.

Though Booth was a supporter of the Conservative Party from the late 1870s, as were most late-Victorian businessmen, he was not a eugenicist. Nevertheless, his great inquiry differentiated between social groups, and Booth, at his most controversial, saw one group in particular, 'Class B', as a problem to be solved by a form of removal. As Booth's research developed, he came to argue that this group, the casual poor of London who worked only intermittently for desperately low daily wages or piece rates, pulled down the remuneration of respectable workers above them in classes C and D, making many otherwise viable and stable

[48] Ibid., 178.
[49] A. H. John, *A Liverpool Merchant House. Being the History of Alfred Booth and Company 1863–1958* (London, 1959), ch. III, 'The Sheepskin Trade. 1865–1890', 38–51.
[50] Charles Booth, *Old Age Pensions and the Aged Poor: A Proposal* (London, 1899).
[51] *Charles Booth's Poverty Maps* (Mary S. Morgan et al.), (Thames & Hudson, London, 2019).
[52] Charles Booth, 'The Inhabitants of Tower Hamlets (School Board Division), their Condition and Occupations', *JRSS*, 50, 1887; 'The Condition and Occupations of the People of East London and Hackney, 1887', *JRSS*, 51, 1888.

working-class families insecure. Booth's solution were what he called 'labour colonies'—work camps in the countryside to which the casuals, by definition capable of work, should be moved. This would remove a source of cheap labour from the London labour market and give a needed uplift to the wages of the respectable working classes. Controversial at the time, labour colonies remain so to this day: Booth's otherwise sympathetic portrait of the poor and their struggles seems to be disfigured by a draconian solution redolent to later generations of the concentration camps of the Boer War, and of yet worse to come in Europe in the 1930s and 1940s.[53] In fact, Booth's ideas were widely welcomed at the time and not only by conservatives: in 1889 the Fabian socialist journal *Today* welcomed labour colonies as an example of the sort of social innovations desired by collectivists.[54]

Booth's sub-classes were defined in strictly social, occupational, and financial ways. What mattered in classifying a household was occupation, income, expenditure, lifestyle, and the impressions of the 'school board visitors', inspectors whose job it was to ensure that all children were sent to school, and whom Booth and his co-workers interviewed across the city. These were not eugenic distinctions based on intelligence or aptitude. Nevertheless, by establishing clear differences between social groups in the metropolis, Booth's inquiry gave Francis Galton an opportunity he could exploit, which he did in his 1901 lecture. Galton may be said to have overlain his own 10-grade segregation of the population according to mental ability on Booth's 8-classes model of London society based on income. In Galton's differentiation there were five groups below the mean of ability, v, u, t, s, r, and then five groups above the mean, R, S, T, U, and V. Galton wrote often of 'civic worth', the criterion by which to assess individuals and groups, though he never adequately defined it. The civic worth of men in classes U and V was 'notably superior to the crowd'. He estimated that one man in 300 belonged to his V-class. As he commented, 'the brains of our nation lie in the higher of our classes'.[55] Galton's ten-class distribution gave an inevitable normal distribution which he then compared to Booth's structure. He contended that at the top, Booth's class H corresponded to his own gradations T, U, and V; in the middle, groups r and R formed the 'mediocre class' of respectable workers in regular employment, equivalent to Booth's Class D. Meanwhile his lowest groups, v, u, and t, corresponded to Booth's class A, criminals and loafers, and class B, the casual poor. Galton defined his classes u and t, and hence Booth's class B, as 'very poor persons who subsist on casual earnings, many of whom are inevitably poor from shiftlessness,

---

[53] John Brown, 'Charles Booth and Labour Colonies 1889–1905', *Economic History Review*, 21, 2, Aug. 1968, 349–60; Trevor Lummis, 'Charles Booth: Moralist or Social Scientist?' and John Brown, 'Social Judgements and Social Policy', *Economic History Review*, 24, 1, 1971, 100–5; 106–13.

[54] Gareth Stedman Jones, *Outcast London. A Study in the Relationship between Classes in Victorian Society* (Oxford, 1971), 314.

[55] Francis Galton, 'The Possible Improvement of the Human Breed under the Existing Conditions of Law and Sentiment', in Galton, *Essays in Eugenics* (London, Eugenics Education Society, 1909), 1–34. See page 11.

idleness or drink'. To Galton, the three groups at the bottom of the ability distribution were without civic worth, what he termed 'undesirables'.[56] To buttress his case, Galton quoted Booth himself in a passage that reflects in its vocabulary of degeneration, and the heritability of anti-social characteristics, the influence of eugenic concepts:

> Their life is the life of savages...From them come the battered figures who slouch through the streets and play the beggar or bully. They render no useful service, they create no wealth; more often they destroy it...[it] is much to be hoped that this class may become less hereditary in its character; there appears to be no doubt that it is hereditary to a very considerable extent.[57]

Galton's lecture went on to make some facile points about coupling young, athletic and talented males with females of equivalent abilities, from good families, and healthy enough to breed profusely. The best of the race should breed the most and be rewarded financially and with public honours. It was crude and distasteful stuff and, for sure, Booth never wrote anything remotely like it. He had been captured and used by the eugenics movement for its own ends, in other words. Galton had assimilated social categories into natural categories. In this way he could argue that his differentiation of society on grounds of inherited mental ability was realized in the actual socio-economic structure of society, which thereby confirmed eugenics. He could also argue that the poor were destined to remain poor because they lacked the mental ability and character to escape their social position. He had turned a social structure, which by its very nature was contingent because reflective of social conditions—and social conditions could change—into an unchangeable natural hierarchy based on inherited characteristics and intelligence.

This was the crucial difference between the thinking of the statistical movement and that of the eugenics movement. The original statisticians looked for the environmental determinants of behaviour, because those might be changed and society thereby improved. And that was always Booth's aim: to assess the full extent of poverty in the city and then take measures to ameliorate it. But to the eugenicists, social categories, possibilities, and opportunities, were fixed by heredity and inherited capacity: people took up the positions and the conditions into which they were born and there were engrained divisions between groups on these principles. By the 1890s it was becoming natural to think in ways that differentiated between social groups, therefore. Both Galton and Booth were responding to the late-Victorian imperative to construct social divisions, enforce social boundaries and isolate the incapable and 'the residuum', a characteristic

---

[56] Ibid., 11.   [57] Ibid., 19.

late-Victorian term.[58] In the process, the new mathematical statistics could be employed to give spurious 'scientific' justification to actions and prejudices that were ideological or simply racist in origin.

14.3 Galton Against Section F: The New versus 'That Old Type of Statistics'

Galton made many crucial contributions to the revolution in statistics.[59] Not the least of them was the application of statistical methods to subjects that had never been thought amenable to ordered and exact scientific inquiry, such as biology, psychology, anthropology, and sociology, where causation is complex and uncertain, rather than direct, as in the physical sciences. He opened up new areas to research by developing new methods for researching. Lacking copious 'anthropological evidence' in the 1860s and 1870s, Galton's inquiries into inheritance led him to the study of something as innocent as the sweet pea. He wanted to know if plants that grew from large seeds went on to produce large seeds themselves. One of the outcomes of the study was his development of the concept of 'regression to the mean'—called by Galton 'regression to mediocrity'—which is the tendency of measured phenomena to revert, over time, to the average of the population being considered, and a key concept of statistical investigations across myriad fields.[60] The discovery of correlation, the measurable link between two sets of data, was another remarkable and transformative by-product of his research into inheritance, a consequence of Galton's need to compute the relationship between characteristics held by one generation and those held by another.

It was a slow and gradual intellectual process that took more than two decades, starting with Galton's first scientific publications in the early 1860s on meteorology 'in which he was clearly feeling his way towards the concepts of correlation and regression'.[61] His study published in 1863, *Meteorographia: Or, Methods of Mapping the Weather*, sought ways in which to synthesise the different types of data taken in weather readings to enable accurate meteorological prediction.[62] In 1870 Galton tried to predict the velocity of winds using measurements

[58] Jose Harris, 'Between Civic Virtue and Social Darwinism: The Concept of the Residuum', in David Englander and Rosemary O'Day (eds), *Retrieved Riches. Social Investigation in Britain 1840–1914* (Aldershot, Hants, 1995), 67–83.

[59] The best account of the historical process by which Galton led this revolution is found in Ruth Schwartz Cowan, 'Francis Galton's Statistical Ideas: The Influence of Eugenics', *Isis*, 63, 4 (Dec. 1972), 509–28.

[60] Francis Galton, 'Regression towards Mediocrity in Hereditary Stature', *Journal of the Anthropological Institute*, 15, 1886, 246–63.

[61] F. N. David, 'Galton, Francis', *Encyclopedia of the Social Sciences* (ed. David Sills) (New York, 1968), vol. 6, 49.

[62] Francis Galton, *Meteorographia: Or, Methods of Mapping the Weather* (London, 1863).

of temperature, humidity and barometric height: 'he was searching for what statisticians now refer to as a multiple regression formula'.[63] His research into hereditary characteristics and traits which could not be measured but which could be ranked in order, such as his suggested rank order of men of distinction in 1874, led him onwards to a method he called 'intercomparison' in 1875.[64] He had first used the term 'correlation' in its statistical sense in *Hereditary Genius* in 1869, and the term 'regression' appeared in 1877 in his paper to the Royal Institution on the inheritance of characteristics in sweet peas entitled 'Typical Laws of Heredity'.[65] This essay showed further technical advances in both fields, but it was not until the 1880s that these methods of analysis were refined and established.

Galton had long been conscious of the 'pressing necessity of obtaining a multitude of exact measurements relating to every measurable faculty of body and mind, for two generations at least, on which to theorise'.[66] He was a very early believer in the idea of the physical degeneration of the race caused by urban living and by the reduction of family size among the intelligent and educated middle classes.[67] He had tried various ways of collecting the human data he required to prove this theory, for example, requesting from headmasters the vital statistics of their pupils for his analysis, and by building up measurements taken from more than 300 families. The stimulus for the 'correlational calculus' was his further collection of more than 9000 sets of vital data—measurements of height, weight, girth, length, and breadth etc.—from volunteers, parents and children, who visited his 'anthropometric laboratory' set up in a corner of the International Health Exhibition in London in 1884.[68] Here 'a man may…get himself and his children weighed, measured and rightly photographed, and have each of their bodily faculties tested, by the best methods known to modern science'.[69] There were more than 150,000 measurements recorded from over 9000 people, and a decade's worth of work to be unravelled and analysed, therefore. As Pearson explained subsequently, 'the need for novel statistical methods, which its problems demanded, led [Galton] to the correlational calculus, the *fons et origo* of that far-reaching ramification—the modern mathematical theory of statistics'.[70] To be precise, the measurement of correlation was established by Galton in the course

[63] Cowan, 'Francis Galton's Statistical Ideas', 513. See Francis Galton, 'Barometric Prediction of Weather', *British Association Reports*, 1870, 31–3.

[64] Francis Galton, 'Statistics by Intercomparison, with Remarks on the Law of Frequency of Error', *London, Edinburgh and Dublin Philosophical Magazine and Journal of Science*, 49, 1875, 33–46.

[65] Francis Galton, 'Typical Laws of Heredity', *Proceedings of the Royal Institution*, 1877, vol. 8, 282–301. Mackenzie, *Statistics in Britain*, 60–3.

[66] Galton, *Memories of My Life*, 244.

[67] Francis Galton, 'Relative Supplies from Town and Country Families to the Population of Future Generations', *JSSL*, 26, March 1873, 19–26.

[68] Pearson, *Life…of Francis Galton*, II, 357.

[69] Galton imagined and described such a laboratory two years before he was enabled to construct it. See 'The Anthropometric Laboratory', *Fortnightly Review*, 31, 1882, 332–8.

[70] Pearson, *Life…of Francis Galton*, II, 357.

of analysing the heights of parents and their offspring, with some help from the Cambridge mathematician J. D. Hamilton Dickson. Hamilton Dickson's contribution was published as an appendix to Galton's paper 'Family Likeness in Stature', delivered to the Royal Society in 1886.[71] Two years later Galton published the seminal essay 'Co-relations and Their Measurement, Chiefly from Anthropometric Data'.[72] Galton had by now come to realize that correlation was not limited to the study of heredity: correlation could be applied to the analysis of the relationship between any two sets of variates, even in circumstances where they were measured in different ways.[73]

According to Pearson, 'Galton's very modest paper of ten pages has led to a revolution in our scientific ideas'.[74] Correlation enabled Pearson's generation, for the first time, 'to submit phenomena to measurement and number – in many branches of inquiry where opinion only had hitherto held sway. It relieved us from the old superstition that where causal relationships could not be traced, there exact or mathematical inquiry was impossible.' It opened up to quantitative analysis 'wide fields of medical, psychological and sociological research'.[75] Continuing the theme of a scientific revolution, one modern historian of science has described the invention of correlation as 'a new view of the universe, both organic and inorganic, which provides all branches of science with a *novum organum*'.[76] Correlation was further developed by Pearson himself. He invented the chi-squared test in 1900 to assess the 'goodness of fit' between two sets of variables, and to calculate the strength of their relationship, whether positive or negative.[77] Correlation, regression, the chi-squared test, and multivariate analysis in general 'have become pillars of modern statistics'.[78] If the artisan statisticians of the London Statistical Society in the 1820s had been able to use tools like these, they might have proved the relationships they were trying to demonstrate between increases in unemployment, pauperism, and crime, and between all three of these variables and falling wages.[79]

---

[71] Francis Galton, 'Family Likeness in Stature', *Proceedings of the Royal Society*, 1886, vol. 40, 42–73. Hamilton Dickson contributed an appendix at 63–73 in this essay.

[72] Francis Galton, 'Co-relations and their Measurement, chiefly from Anthropometric Data, *Proceedings of the Royal Society*, 1889, vol. 45, 135–45.

[73] Cowan, 'Francis Galton's Statistical Ideas', 520; Theodore M. Porter, *The Rise of Statistical Thinking 1820–1900* (Princeton, N. J., 1986), 8.

[74] Pearson, *Life ... of Francis Galton*, IIIa, 56–7.

[75] Pearson, *Life ... of Francis Galton*, II, 357–8.

[76] Victor L. Hilts, 'Statistics and Social Science', in Ronald N. Giere and Richard S. Westfall (eds), *Foundations of Scientific Method. The Nineteenth Century* (Bloomington, Ind., 1973), 207.

[77] Karl Pearson, 'On the criterion that a given system of deviations from the probable in the case of a correlated system of variables is such that it can be reasonably supposed to have arisen from random sampling', in Karl Pearson, *Karl Pearson's Early Statistical Papers* (Cambridge, 1948), 339–57. First published in the *London, Edinburgh and Dublin Philosophical Magazine and Journal of Science*, 5th series, vol. 50, issue 302, 1900, 157–75.

[78] Desrosières, *The Politics of Large Numbers*, 103.

[79] See above, pp. 91–2.

Galton's remarkable insights and innovations moved statistics onto another plane, far beyond the reach of most practitioners. But his achievements were slow to be recognized and this led to his evident frustration with the discipline as it had been up to then, and also with institutions that had yet to understand and register the changes he was making and the enormous analytical potential he was unlocking. It caused him to take public issue with one of the institutional havens for old-style statistics, Section F of the British Association for the Advancement of Science, which had been founded, as we have seen, in Cambridge in 1833 in the presence of Adolphe Quetelet. Now entitled 'Economic Science and Statistics', Section F provided a platform for the mundane discussion of descriptive social statistics, and when that was not the case, for straightforward social and political argument. As Sedgwick and others had feared in 1833, there was not much science to it and plenty of the 'foul Daemon of discord' instead. These reservations attached to Section F throughout its history, and at times to the British Association as a whole.[80] In general, because journalists lacked the scientific or technical background to report on the work of most of the sections, they tended to pay attention to Section F, with the result that the Association as a whole was seen through its unrepresentative subject matter and inexpert discussions. Dickens, who had dubbed the BAAS 'The Mudfog Association for the Advancement of Everything' was only one of many critics, and he was rather more playful than most.[81] As the Scottish physicist, Sir David Brewster, wrote to his friend Lord Brougham in 1861:

> The proceedings of the Association at Manchester were regarded by many as an encroachment on the Social Science Association, and it is a general feeling among scientific members that at the meetings at large manufacturing and commercial towns, the matters of science are overborne by those of a political and social character.[82]

In 1876, the Council of the British Association established a review of its annual meetings which came to focus above all on the inappropriate discussions in Section F. It then received a formal request to abolish Section F which was, and is, assumed to have emanated from Francis Galton.[83]

It appointed a committee, therefore, to report on 'the possibility of excluding unscientific or otherwise unsuitable Papers and Discussions from the Sectional

---

[80] A. D. Orange, 'The Idols of the Theatre: The British Association and its Early Critics', *Annals of Science*, 32, 1975, 277–94.

[81] Charles Dickens, 'Full Report of the First Meeting of the Mudfog Association for the Advancement of Everything', *Bentley's Miscellany*, 2 (1837), 397–413.

[82] Sir David Brewster to Henry Peter, Lord Brougham, 14 Sept. 1861, Brougham Papers, University College, London, B.Mss. 26763.

[83] David Reisman, *Alfred Marshall's Mission* (Toronto, 1990), 185. The assumption is certainly legitimate, but there is no direct evidence for it.

Proceedings of the Association'. In June 1877 Galton submitted to it the case against, a signed paper entitled 'Considerations adverse of the Maintenance of Section F'. William Farr himself replied with 'Considerations...favourable to the Maintenance of Section F'. The papers were published together.[84] It was the new statistics versus the old. Galton began by rehearsing those arguments that had been made in 1833 and subsequently, which were supportive of a statistical section. There was an argument in its favour, he admitted, because statistics have a genuine grounding in the mathematics of probability. They were concerned 'with an important part of human knowledge', therefore. As Babbage had argued in the early 1830s, the British scientific establishment needed connections with both the political elite and with the general public; Section F, with an accessible subject-matter, allowed for these connections to be made and sustained. But to make his counter-argument, Galton set out the subjects of all the papers delivered before Section F in the three years, 1873–75. These included 'Capital and Labour', 'The future of the United States', 'Teaching of hygiene in government schools', 'Comparative mortality of Abstainers and Non-Abstainers from Alcoholic Liquors', 'Acclimatisation of the Silk Worm', 'Domestic Service for Gentlewomen', and many more in a similar random style. As Galton commented, 'It will be observed that not a single memoir treats of the mathematical theory of Statistics, and it can hardly be doubted that if any such paper should be communicated to the Association, the proper place for it would be in Section A [Mathematics]'.[85] He asked whether these and other subjects 'do not depart so widely from the scientific ideal as to make them unsuitable for the British Association' and answered in the affirmative.[86] Section F was detached from the other sections; it had made little impact on the rest of the British Association; it attracted people 'with no scientific training'; but the discredit it brought was 'borne by the whole Association'. Its subjects and its audience belonged elsewhere at the Social Science Association, evidently not an organization that Galton much favoured. He hoped that the Council would give 'the discontinuance of Section F...serious consideration'.[87]

Farr tried a different approach in the defence. He listed all the eminent figures who had presided at the section including Babbage, a host of politicians, and notable political economists like Nassau Senior, Stanley Jevons, and Henry Fawcett. He also linked the work of Section F to mathematicians and savants of the past: Halley, Gompertz, Laplace, Fourier, and Poisson among others, though

[84] For both papers, see 'Economic Science and the British Association', *JSSL*, Sept. 1877, 468–73 (Galton) and 473–6 (Farr).
[85] Galton, 'Considerations Adverse to the Maintenance of Section F (Economic Science and Statistics): submitted by Mr Francis Galton to the Committee appointed by the Council to consider and report on the possibility of excluding unscientific or otherwise unsuitable papers and discussions from the Sectional Proceedings of the Association', 471.
[86] Ibid.
[87] Ibid., 472–3.

this was no sort of defence of the present.[88] He noted especially the number of women who attended its sessions, and the 21 papers delivered by women who included Florence Nightingale, the social and educational reformer Mary Carpenter, the suffragist Lydia Becker, and Maria Georgina Grey, the women's educationist and sister of Emily Shirreff, the friend and biographer of Henry Buckle. Farr picked out for special praise the paper on 'Domestic Service for Gentlewomen', given by Rose Crawshay, another feminist and educationist, who had been the first to raise this important social question.[89] Section F used up very little of the funds of the British Association. Its 'inquiries touch nearly the interests and feelings of the working classes'. And in an oblique reference, surely, to the famous debate at the British Association in 1860 between Huxley and Wilberforce, Farr observed that Section F 'never in any way offends the religious prejudices of the nation'.[90]

Farr incorporated into his submission a defence of Section F written by the joint secretaries of the Statistical Society of London, Robert Giffen and Hammond Chubb. The society argued that to exclude 'the whole subject of the life of man in communities' would degrade the British Association. There was indeed 'a scientific order traceable in that life' and to ignore it would restrict the very definition of science itself. Developing Babbage's original justification, the Society argued that in Section F, public men could be shown the virtues of a scientific approach to the discussion of public affairs. If the British Association existed to advance science it should fulfil its mandate 'in making politicians and philanthropists generally aware of the necessity of scientific method and knowledge in their favourite subjects'.[91]

Farr had received a private letter from Edwin Chadwick in July 1877 in which he had argued that Section F existed to promote the work of the other branches of the British Association: 'questions of economy, of legislation and of administration which serve to advance the progress of the work of the mechanical and other sections; as the patent laws, legislation for scientific instruction...'. Chadwick also saw the sociological benefits of using Section F to attach public life to science:

But then it is to be considered that that section seems to bring in a considerable proportion of non-scientific persons as contributary members; who are not up to most of the discussions which take place in the other sections. Mathematical questions, for example, are in advance of most talkers. Members of Parliament, mayors and local authorities are mostly of the non-scientific classes. Yet their presence and their contributions are useful, as is the general popularity of that section.[92]

---

[88] Farr, 'Considerations...favourable to the maintenance of Section F', 473.
[89] Ibid., 474. Angela John, 'Rose Mary Crawshay 1828–1907', ODNB.
[90] Farr, 'Considerations...favourable to the maintenance of Section F', 476.
[91] The letter was dated 13 July 1877. See ibid., 474–5.
[92] Edwin Chadwick to William Farr, 12 July 1877, William Farr papers, British Library of Political and Economic Science, GB 97 FARR, Vol. 1, Item 19.

The overall defence, which pulled together a range of eclectic, not to say irrelevant arguments, several of them sociological rather than scientific, proved strong enough to see off Galton and other unnamed critics: Section F survived. As one of its later presidents recalled truthfully some years later, 'The attack was able; the defence not particularly brilliant, but the goodness of our cause or the leniency of our judges carried us through, and we were adjudged to have successfully restated the reasons for our existence.'[93] But it survived on sufferance, perhaps because other comparable organizations were no better. As the historical economist T. E. Cliffe Leslie wrote to Jevons in 1878, 'I am by no means desirous of cutting off Section F and relegating it to the Social Science Association, which is ten times worse and less scientific. I had been a member for the last 16 years, but could stand it no longer and retired this year.'[94] Section F remained the same, however. The Cambridge philosopher Henry Sidgwick attended in 1886 and was distinctly unimpressed: 'It seems to me impossible to make the discussions very valuable. The time must necessarily be limited; and there are certain familiar bores who turn up at every meeting and limit still further the fruitful minutes.'[95]

In Karl Pearson's words, the affair of Section F demonstrated 'exactly Galton's feeling with regard to much of what passed for statistics in 1877, that old type of statistics which had no theoretical basis, while Galton was working for a new type.' Pearson added that the criticisms of Section F could have been applied equally to the Statistical Society of London/Royal Statistical Society until it began to admit 'the new type of statisticians' at the beginning of the twentieth century.[96] There are no grounds for thinking Galton insincere or disingenuous in making trouble: there were valid objections to Section F, for sure. But it was also a very obvious way for him to demonstrate the differences between the old and the new, between the mathematical statistics that his brilliance had created and the mere social counting of recent decades. To some extent this was a dispute over personnel, for Section F attracted a diverse range of people, most without any sort of academic training whether in natural science, political economy, or statistics. But Galton was not making an argument for professionalisation or expertise: there is no sense that he wanted the Section to host practitioners as then constituted, rather than the public. Galton's frustration, in fact, was with the practitioners of statistics above all: statisticians who no longer deserved that name because they were unwilling or unable to apply the new mathematical statistics to the study of society. His argument was an intellectual one, designed to draw attention to the

[93] M. E. Grant Duff, 'Address of the President of Section F of the British Association, at the Fifty-first Meeting, held at York, in August 1881', JSSL, xliv, Dec. 1881, 657.

[94] T. E. Cliffe Leslie to W. S. Jevons, 28 Aug. 1878, in R. D. Collison Black and Rosamund Könekamp (eds), Papers and Correspondence of William Stanley Jevons (7 vols, London, 1972–1981), vol. iv, 273.

[95] Henry Sidgwick, Journal entry, 13 Sept. 1886 quoted in Stefan Collini, Public Moralists. Political Thought and Intellectual Life in Britain 1850–1930 (Oxford, 1993), 214.

[96] Pearson, The Life . . . of Francis Galton, II, 347–8.

internal advances within the discipline of statistics which the British Association, the Statistical Society and other institutions, had failed to recognize. He lost this battle, in institutional terms merely a skirmish. But in the intellectual history of Victorian statistics it was a significant juncture: another mark of the decline of the statistical movement after 1870.

By 1885, at the Jubilee Meeting of the Statistical Society of London, the new mathematical statistics were in evidence and in intellectual ascendancy in three papers delivered by Galton himself, Francis Edgeworth and Alfred Marshall. Edgeworth's paper, incorporating 'the best of current statistical theory', was a discussion of the application of significance tests for the comparison of means.[97] It was also a discussion of methodology, distinguishing between 'inductive logic in general' and his use of 'the pure calculus of probabilities'. To show what might now be done using the latter, he reinterpreted some of the most notable statistical findings of the past generation, including William Guy's studies of the influence of a man's profession on his life expectancy, and also more whimsical phenomena such as attendance at London clubs, and Virgilian hexameter.[98] It is not for nothing that Edgeworth has been called 'the poet of statisticians'.[99] His use of probability, the error curve, significance tests, the median, and the tendency of a mean towards normality, left many gasping for air: as one recent historian has commented 'the discussion shows little grasp among his audience of what it has all been about', though the paper was widely noticed subsequently in Britain and abroad.[100] Marshall showed how the interaction of different causes, whether moral, social or economic, could be analysed using new graphical methods.[101] Galton himself demonstrated how to build a probability curve when only a few certain values were known.[102] In all three cases, rather than hide technical issues in footnotes and appendices, the authors expected their audience to follow complex mathematical discussions.[103] It is doubtful if this was possible for many in an audience composed of old-style social statisticians. The evolution of the Royal Statistical Society, renamed as from 1887, into a true learned society was slow, therefore: there remained 'many papers of the old type, but intermingled with a liberal sprinkling of more mathematical papers from Edgeworth, Bowley, Yule

---

[97] Joseph J. Spengler, 'On the Progress of Quantification in Economics', Isis, 52, 2, June 1961, 262n.
[98] F. Y. Edgeworth, 'Methods of Statistics', JSSL, June 1885 (Jubilee volume), 181–217. Stephen M. Stigler, 'Francis Ysidro Edgeworth, Statistician', JRSS, series A, 141 (1978), 296–7; John Aldrich, 'Mathematics in the London/Royal Statistical Society 1834–1934', Electronic Journ@l for History of Probability and Statistics, vol. 6, no. 1, June 2010, 7. http://www.jehps.net/indexang.html
[99] Porter, The Rise of Statistical Thinking, 255.
[100] I. D. Hill, 'Statistical Society of London – Royal Statistical Society. The First 100 years: 1834-1934', JRSS, series A, vol. 147, Pt. 2, 134; Stigler, 'Francis Ysidro Edgeworth', 297.
[101] Alfred Marshall, 'On the Graphic Method of Statistics', Jubilee Volume, June 1885, JSSL, 251–60.
[102] Francis Galton, 'The Application of a Graphic Method to Fallible Measures', ibid., 262–71.
[103] Libby Schweber, Disciplining Statistics. Demography and Vital Statistics in France and England, 1830-1885 (Durham, N.C. and London), 203–5.

and Pearson.[104] A survey of the contributions presented at meetings of the Royal Statistical Society between 1909 and 1933, conducted for its centenary, showed 'some increase in mathematical content' certainly, but the society's *Journal* still focused on 'economic and vital statistics with a sprinkling of mathematical theory.'[105] It took a further generation before the Royal Statistical Society was re-purposed as an academic and professional association specializing in mathematical statistics. The use of trained mathematicians in many different tasks of statistical analysis during and after the Second World War led many of them to join the Society and to permanently change its focus.[106]

14.4 Galton's Moral Inadequacy

Galton's weakness was that 'he pictured life statistically, and with the naiveness of a child spoke the truth.'[107] While there may be very serious doubts about the truth of Galton's views, it is undeniable that he was naïve. The somewhat crude opinions set out in *Hereditary Genius* in 1869 did not win wide acceptance and in that still liberal age 'most reviewers felt that Galton had overstated the case for heredity while insufficiently emphasizing the role of environment.'[108] Galton himself admitted that the book was 'subjected to much criticism' though he dismissed most of this as 'captious or shallow.'[109] So he ploughed on. His essay in 1873 on 'Hereditary Improvement' called for the establishment of a central agency, with local branches, to collect personal data and so identify 'the naturally gifted' and smooth their passage through life. They would be selected and form a leading 'caste' of the most worthy specimens in each social class. The recipients of special favours from the state in starting a home and founding families, these special ones would be expected to intermarry and procreate. Their inferiors, however, would be encouraged to remain celibate and would be 'treated with all kindness' if they did. But should they reproduce as well, the time would come 'when such persons would be considered as enemies of the State, and to have forfeited all claims to kindness.'[110] It is the stuff of dystopian science fiction, a genre given

[104] Hill, 'Statistical Society of London - Royal Statistical Society', 134.
[105] Aldrich, 'Mathematics in the London/Royal Statistical Society 1834–1934', 14. Mackenzie, *Statistics in Britain*, 9. For the survey of papers 1909–33, see J. Bonar and H. W. Macrosty, *Annals of the Royal Statistical Society 1834–1934* (London, 1934), 204–23.
[106] L. H. C. Tippett, 'Annals of the Royal Statistical Society 1934-71', *JRSS*, series A, vol. 135, 4, 545–68. R. L. Plackett, 'Royal Statistical Society: The Last Fifty Years: 1934–84', *JRSS*, series A, vol. 147, 2, 140–50; S. Rosenbaum, 'The Growth of the Royal Statistical Society', *JRSS*, series A, vol. 147, 2, 375–388.
[107] Pearson, *Life…of Francis Galton*, 385.   [108] Gillham, *A Life of Sir Francis Galton*, 171.
[109] Galton, *Memories of My Life*, 290. Brookes, *Extreme Measures*, 167–9, 183.
[110] F. Galton, 'Hereditary Improvement', *Fraser's Magazine*, Jan 1873, vol. vii, 116–30.

motive, cause, and impetus by the eugenics movement.[111] At the end of his life, 35 years later, and in an age far more receptive to opinions like these, Galton was just as outspoken. Hoping that in the future 'social influence will be exerted towards the encouragement of eugenic marriages', he wanted to discourage non-eugenic unions, as well:

> I cannot doubt that our democracy will ultimately refuse consent to that liberty of propagating children which is now allowed to the undesirable classes, but the populace has yet to be taught the true state of these things. A democracy cannot endure unless it be composed of able citizens; therefore it must in self-defence withstand the free introduction of degenerate stock.[112]

Galton's collection of biographical information by questionnaire from Fellows of the Royal Society in 1874 demonstrated how blind he could to contextual factors. The returns apparently 'showed how largely the aptitude for science was an inborn and not an acquired gift, and therefore apt to be hereditary'. But heredity might have little to do with the interests and subsequent direction of a child brought up in a home where scientific discussions were common and where a father was already professionally involved in science.[113] And it would certainly be a *father* in Galton's view, and not a mother, for he evinced a consistently low opinion of women's intellects: 'the virile, independent cast of mind is more suitable to scientific research than the feminine, which is apt to be biased by the emotions and to obey authority'.[114] The misunderstandings—perhaps wilful—continued when he turned his interests to the legal profession and discovered that a relatively high proportion of the relatives of judges 'were father, son, or brother to another judge, a ratio far greater than in the general population'.[115] He believed this also demonstrated the workings of heredity, when common sense would emphasize the significance of familial influence in the choice of career, the attractiveness of the social status and high remuneration of the judiciary, and perhaps also the influence of nepotism: it is an enduring feature of the English bar that its members are indeed related to each other, even to this day. As one critic, the civil servant and economist Herman Merivale, wrote in *The Edinburgh Review*, 'The whole list has the unmistakable character of a snug little family party of jobbers, rather than a galaxy of genius.'[116] Galton always emphasized 'those qualities of intellect and disposition, which urge and qualify a man to perform acts that lead to reputation'.[117]

---

[111] The most obvious example of a novel concerning degeneration and selective breeding was *The Island of Dr. Moreau* by H. G. Wells, published in 1896.

[112] Galton, *Memories of My Life*, 311.   [113] Ibid., 293.   [114] Ibid.

[115] Bernstein, *Against the Gods*, 164.

[116] [Herman Merivale], 'Galton on Hereditary Genius', *Edinburgh Review*, vol. 132, (269), July 1870, 100–25 (quotation at 112).

[117] F. Galton, *Inquiries into Human Faculty and its Development* (London, 1883), 49.

The simpler and more realistic sociological and environmental explanations for eminence were never treated with sufficient interest and respect in his work.

It would be easy enough to convict Galton of malign intent, of deliberately suppressing evidence and explanations that ran counter to eugenics as he had named and established that pseudo-discipline. Galton always believed that moral characteristics were heritable. In his own case, there was certainly *deficiency* of moral sensibility. But it is surely better to see him contextually, in effect to see *his* weaknesses as products of nurture rather than nature. He was an innocent abroad, short of life skills, lacking in emotional range and sensitivity, without experience beyond his own gender and class, cosseted, and cerebral.

Until the 1870s the leading figures in the world of British statistics saw social progress in terms of broad environmental changes for the good of all. Change the conditions of public health and it would be possible for all to live better and more fulfilled lives. To Galton, however, social progress would be attained through the cultivation of the intelligence and other heritable qualities of the individual, and of the few with the highest ability above all. This was eugenics—to improve the race by the improvement of individuals by selective breeding. He did not write or think at the societal level, but at the level of the individual whose characteristics could be measured and assessed. Human life would improve by changing some humans, not by changing the context and environment of all humans. His predecessors in the statistical movement, however, wanted to find natural and social regularities that would help direct broad-scale social reforms. They tried to aggregate humans and determine what they held in common.

This, and a measure of self-conscious intellectual superiority, explains why Galton was dismissive or ignorant of those statisticians who had preceded him, and of Quetelet in particular. According to Pearson, he never studied Laplace or Poisson, nor read the work of Gauss. As for Quetelet, 'he formed no collection of his books, and the few references to Quetelet in Galton's writings are such as might easily arise from indirect sources'.[118] Indeed, he seems to have owned no work at all by the Belgian statistician. He largely disregarded the past history of the problems he studied: he started from scratch and *de novo*. And, in comparison to the statistical movement that preceded him, he was working under the influence of a new paradigm, Darwinism. For these reasons, Galton marks the end of one tradition and the beginning of another in the history of statistics.

This transition is evident in the peculiar but revealing interactions between Florence Nightingale and Francis Galton in the early 1890s. They held to entirely contrasting conceptions of statistics and their social role, as became clear in the course of a tortuous correspondence. There is no better way of demonstrating

---

[118] Pearson, *Life…of F. Galton*, II, 12.

the end of the statistical movement and its supersession by professionalized, mathematical statistics.

14.5 The End of the Statistical Movement: Florence Nightingale versus Francis Galton

Florence Nightingale is probably the most famous of all devotees of Victorian statistics. After her return in triumph from nursing the wounded and sick in the Crimean War, she used her enormous celebrity to reform the living conditions of the British and Indian armies; to improve the sanitary conditions, the internal regime, and the design of British hospitals; and to train nurses, both military and civil. She formed an alliance with the most notable statistician of the era, William Farr, for the reform of the public's health. She wrote papers, which were read for her at learned societies and reforming organizations. She studied blue books— official inquiries—and their statistics. She nursed in wartime and then, in peacetime, fought wars with politicians and bureaucrats in which numbers were her ammunition. The statistics on death and disease in the hospitals of the Crimea, whether accurate or not as she deployed them, had allowed her to make her case about the insanitary conditions of war. She continued to use them thereafter in her struggles with the War Office to improve the conditions of British and imperial soldiers in peace, and in her campaign of reform in civilian hospitals, as well. Hence her interest in the International Statistical Congress in London in 1860.[119]

Like many in the statistical movement, however, she had more than a utilitarian interest in statistics. For Nightingale, numbers transcended the realms of political and sanitary argument to form the basis of a way of understanding not only humankind but society as whole, the natural world, and the deity. To study statistics was to approach the godhead. As Karl Pearson put it somewhat later,

> She held that the universe – including human communities – was evolving in accordance with a divine plan; that it was man's business to endeavour to understand this plan and guide his actions in sympathy with it. But to understand God's thoughts, she held that we must study statistics, for these are the measure of his purpose. Thus the study of statistics was for her a religious duty.[120]

---

[119] The best biographies of Florence Nightingale, each taking a quite different view and approach, are Edward Cook, *The Life of Florence Nightingale* (2 vols, London, 1913); F. B. Smith, *Florence Nightingale. Reputation and Power* (London, 1982); Mark Bostridge, *Florence Nightingale. The Woman and Her Legend* (London, 2008). Anyone who writes about Nightingale and statistics is in debt to Marion Diamond and Mervyn Stone, 'Nightingale on Quetelet', 3 parts, *JRSS* (series A), vol 144, 1981, 66–79; 176–231; 332–51.

[120] Pearson, *The Life...of Francis Galton*, III, 415.

Several of the statisticians we have encountered found a similar transcendence in numbers: Jones and Whewell, Babbage and his expositor Ada Lovelace, Farr himself and also William Guy, were all touched by a fascination with number. But none compares to Nightingale, whose love of order, system, pattern, and symmetry approached the mystical and led her to put her faith in statistics as evidence of the regularities of the natural world and hence, of the existence of God. Whewell and Jones, educated in the natural theological tradition, saw a beneficent God revealed in the order and arrangements of His natural world: it was what led Whewell, by then Master of Trinity College, Cambridge, to ban Darwin's *Origin of Species* from the shelves of the college library.[121] As he wrote to Darwin in early 1860, 'you will not be surprized to be told, that I cannot, yet at least, become a convert to your doctrines'.[122] Other notable Victorians saw revelation in beauty: for John Ruskin the proof of God was the beauty of creation, which, in its turn, informed the beauty of art. But for Nightingale, God dwelt among the numbers, and the numbers showed mankind what God willed.

Her fascination was of longstanding. She had written to her sister in 1847, when she was twenty-seven years old, that 'I can never be sufficiently thankful to Papa for having given me an interest in Statistical & Political matters'.[123] Like many other mid-Victorians, she was excited by Quetelet's works above all. As she wrote to Farr,

I never read Quetelet's *Physique Sociale*[124] (which I have done over & over again) without being astounded at the force of genius and accurate observation which has produced such a work...the world might be reformed and transformed (almost into the fabled Millennium: there will be no heaven unless we make it) if it did but know how to make use of the Statistics which Quetelet has given us in reforming Legislation, Government, Criminal Law, Education, Institutions, Sanitary practice etc. etc. etc. Quetelet's chapter on Medicine alone is a book for a whole Profession to work out.[125]

Receiving copies of the latest editions of Quetelet's *Physique Sociale* and *L'Anthropométrie* from the author himself in 1872, she wrote back in the highest excitement to say that she already knew both works intimately ('que déjà je les connaissais à fond') and could not prevent herself 'each time I leaf through them reading them right to the end'. Quetelet helped her to understand both the

[121] James T. Costa, *Wallace, Darwin and the Origin of Species* (Cambridge, Mass., 2014), 103.

[122] Whewell to Darwin, 2 Jan. 1860, Darwin Correspondence Project, Cambridge, https://www.darwinproject.ac.uk/letter/DCP-LETT-2634.xml

[123] Florence Nightingale to Parthenope Nightingale, 26 Nov. 1847, Florence Nightingale Papers, Wellcome Library, London, 9016/17.

[124] L. A. J. Quetelet, *Physique Sociale: ou Essai sur le développement des facultés de l'homme* (Brussels, 1869). This was an expanded and updated version of Quetelet's 1835 work, *Sur L'Homme*.

[125] Nightingale to Farr, 26 June 1874, Farr papers, Wellcome Library, London 5474/125/1–2.

everyday application of statistics, and also the higher, God-given laws which ordered creation. In regard to the first of these themes, she had explained the uses to which statistics could be put in a letter to Quetelet himself in 1872 in her characteristic French:

Pour moi, cette étude passionnée ne se base pas du tout sur l'amour de la science, à laquelle je ne saurais pretender. Elle provient seulement de ce que j'ai tant vu des souffrances et des misères de l'humanité, des inaptitudes de la legislation et des Gouvernemens de la bêtise, oserai-je le dire?, de notre politique, de lavenglement noir de ceux qui se mêlents de mener notre corps social que ---souvent il me vient comme un éclair à travers l'esprit que seule étude digne du nom est celle don't vous avez si fermement posé les principes.[126]

In regard to the second theme, she wrote to Julius Mohl at the end of 1873 that

Nothing solaces me so much as to write upon the Laws of the Moral World: especially as exemplifying, if possible, the character of a Perfect God, in bringing us to perfection thro' them in eternity. Quetelet, who is an old friend of mine, sent me his new 'Physique Sociale' & 'Anthropométrie'. These especially interested me: because in as far as the laws which register mankind's crime & other social movements *are Laws*; of course, all Legislature & Administration must be based upon them: instead of being just the reverse. Latterly: I have been reading over again the '*Physique Sociale*' with the view of writing as above.[127]

Nightingale recognized, as she wrote to Farr, 'the sacred duty of applying Statistics to reform the world'.[128] But numbers also showed the order in creation; and the order in creation was proof of God's existence.[129] As Nightingale had written to her father much earlier,

Granted that we see signs of universal law all over this world, i.e. law or plan or constant sequences in the moral and intellectual as well as physical phenomena

[126] Florence Nightingale to L. A. J. Quetelet, 18 Nov. 1872, Quetelet Papers, file 1902: 'For me this passionate study is not at all based on a love of science, which I can hardly lay claim to. It comes solely from all that I have seen of the misery and suffering of humanity, of the ineptitude of legislation and of governments, of the stupidity, dare I say, of our politics, of the absolute blindness of those who interfere with our social entity…Often it comes to my mind like a bolt of lightning, that the only study worth its name is that of which you have so firmly set out its principles.'
[127] Florence Nightingale to Julius Mohl, 19 Dec. 1873, quoted in M.Vicinus and B. Nergaard (eds), *Ever Yours, Florence Nightingale. Selected Letters* (London, 1989), 351.
[128] Nightingale to Farr, 26 June 1874, Farr papers, 5474/125/2.
[129] Lynn McDonald, 'Florence Nightingale: Passionate Statistician', *Journal of Holistic Nursing*, 16, 1998, 267–77.

of the world – granted this, we must, in this universal law, find traces of *a* Being who made it, and what is more of the *character* of the Being who made it...[130]

Furthermore, because nature is stable and ordered we have the possibility of achieving moral transcendence for ourselves. God's law, evident in the ordered universe described by statistics, provided 'the road...to universal perfection.'[131] Nightingale knew doubt, of course. A few days after that letter to Mohl, but now worried over news of famine in India, she retracted: 'Sometimes I ask myself, after all my "Laws" and "Moral Worlds": is there a good God after all?'[132] But doubt was the corollary of faith, the necessary process of questioning that encouraged her to keep seeking the evidence of order.

When Quetelet died in 1874 she described him as 'the founder of the most important science in the whole world' and went on to explain both the utilitarian and metaphysical justifications for the study of statistics:

> For upon it depends the practical application of every other [science] & of every other Art: the Science essential to all Political and Social Administration [,] all Education & Organisation based on experience, for *it alone* gives exact results of our experience...Some months ago, I prepared the first sketch of an Essay I meant to publish & dedicate to him on the application of his discoveries to explaining the plan of God in teaching us by these results the laws of our moral progress – to explaining, in short, the path on which we must go if we are to discover the laws of the Divine Government of the Moral World.[133]

Nightingale co-opted Quetelet into her view of God and faith: he, too, became one of her many co-workers and collaborators, though now in divine service. Quetelet himself would have been amazed at the views she foisted onto him for, in his work as a natural scientist as well as a statistician, he never expressed Christian or indeed deistical inspiration. She imputed to Quetelet what she called 'the very highest kind of religion: the seeking in the laws of the Moral World which he had done so much to discover[,] the action or plan of a Supreme Wisdom & Goodness.'[134] She had no doubt that these moral laws could be discovered; doing so would require applying the methodology of natural science to social science. As she wrote on page 156 of her copy of *Physique Sociale*, 'To Discover the Mechanism of the Heavens it was first necessary to collect precise

---

[130] Florence Nightingale to William Nightingale, 6 July 1859 quoted in Cook, *Life of Florence Nightingale*, I, 482.
[131] Ibid., 483.
[132] Nightingale to Mohl, 30 Dec. 1873 in Vicinus and Nergaard (eds), *Ever Yours, Florence Nightingale*, 352.
[133] Nightingale to Farr, 23 Feb. 1874, Farr Papers, Wellcome Library, London, 5474/123/1.
[134] Nightingale to Farr, 4 March 1874, Farr Papers, 5474/124.

observations, to create methods for working them out. So it must be with Social Physics.'[135] Once discerned, these laws would direct the pursuit of human improvement and welfare. According to her inscription on the title page of *Physique Sociale*,

> The Sense of Infinite Power
> The Assurances of Solid Certainty
> & The Endless Vista of Improvement
> From the Principles of *Physique Sociale*
> If only Found Possible to Apply on occasions
> When it is so much wanted.[136]

As Florence Nightingale's biographer put it, Quetelet's book 'was in her eyes a religious work—a revelation of the Will of God.'[137]

Like many devotees, Florence Nightingale wished to proselytise and convert. It was an intrinsic feature of her nature: convinced that she was right she was impelled to convince other Victorians. She fretted after his death that Quetelet was not being given his due in Britain, that he was not read and taught as he should have been and as her admiration for him dictated. As she wrote to Farr,

> I received the other day from the son of our noble and beloved old Quetelet a Sketch of his life & work which I could not help reading through at once. Quetelet's work is always & still the subject of all others which interests me most. But I do not see it make the progress it ought, especially in our places of Education...Quetelet's Physique Sociale & Anthropométrie. These standing classics of his genius are almost now wholly unknown at our Universities & among even highly educated English people: tho' Herschel did so much to popularize them once for us.[138]

At this stage, 1876, she discussed the endowment of a chair in statistics in Oxford with the Master of Balliol College, Benjamin Jowett, a close confidante of hers for many years.[139] Jowett advised a non-resident professorship, appropriate for 'an eminent man' who might give 24 lectures a year and stay in Oxford for a month only at a time. 'The sort of person whom you might hope to get for such a chair would be a young physician', he wrote, perhaps thinking of the doctor-researchers employed by John Simon at the Privy Council Medical Office. Jowett wondered

---

[135] Quoted in Diamond and Stone, 'Nightingale on Quetelet', Pt. ii, 190.
[136] Quoted in Diamond and Stone, 'Nightingale on Quetelet', Pt. ii, 177.
[137] Cook, *Life of Florence Nightingale*, 480.
[138] Nightingale to Farr, 23 March 1875, Farr Papers, Wellcome Library, MS 5474/126.
[139] Mark Bostridge, *Florence Nightingale. The Woman and Her Legend* (London, 2008), 510–11.

'whether you would confine the professorship to social physics, which I suppose to mean reasoning from statistics about the health and social state of man: or include, as an alternative, hygiene & sanitary knowledge generally.'[140] Nothing came of the plan at this time, however, but it arose in their correspondence in 1890–91 when Jowett suggested again that they join together to endow an Oxford chair.[141] Referring to the mass of data now collected and held by government, though put to no good use because the political class lacked an education in statistics, Nightingale did not favour 'an accumulation of facts...but to teach the men who are to govern the country what are the *uses* of facts, of "Statistics"'.[142] In the following year Nightingale also contacted Francis Galton about the Oxford plan and it led to some revealing interactions between them which help us to understand the supersession of the old statistics by the new. Their letters mark, at an actual moment in time, as well as in a symbolic sense, the very end of 'the statistical movement' as it originated in the 1830s.

Nightingale sought Galton's advice 'as to what should be the work and subjects in teaching Social Physics and their practical application in the event of our being able to obtain a Statistical Professorship or Readership at the University of Oxford', which she offered to endow.[143] The use of the term 'social physics', Quetelet's coinage, is revealing and significant, and probably alarmed Galton from the outset. Galton tried to steer her away from the ancient universities, where, as he correctly pointed out, 'unless the subject in which the Professor lectures has a place in the examinations he will get no class at all. His position will be that of a salaried sinecurist, which is proverbially not conducive to activity'.[144] It was an accurate appreciation of the structure and function of what were then two undergraduate, teaching universities, and an early example of the institutional obstructions to the development of the social sciences in them.[145] Instead, Galton suggested that she establish a professorship and a set of annual lectures in statistics at the Royal Institution in London where the audience was 'a mixture of some of the ablest philosophers, of many persons of wide social interests, and of a general public'. For whatever reason, Nightingale dithered and lost commitment to her project; there is a note of impatience in Galton's later letters as this became evident to him, though his lack of enthusiasm for the project may have been the

---

[140]   E. V. Quinn and John Prest (eds), *Dear Miss Nightingale. A Selection of Benjamin Jowett's Letters to Florence Nightingale 1860–93* (Oxford, 1987) , 31 Dec. 1876, letter 362, 275.
[141]   Jowett to Nightingale, 26 Oct. 1890, ibid., letter 422, 314.
[142]   Nightingale to Jowett, 3 Jan. 1891, ibid., letter 423, 315.
[143]   Florence Nightingale to Francis Galton, 7 Feb. 1891, 'Scheme for Social Physics Teaching', in Galton Papers, Special Collections, University College London. The complete correspondence between Nightingale and Galton is included by Pearson in his *Life...of Francis Galton*, II, 414–24, 'The Proposed Professorship of Applied Statistics'.
[144]   Galton to Nightingale, 10 Feb. 1891.
[145]   For the classic argument that the development of British sociology was inhibited by the structures and biases of British universities, see Philip Abrams, *Origins of British Sociology 1834–1914* (Chicago, 1968).

reason that her ardour cooled. That eventually an annual Nightingale Lecture in Statistics was endowed at Oxford suggests that she remained attached to her original scheme. But she may have also recoiled from Galton's agenda of the issues to be addressed by the new professor, for the correspondence captures very well the differences between the outlook of the 1850s, Nightingale's heyday, and the agenda of mathematical statisticians at the end of the century.

In her first letter, Nightingale set out the main questions to be pursued by the appointee, including among many 'hygiene & sanitary work', the results of the new elementary education system, 'the deterrent effects upon crime of being in gaol', the incidence of pauperism and its habitual nature recurring through families down the generations, and the social state of India.[146] It was the agenda of the mid-Victorians, taking us back to the Social Science Association with its focus on the better administration of social institutions, and to the crude, optimistic environmentalism of that era.[147] Indeed, as she wrote to Galton, 'I presume that no one now but understands, however vaguely, that if we change the conditions for the better [,] the evils will diminish accordingly'.[148] To drive home her points she invoked her master, Quetelet: 'You know how Quetelet reduced the most apparently accidental carelessness to ever recurring facts, so that as long as the same conditions exist, the same "accidents" will recur with absolute unfailing regularity.' But this was unlikely to have cut much ice with Galton. Misunderstanding the views and outlook of her correspondent, she urged him to take up this agenda:

> What is wanted is that so high an authority as Mr Francis Galton should jot down other great branches upon which he would wish for statistics, and for some teaching how to use these statistics in order to legislate for and to administer our national life with more precision and experience.

When Galton replied, he courteously acknowledged the social issues Nightingale had raised, and he added three more of his own, but of a different nature and couched in a different idiom:

> In addition to the problems you specify, such may be measured as:
> 1. No. of hours work[ed], and corresponding amounts of value of output in different occupations, whether partly mechanical, partly mental or aesthetic.
> 2. The effect of town life on the offspring, in their number & in their health.
> 3. What are the contributions of the several classes (as to social position & as to residence) to the population of the next generation [?] Who in short are the proletariat?[149]

---

[146] Nightingale to Galton, 7 Feb. 1891, ff. 2–12.
[147] Lawrence Goldman, *Science, Reform and Politics in Victorian Britain: The Social Science Association 1857–1886* (Cambridge, 2002).
[148] Nightingale to Galton, 7 Feb. 1891, f. 16.   [149] Galton to Nightingale, 10 Feb. 1891, f. 1.

This was late-Victorian eugenics answering the institutional reformism that had inspired the social science of the mid-century. Whether the difference in outlook which she encountered at this point was recognized by Nightingale, and whether it contributed to her growing caution over her project, the correspondence with Galton exemplifies the supplanting of early Victorian statistics by something very different.

Galton evidently wanted her to understand that difference. He explained that social statistics, like physics—real physics, not *'physique sociale'*—depended on the insight of 'great men of science' because its problems were intrinsically very difficult. An overt criticism of Nightingale followed because, in Galton's view, she underestimated the degree of difficulty in statistical analysis: 'by no straightforward & expeditious method can the problems in which you – and I may be permitted to add myself – are so much interested, be solved'. Then followed the critical judgment of the present generation on its predecessor, a crucial statement by Galton:

> Quetelet's own history is an example of this. His promise & hopes, and his achievements in 1835-6, remained in *status quo* up to the last edition of his work (*Phys. Sociale*) in 1869. He achieved nothing hardly of real value in all those 33 years. So again Buckle, who started with a flourish of trumpets in the first chapters of his *History of Civilization*, [but] did next to nothing beyond a few flashy applications that have rarely stood others' criticisms. The way in which your object might best be obtained, requires, I think: a man or men conversant with the methods, & especially the *higher methods* of statistics.[150]

Karl Pearson believed that Nightingale's plan to establish some sort of firm university base for the training of applied statisticians was much better than Galton's rather weak suggestion of a course of lectures at the Royal Institution. He put down Galton's hesitation and absence of enthusiasm to his fear that he himself would be asked to take the Oxford chair.[151] But Galton's reluctance may have had an ulterior motive: to prevent Nightingale from institutionalizing the wrong type of statistician and the wrong type of statistical training. She would have endowed the work of social statisticians in the mould of William Farr. To Galton, the future of his subject should be left to practitioners of high mathematical ability and higher mental capacity who would be disengaged from the business of social

---

[150] Galton to Nightingale, 10 Feb. 1891. Even Karl Pearson found this a bit strong and added in a footnote to his account of these letters: 'I venture to think that this is far too sweeping, it overlooks not only what Quetelet achieved in organising official statistics in Belgium, but his great work in unifying international statistics.' It also overlooked the fact that Quetelet suffered a stroke in 1855 which allowed him to continue routine work, but deprived him of creativity. Galton's words are a corrective to Desrosières comment that Galton 'greatly admired' Quetelet. See Alain Desrosières, *The Politics of Large Numbers. A History of Statistical Reasoning* (1993) (Cambridge, MA, and London, 1998), 113.

[151] Pearson, *Life…of Francis Galton*, II, 420–1.

reform. To a eugenicist, Nightingale's reforms would be more than likely to per-petuate existing ills in the vain attempt at improving the lives of the poor. Better no statisticians in Oxford at all than the wrong sort. For these reasons, towards the end of the exchanges Galton tried to guide Nightingale away from Oxford and towards the Royal Statistical Society, now beginning to transform itself into a more academic forum: 'whether the Statistical Society might not appropriately be the body, in whose hands the endowment might be placed, in order to forward your object under the best attainable safeguards. Most statisticians belong to it and a suitable committee of them might be trusted.' Trusted by Galton, we might venture, now that the Society had shown evidence of change.

The correspondence dribbled away. Galton set out some very modest proposals indeed for essay prizes on social questions like education, crime and pauperism that incorporated discussion of statistical sources. Nightingale took a month to answer and suggested he contact relevant authorities to set this up. But Galton replied that the season was already too advanced and that 'there is not now time for doing all this before the vacation begins and people, especially those of the Universities, scatter'. Pearson's commentary on this is touchingly innocent: 'some-how Francis Galton seemed to overlook the very kernel of Florence Nightingale's scheme, and the whole vanished in a trivial essay project'.[152] It is an interpretation that many have since followed, and if ingenuous, it misses the real significance of the exchange.[153] More than a decade later, in 1904, Galton himself endowed the first university home for statistics: a research fellowship in eugenics specifically, at University College, London. In 1911, the year of his death, a full chair in eugenics was established there, held first by Karl Pearson. Since 1907 Pearson had been head of Galton's Eugenics Record Office at UCL.[154]

Galton's exchanges with Nightingale make clear that in his view, hers was an intellectual project and style of social analysis that had run its unsuccessful course by then. Statistics had come of age as a sophisticated branch of higher mathematics; it could substantiate Darwinian natural selection and it was being used to prove the central role in nature played by inheritance. His very lack of enthusiasm for her practical suggestions points to the difference, in Galton's mind, between earlier forms of simple data collection and of analysis used in the attempt to find crude solutions to immediate social ills, and the much more recent use of statistics in building the new science of eugenics—a science of which Florence Nightingale had no inkling. It is for this reason that to understand 'Victorians and Numbers' we must begin in the 1830s before mathematical statistics existed: to dismiss the statistical movement, or ignore it as an unworthy prelude to the real

[152] Ibid., II, 424.
[153] For example, see Richard Stone, *Some British Empiricists in the Social Sciences 1650–1900* (Cambridge, 1997), 336–7.
[154] John M. Eyler, *Victorian Social Medicine. The Ideas and Methods of William Farr* (London and Baltimore, 1979), 208, fn. 8.

history of statistics in Britain as some historians of the discipline have done, is to omit a crucial stage in the story, and overlook the motive for intellectual innovation towards the end of the century.[155] In the abortive meeting of minds between Nightingale and Galton, two different statistical and societal projects were in contention. It marks the symbolic end of the aspirations of the Victorian statistical movement to uncover the patterns that governed social life and which would allow for its improvement.

Nevertheless, we must also recognize a difference between the influence of eugenics itself on public policy, which was very limited, and the innovations in the discipline of statistics that eugenics inspired and stimulated, which were enduring intellectual achievements of the widest importance to the subsequent development of the natural and social sciences, and to the calculations of government and social administrators. Eugenics had a potent impact, usually in early adulthood, on the thinking of many of the most famous founders of the British welfare state, among them John Maynard Keynes, William Beveridge, and Sidney and Beatrice Webb.[156] As a form of social discourse and rhetoric, eugenic tropes and ideas can be heard throughout the late-Victorian, Edwardian, and inter-war eras. We have already encountered them in the language of even a dedicated empiricist and environmentalist like Charles Booth. But as several historians have demonstrated, eugenics never penetrated national politics, social policy, or the protection of the public's health. In this, Britain was different from other western societies, notably the United States, Sweden, and Germany, where eugenic ideas, especially concerning reproduction among those with learning difficulties and other supposedly heritable conditions, led to large-scale programmes of enforced sterilization in the early twentieth century.[157] In Britain, the eugenics movement threw its whole weight behind a scheme of legalised *voluntary* sterilisation only, but still failed to achieve even this relatively modest legislative enactment. In 1913, their attempts to introduce clauses to the Mental Deficiency Act of that year that would have prohibited marriage between, or with, the 'feeble-minded', were defeated. A private members bill in the House of Commons in 1922 to make voluntary sterilization legal also failed.[158]

These failures can be ascribed, in part, to political realities. The Roman Catholic Church opposed any interventions in the field of human reproduction, especially

[155] Mackenzie, *Statistics in Britain*, 8–9.
[156] John Macnicol, 'Eugenics and the Campaign for Voluntary Sterilization in Britain Between the Wars', *Social History of Medicine*, 2, 1989, 149–51.
[157] Dorothy Porter, 'Enemies of the Race: Biologism, Environmentalism, and Public Health in Edwardian England', *Victorian Studies*, 34, 2, Winter 1991, 159–78; Dorothy Porter, 'Eugenics and the Sterilization Debate in Sweden and Britain before World War II', *Scandinavian Journal of History*, 24:2, 145–62; G. R. Searle, *Eugenics and Politics in Britain 1900–1914* (London, 1976); Paul Weindling, *Health, Race, and German Politics between National Unification and Nazism 1870–1945* (Cambridge, 1989).
[158] Porter, 'Eugenics and the Sterilization Debate', 153.

among the most vulnerable, and feared that voluntary sterilization might pave the way for compulsory sterilization.[159] Meanwhile, the labour movement saw eugenics as an ideology espoused by intellectuals—some of them admittedly progressive intellectuals—that was essentially hostile to the working classes. Eugenics, after all, sought to limit the reproduction of the lower classes, however defined. In an age when politics were determined by class and class interest, upper middle-class eugenicists who had only marginal influence on national political parties stood little chance of success.[160] Set against the great issues of the age like war and peace, industrial unrest, unemployment, and poverty, eugenic reform seemed immaterial, even trivial, merely the nostrum of privileged 'faddists' with time and leisure to devote to self-serving ideologies that would punish the weak.[161]

For all its influence over the development of statistics, eugenics was never able to dethrone the environmental caste of thought that had defined liberal Britain in the Victorian age, therefore. Hereditarian explanations of mental handicap never had enough evidence behind them to encourage intellectual support from the medical professions.[162] Physicians, public health professionals, and philanthropic organizations remained resolutely opposed to eugenic solutions. Medical Officers of Health, a cadre of powerful local officials which had emerged in the mid-Victorian era, were 'deeply entrenched' in an environmentalist culture of preventive social medicine.[163] The fundamental assumptions of social policy were never challenged, therefore. Indeed, in the Edwardian era it became all the more evident that the state would assume an expanded role in the establishment of better social conditions, and would also set a minimum standard for those conditions, below which no citizen should fall. The Edwardian foundations of the modern welfare state, laid between 1906 and 1914, were the very antithesis of eugenics, therefore.[164]

Old Age Pensions sustained the elderly. National Insurance provided security for sick as well as unemployed workers. Free school meals, by their very nature, did nothing towards the restriction of working-class fertility. The medical inspection of schoolchildren kept them healthy and moved the burden of healthcare from the family to public authorities. These famous reforms were the outcome of far more potent social forces and ideologies than eugenics. The philosophical idealism, emerging from T. H. Green's Oxford in the 1870s and 1880s, that questioned the social outcomes of a classical laissez-faire economic regime, led to a reconceptualization of the role of the state, and had far greater influence on the collective thought of the governing classes in Britain between 1870 and the

---

[159] Macnicol, 'Eugenics and the Campaign for Voluntary Sterilization', 159.
[160] Ibid., 162–4; Geoffrey Searle, 'Eugenics and Class' in Charles Webster (ed.), *Biology, Medicine and Society 1840–1940* (Cambridge, 1981), 239.
[161] Macnicol, 'Eugenics and the Campaign for Voluntary Sterilization', 168–9.   [162] Ibid., 149.
[163] Porter, 'Enemies of the Race: Biologism, Environmentalism, and Public Health in Edwardian England', 165.
[164] J. R. Hay, *The Origins of the Liberal Welfare Reforms 1906–14* (London, 1975).

Second World War than eugenics.[165] The social investigations of the age, notably those of Booth in London and Seebohm Rowntree in York, provided all the evidence required of stunted lives caused by primary poverty, and gave ammunition and impetus to social reformers. A liberal tradition of emancipatory and protective reform, which had been encouraging social interventions since the 1850s and 1860s—the era of the Social Science Association—already existed and could be extended to new areas of social life.[166] Another liberal tradition that favoured personal liberty against a large and directing central state, which we have encountered at several junctures in the history of Victorian statistics, also prevented the development and implementation of eugenic policies which, by their very nature, infringed fundamental freedoms of the person.

Simon Szreter has argued persuasively that the eugenicists won few ideological battles at this time. The General Register Office, and public health professionals more generally, mounted 'a highly effective counter-attack' in the Edwardian period, starting around 1903, on issues such as infant and child mortality, and successfully upheld the liberal-environmentalist outlook against hereditarian views. Improvements in provisions for welfare and health, introduced by Liberal governments after 1906, were grounded in traditional socio-economic arguments in favour of intervention to reduce risk, to raise remuneration, to improve living conditions, and to enhance life chances.[167] These improvements, moreover, were based on the research and advocacy of reform movements and emerging professions—among them, that of social work itself—which, if anything, signal a reinvigoration of environmental explanations of inequality and poverty at this time.[168] Indeed, Pearson lamented that there were so few positions in universities and government service for the statisticians whom he trained in the principles of eugenics. As he wrote to Galton in 1909, 'At present the biometrician is the man who by calling is medical, botanical or zoological, and he dare not devote all his enthusiasm and energy to our work. The powers that be are against him in this country.'[169] The eugenics movement, which was relatively weak and lacking in institutional bases in the universities and civil society generally in Edwardian Britain, was overwhelmed, as Dorothy Porter has put it, by 'the power of an entrenched public health structure, run by a large, organised occupational group for whom a biologistic social Darwinism had no appeal'.[170]

[165] Jose Harris, 'Political Thought and the Welfare State 1870-1940: An Intellectual Framework for British Social Policy', *Past and Present*, 135 (1992), 116–41; Lawrence Goldman (ed.), *Welfare and Social Policy in Britain since 1870* (Oxford, 2018).

[166] W. L. Burn, *The Age of Equipoise. A Study of the Mid-Victorian Generation* (London, 1964).

[167] Simon Szreter, 'The GRO and the Public Health Movement in Britain, 1837–1914', *Social History of Medicine*, 4, 3, Dec. 1991, 458–461. Simon Szreter, *Fertility, Class and Gender in Britain, 1860-1940* (Cambridge, 1996).

[168] Mackenzie, *Statistics in Britain*, 48.   [169] Pearson, *Life...of Francis Galton*, IIIA, 381.

[170] Porter, 'Enemies of the Race: Biologism, Environmentalism, and Public Health in Edwardian England', 173.

14.6 Conclusion: Galton's Legacy

Until the 1870s the 'politics of statistics' were broadly liberal, and the movement's approach to social questions broadly 'environmental'. The aim was to accumulate social data and analyse it for the patterns and regularities which would explain human behaviour in the mass. From this period the approach changed, however, at least among leading mathematical statisticians, because, under the influence of Social Darwinism and eugenics, behaviour was believed to be biologically and not socially determined. Under the impact of these new paradigms, social structure and social position were believed to be foreordained in human capacity and intelligence; they were not amenable to reform and to change for the good. As the historian of science, Stephen Jay Gould, expressed it, 'Biological determinism is, in its essence, *a theory of limits*. It takes the current status of groups as a measure of where they should and must be (even as it allows some rare individuals to rise as a consequence of their fortunate biology).'[171] It may be compared to the views expressed by the inspiration of the statistical movement, and the paragon of measurement, Alexander von Humboldt, who wrote in 1849 what may stand as the *cri de coeur* of the movement before Galton:

> Whilst we maintain the unity of the human species, we at the same time repel the depressing assumption of superior and inferior races of men. There are nations more susceptible of cultivation than others – but none in themselves nobler than others. All are in like degree designed for freedom.[172]

Writing in 1930, at the end of his vast biography of Galton, Pearson justified eugenics in this way: 'by studying and then applying biological laws to his own species man may step over the corpses of his failures into the hard-won kingdom'.[173] The image is deeply unfortunate and the truth very different, of course: by applying eugenics in Germany after 1933 and in lands occupied by the Third Reich after 1939, corpses were indeed piled up and the reputation of eugenics was destroyed. It is estimated that some 300,000 people with learning difficulties and physical handicaps were murdered by the Nazis, in addition to the millions of Jews, Roma and homosexuals.[174] Writing earlier in 1924 Pearson wondered if the Great War had been enough of a catastrophe to impress on the human race

---

[171] Stephen Jay Gould, *The Mismeasure of Man* (1981) (2nd edn, New York, 1996), 60.

[172] Alexander von Humboldt, *Cosmos: Sketch of a Physical Description of the Universe*, 5 vols (London, 1849–62), I, 368.

[173] Pearson, *Life...of Francis Galton*, III, 434.

[174] Henry Friedlander, *The Origins of Nazi Genocide: From Euthanasia to the Final Solution* (Chapel Hill, N.C., 1995); M. Burleigh, *Death and Deliverance: Euthanasia in Germany c. 1900 to 1945* (Cambridge, 1994); Suzanne E. Evans, *Forgotten Crimes. The Holocaust and People with Disabilities* (Chicago, 2004).

'Galton's teaching'. Perhaps, he mused, yet another crisis would be required to win acceptance for eugenics? He fell to speculating whether 'the immediate social history of Germany may not be uninstructive from this standpoint to the philosophical onlooker'.[175] He was not to know that 'the immediate social history of Germany' did indeed demonstrate what would happen when eugenics took hold of a society, starting with the programme of forced sterilization under the 'Law for the Prevention of Hereditarily Diseased Offspring' of 1933. It is estimated that some 360,000 people were sterilized in the Third Reich in the six years to 1939.[176]

Towards the end of Galton's life, his great-niece and carer, Eva Biggs, worried about her great uncle's associations with the sort of people who had enthusiastically embraced his ideas: 'who knows that some day he may not be made answerable for their actions, for after all he invented Eugenics [?]'[177] It would be factually incorrect and morally wrong to blame Galton for this subsequent history in any direct sense.[178] But the case against him is that he advocated and condoned discrimination on grounds of intelligence, aptitude, moral worth, race, gender and class, and favoured selective breeding. In his naivety he was seemingly unaware of the dangers of an approach to social questions that, almost by definition, first counted, then categorized, and then divided people. Galton used numbers for malign purposes, though he was too lacking in social and political awareness to recognize that. He thus presents a profound irony at the end of this story: that the modern discipline of mathematical statistics was created in a context of ideological opposition to the liberal principles of the original statistical movement. The challenge to explain variation led Galton and his followers to the very heights of intellectual achievement. We may wonder if without this challenge, statistics as a discipline would have become so sophisticated quite so fast, or whether it would have remained as undeveloped as in the era of Quetelet. Many other academic disciplines and modes of thought were similarly influenced by the change of social outlook as Social Darwinism and eugenics took hold in the late-Victorian era; statistics was not alone, by any means. But as a discipline, statistics became associated with peculiarly anti-social views (and *anti*-social is exactly the correct term) in ways that would have surprised and depressed the founders of the statistical movement who saw the exactitude of numbers as liberal and emancipatory.

[175] Pearson, *Life…of Francis Galton*, II, 122.
[176] Richard J. Evans, *The Third Reich in Power* (London, 2005), 507–10.
[177] Eva Biggs to Karl Pearson, quoted in Brookes, *Extreme Measures*, 288.   [178] Ibid., 297.

15

Social Statistics in the 1880s

The Industrial Remuneration Conference, 1885

For three days during the last week of January 1885, a conference was held in the Prince's Hall, Piccadilly, which essayed many of the social issues and problems of the 1880s.[1] It was sponsored by a gentleman from Edinburgh, Robert Miller, an engineer who had made money in Australia.[2] He put up a thousand pounds and requested seven public figures to organize an event to address the following question:

> Is the present system or manner whereby the products of industry are distrib-
> uted between the various persons and classes of the community satisfactory? Or,
> if not, are there any means by which that system could be improved?

Two of the seven were leaders of the working class movement, Thomas Burt, MP, one of the first working men elected to parliament in 1874 as a radical Liberal, and John Burnett, general secretary of the Amalgamated Society of Engineers. The others were Thomas Brassey, the railway promoter and builder; Robert Giffen, statistician in the Board of Trade; Professor Herbert Foxwell, the political economist who collected what became the Goldsmith's Library of economic liter-ature in the University of London; John William Maule Ramsay, the 13th Earl of Dalhousie, Liberal politician; and Frederic Harrison, the Positivist intellectual. Harrison was the moving spirit behind the conference, in fact. It was Harrison to whom Robert Miller went with the idea of a meeting, and it was Harrison who first made public the plan to convene a 'new industrial inquiry' in an article in the *Pall Mall Gazette*.[3] Later, Harrison wrote an obituary of Robert Miller in the *Positivist Review*.[4] One of the aims of the Positivist movement and of its French founder, the philosopher and pioneer sociologist, Auguste Comte, was the cre-ation of industrial and class harmony.[5] Harrison's reputation as a 'friend of labour',

---

[1] *Industrial Remuneration Conference. The Report of Proceedings and Papers Read* (London, 1885) (hereafter *IRC*).
[2] Miller preferred to remain anonymous at the time but was named later by Frederic Harrison as the sponsor. Frederic Harrison, *Autobiographic Memoirs* (2 vols, London, 1911), ii, 296–7.
[3] Frederic Harrison, 'A New Industrial Inquiry', *Pall Mall Gazette*, 8 Sept. 1884.
[4] *Positivist Review*, April 1898, vol. vi, 93.
[5] D. G. Charlton, *Positivist Thought in France During the Second Empire* (Oxford, 1959); Royden Harrison, *Before the Socialists: Studies in Labour and Politics, 1861–1881* (London, 1965).

Victorians and Numbers: Statistics and Society in Nineteenth Century Britain. Lawrence Goldman, Oxford University Press.
© Lawrence Goldman 2022. DOI: 10.1093/oso/9780192847744.003.0015

and his connections across the labour movement and political world, which we have already encountered, made him an obvious choice as convenor of the London meeting.[6]

This group contacted the Statistical Society of London to form a larger organizing committee. The Society nominated a further seven of its members: the statisticians Rawson W. Rawson, then its president, Leone Levi, and Stephen Bourne; Francis Neison, actuary and a founder of the Institute of Actuaries; the economic historian William Cunningham; the industrialist and advocate of industrial arbitration between workers and employers, David Dale of Darlington; and Charles Thomson Ritchie, a Conservative MP and leading minister in future governments. The resulting joint committee invited 125 delegates to the conference in strict proportions: 24% were representatives of business, 40% were trades' unionists; 8% represented Friendly Societies; 12% were Co-operators, and the remaining 16% were economists, statisticians, philanthropists, and various social activists. Remarkably, more than half the participants were from the labour movement. As the conference chairman, Charles Dilke, MP, at that point the 'coming man' in Liberal politics, put it at the conclusion, 'there have been a great many more workmen who have addressed us than capitalists'.[7]

This was in contrast to the situation 30 years before when labour organizations were held in suspicion by mid-Victorian opinion and their spokesmen given little opportunity to participate in the public arena. And it is in even starker contrast to the situation 60 years earlier when the London Statistical Society, founded by artisans, published the *Statistical Illustrations of the British Empire*, discussed above, to draw attention to similar issues of income and remuneration among working people.[8] The statistics of living standards and of social distress were a constant concern in Victorian Britain, as they have been ever since. That, chronologically, this book begins in the 1820s and ends in the 1880s with the consideration of working-class budgets and social conditions in both periods, is not accidental. Throughout the history of statistics, in Britain as elsewhere, their use and their development have been closely associated with the investigation of poverty, and of the standard of living of different classes and groups. This continued to be the case even in an era, after the 1880s, when statistical methods changed and statistics became associated with biological and Darwinian approaches to the study of society. The counting of the poor and the analysis of their condition, though it became more sophisticated in the work of social investigators like Charles Booth in London and Seebohm Rowntree in York, never ceased, though before the First World War it was often inflected with new thinking about the innate and inherited, rather than the contextual and environmental, causes of poverty.

---

[6] See pp. 136–7 and 138 above.   [7] *IRC*, 502.   [8] See Chapter 4 above.

John Mawdsley of the Cotton Spinners; Edward Cowey, Benjamin Pickard, and William Abraham for the Miners; John Burns of the Social Democratic Federation; Lloyd Jones, Edward Greening, and G. J. Holyoake of the co-operative movement; and Hubert Bland of the Fabian Society, were all present.[9] Indeed, the original historian of the Fabians, Edward Pease, wrote later that 'The Conference was the first occasion in which the Fabian Society emerged from its drawing-room obscurity, and the speech of Bernard Shaw on the third day was probably the first he delivered before an audience of more than local importance.'[10] These men met at the IRC Liberal MPs like John Morley, Charles Bradlaugh, and Charles Dilke himself, who had been chairman of the Royal Commission on the Housing of the Working Classes in the previous year. The economists present included Alfred Marshall, William Cunningham, and Foxwell, all from Cambridge. Cunningham and Foxwell were notable proponents of historical economics while Marshall was the great exponent of neo-classical economic theory. J. Shield Nicholson, Professor of Political Economy in Edinburgh University and a true follower of Adam Smith, also participated. The young sociologist, Patrick Geddes, took part alongside the Positivists, Frederic Harrison and E. S. Beesly, the friend of Karl Marx. Entrepreneurs like Brassey and Lowthian Bell, the ironmaster and patron of William Morris, also spoke. Indeed the committee invited Morris and H. M. Hyndman to contribute papers on behalf of the socialist movement, though neither accepted.[11] To give a flavour of the mix, and the eminence of the contributors, in one session of the conference, papers were given for and against land reform as a social panacea by the great naturalist, co-discoverer of natural selection, and social radical, Alfred Russel Wallace, and by the future prime minister A. J. Balfour, while George Bernard Shaw made a contribution to the debate from the floor. Shaw also published an account of the IRC in *The Commonweal* a few weeks later.[12] The distinction of the participants lent a measure of credibility to the IRC then, and lends interest to it today, not least because it has received little attention from historians up to this point.[13]

At first sight the IRC might appear to take a place in the tradition of British corporatism, as one of a series of attempts that each generation has made since the mid-nineteenth century to find a formula for industrial peace and social

---

[9] 'List of Members of the Industrial Remuneration Conference', *IRC*, xiii–xix.

[10] Edward R. Pease, *The History of the Fabian Society* (London, 1916), at http://www.hellenica-world.com/UK/Literature/EdwardRPease/en/TheHistoryOfTheFabianSociety.html#Chapter_III.

[11] *IRC*, xi.

[12] G. B. Shaw, 'The Industrial Remuneration Conference', *The Commonweal*, March 1885, 15.

[13] The notable exceptions to this are John Saville in his admirable introductory essay to the edition of the Conference's proceedings he edited in 1968 and Alon Kadish who has written about popular perceptions of political economy demonstrated in its debates. See John Saville, 'The Background to the Industrial Remuneration Conference of 1885', in *Industrial Remuneration Conference. Reprints of Economic Classics* (Augustus M. Kelley, New York, 1968) and Alon Kadish, *The Oxford Economists in the Late Nineteenth Century* (Oxford, 1982), 127–30.

harmony through various forms of collaboration between key economic and social interests. However, the IRC is not an ancestor of the National Industrial Conference which met at the end of the First World War,[14] nor of the Mond–Turner talks in 1928,[15] nor even of the more recent National Economic Development Council, 'Neddy' as it was known in the 1960s, which was supposed to bring effective planning to the British economy, but which never made its mark, staggering on until abolished without much fanfare or notice in June 1992.[16] The IRC in 1885 was different: it had no representation from the state—politicians and civil servants participated as individuals, not in any official capacity—and it was the state which convened the talks, councils and conferences of the twentieth century, because the state had most to gain from what was later termed in the 1970s, 'the social contract'. The IRC in 1885 was a seminar, not a national plan; it was to share information and prompt creative and consensual thinking rather than act as a bureaucracy implementing policy in itself. It never reconvened, nor set up any apparatus to follow-through agreed approaches to wage bargaining and wage setting, nor was it associated with any conclusions about industrial relations in the future because it came to none. According to Dilke, 'the publication of our proceedings and the careful reading of them which will take place, I believe, by enormous numbers of the most intelligent among the working men and capitalists of this country, cannot but have a most excellent effect for good'.[17] But these were judicious words spoken at the end of sharp debates which had come to no agreed conclusions. The primary significance of the Conference lies in its immediate context as one of several contributions to the debate in the 1880s on living standards and the distribution of wealth in a now mature industrial society. That debate continues, and today as in 1885, it depends on numbers, the statistics of 'the standard of living', so-called.

There were several strands to the 1885 conference. Most of the papers, and the majority of the discussion, focused on industrial relations between capital and labour, on wages and the problem of poverty. The third and last day was devoted to the issue of land reform, however. In another session there were exchanges over trade policy between free traders and protectionists, notably a paper in favour of the re-imposition of tariffs on imported manufactured goods to defend working-class living standards given by the 'fair trader', W. J. Harris, MP, and another advocating tariffs to save British agriculture from the cheaper products of North America and other new sources of food around the globe. They were

---

[14] Rodney Lowe, 'The Failure of Consensus in Britain: The National Industrial Conference, 1919–1921', *The Historical Journal*, 21, 3, Sept. 1978, 649–75.
[15] See Frank Greenaway, 'Mond Family 1867–1973', *ODNB*.
[16] See *Hansard*, House of Lords, 16 June 1992, vol. 538 cc.126–37 for comments on the NEDC at its abolition by several of its former members.
[17] *IRC*, 503.

answered from the floor by the liberal intellectual and biographer of Gladstone, John Morley, MP.[18]

There were two interrelated debates on the standard of living in late-Victorian Britain. One was the straightforward though contentious argument about who earned exactly what, and how far, given existing prices, those real wages stretched. The other was about social justice: after a century of industrialization and in the country which had generated the most wealth in this period, was the distribution of that wealth fair and equitable? Had those who had laboured, received their due? According to Joseph Chamberlain in the 'Unauthorised Programme', his radical Liberal manifesto published a year before the IRC met,

> The vast wealth which modern progress created has run into 'pockets'; individuals and classes have proven rich beyond the dreams of avarice….the majority of the toilers and spinners have derived no proportionate advantage from the prosperity which they have helped to create, while a population equal to that of the whole metropolis has remained constantly in a state of abject destitution and misery.[19]

Was that true? And if so, why?

This latter issue of the historic distribution of the fruits of industry was, in fact, the question that Arnold Toynbee had raised in his lectures to working-class audiences around Britain in the early 1880s which became his posthumous book, *Lectures on the Industrial Revolution in England*, published in 1884. It was the first work to popularize the term 'industrial revolution' and it was used for two generations as a textbook in the adult and workers' education movements. Its success owed much to Toynbee's starting point: that the working communities of Britain had been exploited and historically ill-remunerated since the origins of industrialization. In Toynbee's words, 'the progress in which we believe has been won at the expense of much injustice and wrong'.[20] Toynbee's point was made many times over at the Conference. According to Wallace, 'the actual producers of wealth in the wealthiest country in the world must continue to live without enjoying a fair and adequate share of the wealth which they create'.[21] According to the old co-operator, Lloyd Jones, 'We should frankly acknowledge that as a nation we have not succeeded in distributing equitably the wealth that has come to us so abundantly'.[22] The question remained unanswered and current in the next generation, as well: the

---

[18] W. J. Harris, MP, 'Do any remediable causes influence prejudicially the well-being of the Working Classes?' and S. Harding, 'Home and Foreign Policy, or How to Restore Prosperity to a Distressed and Anxious People', *IRC*, 221–30, 235–40. For Morley's reply see ibid., 246–8.

[19] Joseph Chamberlain, 'Labourers' and Artisans' Dwellings', *The Fortnightly Review*, cciv, 1 Dec. 1883, 761. See C. H. D. Howard, 'Joseph Chamberlain and the "Unauthorised Programme"', *English Historical Review*, 65, 257, Oct. 1950, 447–91; Edith Abbott, 'Charles Booth 1840–1916', *Journal of Political Economy*, 25, 2, Feb. 1917, 196.

[20] A. J. Toynbee, *Lectures on the Industrial Revolution in England* (London, 1884), 58.

[21] *IRC*, 369.   [22] *IRC*, 41.

young William Beveridge, the so-called 'architect of the welfare state' in the 1940s, recalled Edward Caird, the Master of his Oxford college, Balliol, which Beveridge attended between 1900 and 1904, telling the undergraduates that 'one thing that needs doing by some of you is to go and discover why, with so much wealth in Britain, there continues to be so much poverty, and how poverty can be cured'.[23]

Delegates at the IRC were also influenced, though negatively in this case, by another recent publication, the presidential address by Robert Giffen, the civil servant and statistician, to the Statistical Society of London in 1883, which has become known as 'The Progress of the Working Classes', and which attempted a survey of working-class living standards over the previous 50 years, concluding in positive fashion that the workers—all workers—were unquestionably better off now than then.[24] We have met Giffen before in his defence of the Statistical Section of the British Association in 1877.[25] No name was mentioned more often in the debates at the IRC than that of Giffen, almost always in criticism and deprecation. This may explain why the proceedings of the conference include the explanation that Giffen was unable to attend, owing, apparently, to the pressure of official business.[26]

Nothing else Giffen wrote in a long career is of such interest. Gladstone, then prime minister, read the lecture on Boxing Day 1883 and wrote on the following day to the Duke of Bedford

> to recommend to your notice the masterly address of Mr Giffen, of the Board of Trade, in the Statistical Society on the condition of the labouring classes in this country now and 40 or 50 years back. It is, I should think, the best form of answer to Mr. George on *Progress & Poverty*.[27]

Few agreed with Gladstone either then or now. Giffen argued that while wages had risen 'from 20 and in most cases from 50 to 100 per cent'[28] yet 'most articles...have rather diminished in price'.[29] So he concluded that 'the masses of the people are better, immensely better, than they were fifty years ago. This is quite consistent with the fact, which we all lament, that there is a residuum still unimproved.' Beginning with wages, he compared rates and prices in 1883 with those

---

[23]   William Beveridge, *Power and Influence* (London, 1955), 9.

[24]   Robert Giffen, 'The Progress of the Working Classes in the Last Half Century. Being the Inaugural Address of R. Giffen LL.D., President of the Statistical Society, Delivered 20th November 1883', *Journal of the Statistical Society of London*, xlvi, 1883, 593–622.

[25]   See above, p. 276.

[26]   'Mr F. Harrison stated that the Committee deeply regretted that Mr Robert Giffen's official duties have prevented him completing a paper which he was preparing on the subject now before the Conference.' *IRC*, 62.

[27]   H. C. G. Matthew (ed.), *The Gladstone Diaries*, vol. 11, July 1883–Dec. 1886, 26–7 Dec. 1883, 83–4. Gladstone to the Duke of Bedford, 27 Dec. 1883, Gladstone papers, British Library, Add Ms. 44547, f. 21. Colin Matthew includes this letter in the entry for 27 Dec.

[28]   Giffen, 'The Progress of the Working Classes', 598.   [29]   Ibid., 605.

of 50 years ago. Many of his examples were flimsy, fragmentary or very general. His data on rising wages, for example, relied overmuch on only 16 discrete occupations, nine of which were subdivisions of the Yorkshire woollen trade, all of them requiring skilled work.[30] His data on prices, meanwhile, though also very sketchy, demonstrated that they had been generally stable across half a century—indeed, the price of wheat had gone down—and were more stable and dependable than in the 1830s and 1840s, thus assisting working-class budgeting.[31] Even so, these statistics took no account of the peaks and troughs of the trade cycle and the effect of this turbulence on remuneration at specific times. Giffen also ignored the gradations of skill which divided workers into different groups and determined different levels of remuneration.

Giffen tried to prove his point by tabulating the increased consumption of key foodstuffs by volume, from tea and sugar to butter and cheese, though the global figures he provided told nothing about the distribution of the consumption of these foods.[32] He went further to consider what is now sometimes called 'the social wage', counting-in the new costs of sanitary improvements and 'the expend-iture of government for education, for the post office, for the inspection of facto-ries, and for the miscellaneous purposes of civil government'.[33] Writing in 1883, he was able to point to the very recent reduction of the death rate as evidence of progress: in truth, until the mid-1870s, death rates for both sexes had remained stubbornly high.[34] He examined probate returns and income tax data and con-cluded that there was, by 1883, a wider distribution of property than in the 1840s, and a notable increase in the diffusion of wealth.[35] He also concluded that while the real wages of the working classes had doubled approximately, the increase in the return on capital over the previous fifty years had been far smaller. Profits, in other words, had not gone disproportionately to capital, but had been shared with wage earners.[36]

Nevertheless, much of his data was inexact and potentially misleading. Though he could point to a decrease in the number of paupers nationally, that was more likely a consequence of the harsher implementation of the Poor Laws than any fall in overall poverty.[37] Likewise, a growing number of depositors with bank accounts, the number of whom had increased almost ten times between 1831 and 1881, was no measure of overall social improvement when the number of working-class families still making no deposits at all had also risen in line with

---

[30] 'Comparison of Wages Fifty Years Ago and at Present Time', Giffen, 'Progress of the Working Classes', 597.
[31] Ibid., 602.
[32] 'Quantities of the Principal Imported and Excisable Articles retained for Home Consumption, per Head of the Total Population of the United Kingdom 1840–1881', ibid., 608.
[33] Ibid., 605–6.   [34] Ibid., 607.   [35] Ibid., 614–18.   [36] Ibid., 619–20.
[37] Ibid., 610.

the general growth in population.[38] Giffen understood the implications of much of his information: 'that wealth, in certain directions, is becoming more diffused, although it may not be diffusing itself as we would wish', but he was complacent, nevertheless. He admitted himself to be 'one of those who thinks that [this] regime is the best…Surely the lesson is that the nation ought to go on improving on the same lines…Steady progress in the direction maintained for the last fifty years must soon make the English people vastly superior to what they are now.'[39]

This may have soothed Gladstone's conscience but it enraged working men and women who came to the IRC determined to criticize Giffen for failing to take proper account of prices and the purchasing power of real wages. Emma Paterson, the leading woman trade unionist of the era,[40] asked how much

of the meat, bacon, ham, eggs, butter, cocoa, coffee, wine and other articles that Mr Giffen finds the 'masses' now obtain…can be purchased out of 12s a week, when there are also rents, coals, light and clothing to be provided, and perhaps an invalid or aged relative to be helped?[41]

Mrs Ellis of the Huddersfield Pattern Weavers actually held aloft a copy of Giffen's essay

in which he said that pattern weavers in Huddersfield earned 16s per week fifty years ago, and at the present time 25s. To this statement she must give an emphatic denial. Today the men weavers in Huddersfield did not average more than 20s. The women were paid from 15 to 30 per cent less than the men, and did not average more than 15s per week.[42]

Giffen's global averages were given short shrift by people who knew from experience that wages varied according to economic conditions, from trade to trade, from district to district, from factory to factory, and in relation to 'payments in kind'. Many speakers preferred the evidence 'of their own eyes and ears in preference to any number of doubtful tables of figures', a common complaint from workers throughout the three days and whenever living standards were (and are) discussed.[43] As a Mr James Aitkin of the Greenock Chamber of Commerce put it, 'there is a whole army of gentlemen who have had that long experience in the weaving factory, the engine shop, or the building yard. These gentlemen can furnish much more reliable information on this question than it is possible to extract from any statistical tables, however carefully they may be prepared.'[44] Lloyd Jones 'used his own eyes and ears in preference to any number of doubtful tables of

[38] Ibid., 611.   [39] Ibid., 621.
[40] Norbert C. Soldon, 'Emma Anne Paterson 1848-1886', ODNB.   [41] IRC, 205–6.
[42] IRC, 208.   [43] Lloyd Jones, IRC, 83.   [44] IRC, 72.

figures which might be brought forward. What he saw and heard, and had touched, he believed; but were tables of statistics doubled, he would not believe one of them without corroboration.'[45] As G. B. Shaw summarized the exchanges over wages, prices, and the evidence of both,

> The capitalists and their retainers contended that statistics prove that the workman is better off than he has ever been, and that his position is steadily improving. The workers, on the other hand, contended that their personal experience proves that wages, as measured by purchasing power, had fallen. The capitalists objected that the personal experience of one or two workmen proves nothing. The workmen retorted that facts are better than figures; and that workmen agree better on their facts than statisticians on their figures.[46]

The failure to consider the remuneration of different groups of skilled and unskilled workers was noted by many of the workers and their spokesmen. As Lloyd Jones expressed it, 'this class is made up of many classes' and there were 'gradations running into pauperism'.[47] It was pointed out that any official data had to be reduced by 20% or so to take into account the effects on wage earners of ill-health, short time, stoppages, breakdowns and other causes that reduced the hours worked.[48] Some also noted a decline in the quality of life, as opposed to its quantities: a contributor on rural labour in Wiltshire complained about increasing social controls in the countryside and conjured up a picture of playgrounds taken away, games stopped, footpaths closed, nut-gathering and blackberrying prohibited. Even the customs of Merrie England were under attack.[49]

Historians have confirmed what these debates demonstrate: that the often brutal discipline of eking-out a meagre existence and balancing family budgets, and the occasional (or more frequent) flutter at gambling, had made the Victorian workers numerate and highly adept at calculation and mental arithmetic.[50] It could hardly be otherwise: the 'enormous condescension of posterity', as E. P. Thompson put it, should be avoided in this discussion as in so much else concerning the working classes.[51] They were perfectly able to understand not only their own situation but essentially numerical arguments over national questions like tariff reform and its effect on the cost of living, an issue that developed from the 1880s

---

[45] *IRC*, 83.   [46] Shaw, 'The Industrial Remuneration Conference', 15.
[47] Lloyd Jones, 'Profits of Industry and the Workers', *IRC*, 34.
[48] Edith Simcox, 'Loss or Gain of the Working Classes during the Nineteenth Century', *IRC*, 102–3.
[49] W. Saunders, 'Loss or Gain of Labourers in Rural Districts', *IRC*, 109.
[50] James Thompson, 'Printed Statistics and the Public Sphere. Numeracy, Electoral Politics, and the Visual Culture of Numbers, 1880–1914', in Tom Crook and Glen O'Hara (eds), *Statistics and the Public Sphere. Numbers and the People in Modern Britain, c. 1800–2000* (New York and Abingdon, 2011), 135–8.
[51] E. P. Thompson, *The Making of the English Working Class 1780–1830* (London, 1963), 13.

and which erupted into national politics in 1903.[52] And that mental quickness made it possible to alter the way political matters were pictured and represented in the late-Victorian age, now in figures, diagrams, graphs, and pictograms, as well as words.[53]

The more searching criticism of Giffen and other optimists of the age emerged from an ever-present sense of the 'moral economy' which workers had internalized and by which they judged their situation, and which led them to ask Toynbee's question about the fair distribution of the profits of labour and wealth in British society.[54] As Mr G. Sedgwick of the Boot and Shoe Riveters and Finishers' Union put it: 'Whatever the statisticians might say, the workmen would never feel content until they had that portion of the profit on their labour which they had honestly earned.'[55] Things may well have improved, but had the workers received a fair share of that improvement? Not according to John Burnett who contended that the advance in real wages over the past fifty years, Giffen's timespan, was not the point 'but whether there had been that improvement in the position of workmen that they had a right to expect...it would be monstrous if the position of the worker had not improved.'[56] As James Brevitt of the Ironfounders put it,

The real form of the question ought to be, Do the people, the toilers, the millions who, from youth to old age, are engaged in labour obtain anything like an equitable share of the products of their toil, or of those material comforts and social enjoyments which render life tolerable, a blessing and not a curse?[57]

Lloyd Jones agreed: 'The question was not whether the worker was better off than his grandfather in a number of things, but whether he was as well off as the resources of the country entitled him to be?'[58] Or in the words of J. E. Williams of the Social Democratic Federation, the most militant voice at the conference, 'Did they now get a fair share of the product they created by their labour?'[59]

By far the best paper delivered at the conference was a broad summary of changing living standards by Edith Simcox, styled 'anthropologist and political activist' by the *ODNB* and also a trades' unionist, scholar, journalist, feminist campaigner, and acolyte of the author George Eliot, who was the focus of her emotional life.[60] According to G. B. Shaw, her paper carried off 'the honours of

---

[52] Edmund Rogers, 'A "Naked Strength and Beauty". Statistics in the British Tariff Debate, 1880–1914', in Crook and O'Hara (eds.), *Statistics and the Public Sphere*, 224–43.
[53] Thompson, 'Printed Statistics and the Public Sphere', *passim*.
[54] For the meaning of 'moral economy', see the classic essay by E. P. Thompson, 'The Moral Economy of the English Crowd', *Past & Present*, 50, 1, Feb. 1971, 76–136.
[55] *IRC*, 210.   [56] *IRC*, 166.   [57] *IRC*, 62–3.
[58] Lloyd Jones, *IRC*, 83.   [59] J. E. Williams, *IRC*, 79.
[60] Edith Simcox, 'Loss or Gain of the Working Classes during the Nineteenth Century', *IRC*, 84–114. Susanne Stark, 'Edith Simcox 1844–1901', *ODNB*.

the day...the demand for copies outran the supply before she had finished reading it'.[61] In a sophisticated analysis of society as well as incomes, Simcox's central argument was that there had been concentration of wealth in nineteenth-century Britain, with the richest fortunes now much greater than at any time in the past, and all sectors of society more internally differentiated by income and wealth than was the case fifty or a hundred years ago. The differences between great and small industrialists, great and small bankers, great and small building contractors, and between skilled artisans and the unskilled had all been exacerbated.[62] Her critique of the calculations of the supposed experts on living standards, including Giffen, was acute in its arguments and backed by superior information.[63] Her description of 'the prosperous aristocracy of the working classes', who had undoubtedly done well over the decades, sparkles with sharply-observed details. She noted that this was the 'section with which politicians come in contact, and from whence come those whom society is rather over-hasty to welcome as "representative working-men"'. But she reminded her audience of the privations of the more numerous 'less skilled workers, male and female' whose wages 'only suffice for the necessities'. Unlike many speakers she was alive also to the significance of changes in contracts which, acting over time, had left many workers of the 1880s at the mercy of employment by the piece or by the hour only. In discussion she advanced her conclusion 'that a fluctuating fifth of the whole population was to be found permanently on or across the border of pauperism or destitution'.[64]

Many of the speakers at the IRC complained about patchy, inaccurate and inconclusive data and called on the government to address this deficiency. Summing up the Conference, Rawson W. Rawson, a constant presence in statistical circles since the 1830s, hoped 'that one result of this meeting will be that we may be able to represent to Her Majesty's Government the importance of supplying the public with larger and fuller statistics regarding the social condition of the people'.[65] Several speakers, including Thomas Brassey, Charles Bradlaugh, and Arthur Acland, who would soon become a Liberal, and later a Labour MP, drew attention to the better collection of labour statistics by the federal government in the United States, and especially by the Bureau of Labor in Massachusetts.[66] This was the burden of Alfred Marshall's paper also, who admitted to the 'deficiency of our knowledge' and blamed ignorance for both business failure and low wages. Marshall was an optimist over the standard of living, concluding that though the labouring classes 'have now none too much of the necessaries, comforts and luxuries of life', nevertheless at the beginning of the nineteenth century 'they had less

---

[61] Shaw, 'The Industrial Remuneration Conference', 15.
[62] Simcox, 'Loss or Gain', *IRC*, 84–5.
[63] Simcox, 'Loss of Gain', Appendix F: 'The Rate of Wages', *IRC*, 101–3.   [64] *IRC*, 134.
[65] *IRC*, 504.   [66] See *IRC*, 8, 81 (Brassey); 171 (Bradlaugh); 212 (Acland).

than a third of what they have now'.[67] In a speech that reads like a moral tract on how to improve the lives of workers rather than an economic analysis of low wages, Marshall's recipe for industrial peace and stability was to ensure the wide dispersion of accurate knowledge about business and trade so that workers, understanding how business works and the state of their industries, would make realistic wage demands, and employers would pay the going and fair rate. In utopian style, Marshall saw

> no reason why a body of able disinterested men, with a wide range of business knowledge, should not be able to issue predictions of trade storm and of trade weather generally, that would have an appreciable effect in rendering the employment of industry more steady and continuous.[68]

Such a body was not established, perhaps because local arbitration boards of this type to regulate prices and wages had been set up for several industries in the 1860s and 1870s, almost always without success.[69] But the IRC was not without an outcome in this matter. In the autumn of 1886, following a resolution of the House of Commons some months before, John Burnett of the Amalgamated Society of Engineers, who spoke frequently and powerfully at the IRC, was appointed the first Labour Correspondent at the Labour Statistical Bureau in the Board of Trade, charged with the collection of data on wages, earnings, and conditions.[70] By January 1893 this had become the Department of Labour located within the Board, and Burnett was its senior civil servant.

The IRC examined and debated elements of both past and future, both the experiments in co-operation and arbitration from the 1860s and 1870s which Frederic Harrison now wrote off as failures, and the emerging problem of the *lumpen proletariat*, described by Alfred Russel Wallace as 'a vast class of unskilled or little skilled labourers, who, even when in full work, barely earn sufficient to afford them a decent animal subsistence'.[71] The Conference was an important stage in an ongoing national debate over wages, wealth, and the moral economy of the late-Victorian working classes which ran throughout the 1880s and beyond. It was part of the general context in which Charles Booth set to work on what became the multi-volume study of the *Life and Labour of the People in London*. The precise origin of Booth's project 'remains unclear'.[72] Though some of Booth's colleagues later put it down to his reading of *The Bitter Cry of Outcast London*, the

---

[67] Alfred Marshall, 'How far do remediable causes influence prejudicially (a) the continuity of employment, (b) the rate of wages?', IRC, 173–99, quotation at 187.

[68] Ibid., 181.

[69] Lawrence Goldman, *Science, Reform and Politics in Victorian Britain. The Social Science Association 1857–1886* (Cambridge, 2002), 218–23.

[70] Saville, 'The Background to the Industrial Remuneration Conference of 1885', 41.

[71] Wallace, 'How to cause wealth to be more equally distributed', IRC, 369.

[72] Jose Harris, 'Charles Booth 1840–1916', ODNB.

famous pamphlet of 1882 written by the Congregational minister, the Rev. Andrew Mearns, there is no hard evidence for this.[73] There was no blinding reve-lation or crucial event that marked Booth's commitment to the study of the poor, or at least none that we have so far discovered. He became interested in the life and work of the people in 1883 when the first reports based on the 1881 census were published. He joined the Royal Statistical Society in 1885 and published his first sociological essay in its journal in the following year.[74] Booth was also involved in the so-called 'Mansion House Survey' on the condition of the unem-ployed, initiated by the then Lord Mayor of London, Sir R. N. Fowler, in 1884.[75] He lent the Survey one of the clerks from his shipping and trading business.[76] In the summer of 1885, Beatrice Webb found Booth at work on problems of living standards and poverty and contemplating 'personal investigation' into them.[77] He also at this time 'had talks with Mr Hyndman of the Social Democratic Federation, attended the meetings of that body, listened eagerly to addresses, [and] on one occasion giving one himself on the best principles of land ownership'.[78] In Hyndman's autobiography (which is considered an unreliable source) he relates that Booth came to visit him to dispute Hyndman's contention that a quarter of London's population lived in extreme poverty, which Booth believed an exaggera-tion, and Booth told him then that shortly, he would himself be organizing his own survey in London.[79] Booth's biographers give the date of this meeting as February 1886, and in the following month Booth called his friends and collabo-rators together to set his inquiry in motion.[80]

Whatever the precise origins of Booth's great survey, he and Hyndman, along with the speakers at the IRC, were not lone investigators or independent voices setting their research before the public, but participants in a national debate and a process of social investigation which includes the 1885 Industrial Remuneration Conference and much more. We don't have to know the proximate origins of Booth's survey beginning in 1886; it is enough to understand the wider context of this debate about the living standards of the working classes which runs through so much of the social history of the 1880s. Booth was not a heroic social researcher departing from the amateurism of past investigations as presented by some

---

[73] T. S. and M. Simey, *Charles Booth. Social Scientist*, 66.

[74] Charles Booth, 'Occupations of the People of the United Kingdom, 1841–81', *Journal of the Statistical Society of London*, 49, 2, June 1886, 314–444.

[75] *Report of the Mansion House Conference on the Condition of the Unemployed* (London, 1887).

[76] Rosemary O'Day and David Englander, *Mr. Charles Booth's Inquiry. Life and Labour of the People of London Reconsidered* (London, 1993), 29. Belinda Norman-Butler, *Victorian Aspirations: The Life and Labour of Charles and Mary Booth* (London, 1972), 71.

[77] Beatrice Webb's Diary, 22 August 1885: 'Delightful two days with the Booths...Discussed the possibility of social diagnosis...Plenty of workers engaged in the examination of facts collected by others – personal investigation required.' *Beatrice Webb's Dairy*, f. 425: https://digital.library.lse.ac.uk/objects/lse:ros416yur/read/single#page/428/mode/2up.

[78] Mary Booth, *Charles Booth. A Memoir* (London, 1918), 15.

[79] https://booth.lse.ac.uk/learn-more/who-was-charles-booth.

[80] T. S. and M. Simey, *Charles Booth. Social Scientist*, 70n.

historians and sociologists.[81] Rather, he was a wealthy, philanthropic citizen who committed himself to try to solve an empirical dispute which had been troubling British opinion for some years. The Industrial Remuneration Conference helps us to see this dispute in greater detail, and thus helps us better understand what gives thematic unity to a decade focused on the living standards of the workers and the poor—a decade which ended in 1889 with another dispute over living standards, the struggle for the 'dockers' tanner'(6d an hour) at the heart of the London Docks' Strike of that year. That famous labour dispute was co-led by John Burns of the SDF, and he had been one of the loudest and most insistent voices at the IRC four years before where he had vindicated socialism from the floor 'as an artisan, as a follower of Karl Marx, and of Lassalle, as a member of a revolutionary body in England'.[82]

Many of the key figures in the disputes over pay, working conditions, contracts, and economic justice which run through the 1880s were present at the Industrial Remuneration Conference in early 1885. The meeting led to no change in industrial relations and it is unlikely that anyone was better remunerated in consequence of its discussions. But it did host debates on many of these matters, and on other issues of the moment, and is of interest for that reason alone. It also demonstrates that many of the questions associated with specific projects and organizations believed to have definitively changed British social consciousness at the end of the decade, like Booth's 'great inquiry', *Fabian Essays on Socialism* published in 1889, and the 'new unionism' called into being through the strikes of the gas workers and dockers in that year, were already in play. All of these depended, at a fundamental level, on numbers: rates of pay, hours worked, consumer prices, real wages. Indeed, the Royal Statistical Society listened to a large number of empirical analyses of living standards and poverty between the mid-1880s and mid-1890s, during which period it elected Booth as its President, and allowed him to use its rooms as a base for his investigations of London. Booth delivered six papers to the RSS between 1887 and 1894 on subjects including the national employment structure, dock labour, poverty, pauperism, and old age pensions, then an emerging idea.[83] Contributions on poverty were followed in subsequent years by a further group of papers on the statistics of labour, and on the 'labour

[81] Saville, 'The Background to the Industrial Remuneration Conference of 1885', 42–3.
[82] *IRC*, 484–5.
[83] Charles Booth, 'Occupations of the People of the United Kingdom, 1841–81', *Journal of the Statistical Society of London*, 49, 2, June 1886, 314–444; 'The Inhabitants of Tower Hamlets (School Board Division), Their Condition and Occupations', 50, 2, June 1887, 326–401; 'Condition and Occupations of the People of East London and Hackney, 1887', 'Enumeration and Classification of Paupers, and State Pensions for the Aged', 54, 4, Dec. 1891, 600–43; 'The Inaugural Address of Charles Booth, Esq., President of the Royal Statistical Society. Session 1892–93. Delivered 15th November, 1892', 55, 4, Dec. 1892, 521–57 (on Dock Labour); 'Life and Labour of the People in London: First Results of An Inquiry Based on the 1891 Census. Opening Address of Charles Booth, Esq., President of the Royal Statistical Society.Session 1893–94', 56, 4, Dec. 1893, 557–93; 'Statistics of Pauperism in Old Age, 57, 2, June 1894, 235–53.

question' more generally. These included a contribution in fourteen parts, between 1898 and 1906, by G. H. Wood and the social statistician, A. L. Bowley, then teaching at the London School of Economics, on 'The Statistics of Wages in the United Kingdom During the Last Hundred Years', broken down into separate industries.[84]

Two conclusions follow from this. First, across the course of the Victorian period the statistical basis of working-class life, so remote and marginal a subject when the artisans around John Powell and John Gast published their volume of *Statistical Illustrations* in 1825, became a central concern of social statisticians and, indeed, of the nation as a whole. The enfranchisement of millions of workers, the maldistribution of wealth, and what Beatrice Webb famously called 'a new consciousness of sin among men of intellect and men of property' in consequence of that maldistribution, brought the issue of remuneration to the forefront and pulled statisticians with it.[85] But as the debates at the IRC demonstrate, the calculations of statisticians and the numbers they deployed merely became further subjects for contention. The evidence was always contested, of course, whether at the British Association in the 1830s or at the retitled Royal Statistical Society half a century later.

Secondly, the debates at the Industrial Remuneration Conference remind us that even as the discipline of mathematical statistics took off in the same decade in consequence of Francis Galton's brilliant insights and innovations, there remained a far less elevated realm where statistics were the weapons of choice— and often the only weapons—between contending social groups seeking to change or conserve British society. These statistical debates are the very stuff of modern political dispute, 'business and usual', the inevitable consequence of more data being discovered, constructed, and laid before Victorian and subsequent audiences. The Industrial Remuneration Conference has been largely ignored by historians until now, probably because its arguments are so obvious, so 'contemporary', so similar to our own, that it has been overlooked in plain sight. The artisans in Clerkenwell who called themselves the London Statistical Society were overlooked for a different reason, because they were so marginal to the events and the social institutions of the 1820s that they seemed immaterial. But they were among the first to use numbers to improve the lives of the poor and they helped to found a tradition of critical social statistics which, via the Industrial Remuneration Conference and many more forums like it, they have passed on to us.

---

[84] For details, see R. G. D. Allen and R. F. George, 'Professor Sir Arthur Lyon Bowley 1869–1857', *Journal of the Royal Statistical Society*, ser. A, vol. 120, 2, June 1957, 236–41.

[85] Beatrice Webb, *My Apprenticeship* (London, 1926), 179–80.

16

Conclusion

From Statistics to Big Data, 1822–2022

For two generations, from the 1820s until the 1880s, the history of statistics in Victorian Britain was a history of liberal hopes and endeavour. By counting and analysing, statistics offered reformers—Whig potentates redrafting the electoral map, free traders assessing the impact of corn law repeal, Manchester manufacturers fighting for religious equality, physicians and sanitarians struggling to reduce mortality and extend life expectancy—a new way of achieving their aims. Intellectuals like Quetelet, through his influence on his British followers and admirers, provided environmental arguments to buttress these reforms, apparently demonstrating that social behaviour was regular, predictable, even lawbound, and thus amenable to benign intervention and change. National and international organizations pledged themselves to the use of numbers in the pursuit of social welfare and continental amity. Liberal Europe created and then admired the International Statistical Congress, at its best in the 1850s.

Yet it was found that statistics could not explain the origin and spread of diseases; they did not end social and political contestation; they could not unravel the mysteries of history; a powerful tradition in Victorian letters blamed them for introducing iron into the soul. And from the 1870s the history of statistics followed wider international history and the history of ideas. The social concepts of the day became divisive rather than inclusive. Social Darwinism replaced liberal environmentalism; international competition took the place of collaboration. Liberal states, created at mid-century, became conservative states. 'Reform', which had once been associated with liberty, national consolidation, and renewal, became a dangerous term, either associated with insurgent socialism or attempts to buy-off the workers, as in Germany in the 1880s and 1890s. New intellectual paradigms emerged after Darwin that, in their focus on understanding biological and social 'difference', encouraged differentiation by class and race rather than social solidarity. Numbers assisted this process, Galton and the eugenicists who followed him turning them into powerful analytical tools—ironically, more powerful than statistics had been in its liberal, reformist phase—with which to categorize, label, and divide. The arc of statistics, in short, followed that of European and North American political and intellectual development more generally. We should expect nothing else from a discipline and practice that had grown up from

Victorians and Numbers: Statistics and Society in Nineteenth Century Britain. Lawrence Goldman, Oxford University Press.
© Lawrence Goldman 2022. DOI: 10.1093/oso/9780192847744.003.0016

the seventeenth century in association with the modern state. Yet we have seen, as well, that Victorian statistics were part of a vibrant and staunchly independent intellectual culture and were also collected, played with, and analysed for the sheer pleasure and interest of it. For some savants meanwhile, both natural and social scientists as we might now call them, statistics offered a new way of 'doing science' as demonstrated by the Cambridge inductivists: Herschel, Whewell, Babbage, and Jones. Drawn as they were from diverse social and intellectual backgrounds, for many Victorian statisticians the analysis of numbers was a search for order, structure and underlying principle, whether in nature or society.

This raises an ancient philosophical problem in the history of mathematics: whether that discipline is created by human ingenuity and artifice, or whether mathematicians merely discover (or a better word might be *uncover*) ideas, principles, and relationships which are immanent but hidden, ever-present but unnoticed, real and independent of the human mind? The latter view was first advanced by Plato in his 'Theory of Forms' and was intrinsic to Platonism as a body of thought. In short, is the mathematician an inventor or an explorer?[1] For much of the history related here, the statisticians we have encountered were the latter, hoping that the structures and principles they were seeking would unmask themselves once enough data were found, collected, and tabulated. The procedure was Humboldtian: as Alexander von Humboldt himself made measurements and readings of the flora and fauna he collected in the field, and examined them at his leisure, so the statistician collected and then pored over the numbers. By collecting statistics or putting a number on social phenomena, it was hoped that patterns would emerge that could be recognized without the need for complex mathematical procedures and interventions, which in any case, did not then exist. Indeed, for savants and practitioners who believed in a divinely ordered creation and who worked in the tradition of natural theology, the study of statistics should automatically disclose God's method and purpose: the book of God's works could be opened, measured, and counted, and then understood, just like the book of God's words, the Bible. But the 'discoveries' that were made were sometimes false and trivial: that the annual rates of births, marriages, and deaths are steady over time is not the basis of a natural law, nor the foundations for social science and scientific history. Even if the distribution of cases of cholera could be discovered, plotted, and discerned, that in itself would not explain the transmission of the disease and the means to its eradication. To become a genuinely analytical discipline statistics needed invention, construction, and method—the consistent application of mathematical procedures to formal data. This was the contribution

---

[1] The modern *locus classicus* for this problem is its discussion in the memoirs of another mathematician from Trinity College, Cambridge, among several in this book, G. H. Hardy. 'I believe that mathematical reality lies outside us, that our function is to discover or *observe* it, and that the theorems which we prove, and which we describe grandiloquently as our 'creations', are simply our notes of our observations'. G. H. Hardy, *A Mathematician's Apology* (Cambridge, 1940), 63–4.

of Galton and his followers: to invent procedures by which the relations between phenomena, whether natural or social, might be calculated exactly. Remarkable human ingenuity was needed to discern the meaning in the numbers. Although in the 1830s statistics were compared to a new language, they could not speak for themselves. They needed interpretation. The invention of mathematical statistics was required if the natural and social structures buried in the numbers were to be discovered and understood.

One notable historian and historical demographer, looking back to the political arithmetic of the late seventeenth century, has speculated that the appreciation of statistics and statistical order might have led naturally to the loss of religious faith.

> If pattern and meaning can be found by the consideration of *number, weight* and *measure*, as the political arithmeticians succeeded in doing, a belief in the immanent dependence of human action and the social order upon divine agency becomes superfluous, just as the same is true in relation to the motion of the planets or the flight of an arrow.[2]

If true for that period, how much more true should it have been for the mid-Victorian era when so many men and women were subject to religious doubt? To find order revealed through numbers in social life as well as in nature might have suggested that humans had the requisite reason and moral capacity to fend for themselves without a God. It accords with a contemporary sense of ourselves as rational beings capable of understanding creation through science that we should look back on the history of statistics and interpret them as yet another means by which humanity freed itself from ignorance and superstition. Yet there are several figures in this story who saw evidence for the deity reflected back to them from the numbers. After all, according to the *Book of Wisdom*, known as *The Wisdom of Solomon*, part of the Septuagint, had not God 'ordered all things in measure and number and weight'?[3] Arbuthnot, Derham, Whewell, Jones, Malthus, and Florence Nightingale, as well as the older Samuel Taylor Coleridge, can all be so bracketed, though they differed greatly in their sense of who or what 'the deity' might be. Ada Lovelace conceptualized 'mathematical truth as the instrument through which the weak mind of man can most effectually read his creator's works'.[4] Even Charles Babbage accepted 'the existence of a supreme Being' as a

[2] E. A. Wrigley, 'Comments', in Richard Stone, *Some British Empiricists in the Social Sciences 1650–1900* (Cambridge, 1997), 423.
[3] Wisdom of Solomon, ch. xi, 20. https://www.kingjamesbibleonline.org/Wisdom-of-Solomon-Chapter-11/.
[4] Ada Lovelace, 'Sketch of the Analytical Engine Invented by Charles Babbage by L. F. Menabrea. With Notes upon the Memoir by the Translator, Ada Augusta, Countess of Lovelace', *Taylor's Scientific Memoirs*, 3, 1843, 666–731, Note A.

consequence 'of a long life spent in studying the *works* of the Creator'.[5] According to one of his biographers, 'he thought of him as a Programmer'[6] and he explained the possibility of divine miracles using a mathematical analogy: that a mathematical series or sequence might be defined precisely, but still be violated by a rogue number or term that could not be predicted by its formula.[7] Indeed, he would entertain his guests by programming the Difference Engine to suddenly throw out such a random result amidst an otherwise predictable series. As such, miracles were not, according to Babbage, 'a breach of established law' but 'the very circumstances that indicate the existence of far higher laws, which at the appointed time produce their pre-intended results'.[8] A miracle was, in fact, 'a conscious deviation and part of an overall plan' made by the deity.[9] Babbage 'thought it more consonant with reason and exact science, to regard extraordinary events, as the fulfilment of some transcendental laws beyond our reach of observation, than as deviations from the code of Nature'.[10] We might call a miracle, therefore, the planned exception that proved the rule, suddenly introduced by God into a stable system or predictable sequence, but neither random nor inexplicable.[11]

Babbage composed his unofficial *Ninth Bridgewater Treatise* in 1837 because the third volume in that famous series, published three years previously by his friend William Whewell and entitled *Astronomy and General Physics considered with reference to Natural Theology*, had implied that science and mathematics were of no value in understanding the divine origins of the laws of creation.[12] To this Babbage objected strongly, believing 'mathematics as the best preliminary preparation for all other branches of human knowledge, not even excepting

[5] Charles Babbage, *Passages from the Life of a Philosopher* (London, 1864), 403 (his italics).

[6] B. V. Bowden, 'A Brief History of Computation', in B.V. Bowden (ed.), *Faster Than Thought. A Symposium on Digital Computing Machines* (London, 1953), 30.

[7] 'Introduction',*Charles Babbage and his Calculating Engines. Selected Writings by Charles Babbage and Others* (eds Philip Morrison and Emily Morrison) (New York, 1961), xxiv.

[8] Babbage, *Passages in the Life of a Philosopher*, 293.

[9] Imogen Forbes-Macphail, 'The Enchantress of Numbers and the Magic Noose of Poetry: Literature, Mathematics, and Mysticism in the Nineteenth Century', *Journal of Language, Literature and Culture*, 60, 3, 2013, 141. Doron Swade, '"It Will Not Slice a Pineapple": Babbage, Miracles and Machines' in F. Spufford and J. Uglow (eds), *Cultural Babbage. Technology, Time and Invention* (London, 1996), 45–6.

[10] H. W. Buxton, *Memoir of the Life and Labours of the Late Charles Babbage* (1872–80) (ed. Anthony Hyman) (Cambridge, Mass., 1988), 325.

[11] On Babbage's explanation of miracles see Walter Cannon, 'The Problem of Miracles in the 1830s', *Victorian Studies*, 4, 1, Sept. 1960, 23–6; Simon Schaffer, 'Babbage's Dancer and the Impresarios of Mechanism', in Spufford and Uglow (eds.), *Cultural Babbage*, 62–3.

[12] Charles Babbage, *The Ninth Bridgewater Treatise. A Fragment* (London, 1837). The eight treatises were commissioned in the will of the 2nd earl of Bridgewater to demonstrate the 'power, wisdom and goodness of God, as manifested in the Creation'. They were published between 1833 and 1840. The offending passage from Whewell reads: 'We may thus, with the greatest propriety, deny to the mechanical philosophers and mathematicians of recent times any authority with regard to their views of the administration of the Universe...' William Whewell, *Astronomy and General Physics considered with reference to Natural Theology*, (London, 1834) (1841 edn.), 208.

theology'.[13] When he sent a copy of the *Ninth Bridgewater Treatise* to the Duchess of Kent and her daughter, then merely Princess Victoria but on the verge of becoming Queen, 'he had been careful to point out that, while it was a defence of science, it was favourable to religion.'[14] In the history of Victorian statistics as presented here, there is no example on the other side of someone who lost religious faith through working on social statistics—though in the nature of things, Victorians were understandably reticent when discussing religious doubt, and there were many other grounds for coming to the rejection of God. However, the authority of religion and religious ethics were undoubtedly challenged by eugenics, which sought to wrest control of key aspects of life—marriage, procreation, the family, nurture, education, and social welfare—from their traditional locus within Christianity. In its scientific naturalism, as in many other ways, the eugenics movement was a fundamental break with pre-existing traditions. It is hardly surprising that Karl Pearson, who lost his faith as an undergraduate in Cambridge in the 1870s, and who saw himself as a devotee of reason and science only, should have published a collection of his essays entitled *The Ethic of Freethought* in 1888.[15]

Given the diversity of the many Victorian statistical projects and the social range of the groups and individuals involved, we should be wary of automatically associating statistics in the nineteenth century with projects or policies designed to constrain or subvert liberal movements and the interests of ordinary people, what is sometimes called 'social control'. There is a too easy assumption of 'the tyranny of numbers', based on a belief that 'measurement as obsessively practised by our society is about standardization and control'.[16] Counting, it has been asserted, 'means definition and control. To count something, you have to name it and define it.'[17] In that act of definition comes power, it is argued, usually because the definition itself emanates not from those being counted, but from those doing the counting, generally the state. By counting, individuals can be standardized and classified, and classification allows for more efficient management, or worse, for the differentiation or even persecution of some groups. Numbers also take the place of trust: we demand evidence of effort, quality and calibre, and increasingly, we assess these things in quantitative forms only. We start to apply a numerical value to every social transaction or personal interaction, even while we recognize that many of the most valuable human qualities and feelings, to which we all subscribe, cannot be weighed or measured. This was the thrust of the combined critique of numbers by Carlyle, Ruskin, and Dickens.

The coercive relationship between knowledge and power, and the impositions of the 'carceral state' on individuals, are closely associated with the work at the

[13] Buxton, *Memoir of the Life…of Charles Babbage*, 350.
[14] Doris Langley Moore, *Ada, Countess of Lovelace. Byron's Legitimate Daughter* (London, 1977), 94.
[15] Karl Pearson, *The Ethic of Freethought. A Selection of Essays and Lectures* (London, 1888).
[16] David Boyle, *The Tyranny of Numbers. Why Counting Can't Make Us Happy* (London, 2000), 218.
[17] Ibid., 7.

end of the twentieth century of the French theorist and polymath, Michel Foucault. His ideas on the development of the modern 'map of knowledge', what he termed 'the order of things', have been used already to understand the changing conceptualization of statistics, and their changing uses, across the period as a whole.[18] In publications of great variety, intellectual range, and breadth, he traced, as well, the changing nature and consciousness of government in the west—what he termed 'governmentality'—from the early modern period onwards, in a story in which social statistics feature prominently. If medieval forms of government focused on questions of sovereignty and law, from the fifteenth century the 'administrative state' defined itself through the bureaucratic 'apparatus' required for its establishment and growth. Then, according to Foucault's schema, came the management of population in mass society. The key social processes of the eighteenth century, including demographic expansion, agricultural development, and the transformations of the economy, became the new focus of 'governmentality', aided and abetted by the developing subject of statistics. 'Whereas statistics had previously worked within the administrative frame and thus in terms of the functioning of sovereignty, it now gradually reveal[ed] that population has its own regularities, its own rate of deaths and diseases, its cycles of scarcity, and so on.' As the function of government became 'the welfare of the population, the improvement of its condition, the increase of its wealth, longevity, health', so it required 'a range of absolutely new tactics and techniques' for the tasks. Statistics were one of those techniques, providing the knowledge and data required for 'rational' government. Statistics were 'one of the major technical factors of the unfreezing (déblocage) of the art of government'.[19]

To Foucault, this new knowledge and these new techniques were intrinsically related to the development of 'a disciplinary society', a society 'controlled by the apparatuses of security'. In the eighteenth century 'discipline was never more important or more valorised than at the moment when it became important to manage a population'. He describes 'a triangle, sovereignty-discipline-government, which has as its primary target the population and as its essential mechanism the apparatuses of security'.[20] One such apparatus were statistics, providing the data for forms of social control administered through various types of carceral institution— prisons, reformatories, police, the courts, workhouses, hospitals. Knowledge was intrinsically linked to power, its servant and master both: as Foucault expressed it, 'the exercise of power creates and causes to emerge new objects of knowledge and

[18] See above, pp. 192–6.
[19] Michel Foucault, 'Governmentality' (part of a course of lectures on 'Security, Territory, and Population', Collège de France, 1978) in *Power. Essential Works of Foucault 1954–1984: Volume 3* (Colin Gordon, ed.), (Paris, 1994) (2000 edn, London), 201–22. Quotations drawn from pp. 212–17.
[20] Ibid., 219–21.

accumulates new bodies of information...and, conversely, knowledge constantly induces effects of power'.[21]

Stimulating as these ideas may be, as a guide to social institutions as they have functioned historically, Foucault has been found to be empirically unreliable and 'ideological' in the present-minded antagonism he displayed towards actors and social practices in the past of which he evidently disapproved.[22] His explanation of these practices in terms of power and authority—of the control that some groups exercised over others, be they governing elites, social classes, doctors of the insane, penologists, or sanctioned experts of some other description—is both a truism and also a reduction. Power relations are ubiquitous, a feature of any social situation. They must be described and explained, but because of their universality, they are rarely sufficient as descriptions or explanations of any major innovation in social relations.[23] The implication that a single source of centralized social power can be held responsible for all the 'reforms' ascribed to it by Foucault, undervalues the significance of the decisions made by myriad historical actors at local or institutional level, whether they conformed with the dominant values (what Foucault termed an '*episteme*') or challenged them. History is thereby made uniform and deterministic; the institutional and cultural complexity of social institutions is ironed out in ways that diminish the role of groups and individuals who may, or crucially, may not, be servants of the dominant structures of thought. Foucault tended to generalize social history into a history of domination and subordination, arguing that the urge to control could be discerned in all human institutions. Thereby, the scale became too broad and the argument too blunt.

Detailed work on the history of German statistics in the twentieth century, for example, has rejected the easy conclusion that statisticians and statistics functioned 'as a tool of oppressive, dehumanizing reason' when co-opted by the Nazi state. The depiction of 'technocratic totalitarianism' dependent on counting, classifying, objectifying, and then annihilating, has been challenged by Adam Tooze in particular. In reality, Nazi statistical bureaucracy 'was a shambles', and attempts to build a surveillance state failed and were side-lined. 'It was the statistical apparatus created by the Weimar Republic that really provided the underpinnings of the Nazi war effort', he has argued.[24] Germany in the 1930s and 1940s may be an extreme case in every sense, but the argument of this book, reliant on examples

---

[21] Michel Foucault, 'Interview on the Prison: The Book and Its Methods', quoted in Colin Gordon, 'Introduction', ibid., xv.

[22] Andrew Scull, 'Michel Foucault's History of Madness', *The History of the Human Sciences*, iii (1990), 57. See also Lawrence Goldman, *Science, Reform and Politics: The Social Science Association and Nineteenth Century Britain* (Cambridge, 2002), 41–2.

[23] G. Stedman Jones, 'Class Expression versus Social Control? A Critique of Recent Trends in the Social History of Leisure', *History Workshop Journal*, iv (1977), 163–70.

[24] J. Adam Tooze, *Statistics and the German State, 1900–1945. The Making of Modern Economic Knowledge* (Cambridge, 2001), 36–8. Tooze is taking issue with the work of G. Aly and K. H. Roth, *Die restlose Erfassung. Volkszählen, Identifizieren, Aussondernim Nationalsozialismus* (Berlin, 1984).

from that most stable of all modern societies, Victorian Britain, is that the social act of counting is neither intrinsically dehumanizing nor an invitation to social differentiation and dominance. Farr, Guy and Simon—physicians, sanitarians, and statisticians all—were impeccable liberals, reformers in both a political and medical sense, who were at the centre of the statistical movement. By counting the sick and putting a number on disease and its diffusion, they believed that individual health, and the health of the community more widely, might be restored, and illness prevented. The artisans of the London Statistical Society and the businessmen who founded the Manchester Statistical Society were also reformers who shared a contempt for aristocratic government and the 'old corruption' that came with it. They disagreed profoundly on the causes of social distress and the cure for it, however. Counting the rich in the one case, and enumerating the poor in the other, were entirely different approaches, representing profoundly different social outlooks. Yet in both cases, the new technology of statistics seemed to offer a way of making a fresh and powerful arguments against immiseration and religious exclusion, respectively. Remarkably, at a single moment in British history, the early 1830s, when 'Reform' was the cry of every group, Whigs owning broad acres, Manchester millocrats, London professionals, Cambridge dons, Clerkenwell artisans, and Thames shipwrights wanted to harness statistics to change. And all these groups in their different ways subscribed, whether consciously or not, to a broad liberal environmentalism. Though we know how numbers were used to create pseudo-sciences of social division at the end of the century, we must also recognize the liberal promise of statistics at the start of the Victorian period. The history of the modern state might be written in terms of the information required to control population and mass society, and to suppress individualism within it, as Foucault suggested. But it might also be written in terms of the manifold efforts made to free men and women from disease, poverty, and ignorance by statisticians from Babbage to Nightingale and beyond. Techniques of control could also be techniques of liberation.

In her book on the *History of the Modern Fact*, Mary Poovey counterposes two antagonistic conceptualizations of numbers. They are either 'simple descriptors of phenomenal particulars' that 'resist the biases that many people associate with conjecture or theory', or, on the contrary, 'numbers inevitably carry within them the traces of a certain kind of systematic knowledge': they have been chosen and deployed in specific contexts and embody the assumptions, presuppositions, and intent of whoever deploys them.[25] The argument of this book, however, based on a wide variety of examples drawn from the Victorian era, is that numbers are plastic and malleable, tools to be used for good or ill, whose essence cannot be captured by this type of binary division. There is, in fact, a spectrum of uses

---

[25] Mary Poovey, *A History of the Modern Fact. Problems of Knowledge in the Sciences of Wealth and Society* (Chicago and London, 1998), 4.

among the case-studies set out here, from the very self-conscious and instrumental deployment of numbers to sustain a politico-religious cause in Manchester in the 1830s to the belief in the purity, incorruptibility, and transcendence of numbers on the part of Florence Nightingale in the 1870s. Alexander von Humboldt, the inspiration of the statistical movement, used numbers to prove and illustrate the unity of nature. William Whewell and Richard Jones applied them in an academic dispute on the methodology of economic analysis. William Farr hoped that through counting he could plot the distribution of cases of cholera and would discern, thereby, not only the origins and nature of the disease but the means by which it could be defeated. Henry Buckle thought he saw in the statistical predictability of certain key social variables the evidence of historical inevitability. In these, and in other cases, numbers were applied in quite different ways across a wide range of different disciplines so as to defy easy generalization.

According to one historian of modern statistics, 'whether or not statistical facts can claim the status of "truth" or "objectivity" in some metaphysical sense is irrelevant for all practical purposes'.[26] But Florence Nightingale would have demurred at this, for sure. To her, and to many other Victorian statisticians, numbers were without dubiety or ambiguity, and they derived their social authority from this very certainty. As unimpeachable sources they were the only true guides to right conduct and policy. All else was just opinion. Even if we think this hopelessly naïve and an example of Victorian positivism at its most extreme, that people held this view and tried to act upon it is a part of this story, and of the historical fabric in general: it is not irrelevant. To understand the Victorian enthusiasm for statistics, and to fully appreciate the ways in which they were employed in government, politics, and culture, requires sensitivity to all the different views of their utility and essence, from, as Dickens could see, those of Mr Gradgrind to those of Sissy Jupe.

In most of these examples, moreover, there was an admixture of motives that confutes a binary division into just the ideological or non-ideological use of numbers.Whewell and Jones championed a different academic methodology, but they also questioned the desired outcomes of orthodox economic theory, for their goal was the attainment of social solidarity, civic harmony and stability, rather than Ricardian *laissez-faire*. William Farr hoped that statistics *in themselves* would unlock the secrets of cholera, but he was also part of an ongoing debate among sanitarians over theories of disease, whether 'zymotic' or 'miasmatic', whether airborne or waterborne, in which he held views that changed over time and which influenced the way he interpreted the data. Buckle genuinely believed that the social regularities described by statistics were evidence that human history was determined rather than made. This encouraged him to write in a manner that

---

[26] Tooze, *Statistics and the German State, 1900–1945*, 4.

was critical of the narrative and biographical styles of mid-Victorian English historiography which emphasized agency, contingency, and fortune in History. But this was something he was already committed to do, with or without statistics: he was an intellectual outsider with a case to make against orthodox Whig historical writing. In making a case against social inequality and political corruption, the artisan statisticians of the 1820s believed that the numbers they had collected and published in *The Statistical Illustrations* were irrefutable: commentary was unnecessary when faced with these apparently incontestable facts. But the data was deployed to make a strong, open, partisan case in favour of the working classes and against the corruptions of the state and its governing elite. In each of these cases, numbers were used as 'simple descriptors' and were *also* arranged, chosen, and interpreted according to pre-existing theories, ideological positions, and personal dispositions. A binary division between two definitions only, such as Poovey provides, cannot capture the distinctive use of numbers and the combination of ideas and approaches that each separate project encompassed. It is better to take each on its own terms, therefore, showing sensitivity to the use of numbers in each individual instance.

According to one of Foucault's interpreters, he believed that 'nothing, including the exercise of power, is evil in itself – but everything is dangerous'.[27] The historical examples explored here also suggest that the discipline of social statistics is not intrinsically ideological, though it can be made such by abuse, manipulation, error, and conscious or unconscious bias, as can any tool or mode of social analysis. Numbers belong to no one and give solace to no side. Their influence can derive from their conscious embodiment in a deliberate argument, or from the reverse, the sincere belief that the data tells its own story regardless of intent. The history related here does not suggest that numbers, or counting in general, have any intrinsically malign qualities but can be used as circumstances allow or as human will dictates: 'There's nothing either good or bad but thinking makes it so'.[28]

Charles Booth, at the end of this history, could therefore appear both liberal and conservative simultaneously and could be taken to have used numbers both neutrally and also subjectively. No social survey of the nineteenth century came closer to the lives of the poor than Booth's study of *The Life and Labour of the People of London*. Yet even as he demonstrated that poverty was the experience of more than thirty per cent of the population of the city, he sought to assist one group at the expense of another. By differentiating between the respectable and casual poor, the former would be economically advantaged through the social and geographical isolation of the latter in 'labour colonies'. Those who remained in London would be assisted by decanting elsewhere another section of the urban population. It was the policy founded upon the data rather than the data

---

[27] Colin Gordon, 'Introduction' in Foucault, *Power. Essential Works of Foucault 1954–1984*, xv.
[28] *Hamlet*, Act 2, scene 2, lines 1350–1.

themselves that was the issue. The problem with the motto 'Aliis Exterendum' of the Statistical Society of London, of which Booth was the president in the 1890s when it had become the Royal Statistical Society, lay not so much with the information gleaned that was to be 'threshed out by others', but with the identity of the 'others' and the interpretations that they imposed on the data.

Rather overlooked in past studies of the history of statistics, and placed more obviously in the histories of mechanism and computing, it is Babbage who emerges as our central figure. For a year Babbage was at the very heart of the emerging statistical movement: the man who was present at the foundational meetings in Trinity College, Cambridge in June 1833; who then represented the aspiring 'statistical section' to the rest of the British Association for the Advancement Science; at whose house some months later the Statistical Society of London was planned and created; and who led the proceedings in March 1834 when the Society was launched. Babbage was no instinctive organiser or figurehead and he was preoccupied with the Difference and Analytical Engines: by temperament swift to anger and known for his petulance as well as his charm, he retired to his workshop to let others take the helm. But he brought the movement his mathematical skills, his academic distinction in the Lucasian chair, the range of his learning, and his contacts with some of the leading savants of Europe with whom he had been mixing since the 1820s. Quetelet and Humboldt were his friends. Later, Babbage was instrumental in the foundation of the International Statistical Congress in the early 1850s. He published several statistical papers, and all the while he was at work trying to construct and perfect a machine for the processing of 'number' as he called it. If not a technical success, Babbage's engines were a triumph of the mathematical imagination. Properly contextualized, they were also a remarkable outgrowth of the embryonic statistical movement. Babbage's interest in numbers led him to both mechanical and institutional innovations from the 1820s to the 1850s: to help construct organisations for the collection of social statistics and a machine by which to compute and analyse them.

According to the Royal Statistical Society's recent *Data Manifesto*, 'What steam was to the 19th century, and oil has been to the 20th, data is to the 21st.'[29] No one better links the ages of steam and data than Babbage. He was an acknowledged expert in factory organization and technology in the steam era, the author of *On the Economy of Machinery and Manufactures* in 1832.[30] In the same year he was also the author of the article 'On... the Constants of Nature and Art', that curious but tantalizing outline of a project to collect all relevant and significant information in the natural and social orders in numerical form. Babbage envisaged a vast programme in data collection to lay down a store of the most important

---

[29] *Data Manifesto*, Sept. 2014, Royal Statistical Society, London. http://www.statslife.org.uk/images/pdf/rss-data-manifesto-2014.pdf.
[30] Charles Babbage, *On the Economy of Machinery and Manufactures* (London, 1832).

information across all subjects and disciplines, all natural phenomena and human artifice.[31] This included constants of the physical realm, the sort of universal and foundational knowledge in natural science that, once discovered, had to be preserved for all time as the basis of the physical universe and human civilization. But some of the information on social institutions and economic life would be constant in its mutability, requiring to be updated as the data changed. Anyone tempted to dismiss the project as fanciful should heed the judgment of Babbage's friend and biographer, Wilmot Buxton: 'Endowed with a rich and abundant imagination, he seldom gave run to his fancy, or allowed his mind to soar beyond the regions of familiar investigation.'[32]

Babbage provided only an outline of this huge international project that was to be spearheaded by the leading scientific organizations of Britain, France, and Germany. But in doing so he gave his readers and collaborators a sense of what we would call today 'Big Data'. Later, in her notes of 1843 explaining the Analytical Engine, Babbage's collaborator, Ada, Countess of Lovelace, intuited the machine's purpose in processing and retrieving information from among the vast store: it would undertake, at enormous speed, 'processes so lengthy and so complicated, that, although it is possible to arrive at them through great expenditure of time, labour and money, it is yet on these accounts practically almost unattainable.'[33] Of course, to use the term 'Big Data' in this way is anachronistic. Its earliest use, so far as it can be traced, is only as far back as 1997, and it did not enter the technical language of computing until 2005. Its use was later still among the population as a whole.[34] 'Big Data' denotes large data sets which, due to their size and complexity, are impossible to manage and analyse using traditional methods (and traditional methods include, in this case, the use of less powerful computing technology as it existed as recently as the end of the twentieth century).[35] The expansion of computing capacity and data storage since then allows for the analysis of massive amounts of information. Now, clinicians can search through thousands of digitized records, and, with a single computer instruction, 'look for patterns in eye scans that might help the diagnosis or treatment of macular degeneration or

---

[31] 'On the advantage of a collection of numbers to be entitled the Constants of Nature and Art', *Edinburgh Journal of Science*, n.s. xii (1832), 334–40.

[32] Buxton, *Memoir of the Life and Labour of the Late Charles Babbage*, 358.

[33] Lovelace, 'Sketch of the Analytical Engine', Note F.

[34] Gil Press, 'A Very Short History of Big Data', *Forbes Magazine*, 9 May 2013, https://www.forbes.com/sites/gilpress/2013/05/09/a-very-short-history-of-big-data/#9f36f2a65a18.

The first use in 1997 was by Michael Cox and David Ellsworth in 'Application-controlled demand paging for out-of-core visualisation', *Proceedings of the Institute of Electrical and Electronics Engineers*, 8th Conference on Visualisation. Roger Magoulas of O'Reilly Media popularized the term for technical personnel in 2005. See V. Chaorasiya and A. Shrivastava, 'A Survey on Big Data. Techniques and Technologies', in *International Journal of Research and Development in Applied Science and Engineering*, 8, 1, Sept. 2015.

[35] 'What is BIG DATA?' https://www.guru99.com/what-is-big-data.html.

diabetic retinopathy'.[36] The pathology of a single ill patient can now be compared to the pathologies of half a million patients held in data store and clinical conclusions drawn, including the choice of the best treatments. The Manchester Statistical Society emerged from the city-wide efforts of investigators to collect information by going house to house during the cholera epidemic there in 1832. Now, the website 'Patients Like Me' not only brings together those with similar medical conditions but enables them 'to offer up data about their conditions and treatment in an act of 'data altruism' for the good of others.'[37]

Big Data is also defined by the '3Vs': data volume, velocity, and variety.[38] As one author has put it, 'there's a lot of data, it's coming at you very fast and in different forms'.[39] It does not require a great leap of imagination to understand a similar response to the sudden profusion of numbers in the 1830s which underpinned so many of the reforms and social innovations of those years, and which prompted the origins of the statistical movement itself. Contemporaries then remarked on the scale of the new information suddenly available, its ever-changing nature as annual returns became the norm, and the variety of different types of information being kept—too many for the Statistical Department of the Board of Trade which had been established in 1832 as a fledgling statistical service for the whole of British government but which could never keep up, and which was eventually disbanded in the 1850s. Victorian statisticians looked for patterns in the data which might explain what seemed like otherwise random events; indeed, they analysed the numbers in the belief that data would, in itself, demonstrate an ordered universe and society. If Big Data is defined as 'extremely large data sets that may be analysed computationally to reveal patterns, trends, and associations, especially relating to human behaviour and interactions', then there are obvious affinities that link it to the statistical projects of the mid-nineteenth century.[40] Quetelet, Farr, and their collaborators also looked for recurrent patterns in human behaviour that might explain many of the key social variables, from disease and mortality to crime and suicide. Farr would surely have welcomed a recent study of the health of 600,000 people that has revealed the 'intricate links between longevity and lifestyle'. The revelation 'that people cut their life expectancy by an average of nine weeks for every two pounds they are overweight' is in the style and idiom that he made famous in the 1840s and 1850s. If today 'the

[36] Chaorasiya and Shrivastava, 'A Survey on Big Data', 290.
[37] Bradford W. Hesse, Richard P. Moser, and William T. Riley, 'From Big Data to Knowledge in the Social Sciences', *The Annals of the American Academy of Political and Social Science*, 659, May 2015, 29. See https://www.patientslikeme.com/.
[38] The idea was coined by Doug Laney, '3D Data Management: Controlling Data Volume, Velocity, and Variety', 6 Feb. 2001, https://blogs.gartner.com/doug-laney/files/2012/01/ad949-3D-Data-Management-Controlling-Data-Volume-Velocity-and-Variety.pdf.
[39] Timandra Harkness, *Big Data. Does Size Matter?* (London, 2016) (2017 edn), 16.
[40] Definition drawn from *Oxford Languages*, https://www.lexico.com/definition/big_data, 14 September 2020.

power of big data and genetics allow us to compare the effect of different behaviours and diseases in terms of months and years of life lost or gained', we are only using the kind of statistical measures and arresting calculations that made Farr's annual *Letters* on the health of the nation so compelling.[41] For this reason, there are grounds for treating the claims to innovation of the supposedly new concept of 'Big Data' with a degree of scepticism: the quest to understand the order of the world through the analysis of large numerical data sets is not new, even though the technology has changed.

'Big Data' has applications across the whole range of economic, social, and natural scientific disciplines: in the analysis of consumer and market behaviour, flows of finance and shares, the sequencing of the genomes of all living organisms from viruses to humans, the holding of patient data and clinical diagnosis, in the natural sciences from oceanography to astronomy. Yet this list bears some resemblance to the uses to which statistics were being put by the 1840s and 1850s: from the study of tides to the incidence of disease, and from the flow of funds in the City of London, as calculated by Babbage, to the calculation of the workers' 'cost of living'. Whewell spent two decades studying the tides, which required the organization of readings and calculations across the world in an age even before the invention of the telegraph. This was Humboldtian Science, 'physique du globe'. He would surely have appreciated the integration today, by modern oceanographers, of 'signals from remote buoys, satellite telemetry and sensing, oceangoing vessels, airborne weather balloons, and other sources of high-volume, high-velocity data inputs covering large geographic areas' that Big Data now makes possible.[42] It allows for the collection of many more types of information than before, and crucially, allows for their combination and correlation in the pursuit of relational connections. 'The joy of big data is its ability to connect different databases to see a bigger picture.'[43] But even in this regard Big Data may not be entirely new: the artisan statisticians of the London Statistical Society laid out their data in the *Statistical Illustrations* in 1825 in such a way that, without the benefit of correlations, their readers would immediately appreciate the relationships they were exposing between rising pauperism, rising crime, rising taxes, and lower wages.

The wider contention 'that the culture of information systems – knowledge presented efficiently – existed long before the computer, even before the electric telegraph', also seems legitimate.[44] The 'information society' has been set down as

[41] *Daily Telegraph*, 13 Oct. 2017. The article draws on the results of a study undertaken at the Usher Institute, University of Edinburgh. https://www.telegraph.co.uk/science/2017/10/13/nine-weeks-life-expectancy-every-2-lbsoverweightnewreport/#:~:text=People%20cut%20their%20life,links%20between%20longevity%20and%20lifestyle.

[42] Hesse, Moser, and Riley, 'From Big Data to Knowledge in the Social Sciences', 22.

[43] Harkness, *Big Data*, 95.

[44] Daniel R. Headrick, *When Information Came of Age. Technologies of Knowledge in the Age of Reason and Revolution, 1700–1850* (Oxford, 2000), vii.

a product of the Enlightenment: if speeds have increased over time, the principles of data production, storage, retrieval, and display were the same then as now.[45] Think of the aims and import of Diderot's *Encyclopédie*, published between 1751 and 1766, or of the *Encyclopaedia Britannica* which first appeared between 1768 and 1771, and the comparison may be seen to work. Diderot hoped 'to assemble all the knowledge scattered on the surface of the earth, to demonstrate the general system to the people with whom we live, and to transmit it to the people who will come after us'.[46] The *Encyclopédie* was originally conceived as a French version of Ephraim Chambers's *Cyclopaedia, or an Universal Dictionary of Arts and Sciences* published in London in 1728, and *Britannica*, a product of the Scottish Enlightenment, was devised as a conservative answer to the notoriously radical and heterodox *Encyclopédie*. In other words the collection and dissemination of knowledge in London, Paris, and Edinburgh were competitive, emulative, politicized, continuous, and ongoing in the eighteenth century, just as they are today in the internet age. Though the volume and degree of measurement has increased in the modern era, of course, yet we have not, in fact, just lived through 'the first measured century'—the twentieth century—to cite the title of a book published in 2000.[47] The aspiration and the reality of measurement are much older. According to one historian of information, 'Ours is not the first information age in history, for humans have always needed and used information. Yet in certain periods the methods used to handle information changed dramatically. We live in such an age, but it is not the first.'[48] In the late-seventeenth century there had been such a cultural change which manifested itself 'in an increasing interest in information of all sorts – about nature, people, events, business, and other secular topics'.[49] Political Arithmetic and 'Statistik', combinations of just these subjects, take their origin from this transition. As argued in this book, another of these critical ages occurred in the 1830s, and yet another, under the impact of the sophistication and professionalization of mathematical statistics after Galton, came at the end of the nineteenth century. With the full flowering of the 'digital revolution', the beginning of the twenty-first century marks a further 'critical stage'. We should remember that this very term, 'critical stage', was first popularized by the founding sociologist, Henri de Saint Simon, who was an inspiration to Richard Jones and William Whewell when they first read about his ideas in the early 1830s.[50]

---

[45]   Ibid., 217.

[46]   Denis Diderot, 'Encyclopédie', in *Encyclopédie; ou Dictionnaire Raisonné des Sciences, des Arts et des Métiers* (Paris, 1751–1766), http://artflsrv02.uchicago.edu/cgi-bin/philologic/getobject.pl?c.4:1252. encyclopedie0513.

[47]   Theodore Caplow, Louis Hicks, and Ben Wattenberg, *The First Measured Century. An Illustrated Guide to Trends in America, 1900–2000* (Washington D.C., 2000).

[48]   Headrick, *When Information Came of Age*, 217.   [49]   Ibid.

[50]   See above, pp. 134–5.

Two centuries ago, Babbage looked forward from one of these critical periods in the history of data, information, and numbers to another, today. In his correspondence and published work he saw the requirement for greater technical capacity in the future. He recognized that in his obstinate efforts to construct mechanical computational engines he was in advance of the times, as the phrase has it, but he foresaw a stage when the advance of science and mathematics themselves would be frustrated without new technologies for the handling of data and calculation. Hence his prediction as early as 1822 that 'a time will arrive' when 'the useful progress of science' would be impeded without the development of a technology 'for relieving it from the overwhelming encumbrance of numerical detail'.[51] That time has arrived, of course: it is now. Science has moved onwards thanks to the development of the digital computer and its ever greater numerical capacity—memory and processing power, as we would call them today. As Buxton was to put it: 'hitherto theory has been in advance of execution, [but] the Analytical Engine will enable our working powers to keep pace with the march of our theoretical knowledge'.[52]

Babbage saw himself, therefore, as one labourer in the field among many labourers, past, present, and future. He looked forward to the eventual construction, by a later engineer, of a working engine 'embodying in itself the whole of the executive department of mathematical analysis' and he was content to leave 'my reputation in his charge'.[53] He, of all people, would have understood what it meant to be 'a cog in the machine'. After he died, Quetelet, in the last year of *his* life, published a memoir of his friend in which he recalled that Babbage had 'proposed my co-operation in a work which was to contain a register of everything capable of being measured'.[54] This was the 'Constants of Nature and Art', Big Data as Babbage envisaged it in his remarkable essay in 1832, one of 'those works of science which are too large and too laborious for individual efforts...and which would be of the greatest advantage to all classes of the scientific world'.[55] According to Quetelet

The extent of this work, I said, is too vast to be carried out unless by the co-operation of many minds. The outline of what may be necessary for [a] man alone is so great that even with the help of many friends, I could not hope to complete more than a skeleton of the whole.

[51] 'On the Theoretical Principles of the Machinery for Calculating Tables. In a letter to Dr. Brewster', 6 Nov. 1822 ', *Edinburgh Philosophical Journal*, vol. 8, 1823, 128. See above, p. 123.
[52] Buxton, *Memoir of the Life and Labours of the Late Charles Babbage*, 150.
[53] Charles Babbage, *Passages from the Life of a Philosopher* (London, 1864), 450.
[54] Adolphe Quetelet, 'Extracts from a Notice of Charles Babbage, by A. Quetelet of Brussels, translated from the *Annuaire de l'Observatoire Royale de Bruxelles* for 1873', *Annual Report of the Board of Regents of the Smithsonian Institution*, 1873, 184.
[55] 'On the advantage of a collection of numbers to be entitled the Constants of Nature and Art', *Edinburgh Journal of Science*, n.s. xii (1832), 334.

Babbage replied

> that time is an element of the solution which overcomes the greatest difficulties of investigation; and if our efforts are properly directed, our descendants will finish what we have properly begun.[56]

The events, personalities, and subjects of this book may have seemed to the reader, at first sight, to be the components of a distant and unfamiliar history. But if the central figure in this story was correct—and he generally was—we are the intellectual and technical descendants of those who formed the Victorian statistical movement, developing and extending what they began. As Babbage suggested in casting forward in time, and as I hope I have shown, the Victorians who collected, tabulated, analysed and argued over numbers are closer to us than we may have realized.

---

[56] Quetelet, 'Extracts from a Notice of Charles Babbage', 184.

Bibliography

Archival Sources

Charles Babbage Papers, British Library, London.
Charles Booth Collection, British Library of Political and Economic Science, London School of Economics.
Brougham Papers, University College, London.
Buxton Collection, Oxford Museum of the History of Science.
Charles Darwin Correspondence Project, University of Cambridge.
William Farr Papers, Wellcome Library, London.
William Farr Papers, British Library of Political and Economic Science, London School of Economics.
Francis Galton Papers, University College, London.
John Herschel Papers, Royal Society, London.
Edward Jarvis, Mss Autobiography, Houghton Library, Harvard University.
Edward Jarvis, 'European Letters, 1860', 3 vols, Concord Free Public Library, Concord, Mass.
Edward Jarvis Collection, Francis A. Countway Library of Medicine, Harvard Medical School, Boston, Mass.
Minutes of the Manchester Statistical Society, Manchester Central Library.
Florence Nightingale Papers, British Library, London.
Francis Place papers, British Library, London.
L. A. J. Quetelet Papers, Académie Royale de Belgique, Brussels.
Royal Archives, Windsor.
Royal Statistical Society Archives, London.
Edward Henry Stanley papers (15th earl of Derby), Liverpool Record Office.
William Whewell Papers, Trinity College, Cambridge.

Printed Primary Sources

Acton, Lord (John Emerich Edward Dalberg-Acton), 'Mr Buckle's Thesis and Method', and 'Mr. Buckle's Philosophy of History', *Historical Essays and Studies* (eds J. N. Figgis and R. V. Laurence), (London, 1908), 305–23; 324–43.
Albert, Prince Consort, 'Inaugural Address', *Report of the Proceedings of the Fourth Session of the International Statistical Congress Held in London July 16th, 1860, and the Five Following Days* (London, Her Majesty's Stationery Office, 1861), 2–7.
Albert, Prince Consort, 'On Opening the International Statistical Congress. Held in London, July 16th 1860', *The Principal Speeches and Addresses of His Royal Highness the Prince Consort* (ed. Sir Arthur Helps), (London, 1866), 229–46.
Anon, *Select Committee of Artisans, 1823* (London, 1824).
Anon, *Statistical Illustrations of the Territorial Extent and Population, Commerce, Taxation, Consumption, Insolvency, Pauperism and Crime of the British Empire. Compiled for and Published by Order of the London Statistical Society* (London, 1825).

Anon, *Appendix to the First Edition of the Statistical Illustrations* (London, 1826).

Arbuthnot, John, 'Essay on the Usefulness of Mathematical Learning', *Life and Works of John Arbuthnot* (ed. George A. Aitken) (Oxford, 1892).

Arbuthnot, John, 'An Argument for Divine Providence, taken from the constant Regularity observ'd in the Birth of both Sexes', *Philosophical Transactions* (Royal Society), vol. 27, 1710–12, 186–90.

Arnold, Matthew, 'Schools', in T. Humphry Ward (ed.), *The Reign of Queen Victoria: A survey of fifty years of Progress* (2 vols, London, 1887), 238–87.

Aubrey, John, *The Natural History and Antiquity of the County of Surrey* (ed. Richard Rawlinson) (5 vols, 1718–19).

Babbage, Benjamin Herschel, *Babbage's Calculating Machine or Difference Engine* (London, 1872).

Babbage, Charles, *The Works of Charles Babbage* (ed. Martin Campbell-Kelly), 11 vols, (London, 1989).

Babbage, Charles, 'An examination of some questions connected with games of chance', *Transactions of the Royal Society of Edinburgh*, 9, 1821, 153–77.

Babbage, Charles, *On the Application of Machinery to the Purpose of Calculating and Printing Mathematical Tables. A Letter to Sir Humphry Davy* (London, 1822).

Babbage, Charles, 'A Note Respecting the Application of Machinery to the Calculation of Astronomical Tables', *Memoirs of the Astronomical Society*, vol. 1, pt. ii, 1822, 309–14.

Babbage, Charles, 'On the Theoretical Principles of the Machinery for Calculating Tables. In a letter to Dr. Brewster', 6 Nov. 1822', *Edinburgh Philosophical Journal*, vol. 8, 1822–23, 122–8.

Babbage, Charles, *A Comparative View of the Various Institutions for the Assurance of Lives* (London, 1826).

Babbage, Charles, 'On the proportionate number of births of the two sexes under different circumstances', *Edinburgh Journal of Science*, n.s., vol. 1 (1829), 85–104.

Babbage, Charles, 'Sur L'emploi plus ou moins fréquent des mêmes lettres dans les différentes langues', *Correspondence Mathématique et Physique*, vol. 7, 1831, 135–7.

Babbage, Charles, 'On the advantage of a collection of numbers to be entitled the Constants of Nature and Art', *Edinburgh Journal of Science*, n.s. xii (1832), 334–40.

Babbage, Charles, *On the Economy of Machinery and Manufactures* (London, 1832).

Babbage, Charles, 'Une Lettre à M. Quetelet de M. Ch. Babbage relativement à la machine à calculer', Académie royale de Belgique, Bruxelles, *Bulletins*, 2, 1835, 123–6.

Babbage, Charles, *The Ninth Bridgewater Treatise. A Fragment* (London, 1837).

Babbage, Charles [Anon], 'Addition to the Memoir of M. Menabrea on the Analytical Engine', *Philosophical Magazine*, vol. 23 (1843), 235–9.

Babbage, Charles, *Thoughts on the Principles of Taxation with Reference to Property Tax and Its Exceptions* (London, 1848).

Babbage, Charles, *The Exposition of 1851; or, Views of the Industry, the Science and the Government of England* (London, 1851).

Babbage, Charles, 'Sur les constantes de la nature – classe des mammieres', *Compte Rendu des travaux du congrés general de statistique* (Bruxelles, 1853), 222–30.

Babbage, Charles, 'On the Statistics of Lighthouses', *Compte Rendu des travaux du congrés general de statistique* (Bruxelles, 1853), 230–7.

Babbage, Charles, 'Babbage's Note Respecting the Origin of the Statistical Society, Brussels, Sept. 1853', in *JRSS*, 124, 4, 1961, 546.

Babbage, Charles, 'Table of the relative frequency of occurrence of the causes of breaking of plate glass windows', *The Mechanics' Magazine*, vol. 66, Jan.–June 1857, 82.

Babbage, Charles, 'Analysis of the Statistics of the Clearing House during the Year 1839', *JSSL*, 19, 1856, 28–48.

Babbage, Charles, 'On Tables of the Constants of Nature and Art', *Annual Report of the Board of Regents of the Smithsonian Institution for 1856* (Washington D.C., 1857), 289–302.

Babbage, Charles, 'Letter from Charles Babbage, Esq., FRS', *Report of the Proceedings of the Fourth Section of the International Statistical Congress. Held in London, July 16, 1860 and the Five Following Days* (London, 1861), 505–7.

Babbage, Charles, *Passages from the Life of a Philosopher* (London, 1864).

Babbage, Charles, *Charles Babbage and his Calculating Engines. Selected Writings by Charles Babbage and Others* (eds Philip Morrison and Emily Morrison) (New York, 1961).

Babbage, Charles, and John Herschel, 'Barometric Observations at the Fall of the Staubbach. In a Letter from Mr. Babbage to Dr. Brewster', *Edinburgh Philosophical Journal*, 6, 12, 1822, 224–7.

Babbage, Henry Prevost, 'On the Mechanical Arrangements of the Analytical Engine', in Henry Prevost Babbage (ed.), *Babbage's Calculating Engines* (London, 1889), 331–7.

Baker, Thomas, *Memorials of a Dissenting Chapel...Being a Sketch of the Rise of Nonconformity in Manchester* (London, 1884).

Bazard, Amand, *Exposition de la doctrine de St Simon* (2 vols, Paris, 1828–1830).

Bielfeld, J. F. von, *The Elements of Universal Erudition* (trans. W. Hooper) (3 vols, London, 1770).

Booth, Charles, 'Occupations of the People of the United Kingdom, 1841–81', *JSSL*, vol. 49, 2, June 1886, 314–444.

Booth, Charles, 'The Inhabitants of Tower Hamlets (School Board Division), their Condition and Occupations', *JRSS*, 50, 2, 1887, 326–401.

Booth, Charles, 'The Condition and Occupations of the People of East London and Hackney, 1887', *JRSS*, 51, 2, 1888, 276–339.

Booth, Charles, 'Enumeration and Classification of Paupers, and State Pensions for the Aged', *JRSS*, 54, 4, Dec. 1891, 600–43.

Booth, Charles, 'The Inaugural Address of Charles Booth Esq., President of the Royal Statistical Society. Session 1892–93. Delivered 15th November 1892' (on Dock Labour), *JRSS*, 55, 4, Dec. 1892, 521–57.

Booth, Charles, 'Life and Labour of the People in London: First Results of An Inquiry Based on the 1891 Census. Opening Address of Charles Booth, Esq., President of the Royal Statistical Society. Session 1893–94', *JRSS*, 56, 4, Dec. 1893, 557–93.

Booth, Charles, 'Statistics of Pauperism in Old Age', *JRSS*, 57, 2, June 1894, 235–53.

Booth, Charles, *The Life and Labour of the People of London* (17 vols in 3 series: Poverty, Industry, Religion) (London, 1889–1903).

Booth, Charles, *Old Age Pensions and the Aged Poor: A Proposal* (London, 1899).

Booth, Charles, *Charles Booth's Poverty Maps* (Mary S. Morgan et al.), (Thames & Hudson, London, 2019).

Bray, J. F., *Labour's Wrongs and Labour's Remedy, or, the Age of Might and the Age of Right* (Leeds, 1839).

British Association for the Advancement of Science, *Lithographed Signatures of the Members of the British Association for the Advancement of Science, who met at Cambridge, June MDCCCXXXIII with a Report of the Proceedings at the Public Meetings During the Week; and an Alphabetical List of the Members* (Cambridge, 1833).

British Association for the Advancement of Science, *Report of the Third Meeting of the British Association for the Advancement of Science; Held at Cambridge in 1833* (London, 1834).

Brown, Samuel, 'Report on the Sixth International Statistical Congress', *JSSL*, 31, 1, 1868, 11–24.

Brown, Samuel, 'Report of the Seventh International Statistical Congress, held at the Hague, 6–11 September 1869', *JSSL*, 32, 4, 1869, 391–410.

Brown, Samuel, 'Report on the Eighth International Statistical Congress held at St. Petersburg, August 1872', *JSSL*, 35, 4, Dec. 1872, 431–57.

Buckle, Henry Thomas, *History of Civilization in England* (2 vols) (London, 1857, 1861).

Buckle, Henry Thomas, *On Scotland and the Scottish Intellect* (ed. H. J. Hanham) (Chicago and London, 1970).

Burke, Edmund, *Reflections on the Revolution in France* (London, 1790).

Buxton, H. W., *The Life and Labours of the Late Charles Babbage* (1872–80) (ed. Anthony Hyman) (Cambridge, Mass., 1988).

Byron, *Don Juan* (1819–24).

Capper, Benjamin P. *A Statistical Account of the Population and Cultivation, Produce and Consumption of England and Wales* (London, 1801).

Carlyle, Thomas, 'Signs of the Times' (1829) and 'Chartism' (1839) in *Thomas Carlyle: Selected Writings* (ed. Alan Shelston) (Penguin edn., London, 1971), 59–85; 149–232.

Central Society of Education, 'Analysis of the Reports of the Committee of the Manchester Statistical Society on the State of Education in the boroughs of Manchester, Liverpool, and Salford and Bury', in *The First Publication of the Central Society of Education* (London, 1837).

Chadwick, Edwin, *Report of the Sanitary Condition of the Labouring Population of Great Britain* (1842) (ed. M. W. Flinn), (Edinburgh, 1965).

Chamberlain, Joseph, 'Labourers' and Artisans' Dwellings', *The Fortnightly Review*, cciv, 1 Dec. 1883, 761–76.

Child, G. M., 'On the Necessary Limits of the Applicability of the Method of Statistics, with Especial Reference to Sanitary Investigation', *Sessional Proceedings of the National Association for the Promotion of Social Science 1875–6* (London, 1876), 47–56.

Cobbett, William (ed.), *Cobbett's Parliamentary History of England. From the Earliest Period to the Year 1803* (36 vols, London, 1806–20), vol. 14, 1747–1753 (London, 1813).

Coleridge, Samuel Taylor, '128. Statistics', in Robert Southey, *Omniana or Horae Otiosiores*, vol. 1 (London, 1812).

Coleridge, Samuel Taylor, *Biographia Literaria, Or Biographical Sketches of My Literary Life and Opinions* (London, 1817).

Coleridge, Samuel Taylor, *The Notebooks of Samuel Taylor Coleridge* (ed. K. Coburn and A. J. Harding) (5 vols. Abingdon, 2002).

Commission Centrale de Statistique, *Bulletins de Commission Centrale de Statistique*, tomes i–vi (Brussels, 1843–55).

Cooke Taylor, W., 'Objects and Advantages of Statistical Science', *Foreign Quarterly Review*, vol. 16, no. 31, Oct. 1835, 103–16.

Cooke Taylor, W., *The Factory System and the Factory Acts* (London, 1894).

Cunningham, William, 'The Perversion of Economic History', *The Economic Journal*, ii, Sept. 1892, 491–8.

Czoernig, Karl von, *Ethnographie der österreichischen Monarchie* (2 vols, Vienna, 1855, 1857).

Darwin, Charles, *The Voyage of the Beagle* (London, 1839).

Darwin, Charles, *The Origin of Species By Means of Natural Selection, Or the Preservation of Favoured Races in the Struggle for Life* (London, 1859).

Darwin, Charles, *The Autobiography of Charles Darwin 1809–1882* (ed. Nora Barlow) (London, 1958).

Darwin, Charles, *The Life and Letters of Charles Darwin* (ed. Sir Francis Darwin) (3 vols, London, 1887).

Davenant, Charles, *The Political and Commercial Works of that Celebrated Writer Charles D'Avenant* (ed. C. Whitworth) (5 vols, London, 1771).

Davenant, Charles, *Discourses on the Publick Revenues, and on the Trade of England* (1698).

Derham, William, *Physico-Theology; or, a Demonstration of the Being and the Attributes of God from his Works of Creation* (London, 1713).

Dickens, Charles, 'Full Report of the First Meeting of the Mudfog Association for the Advancement of Everything', *Bentley's Miscellany*, 2 (1837), 397–413, reprinted in Charles Dickens, *The Mudfog Papers* (London, 1880).

Dickens, Charles, *Hard Times* (London, 1854).

Dickens, Charles, *Little Dorrit* (London, 1857).

Disraeli, Benjamin, *Coningsby: Or the New Generation* (London, 1844).

Dostoyevsky, Fyodor, *Notes from the Underground* (1864).

[Drinkwater, John, 'Mr Drinkwater's Notes' 1833–4] 'The Royal Statistical Society', *Journal of the Royal Statistical Society* (hereafter *JRSS*), 98, 1, 1935, 140–51.

Eden, Frederick Morton, *The State of the Poor, or, An history of the labouring classes in England from the conquest to the present period; in which are particularly considered their domestic economy with respect to diet, dress, fuel and habitation; and the various plans which, from time to time, have been proposed and adopted for the relief of the poor etc.* (3 vols, London, 1797).

Eden, Frederick Morton, *An Estimate of the Number of the Inhabitants in Great Britain and Ireland, 1800. Written while the Census Bill was before Parliament; partly extracted from The State of the Poor* (London, 1800).

Edgeworth, F. Y., 'Methods of Statistics', *JSSL*, 1885 (Jubilee volume), 181–217.

Eliot, George, *Daniel Deronda* (London, 1876) (New York, Penguin edn., 1967).

Engels, Friedrich, *The Condition of the Working Class in England in 1844* (Leipzig, 1845), translated and edited by W. O. Henderson and W. H. Chaloner (Oxford, 1958).

Ernest II, *Memoirs of Ernest II, Duke of Saxe-Coburg-Gotha* (4 vols, London, 1888–90).

Fairbairn, W., 'A brief memoir of the late John Kennedy, esq.', *Memoirs of the Literary and Philosophical Society of Manchester*, 3rd ser., 1 (1862), 147–57.

Faraday, Michael, *The Correspondence of Michael Faraday* (ed. Frank A. James) (Stevenage, 1996).

Farr, William, 'Vital Statistics; or, the Statistics of Health, Sickness, Diseases and Death', in J. R. McCulloch (ed), *A Statistical Account of the British Empire: Exhibiting its Extent, Physical Capacities, Population, Industry, And Civil and Religious Institutions* (2 vols, London, 1837).

Farr, William, 'On a Method of Determining the Danger and Duration of Diseases at Every Period of Their Progress, Article I', *British Annals of Medicine*, 1, 1837, 72–9.

Farr, William, 'On the Law of Recovery and Dying in Small-Pox, Article II', *British Annals of Medicine*, 1, 1837, 134–43.

Farr, William, 'On Prognosis', *British Medical Almanack*, 1838, 199–216.

Farr, William, 'Note on the Present Epidemic of Small-Pox, and on the Necessity of Arresting Its Ravages', *The Lancet*, 1840–41, i, 353.

[Farr, William], General Register Office, Letters to the Registrar General, *Annual Reports of the Registrar-General of Births, Deaths and Marriages in England* (vols 1–41).

Farr, William, *Report on the Mortality of Cholera in England, 1848–9* (HMSO, London, 1852).

Farr, William, 'Influence of Elevation on the Fatality of Cholera', *JSSL*, 15, 1852, 155–83.

Farr, William, 'Report on the Programme of the Fourth Session of the Statistical Congress', *Programme of the Fourth Session of the International Statistical Congress to be held in London on July 16th and Five Following Days* (London, 1860, Her Majesty's Stationery Office).

Farr, William, 'Reports of the Official Delegates from England at the Meeting of the International Statistical Congress in Berlin', *JSSL*, 26, 4, Dec. 1863, 412–19.

Farr, William, 'Address on Public Health', *Transactions of the National Association for the Promotion of Social Science*, 1866 (Manchester) (London, 1867), 67–83.

Farr, William, 'Report on the Cholera Epidemic of 1866 in England'. *Supplement to the 29th Annual Report of the Registrar General, Parliamentary Papers, 1867–8, xxxvii*.

Farr, William, 'Inaugural Address as President of the Statistical Society of London, delivered 21ˢᵗ Nov. 1871', *JSSL*, vol. 34, 4, 409–23.

Farr, William, 'Presidential Address 1872', *JSSL*, 35, 4, 417–30.

Farr, William, 'Economic Science and the British Association', 'Considerations…favourable to the maintenance of Section F', *JSSL*, vol. 40, 3, Sept. 1877, 473–76.

Farr, William, 'Density or Proximity of Population: Its Advantages and Disadvantages', *Transactions of the National Association for the Promotion of Social Science, 1878* (London, 1879), 530–5.

Fitzmaurice, E. C., *The Life of Sir William Petty, 1623–1687*(London, 1895).

Fletcher, Joseph, 'Moral and Educational Statistics of England and Wales', *JSSL*, 12, 1849, 151–76, 188–335.

Fox, J. J., 'On the Province of the Statistician', *JSSL*, 23, 3, 1860, 330–6.

Galton, Francis, 'Letter addressed by Francis Galton, Esq., to the Secretary', *Journal of the Royal Geographic Society*, vol. 24, 1854, 345–58.

Galton, Francis, *Meteorographia: Or, Methods of Mapping the Weather* (London, 1863).

Galton, Francis, 'Hereditary Talent and Character', *Macmillan's Magazine*, June and August 1865, vol. xii, 157–66, 318–27.

Galton, Francis, *Hereditary Genius* (London, 1869).

Galton, Francis, 'Barometric Prediction of Weather', *British Association Reports*, 1870, 31–3.

Galton, Francis, 'Hereditary Improvement', *Fraser's Magazine*, Jan 1873, vol. vii, 116–30.

Galton, Francis, 'Relative Supplies from Town and Country Families to the Population of Future Generations', *JSSL*, 26, March 1873, 19–26.

Galton, Francis, *English Men of Science, their Nature and Nurture* (London, 1874).

Galton, Francis, 'Nuts and Men', *The Spectator*, 30 May 1874, 689.

Galton, Francis, 'Statistics by Intercomparison, with Remarks on the Law of Frequency of Error', *London, Edinburgh and Dublin Philosophical Magazine and Journal of Science*, 49, 1875, 33–46.

Galton, Francis, 'Typical Laws of Heredity', *Proceedings of the Royal Institution*, 1877, vol. 8, 282–301.

Galton, Francis, 'Economic Science and the British Association', 'Considerations Adverse to the Maintenance of Section F (Economic Science and Statistics)', *JSSL*, Sept. 1877, 468–73.

Galton, Francis, 'The Anthropometric Laboratory', *Fortnightly Review*, 31, 1882, 332–8.

Galton, Francis, *Inquiries Into Human Faculty and Its Development* (London, 1883).

Galton, Francis, 'The Application of a Graphic Method to Fallible Measures', *JSSL*, Jubilee Volume, 1885, 262–71.

Galton, Francis, 'Regression towards Mediocrity in Hereditary Stature', *Journal of the Anthropological Institute*, 15, 1886, 246–63.

Galton, Francis, 'Family Likeness in Stature', *Proceedings of the Royal Society* 1886, vol. 40, 42–63.

Galton, Francis, 'Co-relations and Their Measurement, Chiefly from Anthropometric Data', *Proceedings of the Royal Society*, 1889, vol. 45, 135–45.

Galton, Francis, *Natural Inheritance* (London, 1889).

Galton, Francis, *Memories of My Life* (London, 1908).

Galton, Francis, 'The Possible Improvement of the Human Breed under the Existing Conditions of Law and Sentiment', in Galton, *Essays in Eugenics* (London, Eugenics Education Society, 1909), 1–34.

General Association, *A Narrative and Exposition of the Origin, Progress, Principles, Objects etc. of the General Association, Established in London, for the Purpose of Bettering the Condition of the Manufacturing and Agricultural Labourers* (London, 1827).

Giffen, Robert, 'The Progress of the Working Classes in the Last Half Century. Being the Inaugural Address of R. Giffen LL.D., President of the Statistical Society, Delivered 20th November 1883', *JSSL*, 46, 1883, 593–622.

Gladstone, William E., *The Gladstone Diaries* (eds, M. R. D. Foot and H. C. G. Matthew) (14 vols) (Oxford, 1968–94).

Grant Duff, M. E., 'Address of the President of Section F of the British Association, at the Fifty-first Meeting, held at York, in August 1881', *JSSL*, 44, 4, Dec. 1881, 649–59.

Graunt, John, *Natural and Political Observations mentioned in a following Index, and made upon the Bills of Mortality, by John Graunt, Citizen of London. With reference to the Government, Religion, Trade, Growth, Ayre, Diseases, and several Changes of the said City* (London, 1662).

Greg, S and W. R., *Analysis of the Evidence Taken Before the Factory Commissioners as far as it relates to the Population of Manchester and the Vicinity Engaged in the Cotton Trade* (Manchester, 1834).

Greg, W. R., *Social Statistics of the Netherlands* (Manchester, 1835).

Guy, William, 'On the Value of the Numerical Method as Applied to Science, but especially to Physiology and Medicine', *JSSL*, vol. 2, 1, 1839, 25–47.

Guy, William, *On the best method of collecting and arranging facts: with a proposed new plan of a common-place book* (London, 1840).

Guy, William, 'On the Health of Nightmen, Scavengers and Dustmen', *JSSL*, 11, 1848, 72–81.

Guy, William, 'On the Duration of Life as affected by the Pursuits of Literature, Science and Art: with a summary view of the Duration of Life among the Upper and Middle Classes of Society', *JSSL*, 22, 1859, 337–61.

Guy, William, 'Statistical Methods and Signs', *Report of the Proceedings of the Fourth Session of the International Statistical Congress Held in London July 16th, 1860, and the Five Following Days* (London, Her Majesty's Stationery Office, 1861), 379–83.

Guy, William, 'Dr. Guy's Report on Alleged Fatal Poisoning by Emerald Green; and on the Poisonous Effects of that Substance as used in the Arts', Appendix 3, *Fifth Report of the Medical Officer of the Privy Council*, PP 1863, xxv, 126–62.

Guy, William, 'On the Original and Acquired Meaning of the term "Statistics", and on the Proper Functions of a Statistical Society; also on the Question Whether there be a Science of Statistics; and if so, what are its Nature and Objects, and what is its relation to Political Economy and "Social Science"', *JSSL*, 28, 1865, 478–93.

Guy, William, 'On the Mortality of London Hospitals: and incidentally on the deaths in the Prisons and Public Institutions of the Metropolis', *JSSL*,1867, 30, 293–322.

Guy, William, *Public Health: A Popular Introduction to Sanitary Science* (2 vols, London, 1870, 1874).

Guy, William, 'Inaugural Address as President of the Statistical Society of London, 17th Nov. 1874', *JSSL*, 37, 1874, 411–36.

Guy, William, 'Statistical Development, with Special Reference to Statistics as a Science', *JSSL*, Jubilee Volume (1885), 72–86.

Hallam, Henry, *The constitutional history of England from the accession of Henry VII to the death of George II* (2 vols, London, 1827).

Hallam, Henry, *Introduction to the literature of Europe in the fifteenth, sixteenth, and seventeenth centuries*(4 vols, London, 1837–9).

Hamilton Dickson, J. D. 'Appendix' to F. Galton, 'Family Likeness in Stature', *Proceedings of the Royal Society*1886, vol. 40, 63–73.

Harcourt, E. W. (ed.), *The Harcourt Papers* (1880–1905, Oxford).

Harding, S., 'Home and Foreign Policy, or How to Restore Prosperity to a Distressed and Anxious People', *Industrial Remuneration Conference. The Report of Proceedings and Papers Read* (London, 1885), 235–40.

Harris, W. J., MP, 'Do any remediable causes influence prejudicially the well-being of the Working Classes?', *Industrial Remuneration Conference. The Report of Proceedings and Papers Read* (London, 1885), 221–30.

Harrison, Frederic, 'The Limits of Political Economy', *Fortnightly Review*, vol. I, 1865, 356–76.

Harrison, Frederic, 'A New Industrial Inquiry', *Pall Mall Gazette*, 8 Sept. 1884.

Harrison, Frederic, *Autobiographic Memoirs* (2 vols, London, 1911).

Herschel, J.F.W., *Preliminary Discourse on the Study of Natural Philosophy* (London, 1830).

Herschel, John, and Charles Babbage, 'Barometric Observations at the Fall of the Staubbach. In a Letter from Mr. Babbage to Dr. Brewster', *Edinburgh Philosophical Journal*, 6, 12, 1822, 224–7.

[Herschel, Sir John], 'Quetelet on Probabilities', *Edinburgh Review*, xcii (July–Oct. 1850), 1–56.

Heywood, James, 'Report of an Enquiry, Conducted from House to House, into the State of 176 families in Miles Platting, within the Borough of Manchester in 1837', *JSSL*, 1, 34–6.

Holland, George Calvert, *Inquiry into the Moral, Social and Intellectual Condition of the Industrious Classes of Sheffield* (London, 1839).

Holland, George Calvert, *The Vital Statistics of Sheffield* (London, 1843).

Humboldt, Alexander von, *Personal Narrative of Travels to the Equatorial Regions of the New Continent during the Years 1799–1804* (7 vols, London, 1814–29).

Humboldt, Alexander von, *Cosmos: Sketch of a Physical Description of the Universe*, 5 vols (London, 1849–62).

Huth, Alfred Henry, *The Life and Writings of Henry Thomas Buckle* (2 vols, London, 1880).

Industrial Remuneration Conference.The Report of Proceedings and Papers Read (London, 1885).

Institut International de Statistique, *Bulletin de L'Institut International de Statistique*.

International Statistical Congress, *Comte Rendu des Travaux du Congrès Générale de Statistique, réuni à Bruxelles les 19, 20, 21 et 22 Septembre 1853* (Brussels, 1853).

International Statistical Congress, *Report of the Proceedings of the Fourth Session of the International Statistical Congress Held in London July 16th, 1860, and the Five Following Days* (London, Her Majesty's Stationery Office, 1861).

International Statistical Congress, 'The Permanent Commission of the International Statistical Congress', *JSSL*, 37, 1874, 116.

International Statistical Congress, 'The Permanent Commission of the International Statistical Congress', *JSSL*, 41, 1878, 549–50.

Jacob, William, 'Report on the Agriculture and the Trade in Corn in Some Continental States of Northern Europe', *The Pamphleteer*, 29, no. lviii, 361–456.

Jacob, William, 'Observations and Suggestions Respecting the Collation, Concentration, and Diffusion of Statistical Knowledge Regarding the State of the United Kingdom', *Transactions of the Statistical Society of London*, vol. I, pt. i, 1–2.

Jarvis, Edward, 'Sanitary Reform', *American Journal of Medical Sciences*, vol. 15, April 1848, 419–50.

Jarvis, Edward, 'On the Laws and Practice of Registration in America'; 'On the Crimes of Males and Females'; 'On the further inquiry in the Census as to the Personal Health and Power of each person' and other contributions in *Report of the Proceedings of the Fourth Session of the International Statistical Congress Held in London July 16th, 1860, and the Five Following Days* (London, Her Majesty's Stationery Office, 1861), 51–5; 176; 264–7; 271–2; 277–83; 446–7; 497–9.

Jevons, W. S., *Papers and Correspondence of William Stanley Jevons* (eds. R. D. Collison Black and Rosamund Könekamp) (7 vols, London, 1972–1981).

Jones, Richard, *Literary Remains Consisting of Lectures and Tracts of the late Rev. Richard Jones* (William Whewell ed.), (London, 1859).

Jones, Richard, 'An Introductory Lecture on Political Economy, Delivered at King's College, London, February 27th, 1833', in William Whewell (ed.), *Literary Remains, Consisting of Lectures and Tracts of the Rev. Richard Jones* (London, 1859), 539–79.

Kant, Immanuel, 'Idee zu einer allgemeinen Geschichte in weltbürgerlicher Absicht' ('Idea for a Universal History with a Cosmopolitan Aim'), *Berlinische Monatsschrift*, iv (November 11, 1784). https://www.cambridge.org/core/books/kants-idea-for-a-universal-history-with-a-cosmopolitan-aim/idea-for-a-universal-history-with-a-cosmopolitan-aim/8B2BA346A82FA006AB982E3A941E2A26

Kay, James Phillips, *The Moral and Physical Condition of the Working Classes Employed in the Cotton Manufacture in Manchester* (London, 1832).

Kay, James Phillips, 'On the Establishment of County or District Schools for the training of the Pauper Children Maintained in Union Workhouses, Pt.1, Schools', *JSSL*, 1, May 1838, 14–27.

King, Gregory, 'A Scheme of the Income & Expence of the Several Families of England Calculated for the Year 1688' and 'The General Account of England, France & Holland for the Years 1688 & 1695', in Gregory King, *Natural and Politicall Observations Upon the State and Condition of England* (1696) (ed. G. E. Barnett, Baltimore, 1936).

Lardner, Dionysus, 'Babbage's Calculating Engine', *Edinburgh Review*, 59, 1834, 263–327.

Levi, Leone, *Resume of the Statistical Congress, held at Brussels, September 11th, 1853, for the purpose of introducing unity in the Statistical Documents of all Countries. Read before the Statistical Society of London, 21st November 1853* (London, 1853).

Levi, Leone, 'Resume of the Second Session of the International Statistical Congress held at Paris, September, 1855', *JSSL*, 19, March 1856, 1–11.

Linnean Society, *The Darwin-Wallace Celebration held on Thursday 1st July 1908 by the Linnean Society of London* (London, 1908).

Lovelace, Ada, 'Sketch of the Analytical Engine Invented by Charles Babbage by L. F. Menebrea. With Notes upon the Memoir by the Translator, Ada Augusta, Countess of Lovelace', *Taylor's Scientific Memoirs*, 3, 1843, 666–731.

Ludlow, J. M., 'Account of the West Yorkshire Coal-Strike and Lock-Out of 1858', in *Trades' Societies and Strikes. Report of the Committee on Trades' Societies appointed by the National Association for the Promotion of Social Science* (London, 1860), 11–51.

Lyttelton, Lord, 'Address on Education', *Transactions of the National Association for the Promotion of Social Science*, 1868 (London, 1869), 38–74.

Mailly, Edouard, 'Essai sur la vie et les ouvrages de Quetelet', *Annuaire de L'Académie Royale des Sciences, des Lettres et des Beaux-Arts de Belgique*, 46, (1875), 109–297.

Malthus, T. R., *An Essay on the Principle of Population* (1798) (2nd edn, 1803).

Malthus, T. R., *Principles of Political Economy Considered with a View to their Practical Application* (London, 1820).

Malthus, T. R., *The Works of Thomas Robert Malthus* (eds. E. A. Wrigley and David Soudan) (8 vols. London, 1986).

Manchester Statistical Society, *On the State of Education in Manchester in 1834* (Manchester, 1835).

Manchester Statistical Society, *Report of the Manchester Statistical Society on the State of Education in the Borough of Liverpool* (London, 1836).

Manchester Statistical Society, *Report of a Committee on the Condition of the Working Classes in an Extensive Manufacturing District in 1834, 1835, and 1836* (Manchester, 1838).

Marshall, Alfred, 'How far do remediable causes influence prejudicially (a) the continuity of employment, (b) the rate of wages?', *Industrial Remuneration Conference. The Report of Proceedings and Papers Read* (London, 1885), 173–99.

Marshall, Alfred, 'On the Graphic Method of Statistics', *JSSL*, 1885 (Jubilee volume), 251–60.

Marshall, Alfred, '"The Perversion of Economic History": A Reply', *The Economic Journal*, ii, Sept. 1892, 507–19.

Marx, Karl, *Theories of Surplus Value*, (3 parts), Pt. III (Moscow edn, 1971, trans. J. Cohen and S. W. Ryazanskaya).

Marx, Karl, 'Preface to the First Edition', *Capital* (1867 edn) (Harmondsworth, 1976, ed. Ernest Mandel).

Marx, Karl, *Capital*, vol. III (1894 edn.) https://www.marxists.org/archive/marx/works/1894-c3/

Maxwell, James Clerk, *Scientific Papers of James Clerk Maxwell* (W. D. Niven ed.) (2 vols, Cambridge, 1890).

McCulloch, J. R., *A Discourse on the Rise, Progress, Peculiar Objects and Importance of Political Economy: Containing an Outline of a Course of Lectures on the Principles and Doctrines of that Science* (Edinburgh, 1824).

McCulloch, J. R., 'The State and Defects of British Statistics', *Edinburgh Review*, vol. 61, April 1835, 154–81.

McCulloch, J. R. (ed.), *A Statistical Account of the British Empire: Exhibiting its Extent, Physical Capacities, Population, Industry, And Civil and Religious Institutions* (2 vols, London, 1837).

Menabrea, L. F., 'Notions sur la machine analytique de M. Charles Babbage', *Bibliothèque universelle de Genève*, 82, Oct. 1842, 352–76.

Menabrea, L. F., 'Letter to the Editor of *Cosmos*', *Cosmos*, vol. 6, 1855, 421–2, in Babbage, *Works*, vol. 3, 171–4.

[Merivale, Herman], 'Galton on Hereditary Genius', *Edinburgh Review*, vol. 132, 269, July 1870, 100–25.

Mill, J. S., *Earlier Letters of John Stuart Mill, 1812–1848*, ed. F. E. Mineka (Toronto, 1963).

Mill, J. S., 'On the definition of political economy; and on the method of philosophical investigation in that science', *London and Westminster Review*, iv and xxvi (October 1836), 1–29.

Mill, J. S., *A System of Logic Ratiocinative and Inductive. Being a Connected View of the Principles of Evidence and the Methods of Scientific Investigation* (London, 1843) *Collected Works of John Stuart Mill* (ed. J. M. Robson), vols. 7–8 (Toronto, 1973).

Mill, J. S., *Autobiography* (London, 1873).

Morgan, Sophia De, *Memoir of Augustus De Morgan, by his wife Sophia Elizabeth De Morgan, with selections from his letters* (London, 1882).

Mouat, Frederic J., 'Preliminary Report of the Ninth International Statistical Congress, held at Buda-Pesth, from 1st to 7th September 1876', *JSSL*, vol. 39, Dec. 1876, 628–47.

Mouat, Frederic J., 'Second and Concluding Report of the Ninth International Statistical Congress', *JSSL*, 40, Dec. 1877, 531–56.

Mouat, Frederic J., 'History of the Statistical Society of London', *JRSS*, Jubilee Volume, June 1885, 14–71; 359–71.

Neumann-Spallart, F. X. von, 'Résumé of the Results of the International Statistical Congresses and Sketch of Proposed Plan of an International Statistical Association', *Jubilee Volume of the Statistical Society* (London, 1885), 284–320.

Neumann-Spallart, F. X. von, 'La fondation de l'institut international de statistique', *Bulletin de l'institut international de statistique*, 1, 1886, 1–32.

Newmarch, William, 'The Progress of Economic Science during the Last Thirty Year', *JSSL*, 24, 1861, 451–71.

Newmarch, William, 'The President's Inaugural Address on the Progress and Present Condition of Statistical Inquiry', *JSSL*, 32, Dec. 1869, 359–84.

Nightingale, Florence, 'Proposals for an Uniform Plan of Hospital Statistics', *Report of the Proceedings of the Fourth Session of the International Statistical Congress Held in London July 16th, 1860, and the Five Following Days* (London, Her Majesty's Stationery Office, 1861), 173–4.

Nightingale, Florence, 'Hospital Statistics and Hospital Plans', *Transactions of the National Association for the Promotion of Social Science*, 1861 (London, 1862), 554–60.

Nightingale, Florence, *Dear Miss Nightingale. A Selection of Benjamin Jowett's Letters to Florence Nightingale 1860–93* (eds E. V. Quinn and John Prest), (Oxford, 1987).

Nightingale, Florence, *Ever Yours. Florence Nightingale. Selected Letters* (eds., M. Vicinus and B. Nergaard) (London, 1989).

[Pattison, Mark], 'History of Civilization in England', *Westminster Review*, vol. XII, n.s., Oct. 1857, 375–99.

Pearson, Karl, *The Ethic of Freethought. A Selection of Essays and Lectures* (London, 1888).

[Pearson, Karl], 'Editorial (II.) The Spirit of *Biometrika*', *Biometrika*, 1, 1, 1 Oct. 1901, 3–6.

Pepys, Samuel, *Diary* (London, 1893 edn).

Petty, William, *Several Essays in Political Arithmetic* (1691) (4th edn, London, 1755).

Petty, William, *Political Anatomy of Ireland* (1672).

Petty, William, *Hiberniae Delineatio* (1685).

Petty, William, *A Treatise of Taxes and Contributions* (1662).

Petty, William, *Verbum Sapienti* (1665, published 1691).

Petty, William, *Quantulum conque Concerning Money* (1682, published 1695).

Petty, William, *Essay Concerning the Multiplication of Mankind* (1682, published 1695).

Petty, William, *Political Arithmetick, or a Discourse concerning the extent and value of Lands, People, Buildings; Husbandry, Manufacture, Commerce, Fishery, Artizans, Seamen, Soldiers; Public Revenues, Interest, Taxes* (1677, published 1691).

Peuchet, Jacques, *Statistique Élémentaire de la France. Contenant les Principes de cette Science et leur application a l'analyse de la Richesse, des Forces et de la Puissance de l'Empire Francais* (Paris, 1805).

Playfair, William, *The Commercial and Political Atlas: Representing, by Means of Stained Copper-Plate Charts, the Progress of the Commerce, Revenues, Expenditure and Debts of England during the Whole of the Eighteenth Century* (London, 1787).

Playfair, William, *Lineal Arithmetic applied to show the Progress of the Commerce and Revenue of England During the Present Century* (London, 1798).

Playfair, William, *Statistical Breviary Shewing, on a Principle Entirely New, the Resources of Every State and Kingdom in Europe* (London, 1801).

Playfair, William, *An Inquiry into the Permanent Causes of the Decline and Fall of Powerful and Wealthy Nations* (London, 1805).

Plot, Robert, *Natural History of Oxfordshire, being an Essay toward the Natural History of England* (Oxford, 1676).

Political Economy Club, *Minutes of Proceedings, 1899–1920. Roll of Members and Questions Discussed, 1821–1920, with Documents bearing on the History of the Club*, vi (London, 1921).

Porter, G. R., 'On the Connexion between Crime and Ignorance, as exhibited in Criminal Calandars', *Transactions of the Statistical Society of London*, I, 97–103.

Porter, G. R., *The Progress of the Nation in its Various Social and Economical Relations* (London, 1836) (3rd edn, 1851).

Powell, John, *A Letter Addressed to Edward Ellice, esq., M.P., on the General Influence of Large Establishments of Apprentices in Producing Unfair Competition, Demoralisation of Character, Parish Burthens, Insufficient Workmen, Injured Credit, and Decay of Trade. With Remarks on The Prevailing Theories on Freedom of Trade, and the Justice and Policy of Regulation* (London, 1819).

Powell, John, *A Letter Addressed to Weavers, Shopkeepers, and Publicans, on the Great Value of the Principle of the Spitalfields Acts: in Opposition to the Absurd and Mischievous Doctrines of the Advocates for their Repeal* (London, 1824).

Powell, John, *An Analytical Exposition of the Erroneous Principles and Ruinous Consequences of the Financial and Commercial Systems of Great Britain. Illustrative of their Influence on the Physical, Social and Moral Condition of the People. Founded on the* Statistical Illustrations of the British Empire (London, 1826).

Powell, John, *Plain Reasons for Parliamentary Reform, in familiar letters to a friend; Being the substance of a speech delivered at a meeting of the Clerkenwell Reform Union, September 21, 1831* (London, 1831).

Price, Richard, *An Essay on the Population of England from the Revolution to the Present Time* (London, 1780).

Price, Richard, *Observations on Reversionary Payments: on Schemes for Providing Annuities for Widows, and for Persons in Old Age; on the Method of Calculating the Values of Assurances on Lives; and on the National Debt* (London, 1783).

Quetelet, L. A. J., *Instructions Populaire sur le Calcul des Probabilités* (Brussels, 1828).

Quetelet, L. A. J., *Sur la possibilité de mésurer l'influence des causes qui modifient les Élémens Sociaux* (Brussels, 1832).

Quetelet, L. A. J., *Recherches sur le penchant au crime aux differens ages* (Brussels, 1831).

Quetelet, L. A. J., 'Notes Extraites d'un Voyage en Angleterre, aux moins de Juin et Juillet, 1833', *Correspondance mathematique et physique de l'Observatoire de Bruxelles*,Tome Dieuxième (Brussels, 1835), 1–18.

Quetelet, L. A. J., *Sur L'Homme et le Développement de ses Facultés. Physique Sociale* (Brussels, 1835).

Quetelet, L. A. J., *A Treatise on Man and the Development of his Faculties* (Tr. R. Knox), (Edinburgh, 1842) (1969 edn, Gainesville, Florida).

Quetelet, L. A. J., *Lettres à S.A.R. le Duc Regnant de Saxe-Cobourg et Gotha, sur la Théorie des Probabilités, Appliquée aux Sciences Morales et Politiques* (Brussels, 1846).

Quetelet, L. A. J., 'Sur les indiens O-Jib-Be-Wa's et les proportions de leur corps', *Bulletin de L'Académie Royale des Sciences, des Lettres, et des Beaux-Arts de Belgique*, 15, 1, 1846, 70–6.

Quetelet, L. A. J., *Du Système Social et des Lois qui le Régissent* (Brussels, 1848).

Quetelet, L. A. J., *Letters addressed to HRH the grand-duke of Saxe-Coburg and Gotha on the theory of probabilities as applied to the moral and physical sciences* (London, 1849).

Quetelet, L. A. J., 'Sur les proportions de la race noire', *Bulletin de L'Académie Royale des Sciences, des Lettres, et des Beaux-Arts de Belgique*, 21, 1 (1854), 96–100.

Quetelet, L. A. J., *Anthropométrie, ou mesure des différentes facultés de l'homme* (Brussels, 1870).

Quetelet, L. A. J., 'Extracts from a Notice of Charles Babbage, by A. Quetelet of Brussels, translated from the *Annuaire de l'Observatoire Royale de Bruxelles* for 1873', *Annual Report of the Board of Regents of the Smithsonian Institution*, 1873, 183–7.

Ranke, Leopold von, *History of England, Principally in the Sixteenth and Seventeenth Centuries*, (1859–67) (Eng. tr., 6 vols, Oxford, 1885).

Raper, Henry, and Fitzroy, Robert, 'Hints to Travellers', *Journal of the Royal Geographic Society*, vol. 24, 1854, 328–58.

Rawson, R. W., 'An Enquiry into the Condition of Criminal Offenders in England and Wales, with respect to Education; or, Statistics of Education among the Criminal and General Population of England and other countries', *JSSL*, III, 1841, 331–52.

Rawson, R. W., *Report of the Mansion House Conference on the Condition of the Unemployed* (London, 1887).

[Robertson, G.], 'Transactions of the Statistical Society of London', [vol. 1, pt. 1, 1837]. *The London and Westminster Review*, xxxi, i (April–Aug. 1838), 45–73.

Rogers, James E. Thorold, *Six Centuries of Work and Wages: The History of English Labour* (2 vols, London, 1884).

Ruskin, John, *The Stones of Venice* (3 vols, 1853).

Sadler, Michael, 'The Story of Education in Manchester', in W. H. Brindley (ed.), *The Soul of Manchester* (Manchester, 1929), 39–61.

Saint-Simon, Henri Comte de, *Henri Comte de Saint-Simon 1760–1825: Selected Writings* (ed. F. M. H. Markham) (Oxford, 1952).

Saint-Simon, Henri Comte de, *The Doctrine of Saint-Simon: An Exposition: First Year 1828–9* (ed. G. G. Iggers) (New York, 1958).

[Sanders, T. C.], 'Buckle's History of Civilization', *The Saturday Review*, 11 July 1857, 38–40.

Sanderson, John Burdon, 'Opening Address to Biology Section of Instrument Exhibition', *Nature*, 1 June 1876, 117–19.

Schlözer, August Ludwig von, *Theorie der Statistik nebst Ideen über die Politik überhaupt. Erstes Heft. Einleitung* (Göttingen, 1804).

Senior, Nassau, 'An introductory lecture on political economy, delivered before the University of Oxford, 6th December 1826', *The Pamphleteer*, xxix, no. lvii (London, 1828), 33–48.

Shaw, George Bernard, 'The Industrial Remuneration Conference', *The Commonweal*, March 1885, 15.

Shaw, George Bernard, *George Bernard Shaw: Collected Letters, 1876-1897* (ed. Dan H. Lawrence) (London, 1965).

Shirreff, Emily, 'Biographical Notice', in Helen Taylor (ed.), *Miscellaneous and Posthumous Works of Henry Thomas Buckle* (3 vols, London, 1872), vol. I, ix–lv.

Shuttleworth, James Kay-, *see* Kay, James Phillips

Simcox, Edith, 'Loss or Gain of the Working Classes during the Nineteenth Century', *Industrial Remuneration Conference. The Report of Proceedings and Papers Read* (London, 1885), 84–114.

Simon, John, *English Sanitary Institutions, Reviewed in their Course of Development, and in some of their Political and Social Relations* (London, 1890).

Simon, John, *Personal Recollections of Sir John Simon K. C. B.* [London, 1894, printed privately].

Sinclair, Sir John, *The Statistical Account of Scotland* (21 vols, Edinburgh, 1791–99).

Smith, Edward, *The Present State of the Dietary Question* (London, 1864).

Smith, Edward, *Practical Dietary for Families, Schools and the Working Classes* (London, 1864).

Southwood Smith, T., *The Philosophy of Health* (2 vols, London, 1835).

Sprat, Thomas, *The History of the Royal Society of London, for the Improving of Natural Knowledge* (London, 1667).

Stanley, Lord, *Disraeli, Derby and the Conservative Party: Journals and Memoirs of Edward Henry, Lord Stanley 1849–1869* (ed. John Vincent) (Hassocks, Sussex, 1978).

[Stead, W. T.], 'The Labour Party and the Books that Helped to Make It', *Review of Reviews*, 33, June 1906, 568–82.

[Stephen, James Fitzjames], 'Buckle's History of Civilization in England', *Edinburgh Review*, April 1858, vol. cvii, 465–512.

Stephens, W. R. W., *A Memoir of the Rt. Hon. William Page Wood, Baron Hatherley* (London, 1883).

Süssmilch, Johann Peter, *Die Göttliche Ordnung in den Veränderungen des Menschlichen Geschlechts, aus der Geburt, dem Tode und der Fortpflanzung Desselben Erwiesen* (1741).

Thomas, E. L. H., *Illustrations of Cross Street Chapel* (Manchester, 1917).

Tocqueville, Alexis de, *Journeys to England and Ireland* (tr. J. P. Mayer) (New Haven, Ct., 1958).

Tocqueville, Alexis de, *Democracy in America* (2 vols, 1835, 1840), ed. Phillips Bradley (New York, 1945).

Tocqueville, Alexis de, *Selected Letters on Politics and Society* (ed. R. Boesche) (Berkeley and London, 1985).

Todhunter, Isaac, *William Whewell D.D.: An account of his writings with selections from his library and scientific correspondence* (2 vols, London, 1876).

Toynbee, A. J., *Lectures on the Industrial Revolution in England* (London, 1884).

Trollope, Anthony, *Phineas Redux* (London, 1874).

Twain, Mark, 'Chapters from my Autobiography' (Ch. XX of XXV), *North American Review*, 185, 618, 5 July 1907, 465–74.

Tylor, E. B., 'Primitive Society', *Contemporary Review*, Pt i, vol. 21, Dec. 1872, 701–18; Pt. ii, vol. 22, June 1873, 53–72.

Tylor, E. B., 'On a Method of Investigating the Development of Institutions: Applied to Laws of Marriage and Descent', *Journal of the Anthropological Institute*, 18, 1889, 245–72.

Venn, John, *The Logic of Chance. An Essay on the foundations and province of the Theory of Probability, with especial reference to its logical bearings on its applications to Moral and Social Science and Statistics* (London, 1866).

Wade, John, *The Extraordinary Black Book…presenting a complete view of the expenditure, patronage, influence, and abuses of the Government, in Church, State, Law and Representation* (London, 1819).

Webb, Beatrice, *My Apprenticeship* (London, 1926).

Wells, H. G., *The Island of Dr. Moreau* (London, 1896).

Whewell, William, 'Address', *Report of the Third Meeting of the British Association for the Advancement of Science, Held at Cambridge, 1833* (London, 1834).

[Whewell, William], 'An Essay on the Distribution of Wealth and on the Sources of Taxation', *The British Critic and Quarterly Theological Review*, vol. x, no. xix, July 1831, 41–61.

Whewell, William, *Mathematical Exposition of Some Doctrines of Political Economy* (Cambridge, 1829) reprinted in *Transactions of the Cambridge Philosophical Society*, vol. iii, 1831.

Whewell, William, 'Mathematical Exposition of some of the Leading Doctrines in Mr. Ricardo's *Principles of Political Economy and Taxation*', *Transactions of the Cambridge Philosophical Society*, (Cambridge, 1833),vol. iv, 1833, 155–198.

Whewell, William, *Astronomy and General Physics considered with reference to Natural Theology* (London, 1834) (The 3rd Bridgewater Treatise).

Whewell, William, *The History of the Inductive Sciences, from the Earliest to the Present Times* (3 vols, London, 1837).

Whewell, William, *On the Plurality of Worlds* (London, 1853).

Whewell, William, 'Prefatory Notice' in Whewell (ed.), *Literary Remains, Consisting of Lectures and Tracts on Political Economy of the late Rev. Richard Jones* (London, 1859), ix–xl.

Whewell, William, 'Comte and Positivism', *Macmillan's Magazine*, vol. 13, March 1866, 353–62.

Wilkinson, Thomas Read, 'On the Origin and History of the Manchester Statistical Society', 17 Nov. 1875, *Transactions of the Manchester Statistical Society*, 1875–6, 9–17.

Young, Arthur, *Political Arithmetic: Containing Observations on the Present State of Great Britain and the Principles of her Policy in Encouragement of Agriculture* (London, 1774).

Secondary Sources

Abbott, Edith, 'Charles Booth 1840–1916', *Journal of Political Economy*, Feb. 1917, 25, 2, 195–200.

Abrams, Philip, *The Origins of British Sociology 1834–1914* (Chicago and London, 1968).

Adams, Maeve E., 'Numbers and Narratives. Epistemologies of Aggregation in British Statistics and Social Realism, c. 1790–1880', in Tom Crook and Glen O'Hara (eds), *Statistics and the Public Sphere. Numbers and the People in Modern Britain, c. 1800–2000* (New York and Abingdon, 2011), 103–20.

Adelman, Paul, 'Frederic Harrison and the "positivist" attack on orthodox political economy', *History of Political Economy*, iii, 1971, 170–89.

Alborn, Timothy L., 'A Calculating Profession: Victorian Actuaries among the Statisticians', *Science in Context*, 7, 3 (1994), 433–68,

Alborn, Timothy L., *Regulated Lives: Life Insurance and British Society, 1800–1914* (Toronto, 2009).

Aldrich, John, Mathematics in the London/Royal Statistical Society 1834–1934, *Electronic Journal for History of Probability and Statistics*, 6, 1, June 2010, 7. http://www.jehps.net/indexang.html

Allen, R. G. D., and George, R. F., 'Professor Sir Arthur Lyon Bowley 1869–1957', *JRSS*, ser. A, vol. 120, 2, June 1957, 236–41.

Anderson, Margo, 'The US Bureau of the Census in the Nineteenth Century', *Social History of Medicine*, 4, 3, Dec. 1991, 497–513.

Anderson, Perry, 'Components of the National Culture', *New Left Review*, 50 (1968), 1–57.

Aron, Raymond, *Main Currents in Sociological Thought* (2 vols) (Harmondsworth, 1967 edn).

Ashton, T. S., *Economic and Social Investigation in Manchester, 1833–1933. A Centenary History of the Manchester Statistical Society* (London, 1934).

Aylmer, Gerald, *The King's Servants: The Civil Service of Charles I, 1625–42* (London, 1961).

Aylmer, Gerald, *The State's Servants: The Civil Service of the English Republic, 1649–1660* (London, 1973).

Aylmer, Gerald, *The Crown's Servants: Government and Civil Service under Charles II, 1660–1685* (Oxford, 2002).

Bailey, S. D., 'Parliament and the Prying Proclivities of the Registrar-General', *History Today*, 31, 4, April 1981.

Baker, K. M., 'The Early History of the Term "Social Science"', *Annals of Science*, XX, 1964, 211–26.

Bayly, Christopher, *Imperial Meridian. The British Empire and the World 1780–1830* (London, 1989).

Becher, Harvey W., 'William Whewell's Odyssey. From Mathematics to Moral Philosophy', in M. Fisch and S. Schaffer (eds), *William Whewell. A Composite Portrait* (Oxford, 1991), 1–30.

Berg, Maxine, *The Machinery Question and the Making of Political Economy 1815–48* (Cambridge, 1980).

Bernstein, Peter L., *Against the Gods. The Remarkable Story of Risk* (New York, 1996).

Beveridge, William, *Power and Influence* (London, 1955).

Bevis, James, 'Kay, Heywood & Langton: 19th Century Statisticians and Social Reformer', *Transactions of the Manchester Statitistical Society*, 2010–11, 65–77.

Blake, Robert, *Disraeli* (1966).

Blaug, Mark, *Ricardian Economics: A Historical Study* (New Haven, 1958).

Bonar, J., and Macrosty, H. W., *Annals of the Royal Statistical Society 1834–1934* (London, 1934).

Booth, Mary, *Charles Booth. A Memoir* (London, 1918).

Bostridge, Mark, *Florence Nightingale. The Woman and Her Legend* (London, 2008).

Bostridge, Mark, 'Was Ada Lovelace the true founder of Silicon Valley?', *The Spectator*, 17 March 2018.

Botting, Douglas, *Humboldt and the Cosmos* (New York, 1973).

Bowden, B. V., 'A Brief History of Computation', in B.V. Bowden (ed.), *Faster Than Thought. A Symposium on Digital Computing Machines* (London, 1953), 1–34.

Boyle, David, *The Tyranny of Numbers. Why Counting Can't Make Us Happy* (London, 2000).

Brewer, John, *The Sinews of Power. War, Money and the English State 1688–1783* (London, 1989).

Briggs, Asa, 'The Language of "Class" in Early Nineteenth-Century England', in A. Briggs and J. Saville (eds.), *Essays in Labour History* (London, 1960), 43–73.

Brock, W. H., 'Humboldt and the British: A Note on the Character of British Science', *Annals of Science*, vol. 50, 4, 365–72.

Brockington, C. Fraser, 'Public Health at the Privy Council 1858–71', *Medical Officer*, xci, 1959, 176–7; 211–14; 243–6; 259–60; 278–80; 287–90.

Brockington, C. Fraser, *Public Health in the Nineteenth Century* (Edinburgh and London, 1965).

Bromley, Alan G., 'Table Making and Calculating Engines', in 'General Introduction', *The Works of Charles Babbage* (ed. Martin Campbell-Kelly), 11 vols (London, 1989) vol. 1, 22–7.

Brookes, Martin, *Extreme Measures. The Dark Visions and Bright Ideas of Francis Galton* (London, 2004).

Brown, John, 'Charles Booth and Labour Colonies 1889–1905', *Economic History Review*, 21, 2, Aug. 1968, 349–60.

Brown, John, 'Social Judgements and Social Policy', *Economic History Review*, vol. 24, 1, 1971, 106–13.

Brown, Lucy, *The Board of Trade and the Free-Trade Movement 1830–42* (Oxford, 1958).

Brundage, Anthony, *England's "Prussian Minister". Edwin Chadwick and the Politics of Government Growth, 1832–1854* (London, 1988).

Buck, Peter, 'Seventeenth-Century Political Arithmetic: Civil Strife and Vital Statistics', *Isis*, 68, 1, March 1977, 67–84.

Buck, Peter, 'People Who Counted: Political Arithmetic in the Eighteenth Century', *Isis*, 73, 1, March 1982, 28–45.

Burleigh, Michael, *Death and Deliverance: Euthanasia in Germany c. 1900 to 1945* (Cambridge, 1994).

Burn, W. L., *The Age of Equipoise. A Study of the Mid-Victorian Generation* (London, 1964).

Burns, J. H., 'J. S. Mill and the Term "Social Science"', *Journal of the History of Ideas*, xx, 3, 1959, 431–2.

Burrow, J. W., *The Crisis of Reason. European Thought, 1848–1914* (New Haven and London, 2000).

Campion, H., 'International Statistics', *JRSS*, series A, 1949, no. 2, 105–43.

Cannon, Susan Faye, *Science in Culture. The Early Victorian Period* (New York, 1978).

Cannon, Walter, 'The Problem of Miracles in the 1830s', *Victorian Studies*, 4, 1, Sept. 1960, 4–32.

Caplow, Theodore, Hicks, Louis, and Wattenberg, Ben, *The First Measured Century. An Illustrated Guide to Trends in America, 1900–2000* (Washington D.C., 2000).

Cassedy, James H., *Demography in Early America* (Cambridge, MA., 1969).

Chaorasiya, V., and A. Shrivastava, 'A Survey on Big Data. Techniques and Technologies', *International Journal of Research and Development in Applied Science and Engineering*, 8, 1, Sept. 2015. http://ijrdase.com/ijrdase/wp-content/uploads/2015/10/A-survey-on-Big-Data-Techniques-and-Technologies-Vinay-Chourasia.pdf

Charlton, D. G., *Positivist Thought in France During the Second Empire* (Oxford, 1959).

Clark, J. C. D., *English Society 1660–1832. Religion, Society and Politics during the Ancien Regime* (Cambridge, 1985).

Clokie, H. M., and J. William Robinson, *Royal Commissions of Inquiry* (Stanford University Press, 1937).

Coats, A. W., 'The Historist Reaction in English Political Economy 1870–1890', *Economica*, 21, 82, May 1954, 143–53.

Cohen, I. Bernard, 'Foreword' in 'General Introduction', *The Works of Charles Babbage* (ed. Martin Campbell-Kelly), 11 vols (London, 1989) vol. 1, 7–18.

Cole, G. D. H., 'A Study in British Trade Union History. Attempts at a "General Union" 1829–1834', *International Review for Social History*, 1938, vol. 4, 359–462.

Cole, Stephen, 'Continuity and Institutionalisation in Science: A Case Study of Failure', in Anthony Oberschall (ed.), *The Establishment of Empirical Sociology: Studies in Continuity, Discontinuity and Institutionalization* (New York, 1972), 73–129.

Collier, Bruce and MacLauchlan, James, *Charles Babbage and the Engines of Perfection* (Oxford, 1998).

Collini, Stefan, *Public Moralists. Political Thought and Intellectual Life in Britain 1850–1930* (Oxford, 1993).

Comrie, L. J., 'Babbage's Dream Comes True', *Nature*, Oct. 26, 1946, 567–8.

Cook, Edward, *The Life of Florence Nightingale* (2 vols, London, 1913).

Costa, James T., *Wallace, Darwin and the Origin of Species* (Cambridge, Mass., 2014).

Cowan, Ruth Schwartz, 'Francis Galton's Statistical Ideas: The Influence of Eugenics', *Isis*,. 63, 4 (Dec. 1972), 509–28.

Crellin, J. K., 'The Dawn of the Germ Theory: Particles, Infection and Biology', in F. N. L. Poynter (ed.), *Medicine and Society in the 1860s* (London, 1968), 57–76.

Crook, Tom and O'Hara, Glen, 'The "Torrent of Numbers": Statistics and the Public Sphere in Britain, c.1800–2000', in Tom Crook and Glen O'Hara (eds), *Statistics and the Public Sphere. Numbers and the People in Modern Britain, c. 1800–2000* (New York and Abingdon, 2011).

Cullen, Michael J., 'The Making of the Civil Registration Act of 1836', *Journal of Ecclesiastical History*, 25, 1974, 39–59.

Cullen, Michael J., *The Statistical Movement in Early Victorian Britain. The Foundations of Empirical Social Research* (Hassocks, Sussex, 1975).

Daston, Lorraine J., 'Introduction', in Lorenz Krüger, Lorraine Daston, and Michael Heidelberger (eds), *The Probabilistic Revolution*, (2 vols), vol. 1, *Ideas in History* (1987) (Cambridge, Mass. 1990 edn.), 1–4.

Daston, Lorraine J., 'The Domestication of Risk: Mathematical Probability and Insurance 1650–1830', in L Krüger, L. Daston, and M. Heidelberger (eds), *The Probabilistic Revolution*, vol. 1, 237–60.

Daston, Lorraine J., 'Rational Individuals versus Laws of Society: From Probability to Statistics', in L. Krüger, L. J. Daston, and M. Heidelberger (eds), *The Probabilistic Revolution*, vol. 1, 295–304.

Daston, Lorraine J., *Classical Probability in the Enlightenment* (Princeton, NJ, 1988).

David, F. N., 'Galton, Francis', *Encyclopedia of the Social Sciences* (ed. David Sills) (19 vols, New York, 1968), vol. 6, 48–53.

Desrosières, Alain, *The Politics of Large Numbers. A History of Statistical Reasoning* (1993) (Cambridge and London, 1998).

Desrosières, Alain, 'Official Statistics and Medicine in Nineteenth-Century France: The SGF as a Case Study', *Social History of Medicine*, 4, 3, Dec. 1991, 515–37.

Dettelbach, Michael, 'Humboldtian Science', in N. Jardine, J. A. Secord, and E. C. Spary (eds), *Cultures of Natural History* (Cambridge, 1996), 287–304.

Diamond, Marion, and Mervyn Stone, 'Nightingale on Quetelet' (3 pts), *Journal of the Royal Statistical Society* (series A), vol. 144, 1981, 66–79; 176–231; 332–5.

Diamond, Solomon 'Introduction', in L. A. J. Quetelet, *A Treatise on Man and the Development of his Faculties* (Tr. R. Knox), (Edinburgh, 1842) (1969 edn, Gainesville, Florida), v–xii.

Dicey, A.V., *Lectures on the Relation between Law and Public Opinion in England during the Nineteenth Century* (London, 1905).

Donnelly, Kevin, *Adolphe Quetelet. Social Physics and the Average Men of Science 1796–1874* (London, 2015).

Draper, Nicholas, *The Price of Emancipation. Slave-Ownership, Compensation and British Society at the End of Slavery* (Cambridge, 2010).

Drolet, Michael, 'Tocqueville's Interest in the Social: Or How Statistics Informed His "New Science of Politics"', *History of European Ideas*, 31, 2005, 451–71.

Dubbey, John M., 'Mathematical Papers' in 'General Introduction', *The Works of Charles Babbage* (ed. Martin Campbell-Kelly), (11 vols, London, 1989), vol. 1, 18–22.

Eastwood, David, '"Amplifying the Province of the Legislature"': The Flow of Information and the English State in the Early Nineteenth Century', *Historical Research*, vol. 62, no. 149, October 1989, 276–94.

Edwards, A.W. F., 'Statisticians in Stained Glass Houses', *Significance* (Royal Statistical Society), 3 (4), 182–3, https://rss.onlinelibrary.wiley.com/doi/pdf/10.1111/j.1740–9713. 2006.00203.x

Elesh, David, 'The Manchester Statistical Society: A Case Study of Discontinuity in the History of Empirical Social Research', in Anthony Oberschall (ed.), *The Establishment of Empirical Sociology: Studies in Continuity, Discontinuity and Institutionalization* (New York, 1972), 31–72.

Elton, Geoffrey, *The Tudor Revolution in Government. Administrative Changes in the Reign of Henry VIII* (Cambridge, 1953).

Emmett, Frank, 'Classic Text Revisited. The Moral and Physical Condition of the Working Classes (1832)', *Transactions of the Historic Society of Lancashire and Cheshire*, 162 (2013), 185–219.

Evans, Richard J., *The Third Reich in Power* (London, 2005).

Evans, Richard J., 'R A Fisher and the Science of Hatred', *New Statesman*, 28 July 2020, https://www.newstatesman.com/international/science-tech/2020/07/ra-fisher-and-science-hatred

Evans, Suzanne E., *Forgotten Crimes. The Holocaust and People with Disabilities* (Chicago, 2004).

Eyler, John M., 'William Farr on the Cholera. The Sanitarian's Disease Theory and the Statistician's Method', *Journal of the History of Medicine and Allied Sciences*, 28, April 1973, 79–100.

Eyler, John M., *Victorian Social Medicine. The Ideas and Methods of William Farr* (Baltimore and London, 1979).

Fairlie, Susan, 'The 19th Century Corn Laws Reconsidered', *Economic History Review*, 2nd ser., 18, 1965, 562–75.

Fetter, F. W., 'The Influence of Economists in Parliament on British Legislation from Ricardo to John Stuart Mill', *Journal of Political Economy*, lxxxiii, 5, 1975, 1051–60.

Finer, S. E., *The Life and Times of Sir Edwin Chadwick* (London and New York, 1952).

Finer, S. E., 'The Transmission of Benthamite Ideas 1820–50', in G. Sutherland (ed.), *Studies in the Growth of Nineteenth-Century Government* (London, 1972), 11–32.

Finlayson, G. B. A. M., *The Seventh Earl of Shaftesbury* (London, 1981).

Fisch, Menachem and Schaffer, Simon (eds), *William Whewell. A Composite Portrait* (Oxford, 1991).

Fontana, Biancamaria, *Rethinking the Politics of Commercial Society: the Edinburgh Review 1802–1832* (Cambridge, 1985).

Forbes-Macphail, Imogen, 'The Enchantress of Numbers and the Magic Noose of Poetry: Literature, Mathematics, and Mysticism in the Nineteenth Century', *Journal of Language, Literature and Culture*, 2013, vol. 60, 3, 138–56.

Foreman, Amanda, *A World on Fire. An Epic History of Two Nations Divided* (London, 2010).

Forster, John, *The Life of Charles Dickens* (London, 1928).

Foucault, Michel, 'The Birth of Social Medicine' (1974) and 'Governmentality' (part of a course of lectures on 'Security, Territory, and Population', Collège de France, 1978) in Foucault, *Power. Essential Works of Foucault 1954–1984: Volume 3* (ed., Colin Gordon) (Paris, 1994) (2000 edn, London).

Frazer, W. M., *Duncan of Liverpool. Being an Account of the Work of Dr. W. H. Duncan, Medical Officer of Health of Liverpool 1847–63* (London, 1947).

Friedlander, Henry, *The Origins of Nazi Genocide: From Euthanasia to the Final Solution* (Chapel Hill, N.C., 1995).

Gatrell, V. A. C., 'Incorporation and the Pursuit of Liberal Hegemony in Manchester 1790–1839', in Derek Fraser (ed.) *Municipal Reform and the Industrial City* (Leicester and New York, 1982), 16–60.

Geyer, Martin H., 'One Language for the World: The Metric System, International Coinage, Gold Standard, and the Rise of Internationalism, 1850–1900' in Martin H. Geyer and Johannes Paulmann (eds) *The Mechanics of Internationalism. Culture, Society, and Politics from the 1840s to the First World War* (Oxford, 2001), 55–92.

Gibson, William, and Bruce Sterling, *The Difference Engine* (London, 1990).

Gillham, Nicholas Wright, *A Life of Sir Francis Galton. From African Exploration to the Birth of Eugenics* (Oxford, 2001).

Glass, D. V., *Numbering the People: The Eighteenth Century Population Controversy and the Development of Census and Vital Statistics in Britain* (Farnborough, Hants, 1973).

Glass, D. V., M. Ogborn, and I. Sutherland, 'John Graunt and His Natural and Political Observations [and Discussion]'. *Proceedings of the Royal Society of London. Series B, Biological Sciences*, 159, 974 (1963), 2–37.

Goldman, Lawrence, 'The Origins of British "Social Science": Political Economy, Natural Science and Statistics, 1830–35', *The Historical Journal*, 26, 3 (1983), 587–616.

Goldman, Lawrence, '"A Peculiarity of the English? The Social Science Association and the Absence of Sociology in Nineteenth Century Britain', *Past & Present*, 114 (Feb. 1987), 133–71.

Goldman, Lawrence (ed.), *The Blind Victorian. Henry Fawcett and British Liberalism* (Cambridge, 1989).

Goldman, Lawrence, 'Statistics and the Science of Society in Early Victorian Britain: An Intellectual Context for the General Register Office', *Social History of Medicine*, 4, 3, Dec. 1991, 415–34.

Goldman, Lawrence, 'Experts, Investigators and The State in 1860: British Social Scientists through American Eyes', in M. Lacey and M. O. Furner (eds), *The State and Social Investigation in Britain and the United States* (Cambridge, 1993), 95–126.

Goldman, Lawrence, *Dons and Workers. Oxford and Adult Education Since 1850* (Oxford, 1995).

Goldman, Lawrence, 'Ruskin, Oxford and the British Labour Movement 1880–1914', in Dinah Birch (ed.), *Ruskin and the Dawn of the Modern* (Oxford, 1999), 57–86.

Goldman, Lawrence, *Science, Reform and Politics: The Social Science Association 1857–1886* (Cambridge, 2002).

Goldman, Lawrence, 'Victorian Social Science: From Singular to Plural', in Martin Daunton (ed.), *The Organisation of Knowledge in Victorian Britain* (Oxford, 2005), 87–114.

Goldman, Lawrence, 'The Defection of the Middle Class: The Endowed Schools Act, the Liberal Party, and the 1874 Election', in Peter Ghosh and Lawrence Goldman (eds), *Politics and Culture in Victorian Britain. Essays in Memory of Colin Matthew* (Oxford, 2006), 118–35.

Goldman, Lawrence, 'Virtual Lives: History and Biography in an Electronic Age', *Australian Book Review*, June 2007, 37–44.

Goldman, Lawrence, 'Foundations of British Sociology 1880–1930. Contexts and Biographies', *The Sociological Review*, 55, 3, August 2007, 431–40.

Goldman, Lawrence, '"A Total Misconception". Lincoln, the Civil War, and the British, 1860–1865', in Richard Carwardine and Jay Sexton (eds), *The Global Lincoln* (Oxford, 2011), 107–22.

Goldman, Lawrence, 'Conservative Political Thought from the Revolutions of 1848 until the Fin de Siècle', in Gareth Stedman Jones and Gregory Claeys (eds), *Cambridge History of Political Thought vol. vi: The Nineteenth Century* (Cambridge University Press, 2012), 691–719.

Goldman, Lawrence, 'Social Reform and the Pressure of "Progress" on Parliament, 1660-1914' in Richard Huzzey (ed.), *Pressure and Parliament. From Civil War to Civil Society, Parliamentary History: Texts and Studies*, 13, 2018, 72–88.

Goldman, Lawrence, 'From Art to Politics: William Morris and John Ruskin', in John Blewitt (ed.), *William Morris and John Ruskin. A New Road on which the World Should Travel* (Exeter, 2019), 123–42.

Goldman, Lawrence, 'Victorians and Numbers: Statistics and Social Science in Nineteenth-Century Britain', in Plamena Panayotova (ed.), *The History of Sociology in Britain. New Research and Revaluation* (Cham, Switzerland, 2019), 71–100.

Goldman, Lawrence, (ed.), *Welfare and Social Policy in Britain since 1870* (Oxford, 2019).

Gordon, Colin, 'Introduction', in *Power. Essential Works of Foucault 1954–1984: Volume 3* (ed., Colin Gordon) (Paris, 1994) (2000 edn, London), xi–xli.

Gossart, M., 'Adolphe Quetelet et le Prince Albert de Saxe-Cobourg (1836–1861)', *Bulletins de l'Académie Royale de Belgique, Classe des lettres et des sciences morales et politique* (Brussels, 1919), 211–54.

Gould, Stephen Jay, *The Mismeasure of Man* (New York, 1981).

Grimley, Matthew, 'You Got an Ology? The Backlash against Sociology in Britain, c. 1945–90', in Lawrence Goldman (ed.), *Welfare and Social Policy in Britain since 1870* (Oxford, 2019), 178–93.

Grob, Gerald N., *Edward Jarvis and the Medical World of Nineteenth Century America* (Knoxville, TN, 1978).

Hacking, Ian, 'How Should We Do the History of Statistics?', *I & C*, Spring 1981, no. 8, 15–26.

Hacking, Ian, 'Nineteenth Century Cracks in the Concept of Determinism', *Journal of the History of Ideas*, 44, 3, 1983, 455–75.

Hacking, Ian, 'Prussian Numbers 1860–1882', in Lorenz Krüger, Lorraine J. Daston, and Michael Heidelberger (eds), *The Probabilistic Revolution* (2 vols), vol. 1, *Ideas in History* (Cambridge, Mass., 1987), 377–94.

Hacking, Ian, 'Was there a Probabilistic Revolution 1800–1930?', in Krüger, Daston, and Heidelberger (eds), *The Probabilistic Revolution* vol. 1, 45–55.

Hacking, Ian, *The Taming of Chance* (Cambridge, 1990).

Halbwachs, Maurice, *La Théorie de l'Homme Moyen: Essai sur Quetelet et la Statistique Morale* (Paris, 1913).

Hammond, J. L., and Barbara Hammond, *The Town Labourer 1760–1832* (London and New York, 1917).

Hand, David J., *Statistics. A Very Short Introduction* (Oxford, 2008).

Hankins, F. H., 'Adolphe Quetelet as Statistician', *Columbia University Studies in History, Economics and Public Law*, no. xxxi (New York, 1908).

Hardy, G. H., *A Mathematician's Apology* (Cambridge, 1940).

Harkness, Timandra, *Big Data. Does Size Matter?* (London, 2016) (2017 edn).

Harris, Jose, *Private Lives, Public Spirit: Britain 1870–1914* (Harmondsworth, 1993).

Harris, Jose, 'Political Thought and the Welfare State 1870–1940: An Intellectual Framework for British Social Policy', *Past & Present*, 135 (1992), 116–41.

Harris, Jose, 'Between Civic Virtue and Social Darwinism: The Concept of the Residuum', in David Englander and Rosemary O'Day (eds), *Retrieved Riches. Social Investigation in Britain 1840–1914* (Aldershot, Hants, 1995), 67–83.

Harris, Robert, *Politics and the Nation: Britain in the Mid-Eighteenth Century* (Oxford, 2002).

Harrison, Brian, 'Finding Out How the Other Half Live: Social Research and British Government since 1780', in Harrison, *Peaceable Kingdom. Stability and Change in Modern Britain* (Oxford, 1982), 260–308.

Harrison, Royden, *Before the Socialists: Studies in Labour and Politics, 1861–1881* (London, 1965).

Hart, Jennifer, 'Nineteenth Century Social Reform: A Tory Interpretation of History', *Past & Present*, 31, 1965, 39–61.

Hart, Jennifer, *Proportional Representation. Critics of the British Electoral System 1820–1945* (Oxford, 1992).

Hartwell, R. M., 'Cunningham, William', https://www.encyclopedia.com/social-sciences/applied-and-social-sciences-magazines/cunningham-william

Hay, J. R., *The Origins of the Liberal Welfare Reforms 1906–14* (London, 1975).

Head, Brian W., 'The Origins of "La Science Sociale" in France', *Australian Journal of French Studies*, vol. xix, 2, May–Aug 1982, 115–32.

Headrick, Daniel R., *When Information Came of Age. Technologies of Knowledge in the Age of Reason and Revolution, 1700–1850* (Oxford, 2000).

Henderson, W. O., *The Life of Friedrich Engels* (2 vols, London, 1976).

Hesketh, Ian, *The Science of History in Victorian Britain. Making the Past Speak* (London, 2011).

Hesse, Bradford W., Richard P. Moser, and William T. Riley, 'From Big Data to Knowledge in the Social Sciences', *The Annals of the American Academy of Political and Social Science*, 659, May 2015, 16–32.

Hevly, Bruce, 'The Heroic Science of Glacier Motion', *Osiris*, 11, 1996, 66–86.

Hewitt, Martin, *Making Social Knowledge in the Victorian City. The Visiting Mode in Manchester 1832–1914* (Abingdon, Oxfordshire, 2020).

Heyck, T. W., *The Transformation of Intellectual Life in Victorian England* (London, 1982).

Higgs, Edward, *Making Sense of the Census. The Manuscript Returns 1801–1901* (London, 1989).

Higgs, Edward, 'Disease, Febrile Poisons, and Statistics: The Census as a Medical Survey, 1841–1911', *Social History of Medicine*, 4, 3, Dec. 1991, 465–78.

Higgs, Edward, *A Clearer Sense of the Census: The Victorian Censuses and Historical Research* (London, 1996).

Higgs, Edward, 'A Cuckoo in the Nest? The Origins of Civil Registration and State Medical Statistics in England and Wales', *Continuity and Change*, 11, 1, 1996, 115–34.

Higgs, Edward, *Life, Death and Statistics: Civil Registration, Censuses and the Work of the General Register Office 1837–1952* (Hatfield, 2004).

Hill, I. D., 'Statistical Society of London – Royal Statistical Society. The First 100 Years: 1834–1934', *Journal of the Royal Statistical Society*, ser. A, 147 (1984), 130–9.

Hilts, Victor L., 'William Farr (1807–1883) and the "Human Unit"', *Victorian Studies*, 14, 2, Dec. 1970, 143–50.

Hilts, Victor L., 'Statistics and Social Science', in Ronald N. Giere and Richard S. Westfall (eds), *Foundations of Scientific Method. The Nineteenth Century* (Bloomington, Ind., 1973).

Hilts, Victor L., '*Aliis exterendum*, or, the Origins of the Statistical Society of London', *Isis*, 1978, 69, 21–43.

Hirst, Derek, *Authority and Conflict: England 1603–1658* (London, 1986).

Hobsbawm, Eric, *The Age of Capital. Europe 1848–1875* (London, 1975).

Hodder, E., *The Life and Work of the Seventh Earl of Shaftesbury* (3 vols, London, 1887).

Hollings, Christopher, Ursula Martin, Adrian Rice, 'The Early Mathematical Education of Ada Lovelace', *BSHM Bulletin: Journal of the British Society for the History of Mathematics*, 32, 3, 2017, 221–34.

Holmes, G. S., 'Gregory King and the Social Structure of Pre-Industrial England', *Transactions of the Royal Historical Society*, 5th ser., 27 (1977), 41–68.

Howard, C. H. D., 'Joseph Chamberlain and the "Unauthorised Programme"', *English Historical Review*, 65, 257, Oct. 1950, 447–91.

Huzzey, Richard, *Freedom Burning. Anti-Slavery and Empire in Victorian Britain* (Ithaca NY and London, 2012).

Hyman, Anthony, *Charles Babbage. Pioneer of the Computer* (Oxford, 1982).

Iggers, G. G., 'Further Remarks about the Early Uses of the Term "Social Science"', *Journal of the History of Ideas*, xx, 1959, 433–6.

Innes, Joanna, 'Jonathan Clark, Social History and England's "Ancien Regime"', *Past & Present*, 115, 1, 1987, 165–200.

Innes, Joanna, 'Power and Happiness: Empirical Social Enquiry in Britain, from "Political Arithmetic" to "moral statistics"', in Innes, *Inferior Politics. Social Problems and Social Policies in Eighteenth-Century Britain* (Oxford, 2009), 109–75.

Jahoda, Gustav, 'Quetelet and the Emergence of the Behavioral Sciences', *SpringerPlus*, 4, 473 (2015). https://doi.org/10.1186/s40064-015-1261-7

Jensen, J. Vernon, *Thomas Henry Huxley: Communicating for Science* (Cranbury, NJ, 1991).

John, A. H., *A Liverpool Merchant House. Being the History of Alfred Booth and Company 1863–1958* (London, 1959).

Johnson, L. G., *Richard Jones Reconsidered* (London, 1955).

Johnson, Paul, *The Birth of the Modern. World Society 1815–1830* (New York, 1991).

Johnson, Richard, 'Administrators in Education before 1870: Patronage, Social Position and Role', in Gillian Sutherland (ed.), *Studies in the Growth of Nineteenth Century Government* (London, 1972), 110–38.

Jones, Greta, *Social Darwinism and English Thought. The Interaction between Biological and Social Theory* (Brighton, Sussex, 1980).

Kadish, Alon, *The Oxford Economists in the Late Nineteenth Century* (Oxford, 1982).

Karder, Lawrence, *The Asiatic Mode of Production. Sources, Development and Critique in the Writing of Karl Marx* (Assen, 1975).

King, Steven, '"In These You May Trust". Numerical Information, Accounting Practices, and the Poor Law, c. 1790 to 1840', in Tom Crook and Glen O'Hara (eds), *Statistics and the Public Sphere. Numbers and the People in Modern Britain, c. 1800-2000* (New York and Abingdon, 2011), 51-66.

Kitson Clark, G., '"Statesmen in Disguise": Reflexions on the History of the Neutrality of the Civil Service', *Historical Journal*, 2, 1959, 19-39.

Klein, Judy L., *Statistical Visions in Time: A History of Time Series Analysis 1662-1938* (Cambridge, 1997).

Kline, Morris, *Mathematics. The Loss of Certainty* (Oxford, 1980).

Klingender, Francis D., *Art and the Industrial Revolution* (London, 1972).

Knight, Roger, *Britain Against Napoleon. The Organization of Victory 1793-1815* (London, 2014).

Koot, G. M., 'T. E. Cliffe-Leslie, Irish Social Reform and the Origins of the English Historical School of Economics', *History of Political Economy*, 7, 3, 1975, 312-36.

Koot, G. M., *English Historical Economics, 1870-1926. The Rise of Economic History and Neomercantilism* (Cambridge, 1988).

Korski, Daniel, 'Universities Are About to Enter a New Era', *Daily Telegraph*, 24 Jan. 2017.

Krüger, Lorenz, 'Preface', in Lorenz Krüger, Lorraine J. Daston, and Michael Heidelberger (eds), *The Probabilistic Revolution* (2 vols), vol. 1, *Ideas in History* (1987) (Cambridge, Mass., 1990 edn.), xv.

Kuhn, Thomas, 'The Function of Measurement in Modern Physical Science', *Isis*, 52, 2, June 1961, 161-90.

Lambert, Royston, *Sir John Simon 1816-1904 and English Social Administration* (London, 1963).

Landau, D., and Lazarsfeld, Paul F., 'Quetelet', *International Encyclopedia of the Social Sciences*, xiii (New York, 1968), 247-56.

Laslett, Peter, 'Comments' in Richard Stone, *Some British Empiricists in the Social Sciences 1650-1900* (Cambridge, 1997), 399-404.

Latter, R., *The International Statistical Institute 1885-1985* (Voorburg, Netherlands, 1985).

Lawton, R., *The Census and Social Structure* (London, 1978).

Lazarsfeld, Paul F., 'Notes on the History of Quantification in Sociology - Trends, Sources and Problems', *Isis*, 52, 2, June 1961, 277-333.

Lazarsfeld, Paul F., 'Toward a History of Empirical Sociology', in E. Privat (ed.) *Mélanges en l'honneur de Fernand Braudel. Methodologie de l'Histoire et des Sciences Humaines* (Toulouse, 1973), 289-303.

Lécuyer, Bernard-Pierre, 'Probability in Vital and Social Statistics: Quetelet, Farr, and the Bertillons', in Lorenz Krüger, Lorraine J. Daston, and Michael Heidelberger (eds), *The Probabilistic Revolution* (2 vols), vol. 1, *Ideas in History* (1987) (Cambridge, Mass., 1990 edn.), 317-35.

Lee, C. H., *A Cotton Enterprise. A History of M'Connel and Kennedy, Fine Cotton Spinners* (Manchester, 1972).

Lewes, Fred, 'William Farr and the Cholera', *Population Trends*, Spring 1983, 8-12.

Lewes, Fred, 'The Letters Between Adolphe Quetelet and William Farr 1852-1874', Académie Royale de Belgique, *Bulletin de la Classe des Lettres et des Sciences Morales et Politique*, 5 series—Tome 69, 1983, 417-28.

Lewes, Fred, 'Dr Marc d'Espine's statistical nosology', *Medical History*, 32, 1988, 301-13.

Lindenfeld, David F., *The Practical Imagination. The German Sciences of State in the Nineteenth Century* (Chicago, 1997).

Lindert. P. H., and J. G. Williamson, 'Revising England's Social Tables, 1688-1812' and 'Reinterpreting Britain's Social Tables 1688-1913', *Explorations in Economic History*, vol. 19, 1982, 358-408, and vol. 20, 1983, 94-109.

Linebaugh, Peter, 'Labour History without the Labour Process: A Note on John Gast and His Times', *Social History*, 7, 3, Oct. 1982, 319–28.

Lowe, Philip, 'The British Association and the Provincial Public', in Roy MacLeod and Peter Collins (eds), *The Parliament of Science. The British Association for the Advancement of Science 1831–1981* (London, 1981), 118–44.

Lowe, Rodney, 'The Failure of Consensus in Britain: The National Industrial Conference, 1919–1921', *The Historical Journal*, 21, 3, Sept. 1978, 649–75.

Lubenow, William, *Liberal Intellectuals and Public Culture in Modern Britain 1815–1914* (Woodbridge, Suffolk, 2010).

Lummis, Trevor, 'Charles Booth: Moralist or Social Scientist?', *Economic History Review*, 24, 1, 1971, 100–5.

MacDonagh, Oliver, 'The Nineteenth Century Revolution in Government: A Reappraisal', *Historical Journal*, 1, 1958, 52–67.

Mack, Mary P., *Jeremy Bentham. An Odyssey of Ideas* (London, 1962).

Mackenzie, Donald A., *Statistics in Britain, 1865–1930: The Social Construction of Scientific Knowledge* (Edinburgh, 1981).

Macleod, Roy M., 'The Frustration of State Medicine 1880–1899', *Medical History*, 11, 1967, 15–40.

Macleod, Roy M., 'Introduction. On the Advancement of Science', in Roy MacLeod and Peter Collins (eds), *The Parliament of Science. The British Association for the Advancement of Science 1831–1981* (London, 1981), 17–42.

Macnicol, John, 'Eugenics and the Campaign for Voluntary Sterilization in Britain between the Wars', *Social History of Medicine*, 2, 1989, 147–69.

Magnello, M. Eileen, 'Darwinian Variation and the Creation of Mathematical Statistics', in *Mathematics in Victorian Britain* (eds Raymond Flood, Adrian Rice, and Robin Wilson) (Oxford, 2011), 283–300.

Maloney, John, 'Marshall, Cunningham, and the Emerging Economics Profession', *Economic History Review*, n.s., 29, 3, Aug. 1976, 440–51.

Maltby, S. E., *Manchester and the Movement for National Elementary Education* (Manchester, 1918).

Manuel, Frank E., *The Prophets of Paris: Turgot, Condorcet, Saint-Simon, Fourier and Comte* (Oxford, 1962).

Marchi, N. B. de, and R. P. Sturges, 'Malthus and Ricardo's Inductivist Critics: Four Letters to William Whewell', *Economica*, 1973, 379–93.

Marcus, Steven, *Engels, Manchester, and the Working Class* (New York, 1974).

Martin, Theodore, *The Life of His Royal Highness The Prince Consort* (5 vols, London, 1879).

McCord, Norman, 'Cobden and Bright in Politics, 1846–1857', in R. Robson (ed.), *Ideas and Institutions in Victorian Britain* (London, 1967), 87–114.

McDonald, Lynn, 'Florence Nightingale: Passionate Statistician', *Journal of Holistic Nursing*, 16, 1998, 267–77.

McLachlan, H., *The Unitarian Movement in the Religious Life of England* (London, 1934).

Miller, Ian, and Chris Wild, *A & G Murray and the Cotton Mills of Ancoats* (Lancaster, 2007).

Miller, W. L., 'Richard Jones: A Case Study in Methodology', *History of Political Economy*, iii, 1971, 198–207.

Mitchison, Rosalind, *Agricultural Sir John: The Life of Sir John Sinclair of Ulbster* (London, 1962).

Moore, Doris Langley, *Ada, Countess of Lovelace. Byron's Legitimate Daughter* (London, 1977).

Morrell, J., and A. Thackray, *Gentlemen of Science: Early Years of the British Association for the Advancement of Science* (Oxford, 1981).

Moseley, Maboth, *Irascible Genius. A Life of Charles Babbage, Inventor* (London, 1964).

New, Chester, *The Life of Henry Brougham to 1831* (Oxford, 1961).

Nicolson, Malcolm, 'Alexander von Humboldt, Humboldtian Science and the Origins of the Study of Vegetation', *History of Science*, 25, 2, June 1987, 167–94.

Nissel, M., *People Count. A History of the General Register Office* (London, 1987).

Nixon, J. W. *A History of the International Statistical Institute 1885–1960* (The Hague, 1960).

Norman-Butler, Belinda, *Victorian Aspirations: The Life and Labour of Charles and Mary Booth* (London, 1972).

Oberschall, Anthony, *Empirical Social Research in Germany 1848–1914* (The Hague, 1965).

Oberschall, Anthony, 'The Two Empirical Roots of Social Theory and the Probability Revolution', in Lorenz Krüger, Gerd Gigerenzer, and Mary S. Morgan (eds), *The Probabilistic Revolution* (2 vols), *Vol. 2: Ideas in the Sciences* (1987) (Cambridge, Mass., 1990 edn.), 103–31.

O'Brien, Christopher, 'The Origins and Originators of Early Statistical Societies: A Comparison of Liverpool and Manchester', *JRSS*, series A, vol. 174, 1, Jan. 2011, 51–62.

O'Brien, Christopher, 'The Manchester Statistical Society. Civic Engagement in the 19th Century', *Transactions of the Manchester Statistical Society*, 2011–12, 96–104.

O'Brien, D. P., *J. R. McCulloch: A Study in Classical Economics* (London, 1970).

O'Day, Rosemary, and David Englander, *Mr. Charles Booth's Inquiry. Life and Labour of the People of London Reconsidered* (London, 1993).

Orange, A. D., 'The Idols of the Theatre: The British Association and its Early Critics', *Annals of Science*, 32, 1975, 277–94.

Palmer, Stanley H., *Economic Arithmetic. A Guide to the Statistical Sources of English Commerce, Industry, and Finance 1700–1850* (New York, 1977).

Parker, Christopher, 'English Historians and the Opposition to Positivism', *History and Theory*, 22, 2, May 1983, 120–45.

Pearson, Karl, 'On the Criterion that a Given System of Deviations from the Probable in the Case of a Correlated System of Variables Is Such that It Can Be Reasonably Supposed to Have Arisen from Random Sampling', *London, Edinburgh and Dublin Philosophical Magazine and Journal of Science*, 5th series, vol. 50, issue 302, 1900, 157–75.

Pearson, Karl, *The Life, Letters and Labours of Francis Galton* (4 vols, Cambridge, 1914–30).

Pearson, Karl, *Karl Pearson's Early Statistical Papers* (Cambridge, 1948).

Pease, Edward R., *The History of the Fabian Society* (London, 1916).

Pettigrew, William A., *Freedom's Debt. The Royal African Company and the Politics of the Atlantic Slave Trade, 1672–1752* (Chapel Hill, NC, 2013).

Pick, Daniel, *Faces of Degeneration. A European Disorder c. 1848–c. 1918* (Cambridge, 1989).

Pickstone, John V., 'Manchester's History and Manchester's Medicine', *British Medical Journal*, 295, 19–26 Dec. 1987, 1604–8.

Pickstone, John V., *Ways of Knowing. A New History of Science, Technology and Medicine* (Manchester, 2000).

Pigou, A.C. (ed.), *Memorials of Alfred Marshall* (1925) (New York, 1956 edn).

Plackett, R. L., 'Royal Statistical Society: The Last Fifty Years: 1934–84', *JRSS*, series A, 147, 2, 140–50.

Poovey, Mary, 'Figures of Arithmetic, Figures of Speech: The Discourse of Statistics in the 1830s', *Critical Inquiry*, 19, 2, Winter 1993, 256–76.

Poovey, Mary, *The History of the Modern Fact. Problems of Knowledge in the Sciences of Wealth and Society* (Chicago and London, 1998).

Popkin, R. H., *The Philosophy of the Sixteenth and Seventeenth Centuries* (New York, 1966).

Porter, Dorothy, 'Enemies of the Race: Biologism, Environmentalism, and Public Health in Edwardian England', *Victorian Studies*, 34, 2, Winter 1991, 159–78.

Porter, Dorothy, 'Eugenics and the Sterilization Debate in Sweden and Britain before World War II', *Scandinavian Journal of History*, 24, 2, 145–62.

Porter, Roy, *The Creation of the Modern World. The Untold Story of the British Enlightenment* (New York, 2000).

Porter, Theodore M., 'A Statistical Survey of Gases: Maxwell's Social Physics', *Historical Studies in the Physical Sciences*, 8, 1981, 77–116.

Porter, Theodore M., 'The Mathematics of Society: Variation and Error in Quetelet's Statistics', *British Journal for the History of Science*, 18, 1, March 1985, 56.

Porter, Theodore M., *The Rise of Statistical Thinking, 1820–1900* (Princeton, NJ, 1986).

Porter, Theodore M., 'Lawless Society: Social Science and the Reinterpretation of Statistics in Germany, 1850–1880', in Lorenz Krüger, Lorraine J. Daston, and Michael Heidelberger (eds), *The Probabilistic Revolution* (2 vols), vol. 1, *Ideas in History* (Cambridge, Mass., 1987), 351–75.

Porter, Theodore M., *Karl Pearson: The Scientific Life in a Statistical Age* (Princeton, NJ, 2004).

Porter, Theodore M., 'Statistics and the Career of Public Reason. Engagement and Detachment in a Quantified World', in Tom Crook and Glen O'Hara (eds), *Statistics and the Public Sphere. Numbers and the People in Modern Britain, c. 1800–2000* (New York and Abingdon, 2011), 32–47.

Poynter, J. R., *Society and Pauperism: English Ideas on Poor Relief, 1795–1834* (London, 1969).

Press, Gil, 'A Very Short History of Big Data', *Forbes Magazine*, 9 May 2013. https://www.forbes.com/sites/gilpress/2013/05/09/a-very-short-history-of-big-data/#9f36f2a65a18

Prothero, I. J., *Artisans and Politics in Early Nineteenth Century London. John Gast and his Times* (London, 1979).

Randeraad, Nico, *States and Statistics in the Nineteenth Century. Europe by Numbers* (Manchester, 2010).

Randeraad, Nico, 'The International Statistical Congress (1853–1876): Knowledge Transfers and their Limits', *European History Quarterly*, 41, 1, Jan. 2011, 50–65.

Rashid, Salim, 'Richard Jones and Baconian Historicism at Cambridge', *Journal of Economic Issues*, xiii, 1, March 1979, 159–73.

Read, Donald, *Cobden and Bright. A Victorian Political Partnership* (London, 1967).

Reisman, David, *Alfred Marshall's Mission* (Toronto, 1990).

Richter, Melvin, *The Politics of Conscience: T. H. Green and His Age* (London, 1964).

Robertson, John Mackinnon, *Buckle and His Critics. A Study in Sociology* (London, 1895).

Rogers, Edmund, 'A "Naked Strength and Beauty". Statistics in the British Tariff Debate, 1880–1914', in Tom Crook and Glen O'Hara (eds), *Statistics and the Public Sphere. Numbers and the People in Modern Britain, c. 1800–2000* (New York and Abingdon, 2011), 224–43.

Roll, Eric, *A History of Economic Thought* (1938) (London, 4th edn, 1973).

Rollin, Frank A., *Life and Public Services of Martin R. Delany* (Boston, 1868).

Romano, Richard M., 'The Economic Ideas of Charles Babbage', *History of Political Economy*, 14, 3, 1982, 385–405.

Romano, Terrie M., *Making Medicine Scientific. John Burdon Sanderson and the Culture of Victorian Science* (Baltimore and London, 2002).

Rose, Todd, 'How the Idea of a 'Normal' Person Got Invented', *The Atlantic*, 18 Feb. 2016, https://www.theatlantic.com/business/archive/2016/02/the-invention-of-the-normal-person/463365/

Rose, Todd, *The End of Average: How We Succeed in a World that Values Sameness* (London, 2017).

Rosenbaum, S., 'The Growth of the Royal Statistical Society', *JRSS*, series A, vol. 147, 2, 375–88.

Royal Statistical Society, London, *Data Manifesto*, Sept. 2014. http://www.statslife.org.uk/images/pdf/rss-data-manifesto-2014.pdf

Ruse, Michael, 'William Whewell: Omniscientist', in M. Frisch and S. Schaffer (eds), *William Whewell. A Composite Portrait* (Oxford, 1991), 87–116.

Ruskin, Steven, *John Herschel's Cape Voyage. Private Science, Public Imagination and the Ambitions of Empire* (London, 2004).

Sarton, G., *Sarton on the History of Science. Essays by George Sarton* (D. Stimson, ed.) (Cambridge, Mass., 1962).

Saville, John, 'The Background to the Industrial Remuneration Conference of 1885', in *Industrial Remuneration Conference. Reprints of Economic Classics* (1885) (Augustus M. Kelley, New York, 1968), 5–44.

Schaffer, Simon, 'The History and Geography of the Intellectual World: Whewell's Politics of Language', in M. Frisch and S. Schaffer (eds.), *William Whewell. A Composite Portrait* (Oxford, 1991), 201–31.

Schaffer, Simon, 'Babbage's Intelligence: Calculating Engines and the Factory System', *Critical Inquiry*, vol 21, Autumn 1994, 203–27.

Schaffer, Simon, 'Babbage's Dancer and the Impresarios of Mechanism' in F. Spufford and J. Uglow (eds.), *Cultural Babbage. Technology, Time and Invention* (London, 1996), 53–80.

Schofield, Robert E., *The Lunar Society of Birmingham. A Social History of Provincial Science and Industry in Eighteenth Century England* (Oxford, 1963).

Schumpeter, J. A., *A History of Economic Analysis* (New York, 1954).

Schweber, Libby, *Disciplining Statistics. Demography and Vital Statistics in France and England, 1830–1885* (Durham, N. C. and London, 2006).

Scull, Andrew, 'Michel Foucault's History of Madness', *The History of the Human Sciences*, iii, 1, Feb 1990, 57–67.

Searle, G. R., *Eugenics and Politics in Britain 1900–1914* (London, 1976).

Searle, G. R., 'Eugenics and Class', in Charles Webster (ed.), *Biology, Medicine and Society 1840–1940* (Cambridge, 1981), 217–42.

Selleck, Richard W., 'The Manchester Statistical Society and the Foundation of Social Science Research', *Australian Educational Researcher*, 16, 1, 1989, 1–14.

Selleck, Richard W., *James Kay-Shuttleworth: Journey of an Outsider* (London, 1995).

Seymour, Miranda, *In Byron's Wake. The Turbulent Lives of Lord Byron's Wife and Daughter: Annabella Milbanke and Ada Lovelace* (London, 2018).

Shaw, Martin and Miles, Ian, 'The Social Roots of Statistical Knowledge', in *Demystifying Social Statistics* (eds. John Irvine, Ian Miles, and Jeff Evans) (London, 1979), 27–38.

Sheynin, Oscar, 'Quetelet as Statistician', *Archive for History of Exact Sciences*, 36, 4, Dec. 1986, 281–325.

Shils, Edward, 'Tradition, Ecology and Institution in the History of Sociology', *Daedalus*, 99, 4, 1970, 760–825.

Shoen, Harriet H., 'Prince Albert and the Application of Statistics to Problems of Government, *Osiris*, 5, 1938, 276–318.

Simey, T. S. and Simey, M. B., *Charles Booth. Social Scientist* (Oxford, 1960).

Simon, Brian, *Studies in the History of Education 1780–1870* (London, 1960) (1974 edn).

Skinner, Quentin, 'Meaning and Understanding in the History of Ideas', *History and Theory*, viii, no. 1, 1969, 3–53.

Smith, F. B., *Florence Nightingale. Reputation and Power* (London, 1982).

Smith, Frank, *The Life and Work of Sir James Kay-Shuttleworth* (1923) (1974 edn, Trowbridge, Wilts).

Snyder, Laura J., *The Philosophical Breakfast Club. Four Remarkable Friends Who Transformed Science and Changed the World* (New York, 2011).

Spengler, Joseph J., 'On the Progress of Quantification in Economics', *Isis*, 52, 2, June 1961, 258–76.

Spufford, F. and J. Uglow, 'The Difference Engine and *The Difference Engine*' in F. Spufford and J. Uglow (eds), *Cultural Babbage. Technology, Time and Invention* (London, 1996), 266–90.

Spychal, Martin, 'Crown and Country: Research Profiles in the IHR', Martin Spychal in conversation with Lawrence Goldman, *Past and Future: The Magazine of the Institute of Historical Research*, Issue 21, Spring/Summer 2017, 12–13.

Stedman Jones, Gareth, *Outcast London. A Study in the Relationship Between Classes in Victorian Society* (Oxford, 1971).

Stedman Jones, Gareth, 'Class Expression versus Social Control? A Critique of Recent Trends in the Social History of Leisure', *History Workshop Journal*, iv (1977), 163–70.

Stern, Fritz, *Bismarck, Bleichröder and the building of the German Empire* (New York, 1977).

Stewart, Ian, *Nature's Numbers. Discovering Order and Pattern in the Universe* (London, 1995).

Stewart, Ian, *Life's Other Secret. The New Mathematics of the Living World* (New York, 1998).

Stigler, Stephen, 'Francis Ysidro Edgeworth, Statistician', *JRSS*, series A, 141 (1978), 287–322.

Stigler, Stephen, *The History of Statistics: The Measurement of Uncertainty before 1900* (Cambridge, MA, 1986).

Stigler, Stephen, 'The Measurement of Uncertainty in Nineteenth-Century Social Science', in Lorenz Krüger, Lorraine J. Daston, and Michael Heidelberger (eds), *The Probabilistic Revolution* (2 vols), vol. 1, *Ideas in History* (1987) (Cambridge, Mass., 1990 edn), 287–92.

Stone, Richard, *Some British Empiricists in the Social Sciences 1650–1900* (Cambridge, 1997).

Supple, Barry, *The Royal Exchange Assurance: A History of British Insurance 1720–1970* (Cambridge, 1970).

Swade, Doron, *Charles Babbage and his Calculating Engines* (Science Museum, London, 1991).

Swade, Doron, '"It will not slice a pineapple": Babbage, Miracles and Machines', in F. Spufford and J. Uglow (eds), *Cultural Babbage. Technology, Time and Invention* (London, 1996), 34–52.

Swade, Doron, *The Cogwheel Brain. Charles Babbage and the Quest to Build the First Computer* (London, 2000).

Swade, Doron, 'Calculating Engines. Machines, Mathematics and Misconceptions', in R. Flood, A. Rice, and R. Wilson (eds.), *Mathematics in Victorian Britain* (Oxford, 2011), 239–51.

Szreter, Simon, 'The Genesis of the Registrar-General's Social Classification of Occupations', *British Journal of Sociology*, 35, 4, Dec. 1984, 522–46.

Szreter, Simon, 'Introduction: The GRO and the Historians', in Simon Szreter (ed.), *The General Register Office of England and Wales and the Public Health Movement 1837–1914. A Comparative Perspective*, Special Issue, *Social History of Medicine*, 4, 3, Dec. 1991, 401–14.

Szreter, Simon, *Fertility, Class and Gender in Britain 1860–1940* (Cambridge, 1996).

Thackray, Arnold, 'Natural Knowledge in Cultural Context: The Manchester Mode', *American Historical Review*, 79, 3, June 1974, 627–709.

Tholfsen, Trygve R., *Sir James Kay-Shuttleworth on Popular Education* (London, 1974).

Thompson, D'Arcy, *On Growth and Form* (London, 1917).

Thompson, E. P., *The Making of the English Working Class 1780–1830* (London, 1963).

Thompson, E. P., 'The Moral Economy of the English Crowd', *Past & Present*, 50, 1, Feb. 1971, 76–136.

Thompson, James, 'Printed Statistics and the Public Sphere. Numeracy, Electoral Politics, and the Visual Culture of Numbers, 1880–1914', in Tom Crook and Glen O'Hara (eds), *Statistics and the Public Sphere. Numbers and the People in Modern Britain, c. 1800–2000* (New York and Abingdon, 2011), 121–43.

Thompson, S. J., '"Population Combined with Wealth and Taxation". Statistics, Representation and the Making of the 1832 Reform Act', in Tom Crook and Glen O'Hara (eds), *Statistics and the Public Sphere. Numbers and the People in Modern Britain, c. 1800–2000* (New York and Abingdon, 2011), 205–23.

Tippett, L. H. C., 'Annals of the Royal Statistical Society 1934–71', *JRSS*, series A, 135, 4, 545–68.

Tooze, J. Adam, *Statistics and the German State, 1900–1945. The Making of Modern Economic Knowledge* (Cambridge, 2001).

Trebilcock, Clive, *Phoenix Assurance and the Development of British Assurance. Volume 1: 1780–1870; Volume 2: The Era of Insurance Giants, 1870–1984* (Cambridge, 1985, 1998).

Tribe, Keith, *Land, Labour and Economic Discourse* (Cambridge, 1978).

Uglow, Jenny, 'Introduction', in F. Spufford and J. Uglow (eds), *Cultural Babbage. Technology, Time and Invention* (London, 1996).

Uglow, Jenny, *The Lunar Men: Five Friends whose Curiosity Changed the World* (London, 2002).

Ullman, Victor, *Martin R. Delany: The Beginnings of Black Nationalism* (Boston, 1971).

Van Sinderen, Alfred W., 'Babbage's Letter to Quetelet', *Annals of the History of Computing*, 5, 3, July 1983, 263–7.

Waddell, D. A. G., 'Charles Davenant 1656–1714 – a Biographical Sketch', *Economic History Review*, 2nd series, 11, 1958–9, 279–88.

Walton, John, *Lancashire. A Social History 1558–1939* (Manchester, 1987).

Weindling, Paul, *Health, Race, and German Politics between National Unification and Nazism 1870–1945* (Cambridge, 1989).

Westergaard, Harald, *Contributions to the History of Statistics* (London, 1932).

Wheeler, Michael, *Lies, Damned Lies, and Statistics: The Manipulation of Public Opinion in America* (New York, 1976).

Willcox, Walter F., 'Note on the Chronology of Statistical Societies', *Journal of the American Statistical Association*, 29, 188, Dec. 1934, 418–20.

Willcox, Walter F., 'Development of International Statistics', *Milbank Memorial Fund Quarterly*, 27, 2, 1949, 143–53.

Williams, Orlo, *Lamb's Friend the Census-Taker: Life and Letters of John Rickman* (London,1912).

Williams, Perry, 'Passing on the Torch. Whewell's Philosophy and the Principles of English University Education', in M. Fisch and S. Schaffer (eds), *William Whewell. A Composite Portrait* (Oxford, 1991), 117–48.

Williams, Raymond, *Politics and Letters* (London, 1979).

Wolmar, Christian, *Fire and Steam* (London, 2007).

Woodward, Llewellyn, *The Age of Reform 1815–1870* (Oxford, 1963).

Wormell, Deborah, *Sir John Seeley and the Uses of History* (Cambridge, 1980).

Wright, T. W., *The Religion of Humanity. The Impact of Comtean Positivism on Victorian Britain* (Cambridge, 1986).

Wrigley, E. A., 'Comments', in Richard Stone, *Some British Empiricists in the Social Sciences 1650–1900* (Cambridge, 1997), 423–8.

Wulf, Andrea, *The Invention of Nature. The Adventures of Alexander von Humboldt. The Lost Hero of Science* (London, 2015).

Yeo, Richard, 'William Whewell, Natural Theology and the Philosophy of Science in Mid Nineteenth Century Britain', *Annals of Science*, 36, 5, 1979.

Yeo, Richard, 'Scientific Method and the Image of Science 1831–1891', in R. Macleod and P. Collins (eds), *The Parliament of Science: The British Association for the Advancement of Science 1831–1981* (London, 1981), 65–88.

Yeo, Richard, 'William Whewell's Philosophy of Knowledge and its Reception'. in M. Fisch and S. Schaffer (eds), *William Whewell. A Composite Portrait* (Oxford, 1991), 175–99.

Yeo, Richard, *Defining Science. William Whewell, Natural Knowledge and Public Debate in Early Victorian England* (Cambridge, 1993).

Young, R. M., 'The Historiographical and Ideological Contexts of the Nineteenth-Century Debate on Man's Place in Nature', in M. Teich and R. M. Young (eds), *Changing Perspectives in the History of Science* (London, 1973), 344–438.

Zahn, Friedrich, *50 Année de L'Institut International de Statistique* (Munich, 1934).

Zelizer, Viviana A. Rotman, *Morals and Markets. The Development of Life Insurance in the United States* (1979) (New York, 2017 edn).

Official Publications

Hansard's Parliamentary Debates

Parliamentary Papers

Report from the Select Committee on Parochial Registration, 1833, vol. xiv, 119–22.

Report from the Select Committee on Education in England and Wales: Together with the Minutes of Evidence, Appendix, and Index (House of Commons, 1835).

Papers Relating to the Sanitary State of the People of England, 1857–8, vol. xxiii, i–xlvi, 1–164.

Reports of the Medical Officer of the Privy Council, i–xiii, Parliamentary Papers, 1859–71.

Census of Great Britain, 1851. Population Tables. 1. Numbers of Inhabitants. Report and Summary Tables (London, HMSO, 1852).

Census of Great Britain 1851. Religious Worship in England and Wales (London, HMSO, 1854).

Census of Great Britain 1851. Education. England and Wales. Report and Tables (London, HMSO, 1854).

Websites

'Sketch of the Analytical Engine', L. Menabrea, Tr. Ada Lovelace (1843) www.fourmilab.ch/babbage/sketch.html

'What Is BIG DATA?' https://www.guru99.com/what-is-big-data.html

Who Was Charles Booth? (London School of Economics) https://booth.lse.ac.uk/learn-more/who-was-charles-booth

Cambridge Core, Cambridge University Press

https://www.cambridge.org/core/books/kants-idea-for-a-universal-history-with-a-cosmopolitan-aim/idea-for-a-universal-history-with-a-cosmopolitan-aim/8B2BA346A82FA006AB982E3A941E2A26

https://www.census.gov/history/www/census_then_now/director_biographies/directors_1840_-_1865.html

Hartwell, R. M., 'Cunningham, William', https://www.encyclopedia.com/social-sciences/applied-and-social-sciences-magazines/cunningham-william.

https://www.kingjamesbibleonline.org/Wisdom-of-Solomon-Chapter-11/

Legacies of British Slavery website, University College, London. https://www.ucl.ac.uk/lbs/

'Lies, Damned Lies and Statistics', Department of Mathematics, University of York: https://www.york.ac.uk/depts/maths/histstat/lies.htm

'Maine at Gettysburg: Report of Maine Commissioners' (Portland, Maine, The Lakeside Press, 1898) at www.segtours.com/files/maineatgettysburg.pdf

Marx, Karl, *Capital*, vol. III, ch. 47. https://www.marxists.org/archive/marx/works/1894-c3/ch47.htm

Oxford Languages, https://www.lexico.com/definition/big_data

Royal Society, London
https://royalsociety.org/about-us/history/

Royal Statistical Society, London*, Data Manifesto*, Sept. 2014. http://www.statslife.org.uk/images/pdf/rss-data-manifesto-2014.pdf

Steamindex: https://www.steamindex.com/people/kennedy.htm

Beatrice Webb's Diary (London School of Economics) https://digital.library.lse.ac.uk/objects/lse:ros416yur/read/single#page/428/mode/2up

World Health Organisation, *International Classification of Diseases* (11th revision, 18 June 2018) https://www.who.int/classifications/icd/en/

Unpublished Dissertations

Downing, Arthur, 'The Friendly Planet. Friendly societies and fraternal associations around the English-speaking world, 1840–1925', Unpublished D. Phil thesis, University of Oxford, 2015.

Morewood, John, 'Henry Peter Brougham and Anti-Slavery 1802–1843', Unpublished PhD thesis, Institute of Historical Research, University of London, 2021.

Ryan, Rebecca, '"Chaos vom Fabriken": A German insight into the Industrial Revolution in Britain in 1840', unpublished BA dissertation, Oxford History Schools, 2010 (St. Peter's College, Oxford).

Spychal, Martin, 'Constructing England's Electoral Map. Parliamentary Boundaries and the 1832 Reform Act', Unpublished PhD thesis, Institute of Historical Research, University of London, 2017.

Index

For the benefit of digital users, indexed terms that span two pages (e.g., 52–53) may, on occasion, appear on only one of those pages.